STUDENT SOLUTIONS MANUAL T

ADVANCED
Engineering
Mathematics

SEVENTH EDITION
by Dennis G. Zill

Warren S. Wright
Roberto Martinez

JONES & BARTLETT
LEARNING

World Headquarters
Jones & Bartlett Learning
5 Wall Street
Burlington, MA 01803
978-443-5000
info@jblearning.com
www.jblearning.com

Jones & Bartlett Learning books and products are available through most bookstores and online booksellers. To contact Jones & Bartlett Learning directly, call 800-832-0034, fax 978-443-8000, or visit our website, www.jblearning.com.

Substantial discounts on bulk quantities of Jones & Bartlett Learning publications are available to corporations, professional associations, and other qualified organizations. For details and specific discount information, contact the special sales department at Jones & Bartlett Learning via the above contact information or send an email to specialsales@jblearning.com.

Production Credits
Director of Product Management: Laura Pagluica
Product Manager: Edward Hinman
Content Strategist: Melissa Duffy
Content Coordinator: Paula Gregory
Manager, Project Management: Lori Mortimer
Senior Project Specialist: Jennifer Risden
Director of Marketing: Andrea DeFronzo
Product Fulfillment Manager: Wendy Kilborn
Composition: Exela Technologies
Project Management: Exela Technologies
Cover Design: Scott Moden
Cover Image (Title Page): © Pol.mch/Shutterstock
Printing and Binding: McNaughton & Gunn

ISBN: 978-1-284-20626-5

6048

Printed in the United States of America
24 23 22 21 20   10 9 8 7 6 5 4 3 2 1

Contents

Chapter 1

Introduction to Differential Equations

1.1 | Definitions and Terminology

3. Fourth order; linear

6. Second order; nonlinear because of R^2

9. First order; nonlinear because of $\sin\left(\dfrac{dy}{dx}\right)$

12. Writing the differential equation in the form $u(dv/du) + (1 + u)v = ue^u$, we see that it is linear in v. However, writing it in the form $(v + uv - ue^u)(du/dv) + u = 0$, we see that it is nonlinear in u.

15. From $y = e^{3x}\cos 2x$ we obtain $y' = 3e^{3x}\cos 2x - 2e^{3x}\sin 2x$ and $y'' = 5e^{3x}\cos 2x - 12e^{3x}\sin 2x$, so that $y'' - 6y' + 13y = 0$.

18. Since $\tan x$ is not defined for $x = \pi/2 + n\pi$, n an integer, the domain of $y = 5\tan 5x$ is $\{x \mid 5x \neq \pi/2 + n\pi\}$
or $\{x \mid x \neq \pi/10 + n\pi/5\}$. From $y' = 25\sec^2 5x$ we have

$$y' = 25(1 + \tan^2 5x) = 25 + 25\tan^2 5x = 25 + y^2.$$

An interval of definition for the solution of the differential equation is $(-\pi/10, \pi/10)$. Another interval is $(\pi/10, 3\pi/10)$, and so on.

21. Writing $\ln(2X - 1) - \ln(X - 1) = t$ and differentiating implicitly we obtain

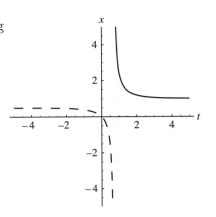

$$\frac{2}{2X - 1}\frac{dX}{dt} - \frac{1}{X - 1}\frac{dX}{dt} = 1$$

$$\left(\frac{2}{2X - 1} - \frac{1}{X - 1}\right)\frac{dX}{dt} = 1$$

$$\frac{2X - 2 - 2X + 1}{(2X - 1)(X - 1)}\frac{dX}{dt} = 1$$

$$\frac{dX}{dt} = -(2X - 1)(X - 1) = (X - 1)(1 - 2X).$$

Exponentiating both sides of the implicit solution we obtain

$$\frac{2X - 1}{X - 1} = e^t$$

$$2X - 1 = Xe^t - e^t$$

$$(e^t - 1) = (e^t - 2)X$$

$$X = \frac{e^t - 1}{e^t - 2}.$$

Solving $e^t - 2 = 0$ we get $t = \ln 2$. Thus the solution is defined on $(-\infty, \ln 2)$ or on $(\ln 2, \infty)$. The graph of the solution defined on $(-\infty, \ln 2)$ is dashed, and the graph of the solution defined on $(\ln 2, \infty)$ is solid.

24. Differentiating $y = 2x^2 - 1 + c_1 e^{-2x^2}$ we obtain $\dfrac{dy}{dx} = 4x - 4xc_1 e^{-2x^2}$, so that

$$\frac{dy}{dx} + 4xy = 4x - 4xc_1 e^{-2x^2} + 8x^3 - 4x + 4c_1 xe^{-x^2} = 8x^3$$

In Problems 27–30, we use the Product Rule and the derivative of an integral ((12) of this section): $\dfrac{d}{dx}\displaystyle\int_a^x g(t)\,dt = g(x).$

27. Differentiating $y = e^{3x}\displaystyle\int_1^x \frac{e^{-3t}}{t}\,dt$ we obtain $\dfrac{dy}{dx} = e^{3x}\displaystyle\int_1^x \frac{e^{-3t}}{t}\,dt + \frac{e^{-3t}}{x}\cdot e^{3x}$ or

$\dfrac{dy}{dx} = e^{3x}\displaystyle\int_1^x \frac{e^{-3t}}{t}\,dt + \frac{1}{x}$, so that

$$x\frac{dy}{dx} - 3xy = x\left(e^{3x}\int_1^x \frac{e^{-3t}}{t}\,dt + \frac{1}{x}\right) - 3x\left(e^{3x}\int_1^x \frac{e^{-3t}}{t}\,dt\right)$$

$$= xe^{3x}\int_1^x \frac{e^{-3t}}{t}\,dt + 1 - 3xe^{3x}\int_1^x \frac{e^{-3t}}{t}\,dt = 1$$

30. Differentiating $y = e^{-x^2} + e^{-x^2} \int_0^x e^{t^2}\, dt$ we obtain $\dfrac{dy}{dx} = -2xe^{-x^2} - 2xe^{-x^2} \int_0^x e^{t^2}\, dt + e^{x^2} \cdot e^{-x^2}$

or $\dfrac{dy}{dx} = -2xe^{-x^2} - 2xe^{-x^2} \int_0^x e^{t^2}\, dt + 1$, so that

$$\frac{dy}{dx} + 2xy = \left(-2xe^{-x^2} - 2xe^{-x^2} \int_0^x e^{t^2}\, dt + 1 \right) + 2x \left(e^{-x^2} + e^{-x^2} \int_0^x e^{t^2}\, dt \right)$$

$$= -2xe^{-x^2} - 2xe^{-x^2} \int_0^x e^{t^2}\, dt + 1 + 2xe^{-x^2} + 2xe^{-x^2} \int_0^x e^{t^2}\, dt = 1$$

33. Force the function $y = e^{mx}$ into the equation $y' + 2y = 0$ to get

$$(e^{mx})' + 2(e^{mx}) = 0$$

$$me^{mx} + 2e^{mx} = 0$$

$$e^{mx}(m + 2) = 0$$

Now since $e^{mx} > 0$ for all values of x, we must have $m = -2$ and so $y = e^{-2x}$ is a solution.

36. Force the function $y = e^{mx}$ into the equation $2y'' + 9y' - 5y = 0$ to get

$$2(e^{mx})'' + 9(e^{mx})' - 5(e^{mx}) = 0$$

$$2m^2 e^{mx} + 9me^{mx} - 5e^{mx} = 0$$

$$e^{mx}(2m^2 + 9m - 5) = 0$$

$$e^{mx}(m + 5)(2m - 1) = 0$$

Now since $e^{mx} > 0$ for all values of x, we must have $m = -5$ and $m = 1/2$ therefore $y = e^{-5x}$ and $y = e^{x/2}$ are solutions.

39. Force the function $y = x^m$ into the equation $x^2 y'' - 7xy' + 15y = 0$ to get

$$x^2 \cdot (x^m)'' - 7x \cdot (x^m)' + 15(x^m) = 0$$

$$x^2 \cdot m(m - 1)x^{m-2} - 7x \cdot mx^{m-1} + 15x^m = 0$$

$$(m^2 - m)x^m - 7mx^m + 15x^m = 0$$

$$x^m[m^2 - 8m + 15] = 0$$

$$x^m[(m - 3)(m - 5)] = 0$$

The last line implies that $m = 3$ and $m = 5$ therefore $y = x^3$ and $y = x^5$ are solutions.

In Problems 41–44, we substitute $y = c$ into the differential equations and use $y' = 0$ and $y'' = 0$

42. Solving $c^2 + 2c - 3 = (c + 3)(c - 1) = 0$ we see that $y = -3$ and $y = 1$ are constant solutions.

45. From $y = (x + c_1)^2$ we obtain

$$\left(\frac{dy}{dx}\right)^2 = (2\,(x + c_1))^2 = 4\,(x + c_1)^2\,.$$

Then

$$4y = 4(x + c_1)^2.$$

Inspection of the differential equation reveals that $y = 0$ is a solution of the differential equation but is not a member of the one-parameter family $y = (x + c_1)^2$.

48. From $y = x - (x - c_1)^2$ we obtain

$$\left(\frac{dy}{dx}\right)^2 = (1 - 2\,(x - c_1))^2 = 1 - 4\,(x - c_1) + 4\,(x - c_1)^2 = 1 - 4x + 4c_1 + 4x^2 - 8xc_1 + 4c_1^2$$

$$-2\frac{dy}{dx} = -2\,(1 - 2\,(x - c_1)) = -2 + 4x - 4c_1$$

$$4y = 4\left(x - (x - c_1)^2\right) = 4\left(x - x^2 + 2xc_1 - c_1^2\right) = 4x - 4x^2 + 8xc_1 - 4c_1^2.$$

Then

$$\left(\frac{dy}{dx}\right)^2 - 2\frac{dy}{dx} + 4y = 4x - 1.$$

Inspection of the differential equation reveals that $y = x$ is a solution of the differential equation but not a member of the one-parameter family $y = x - (x - c_1)^2$.

1.2 | Initial-Value Problems

3. Letting $x = 2$ and solving $1/3 = 1/(4 + c)$ we get $c = -1$. The solution is $y = 1/(x^2 - 1)$. This solution is defined on the interval $(1, \infty)$.

6. Letting $x = 1/2$ and solving $-4 = 1/(1/4 + c)$ we get $c = -1/2$. The solution is $y = 1/(x^2 - 1/2) = 2/(2x^2 - 1)$. This solution is defined on the interval $(-1/\sqrt{2}\,, 1/\sqrt{2}\,)$.

In Problems 7–10, we use $x = c_1 \cos t + c_2 \sin t$ and $x' = -c_1 \sin t + c_2 \cos t$ to obtain a system of two equations in the two unknowns c_1 and c_2.

9. From the initial conditions we obtain

$$\frac{\sqrt{3}}{2}c_1 + \frac{1}{2}c_2 = \frac{1}{2} - \frac{1}{2}c_2 + \frac{\sqrt{3}}{2} = 0$$

Solving, we find $c_1 = \sqrt{3}/4$ and $c_2 = 1/4$. The solution of the initial-value problem is

$$x = (\sqrt{3}/4)\cos t + (1/4)\sin t.$$

In Problems 11–14, we use $y = c_1 e^x + c_2 e^{-x}$ and $y' = c_1 e^x - c_2 e^{-x}$ to obtain a system of two equations in the two unknowns c_1 and c_2.

12. From the initial conditions we obtain

$$ec_1 + e^{-1}c_2 = 0$$

$$ec_1 - e^{-1}c_2 = e.$$

Solving, we find $c_1 = \frac{1}{2}$ and $c_2 = -\frac{1}{2}e^2$. The solution of the initial-value problem is

$$y = \frac{1}{2}e^x - \frac{1}{2}e^2 e^{-x} = \frac{1}{2}e^x - \frac{1}{2}e^{2-x}.$$

15. Two solutions are $y = 0$ and $y = x^3$.

18. For $f(x,y) = \sqrt{xy}$ we have $\partial f/\partial y = \frac{1}{2}\sqrt{x/y}$. Thus the differential equation will have a unique solution in any region where $x > 0$ and $y > 0$ or where $x < 0$ and $y < 0$.

21. For $f(x,y) = x^2/(4 - y^2)$ we have $\partial f/\partial y = 2x^2 y/(4 - y^2)^2$. Thus the differential equation will have a unique solution in any region where $y < -2$, $-2 < y < 2$, or $y > 2$.

24. For $f(x,y) = (y + x)/(y - x)$ we have $\partial f/\partial y = -2x/(y - x)^2$. Thus the differential equation will have a unique solution in any region where $y < x$ or where $y > x$.

In Problems 25–28, we identify $f(x,y) = \sqrt{y^2 - 9}$ and $\partial f/\partial y = y/\sqrt{y^2 - 9}$. We see that f and $\partial f/\partial y$ are both continuous in the regions of the plane determined by $y < -3$ and $y > 3$ with no restrictions on x.

27. Since $(2, -3)$ is not in either of the regions defined by $y < -3$ or $y > 3$, there is no guarantee of a unique solution through $(2, -3)$.

30. (a) Since $\dfrac{d}{dx}\tan(x + c) = \sec^2(x + c) = 1 + \tan^2(x + c)$, we see that $y = \tan(x + c)$ satisfies the differential equation.

(b) Solving $y(0) = \tan c = 0$ we obtain $c = 0$ and $y = \tan x$. Since $\tan x$ is discontinuous at $x = \pm\pi/2$, the solution is not defined on $(-2, 2)$ because it contains $\pm\pi/2$.

(c) The largest interval on which the solution can exist is $(-\pi/2, \pi/2)$.

33. (a) Differentiating $3x^2 - y^2 = c$ we get $6x - 2yy' = 0$ or $yy' = 3x$.

(b) Solving $3x^2 - y^2 = 3$ for y we get

$$y = \phi_1(x) = \sqrt{3(x^2 - 1)}, \qquad 1 < x < \infty,$$

$$y = \phi_2(x) = -\sqrt{3(x^2 - 1)}, \qquad 1 < x < \infty,$$

$$y = \phi_3(x) = \sqrt{3(x^2 - 1)}, \qquad -\infty < x < -1,$$

$$y = \phi_4(x) = -\sqrt{3(x^2 - 1)}, \qquad -\infty < x < -1.$$

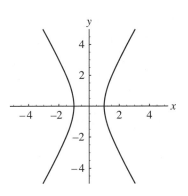

(c) Only $y = \phi_3(x)$ satisfies $y(-2) = 3$.

In Problems 35–38, we consider the points on the graphs with x-coordinates $x_0 = -1$, $x_0 = 0$, and $x_0 = 1$. The slopes of the tangent lines at these points are compared with the slopes given by $y'(x_0)$ in (a) through (f).

36. The graph satisfies the conditions in (e).

39. Using the function $y = c_1 \cos 3x + c_2 \sin 3x$ and the first boundary condition we get

$$y(0) = c_1 \cos 0 + c_2 \sin 0 = 0$$

Therefore $c_1 = 0$. Similarly, for the second boundary condition we get

$$y(\pi/6) = c_2 \sin 3 \left(\pi/6 \right) = -1$$

Therefore $c_2 = -1$. The solution to the boundary value problem is $y(x) = -\sin 3x$.

42. The derivative of the function $y = c_1 \cos 3x + c_2 \sin 3x$ is $y' = -3c_1 \sin 3x + 3c_2 \cos 3x$ and using the two boundary conditions we get

$$y(0) = c_1 + 0 = 1$$

Therefore $c_1 = 1$. In addition

$$y'(\pi) = 0 - 3c_2 = 5$$

Therefore $c_2 = -5/3$. The solution to this boundary value problem is $y(x) = \cos 3x - \dfrac{5}{3} \sin 3x$.

1.3   Differential Equations as Mathematical Models

3. Let b be the rate of births and d the rate of deaths. Then $b = k_1 P$ and $d = k_2 P^2$. Since $dP/dt = b - d$, the differential equation is $dP/dt = k_1 P - k_2 P^2$.

6. By inspecting the graph in the text we take T_m to be $T_m(t) = 80 - 30 \cos \left(\pi t/12 \right)$. Then the temperature of the body at time t is determined by the differential equation

$$\frac{dT}{dt} = k \left[T - \left(80 - 30 \cos \left(\frac{\pi}{12} t \right) \right) \right], \quad t > 0.$$

9. The rate at which salt is leaving the tank is

$$R_{out} \ (3 \ \text{gal/min}) \cdot \left(\frac{A}{300} \ \text{lb/gal} \right) = \frac{A}{100} \ \text{lb/min}.$$

Thus $dA/dt = A/100$. The initial amount is $A(0) = 50$.

12. The rate at which salt is entering the tank is

$$R_{in} = (c_{in} \text{ lb/gal}) \cdot (r_{in} \text{ gal/min}) = c_{in}r_{in} \text{ lb/min.}$$

Now let $A(t)$ denote the number of pounds of salt and $N(t)$ the number of gallons of brine in the tank at time t. The concentration of salt in the tank as well as in the outflow is $c(t) = x(t)/N(t)$. But the number of gallons of brine in the tank remains steady, is increased, or is decreased depending on whether $r_{in} = r_{out}$, $r_{in} > r_{out}$, or $r_{in} < r_{out}$. In any case, the number of gallons of brine in the tank at time t is $N(t) = N_0 + (r_{in} - r_{out})t$. The output rate of salt is then

$$R_{out} = \left(\frac{A}{N_0 + (r_{in} - r_{out})t} \text{ lb/gal} \right) \cdot (r_{out} \text{ gal/min}) = r_{out} \frac{A}{N_0 + (r_{in} - r_{out})t} \text{ lb/min.}$$

The differential equation for the amount of salt, $dA/dt = R_{in} - R_{out}$, is

$$\frac{dA}{dt} = c_{in}r_{in} - r_{out} \frac{A}{N_0 + (r_{in} - r_{out})t} \quad \text{or} \quad \frac{dA}{dt} + \frac{r_{out}}{N_0 + (r_{in} - r_{out})t} A = c_{in}r_{in}.$$

15. Since $i = dq/dt$ and $L\, d^2q/dt^2 + R\, dq/dt = E(t)$, we obtain $L\, di/dt + Ri = E(t)$.

18. Since the barrel in Figure 1.3.17(b) in the text is submerged an additional y feet below its equilibrium position, the number of cubic feet in the additional submerged portion is the volume of the circular cylinder: $\pi \times (\text{radius})^2 \times \text{height}$ or $\pi(s/2)^2 y$. Then we have from Archimedes' principle

$$\text{Upward force of water on barrel} = \text{Weight of water displaced}$$

$$= (62.4) \times (\text{Volume of water displaced})$$

$$= (62.4)\pi(s/2)^2 y = 15.6\pi s^2 y.$$

It then follows from Newton's second law that

$$\frac{w}{g}\frac{d^2y}{dt^2} = -15.6\pi s^2 y \quad \text{or} \quad \frac{d^2y}{dt^2} + \frac{15.6\pi s^2 g}{w} y = 0,$$

where $g = 32$ and w is the weight of the barrel in pounds.

21. As the rocket climbs (in the positive direction), it spends its amount of fuel and therefore the mass of the fuel changes with time. The air resistance acts in the opposite direction of the motion and the upward thrust R works in the same direction. Using Newton's second law we get

$$\frac{d}{dt}(mv) = -mg - kv + R$$

Now because the mass is variable, we must use the product rule to expand the left side of the equation. Doing so gives us the following:

$$\frac{d}{dt}(mv) = -mg - kv + R$$

$$v \times \frac{dm}{dt} + m \times \frac{dv}{dt} = -mg - kv + R$$

The last line is the differential equation we wanted to find.

24. The gravitational force on m is $F = -kM_r m/r^2$. Since $M_r = 4\pi\delta r^3/3$ and $M = 4\pi\delta R^3/3$, we have $M_r = r^3 M/R^3$ and

$$F = -k\frac{M_r m}{r^2} = -k\frac{r^3 Mm/R^3}{r^2} = -k\frac{mM}{R^3}r.$$

Now from $F = ma = d^2r/dt^2$ we have

$$m\frac{d^2r}{dt^2} = -k\frac{mM}{R^3}r \quad \text{or} \quad \frac{d^2r}{dt^2} = -\frac{kM}{R^3}r.$$

27. The differential equation is $x'(t) = r - kx(t)$ where $k > 0$.

Chapter 1 in Review

3. $\dfrac{d}{dx}(c_1 \cos kx + c_2 \sin kx) = -kc_1 \sin kx + kc_2 \cos kx;$

$$\frac{d^2}{dx^2}(c_1 \cos kx + c_2 \sin kx) = -k^2 c_1 \cos kx - k^2 c_2 \sin kx = -k^2(\overbrace{c_1 \cos kx + c_2 \sin kx}^{y});$$

$$\frac{d^2y}{dx^2} = -k^2 y \quad \text{or} \quad \frac{d^2y}{dx^2} + k^2 y = 0$$

6. $y' = -c_1 e^x \sin x + c_1 e^x \cos x + c_2 e^x \cos x + c_2 e^x \sin x;$

$y'' = -c_1 e^x \cos x - c_1 e^x \sin x - c_1 e^x \sin x + c_1 e^x \cos x - c_2 e^x \sin x + c_2 e^x \cos x + c_2 e^x \cos x + c_2 e^x \sin x$

$\qquad = -2c_1 e^x \sin x + 2c_2 e^x \cos x;$

$y'' - 2y' = -2c_1 e^x \cos x - 2c_2 e^x \sin x = -2y; \qquad y'' - 2y' + 2y = 0$

9. b **12.** a, b, d

15. The slope of the tangent line at (x, y) is y', so the differential equation is $y' = x^2 + y^2$.

18. **(a)** Differentiating $y^2 - 2y = x^2 - x + c$ we obtain $2yy' - 2y' = 2x - 1$ or $(2y - 2)y' = 2x - 1$.

(b) Setting $x = 0$ and $y = 1$ in the solution we have $1 - 2 = 0 - 0 + c$ or $c = -1$. Thus, a solution of the initial-value problem is $y^2 - 2y = x^2 - x - 1$.

(c) Solving the equation $y^2 - 2y - (x^2 - x - 1) = 0$ by the quadratic formula we get $y = (2 \pm \sqrt{4 + 4(x^2 - x - 1)})/2 = 1 \pm \sqrt{x^2 - x} = 1 \pm \sqrt{x(x - 1)}$. Since $x(x - 1) \geq 0$ for $x \leq 0$ or $x \geq 1$, we see that neither $y = 1 + \sqrt{x(x - 1)}$ nor $y = 1 - \sqrt{x(x - 1)}$ is differentiable at $x = 0$. Thus, both functions are solutions of the differential equation, but neither is a solution of the initial-value problem.

21. (a)

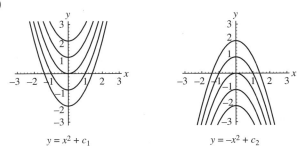

$y = x^2 + c_1$ $y = -x^2 + c_2$

(b) When $y = x^2 + c_1$, $y' = 2x$ and $(y')^2 = 4x^2$. When $y = -x^2 + c_2$, $y' = -2x$ and $(y')^2 = 4x^2$.

(c) Pasting together x^2, $x \geq 0$, and $-x^2$, $x \leq 0$, we get $y = \begin{cases} -x^2, & x \leq 0 \\ x^2, & x > 0. \end{cases}$

24. Differentiating $y = x \sin x + (\cos x) \ln(\cos x)$ we get

$$y' = x \cos x + \sin x + \cos x \left(\frac{-\sin x}{\cos x} \right) - (\sin x) \ln (\cos x)$$

$$= x \cos x + \sin x - \sin x - (\sin x) \ln (\cos x)$$

$$= x \cos x - (\sin x) \ln (\cos x)$$

and

$$y'' = -x \sin x + \cos x - \sin x \left(\frac{-\sin x}{\cos x} \right) - (\cos x) \ln (\cos x)$$

$$= -x \sin x + \cos x + \frac{\sin^2 x}{\cos x} - (\cos x) \ln (\cos x)$$

$$= -x \sin x + \cos x + \frac{1 - \cos^2 x}{\cos x} - (\cos x) \ln (\cos x)$$

$$= -x \sin x + \cos x + \sec x - \cos x - (\cos x) \ln (\cos x)$$

$$= -x \sin x + \sec x - (\cos x) \ln (\cos x).$$

Thus

$$y'' + y = -x \sin x + \sec x - (\cos x) \ln(\cos x) + x \sin x + (\cos x) \ln (\cos x) = \sec x.$$

To obtain an interval of definition we note that the domain of $\ln x$ is $(0, \infty)$, so we must have $\cos x > 0$. Thus, an interval of definition is $(-\pi/2, \pi/2)$.

In Problems 27–30 we use (12) of Section 1.1 and the Product Rule.

27.

$$y = e^{\cos x} \int_0^x t e^{-\cos t} \, dt$$

$$\frac{dy}{dx} = e^{\cos x} \left(x e^{-\cos x} \right) - \sin x e^{\cos x} \int_0^x t e^{-\cos t} \, dt$$

$$\frac{dy}{dx} + (\sin x)\, y = e^{\cos x} x e^{-\cos x} - \sin x e^{\cos x} \int_0^x t e^{-\cos t}\, dt + \sin x \left(e^{\cos x} \int_0^x t e^{-\cos t}\, dt \right)$$

$$= x - \sin x e^{\cos x} \int_0^x t e^{-\cos t}\, dt + \sin x e^{\cos x} \int_0^x t e^{-\cos t}\, dt = x$$

30.

$$y = \sin x \int_0^x e^{t^2} \cos t\, dt - \cos x \int_0^x e^{t^2} \sin t\, dt$$

$$y' = \sin x \left(e^{x^2} \cos x \right) + \cos x \int_0^x e^{t^2} \cos t\, dt - \cos x \left(e^{x^2} \sin x \right) + \sin x \int_0^x e^{t^2} \sin t\, dt$$

$$= \cos x \int_0^x e^{t^2} \cos t\, dt + \sin x \int_0^x e^{t^2} \sin t\, dt$$

$$y'' = \cos x \left(e^{x^2} \cos x \right) - \sin x \int_0^x e^{t^2} \cos t\, dt + \sin x \left(e^{x^2} \sin x \right) + \cos x \int_0^x e^{t^2} \sin t\, dt$$

$$= e^{x^2} \left(\cos^2 x + \sin^2 x \right) - \left(\overbrace{\sin x \int_0^x e^{t^2} \cos t\, dt - \cos x \int_0^x e^{t^2} \sin t\, dt}^{y} \right)$$

$$= e^{x^2} - y$$

$$y'' + y = e^{x^2} - y + y = e^{x^2}$$

33. Using implicit differentiation we get

$$y^3 + 3y = 2 - 3x$$

$$3y^2 y' + 3y' = -3$$

$$y^2 y' + y' = -1$$

$$(y^2 + 1)y' = -1$$

$$y' = \frac{-1}{y^2 + 1}$$

Differentiating the last line and remembering to use the quotient rule on the right side leads to

$$y'' = \frac{2yy'}{(y^2 + 1)^2}$$

Now since $y' = -1/(y^2 + 1)$ we can write the last equation as

$$y'' = \frac{2y}{(y^2 + 1)^2} y' = \frac{2y}{(y^2 + 1)^2} \frac{-1}{(y^2 + 1)} = 2y \left(\frac{-1}{y^2 + 1} \right)^3 = 2y(y')^3$$

which is what we wanted to show.

36. Substituting $y = c_1 + c_2 x$ and $y' = c_2$ into the left-hand side of the differential equation gives

$$y' + 2y = c_2 + 2(c_1 + c_2 x) = c_2 + 2c_1 + 2c_2 x.$$

Setting the result equal to the right-hand side of the differential equation yields

$$c_2 + 2c_1 + 2c_2 x = 3x$$

$$(c_2 + 2c_1) + 2c_2 x = 0 + 3x.$$

Therefore, $2c_2 = 3$ or $c_2 = \dfrac{3}{2}$, and $\dfrac{3}{2} + 2c_1 = 0$ or $c_1 = -\dfrac{3}{2} \cdot \dfrac{1}{2} = -\frac{3}{4}$. Thus $y = -\dfrac{3}{4} + \dfrac{3}{2} x$.

In Problem 39–42, $y = c_1 e^{-3x} + c_2 e^x + 4x$ is given as a two-parameter family of solutions of the second-order differential equation $y'' + 2y' - 3y = -12x + 8$.

39. If $y(0) = 0$ and $y'(0) = 0$, then

$$c_1 + c_2 = 0$$

$$-3c_1 + c_2 = -4$$

subtracting the second equation from the first gives us $4c_1 = 4$ or $c_1 = 1$, and thus $c_2 = -1$. Therefore $y = e^{-3x} - e^x + 4x$.

42. If $y(-1) = 1$ and $y'(-1) = 1$, then

$$c_1 e^3 + c_2 e^{-1} = 5$$

$$-3c_1 e^3 + c_2 e^{-1} = -3$$

subtracting the second equation from the first gives us $4c_1 = 8$ or $c_1 = 2e^{-3}$, and thus $c_2 = 3e$. Therefore $y = 2e^{-3x-3} + 3e^{x+1} + 4x$.

45. From the graph we see that estimates for y_0 and y_1 are $y_0 = -3$ and $y_1 = 0$.

Chapter 2

First-Order Differential Equations

2.1 | Solution Curves Without a Solution

3.

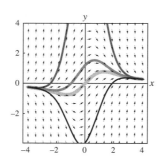

6.

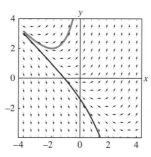

9.

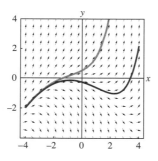

12.

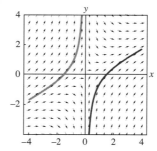

15. (a) The isoclines have the form $y = -x + c$, which are straight
lines with slope -1.

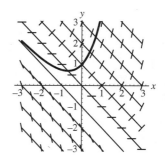

(b) The isoclines have the form $x^2 + y^2 = c$, which are circles
centered at the origin.

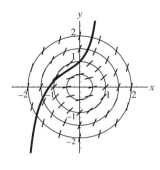

21. Solving $y^2 - 3y = y(y - 3) = 0$ we obtain the critical points 0 and 3. From the
phase portrait we see that 0 is asymptotically stable (attractor) and 3 is unstable
(repeller).

24. Solving $10 + 3y - y^2 = (5 - y)(2 + y) = 0$ we obtain the critical points -2 and 5.
From the phase portrait we see that 5 is asymptotically stable (attractor) and -2 is
unstable (repeller).

27. Solving $y \ln(y+2) = 0$ we obtain the critical points -1 and 0. From the phase portrait
we see that -1 is asymptotically stable (attractor) and 0 is unstable (repeller).

30. The critical points are approximately at $-2, 2$, 0.5, and 1.7. Since $f(y) > 0$ for $y < -2.2$
and $0.5 < y < 1.7$, the graph of the solution is increasing on the y-intervals $(-\infty, -2.2)$ and
$(0.5, 1.7)$. Since $f(y) < 0$ for $-2.2 < y < 0.5$ and $y > 1.7$, the graph is decreasing on the
y-interval $(-2.2, 0.5)$ and $(1.7, \infty)$.

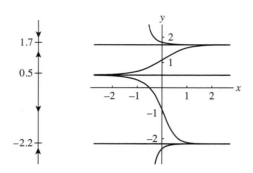

39. From the equation $dP/dt = k\left(P - h/k\right)$ we see that the only critical point of the autonomous differential equation is the positive number h/k. A phase portrait shows that this point is unstable, that is, h/k is a repeller. For any initial condition $P(0) = P_0$ for which $0 < P_0 < h/k$, $dP/dt < 0$ which means $P(t)$ is monotonic decreasing and so the graph of $P(t)$ must cross the t-axis or the line $P - 0$ at some time $t_1 > 0$. But $P(t_1) = 0$ means the population is extinct at time t_1.

42. (a) From the phase portrait we see that critical points are α and β. Let $X(0) = X_0$. If $X_0 < \alpha$, we see that $X \to \alpha$ as $t \to \infty$. If $\alpha < X_0 < \beta$, we see that $X \to \alpha$ as $t \to \infty$. If $X_0 > \beta$, we see that $X(t)$ increases in an unbounded manner, but more specific behavior of $X(t)$ as $t \to \infty$ is not known.

(b) When $\alpha = \beta$ the phase portrait is as shown. If $X_0 < \alpha$, then $X(t) \to \alpha$ as $t \to \infty$. If $X_0 > \alpha$, then $X(t)$ increases in an unbounded manner. This could happen in a finite amount of time. That is, the phase portrait does not indicate that X becomes unbounded as $t \to \infty$.

(c) When $k = 1$ and $\alpha = \beta$ the differential equation is $dX/dt = (\alpha - X)^2$. For $X(t) = \alpha - 1/(t+c)$ we have $dX/dt = 1/(t+c)^2$ and

$$(\alpha - X)^2 = \left[\alpha - \left(\alpha - \frac{1}{t+c}\right)\right]^2 = \frac{1}{(t+c)^2} = \frac{dX}{dt} \ .$$

For $X(0) = \alpha/2$ we obtain

$$X(t) = \alpha - \frac{1}{t + 2/\alpha} \ .$$

For $X(0) = 2\alpha$ we obtain

$$X(t) = \alpha - \frac{1}{t - 1/\alpha}.$$

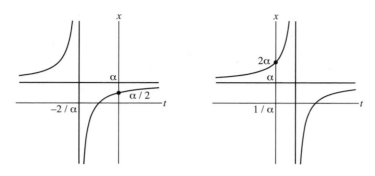

For $X_0 > \alpha$, $X(t)$ increases without bound up to $t = 1/\alpha$. For $t > 1/\alpha$, $X(t)$ increases but $X \to \alpha$ as $t \to \infty$.

2.2 | Separable Equations

In many of the following problems we will encounter an expression of the form $\ln |g(y)| = f(x)+c$. To solve for $g(y)$ we exponentiate both sides of the equation. This yields $|g(y)| = e^{f(x)+c} = e^c e^{f(x)}$ which implies $g(y) = \pm e^c e^{f(x)}$. Letting $c_1 = \pm e^c$ we obtain $g(y) = c_1 e^{f(x)}$.

3. From $dy = -e^{-3x}\, dx$ we obtain $y = \frac{1}{3}e^{-3x} + c$.

6. From $\frac{1}{y^2}\, dy = -2x\, dx$ we obtain $-\frac{1}{y} = -x^2 + c$ or $y = \frac{1}{x^2 + c_1}$.

9. From $\left(y + 2 + \frac{1}{y}\right) dy = x^2 \ln x\, dx$ we obtain $\frac{y^2}{2} + 2y + \ln |y| = \frac{x^3}{3} \ln |x| - \frac{1}{9}x^3 + c$.

12. From $2y\, dy = -\frac{\sin 3x}{\cos^3 3x}\, dx$ or $2y\, dy = -\tan 3x \sec^2 3x\, dx$ we obtain $y^2 = -\frac{1}{6}\sec^2 3x + c$.

15. From $\frac{1}{S}\, dS = k\, dr$ we obtain $S = ce^{kr}$.

18. From $\frac{1}{N}\, dN = \left(te^{t+2} - 1\right) dt$ we obtain $\ln |N| = te^{t+2} - e^{t+2} - t + c$ or $N = c_1 e^{te^{t+2} - e^{t+2} - t}$.

21. From $x\, dx = \frac{1}{\sqrt{1 - y^2}}\, dy$ we obtain $\frac{1}{2}x^2 = \sin^{-1} y + c$ or $y = \sin\left(\frac{x^2}{2} + c_1\right)$.

24. From $\frac{1}{y^2 - 1}\, dy = \frac{1}{x^2 - 1}\, dx$ or $\frac{1}{2}\left(\frac{1}{y - 1} - \frac{1}{y + 1}\right) dy = \frac{1}{2}\left(\frac{1}{x - 1} - \frac{1}{x + 1}\right) dx$ we obtain

$\ln |y - 1| - \ln |y + 1| = \ln |x - 1| - \ln |x + 1| + \ln c$ or $\frac{y - 1}{y + 1} = \frac{c(x - 1)}{x + 1}$. Using $y(2) = 2$ we

find $c = 1$. A solution of the initial-value problem is $\frac{y - 1}{y + 1} = \frac{x - 1}{x + 1}$ or $y = x$.

27. Separating variables and integrating we obtain

$$\frac{dx}{\sqrt{1-x^2}} - \frac{dy}{\sqrt{1-y^2}} = 0 \quad \text{and} \quad \sin^{-1} x - \sin^{-1} y = c.$$

Setting $x = 0$ and $y = \sqrt{3}/2$ we obtain $c = -\pi/3$. Thus an implicit solution of the initial-value problem is $\sin^{-1} x - \sin^{-1} y = \pi/3$. Solving for y and using an addition formula from trigonometry, we get

$$y = \sin\left(\sin^{-1} x + \frac{\pi}{3}\right) = x \cos\frac{\pi}{3} + \sqrt{1-x^2}\sin\frac{\pi}{3} = \frac{x}{2} + \frac{\sqrt{3}\sqrt{1-x^2}}{2}.$$

30. From $\dfrac{\sinh y}{\cosh y} = \dfrac{dx}{x}$ we obtain

$$\ln\left(\cosh y\right) = \ln x + \ln c.$$

Using $y(1) = 0$ we find $0 = 0 + \ln c$ or $c = 1$. Thus an implicit solution of the initial-value problem is $\cosh y = x$. Solving for y we get

$$y = \cosh^{-1} x$$

$$y = \ln\left(x + \sqrt{x^2 - 1}\right), \ x \ge 1.$$

33. Separating variables and then proceeding as in Example 6 we get

$$\frac{dy}{dx} = \left(1 + y^2\right)\sqrt{1 + \cos x^3}$$

$$\frac{1}{1+y^2}\frac{dy}{dx} = \sqrt{1 + \cos x^3}$$

$$\int_1^x \frac{1}{1+y^2(t)}\frac{dy}{dt}\, dt = \int_1^x \sqrt{1 + \cos t^3}\, dt$$

$$\tan^{-1} y(t)\Big|_1^x = \int_1^x \sqrt{1 + \cos t^3}\, dt$$

$$\tan^1 y(x) - \tan^{-1} y(1) = \int_1^x \sqrt{1 + \cos t^3}\, dt$$

$$\tan^1 y(x) - \frac{\pi}{4} = \int_1^x \sqrt{1 + \cos t^3}\, dt$$

$$\tan^1 y(x) = \frac{\pi}{4} + \int_1^x \sqrt{1 + \cos t^3}\, dt$$

$$y(x) = \tan\left(\frac{\pi}{4} + \int_1^x \sqrt{1 + \cos t^3}\, dt\right)$$

36. Separating variables we get

$$(2y - 2)\,\frac{dy}{dx} = 3x^2 + 4x + 2$$

$$(2y - 2)\,dy = \left(3x^2 + 4x + 2\right)\,dx$$

$$\int (2y - 2)\,dy = \int \left(3x^2 + 4x + 2\right)\,dx$$

$$\int 2\,(y - 1)\,dy = \int \left(3x^2 + 4x + 2\right)\,dx$$

$$(y - 1)^2 = x^3 + 2x^2 + 2x + c$$

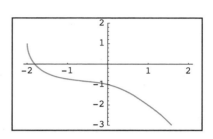

The condition $y(1) = -2$ implies $c = 4$. Thus $y = 1 - \sqrt{x^3 + 2x^2 + 2x + 4}$ where the minus sign is indicated by the initial condition. Now $x^3 + 2x^2 + 2x + 4 = (x + 2)\left(x^2 + 1\right) > 0$ implies $x > -2$, so the interval of definition is $(-2, \infty)$.

39. (a) The equilibrium solutions $y(x) = 2$ and $y(x) = -2$ satisfy the initial conditions $y(0) = 2$ and $y(0) = -2$, respectively. Setting $x = \frac{1}{4}$ and $y = 1$ in $y = 2(1 + ce^{4x})/(1 - ce^{4x})$ we obtain

$$1 = 2\frac{1 + ce}{1 - ce}, \quad 1 - ce = 2 + 2ce, \quad -1 = 3ce, \quad \text{and} \quad c = -\frac{1}{3e}.$$

The solution of the corresponding initial-value problem is

$$y = 2\frac{1 - \frac{1}{3}e^{4x-1}}{1 + \frac{1}{3}e^{4x-1}} = 2\frac{3 - e^{4x-1}}{3 + e^{4x-1}}.$$

(b) Separating variables and integrating yields

$$\frac{1}{4}\ln|y - 2| - \frac{1}{4}\ln|y + 2| + \ln c_1 = x$$

$$\ln|y - 2| - \ln|y + 2| + \ln c = 4x$$

$$\ln\left|\frac{c(y - 2)}{y + 2}\right| = 4x$$

$$c\,\frac{y - 2}{y + 2} = e^{4x}.$$

Solving for y we get $y = 2(c + e^{4x})/(c - e^{4x})$. The initial condition $y(0) = -2$ implies $2(c + 1)/(c - 1) = -2$ which yields $c = 0$ and $y(x) = -2$. The initial condition $y(0) = 2$ does not correspond to a value of c, and it must simply be recognized that $y(x) = 2$ is a solution of the initial-value problem. Setting $x = \frac{1}{4}$ and $y = 1$ in $y = 2(c + e^{4x})/(c - e^{4x})$ leads to $c = -3e$. Thus a solution of the initial-value problem is

$$y = 2\frac{-3e + e^{4x}}{-3e - e^{4x}} = 2\frac{3 - e^{4x-1}}{3 + e^{4x-1}}.$$

42. Differentiating $\ln{(x^2 + 10)} + \csc y = c$ we get

$$\frac{2x}{x^2 + 10} - \csc y \, \cot y \, \frac{dy}{dx} = 0,$$

$$\frac{2x}{x^2 + 10} - \frac{1}{\sin y} \cdot \frac{\cos y}{\sin y} \frac{dy}{dx} = 0,$$

or

$$2x \sin^2 y \, dx - (x^2 + 10) \cos y \, dy = 0.$$

Writing the differential equation in the form

$$\frac{dy}{dx} = \frac{2x \sin^2 y}{(x^2 + 10) \cos y}$$

we see that singular solutions occur when $\sin^2 y = 0$, or $y = k\pi$, where k is an integer.

45. Separating variables we obtain $\dfrac{dy}{(y-1)^2 + 0.01} = dx$. Then

$$10 \tan^{-1} 10(y-1) = x + c \quad \text{and} \quad y = 1 + \frac{1}{10} \tan \frac{x+c}{10}.$$

Setting $x = 0$ and $y = 1$ we obtain $c = 0$. The solution is

$$y = 1 + \frac{1}{10} \tan \frac{x}{10}.$$

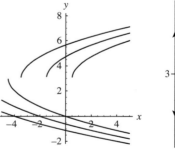

48. (a) The second derivative of y is

$$\frac{d^2y}{dx^2} = -\frac{dy/dx}{(y-1)^2} = -\frac{1/(y-3)}{(y-3)^2} = -\frac{1}{(y-3)^3}.$$

The solution curve is concave down when $d^2y/dx^2 < 0$ or $y > 3$, and concave up when $d^2y/dx^2 > 0$ or $y < 3$. From the phase portrait we see that the solution curve is decreasing when $y < 3$ and increasing when $y > 3$.

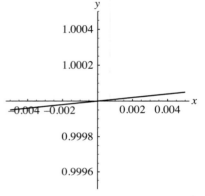

(b) Separating variables and integrating we obtain

$$(y - 3) \, dy = dx$$

$$\frac{1}{2}y^2 - 3y = x + c$$

$$y^2 - 6y + 9 = 2x + c_1$$

$$(y - 3)^2 = 2x + c_1$$

$$y = 3 \pm \sqrt{2x + c_1}.$$

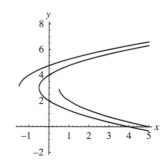

The initial condition dictates whether to use the plus or minus sign.

When $y_1(0) = 4$ we have $c_1 = 1$ and $y_1(x) = 3 + \sqrt{2x+1}$.

When $y_2(0) = 2$ we have $c_1 = 1$ and $y_2(x) = 3 - \sqrt{2x+1}$.

When $y_3(1) = 2$ we have $c_1 = -1$ and $y_3(x) = 3 - \sqrt{2x-1}$.

When $y_4(-1) = 4$ we have $c_1 = 3$ and $y_4(x) = 3 + \sqrt{2x+3}$.

51. Separating variables, we have

$$\frac{dy}{\sqrt{y}+y} = \frac{dx}{\sqrt{x}+x}.$$

Integrate $\dfrac{dx}{\sqrt{x}+x}$ using u-substitution with $u^2 = x$. So $2u\,du = dx$.

$$\int \frac{dx}{\sqrt{x}+x} = \int \frac{2u\,du}{u+u^2} = 2\int \frac{du}{1+u} = 2\ln(1+u) + c_0 = 2\ln\left(1+\sqrt{x}\right) + c_0.$$

Similarly,

$$\int \frac{dy}{\sqrt{y}+y} = 2\ln\left(1+\sqrt{y}\right) + c_1.$$

Thus

$$2\ln\left(1+\sqrt{y}\right) + c_1 = 2\ln\left(1+\sqrt{x}\right) + c_0$$

$$2\ln\left(1+\sqrt{y}\right) = 2\ln\left(1+\sqrt{x}\right) + (c_0 - c_1)$$

$$\ln\left(1+\sqrt{y}\right) = \ln\left(1+\sqrt{x}\right) + 2(c_0 - c_1) = \ln\left(1+\sqrt{x}\right) + c_2$$

$$1 + \sqrt{y} = c_3\left(1+\sqrt{x}\right)$$

$$\sqrt{y} = c_3\left(1+\sqrt{x}\right) - 1$$

$$y = \left[c_3\left(1+\sqrt{x}\right) - 1\right]^2.$$

54. Separating variables, we have

$$y\,dy = x\tan^{-1}x\,dx.$$

Using integration by parts on the right-hand side yields

$$\int y\,dy = \int x\tan^{-1}x\,dx$$

$$\frac{1}{2}y^2 = \frac{1}{2}x^2\tan^{-1}x - \frac{1}{2}x + \frac{1}{2}\tan^{-1}x + c_1$$

$$y^2 = x^2\tan^{-1}x - x + \tan^{-1}x + c_2$$

$$y = \sqrt{x^2\tan^{-1}x - x + \tan^{-1}x + c_2}.$$

Solve for c_2 using $y(0) = 3$.

$$3 = \sqrt{c_2}$$

$$9 = c_2.$$

Hence $y = \sqrt{x^2 \tan^{-1} x - x + \tan^{-1} x + 9}$.

63. Separating variables of $1 + \left(\dfrac{dx}{dy}\right)^2 = \dfrac{a}{y}$, we have

$$\left(\frac{dx}{dy}\right)^2 = \frac{a}{y} - 1 = \frac{a - y}{y}$$

$$\frac{dx}{dy} = \sqrt{\frac{a - y}{y}}$$

$$dx = \sqrt{\frac{a - y}{y}}\, dy$$

To integrate the right-hand side we use the substitution $y = a \sin^2 \phi$. So $dy = 2a \sin \phi \cos \phi\, d\phi$.

$$\int \sqrt{\frac{a - y}{y}}\, dy = \int \sqrt{\frac{a - a \sin^2 \phi}{a \sin^2 \phi}}\, 2a \sin \phi \cos \phi\, d\phi = \int \sqrt{\frac{a \cos^2 \phi}{a \sin^2 \phi}}\, 2a \sin \phi \cos \phi\, d\phi$$

$$= 2a \int \frac{\cos \phi}{\sin \phi} \sin \phi \cos \phi\, d\phi = 2a \int \cos^2 \phi\, d\phi = 2a \int \frac{1 + \cos 2\phi}{2}\, d\phi$$

$$= a \int (1 + \cos 2\phi)\, d\phi = a \left(\phi + \frac{1}{2} \sin 2\phi\right) + c_1 = \frac{a}{2}(2\phi + \sin 2\phi) + c_1.$$

Therefore the solution to the differential equation is

$$\int dx = \int \sqrt{\frac{a - y}{y}}\, dy$$

$$x = \frac{a}{2}(2\phi + \sin 2\phi) + c_1.$$

When $y = 0$, we get $0 = a \sin^2 \phi$ or $\phi = 0$. Since the curve passes through $(0, 0)$,

$$0 = \frac{a}{2}(0 + \sin 0) + c_1$$

$$0 = c_1$$

Using the identity $\sin^2 \phi = \dfrac{1}{2}(1 - \cos 2\phi)$, parametric equations for the curve are

$$x = \frac{a}{2}(2\phi + \sin 2\phi)$$

$$y = a \sin^2 \phi = \frac{a}{2}(1 - \cos 2\phi).$$

It can be shown these are parametric equations of a cycloid.

2.3 Linear Equations

3. For $y' + y = e^{3x}$ an integrating factor is $e^{\int dx} = e^x$ so that $\dfrac{d}{dx}\left[e^x y\right] = e^{4x}$ and $y = \frac{1}{4}e^{3x} + ce^{-x}$

for $-\infty < x < \infty$. The transient term is ce^{-x}.

6. For $y' + 2xy = x^3$ an integrating factor is $e^{\int 2x\,dx} = e^{x^2}$ so that $\dfrac{d}{dx}\left[e^{x^2} y\right] = x^3 e^{x^2}$ and

$y = \frac{1}{2}x^2 - \frac{1}{2} + ce^{-x^2}$ for $-\infty < x < \infty$. The transient term is ce^{-x^2}.

9. For $y' - \dfrac{1}{x}y = x\sin x$ an integrating factor is $e^{-\int (1/x)\,dx} = \dfrac{1}{x}$ so that $\dfrac{d}{dx}\left[\dfrac{1}{x}y\right] = \sin x$ and

$y = cx - x\cos x$ for $0 < x < \infty$.

12. For $y' - \dfrac{x}{(1+x)}y = x$ an integrating factor is $e^{-\int [x/(1+x)]dx} = (x+1)e^{-x}$ so that

$\dfrac{d}{dx}\left[(x+1)e^{-x}y\right] = x(x+1)e^{-x}$ and $y = -x - \dfrac{2x+3}{x+1} + \dfrac{ce^x}{x+1}$ for $-1 < x < \infty$.

15. For $\dfrac{dx}{dy} - \dfrac{4}{y}x = 4y^5$ an integrating factor is $e^{-\int (4/y)\,dy} = e^{\ln y^{-4}} = y^{-4}$ so that $\dfrac{d}{dy}\left[y^{-4}x\right] = 4y$

and $x = 2y^6 + cy^4$ for $0 < y < \infty$.

18. For $y' + (\cot x)y = \sec^2 x\csc x$ an integrating factor is $e^{\int \cot x\,dx} = e^{\ln|\sin x|} = \sin x$ so that

$\dfrac{d}{dx}\left[(\sin x)\,y\right] = \sec^2 x$ and $y = \sec x + c\csc x$ for $0 < x < \pi/2$.

21. For $\dfrac{dr}{d\theta} + r\sec\theta = \cos\theta$ an integrating factor is $e^{\int \sec\theta\,d\theta} = e^{\ln|\sec x + \tan x|} = \sec\theta + \tan\theta$ so

that $\dfrac{d}{d\theta}\left[(\sec\theta + \tan\theta)r\right] = 1 + \sin\theta$ and $(\sec\theta + \tan\theta)r = \theta - \cos\theta + c$ for $-\pi/2 < \theta < \pi/2$.

24. For $y' + \dfrac{2}{x^2 - 1}y = \dfrac{x+1}{x-1}$ an integrating factor is $e^{\int [2/(x^2-1)]dx} = \dfrac{x-1}{x+1}$

so that $\dfrac{d}{dx}\left[\dfrac{x-1}{x+1}y\right] = 1$ and $(x-1)y = x(x+1) + c(x+1)$ for $-1 < x < 1$.

27. For $\dfrac{di}{dt} + \dfrac{R}{L}i = \dfrac{E}{L}$ an integrating factor is $e^{\int (R/L)\,dt} = e^{Rt/L}$ so that $\dfrac{d}{dt}\left[e^{Rt/L}\,i\right] = \dfrac{E}{L}e^{Rt/L}$

and $i = \dfrac{E}{R} + ce^{-Rt/L}$ for $-\infty < t < \infty$. If $i(0) = i_0$ then $c = i_0 - E/R$ and $i = \dfrac{E}{R} +$

$\left(i_0 - \dfrac{E}{R}\right)e^{-Rt/L}$.

30. For $y' + (\tan x)y = \cos^2 x$ an integrating factor is $e^{\int \tan x \, dx} = e^{\ln|\sec x|} = \sec x$ so that

$$\frac{d}{dx}\left[(\sec x)\,y\right] = \cos x \text{ and } y = \sin x \cos x + c \cos x \text{ for } -\pi/2 < x < \pi/2. \text{ If } y(0) = -1$$

then $c = -1$ and $y = \sin x \cos x - \cos x$.

33. For $y' + 2y = f(x)$ an integrating factor is e^{2x} so that

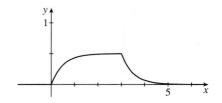

$$ye^{2x} = \begin{cases} \dfrac{1}{2}e^{2x} + c_1, & 0 \le x \le 3 \\[2mm] c_2, & x > 3. \end{cases}$$

If $y(0) = 0$ then $c_1 = -1/2$ and for continuity we must have $c_2 = \frac{1}{2}e^6 - \frac{1}{2}$ so that

$$y = \begin{cases} \dfrac{1}{2}(1 - e^{-2x}), & 0 \le x \le 3 \\[2mm] \dfrac{1}{2}(e^6 - 1)e^{-2x}, & x > 3. \end{cases}$$

36. For

$$y' + \frac{2x}{1+x^2}y = \begin{cases} \dfrac{x}{1+x^2}, & 0 \le x \le 1 \\[2mm] \dfrac{-x}{1+x^2}, & x > 1, \end{cases}$$

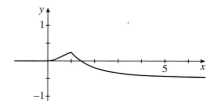

an integrating factor is $1 + x^2$ so that

$$(1 + x^2)\,y = \begin{cases} \dfrac{1}{2}x^2 + c_1, & 0 \le x \le 1 \\[2mm] -\dfrac{1}{2}x^2 + c_2, & x > 1. \end{cases}$$

If $y(0) = 0$ then $c_1 = 0$ and for continuity we must have $c_2 = 1$ so that

$$y = \begin{cases} \dfrac{1}{2} - \dfrac{1}{2\,(1+x^2)}, & 0 \le x \le 1 \\[2mm] \dfrac{3}{2\,(1+x^2)} - \dfrac{1}{2}, & x > 1. \end{cases}$$

39. We rewrite equation $dx = \left(x + y^2\right) dy$ as

$$\frac{dx}{dy} = x + y^2$$

$$\frac{dx}{dy} - x = y^2.$$

This equation is linear in x, so an integrating factor is $e^{\int -1\,dy} = e^{-y}$, so

$$\frac{d}{dy}\left[e^{-y}x\right] = y^2 e^{-y}$$

$$e^{-y}x = \int y^2 e^{-y}\,dy = -y^2 e^{-y} - 2y e^{-y} - 2e^{-y} + c$$

$$x = -y^2 - 2y - 2 + ce^y.$$

42. An integrating factor for $y' - 2xy = -1$ is e^{-x^2}. Thus

$$\frac{d}{dx}[e^{-x^2}y] = -e^{-x^2}$$

$$e^{-x^2}y = -\int_0^x e^{-t^2}\,dt = -\frac{\sqrt{\pi}}{2}\operatorname{erf}(x) + c.$$

From $y(0) = \sqrt{\pi}/2$, and noting that $\operatorname{erf}(0) = 0$, we get $c = \sqrt{\pi}/2$. Thus

$$y = e^{x^2}\left(-\frac{\sqrt{\pi}}{2}\operatorname{erf}(x) + \frac{\sqrt{\pi}}{2}\right) = \frac{\sqrt{\pi}}{2}e^{x^2}(1 - \operatorname{erf}(x)) = \frac{\sqrt{\pi}}{2}e^{x^2}\operatorname{erfc}(x).$$

45. For $y' + xy = \cos x$ an integrating factor is $e^{x^2/2}$. Thus

$$\frac{d}{dx}\left[e^{x^2/2}y\right] = e^{x^2/2}\cos x \quad\text{and}\quad e^{x^2/2}y = \int_0^x e^{t^2/2}\cos t\,dt + c.$$

From $y(0) = 2$ we get $c = 2$, so $y = 2e^{-x^2/2} + e^{-x^2/2}\int_0^x e^{t^2/2}\cos t\,dt$.

48. The integrating factor for $y' - (\sin x^2)\,y = 0$ is $e^{-\int_0^x \sin t^2\,dt}$. Then

$$\frac{d}{dx}\left[e^{-\int_0^x \sin t^2\,dt}y\right] = 0$$

$$e^{-\int_0^x \sin t^2\,dt}y = c_1$$

$$y = c_1 e^{\int_0^x \sin t^2\,dt}$$

Letting $t = \sqrt{\pi/2}\,u$ we have $dt = \sqrt{\pi/2}\,du$ and

$$\int_0^x \sin t^2\,dt = \sqrt{\frac{\pi}{2}}\int_0^{\sqrt{2/\pi}\,x} \sin\left(\frac{\pi}{2}u^2\right)\,du = \sqrt{\frac{\pi}{2}}\,S\left(\sqrt{\frac{2}{\pi}}\,x\right)$$

so $y = c_1 e^{\sqrt{\pi/2}\,S\left(\sqrt{2/\pi}\,x\right)}$. Using $S(0) = 0$ and $y(0) = c_1 = 5$ we have $y = 5e^{\sqrt{\pi/2}\,S\left(\sqrt{2/\pi}\,x\right)}$.

57. Writing the differential equation as $\dfrac{dE}{dt} + \dfrac{1}{RC}E = 0$ we see that an integrating factor is $e^{t/RC}$. Then

$$\frac{d}{dt}\left[e^{t/RC}E\right] = 0$$

$$e^{t/RC}E = c$$

$$E = ce^{-t/RC}$$

From $E(4) = ce^{-4/RC} = E_0$ we find $c = E_0 e^{4/RC}$. Thus the solution of the initial-value problem is

$$E = E_0 e^{4/RC} e^{-t/RC} = E_0 e^{-(t-4)/RC}.$$

2.4 | Exact Equations

3. Let $M = 5x + 4y$ and $N = 4x - 8y^3$ so that $M_y = 4 = N_x$. From $f_x = 5x + 4y$ we obtain $f = \frac{5}{2}x^2 + 4xy + h(y)$, $h'(y) = -8y^3$, and $h(y) = -2y^4$. A solution is $\frac{5}{2}x^2 + 4xy - 2y^4 = c$.

6. Let $M = 4x^3 - 3y\sin 3x - y/x^2$ and $N = 2y - 1/x + \cos 3x$ so that $M_y = -3\sin 3x - 1/x^2$ and $N_x = 1/x^2 - 3\sin 3x$. The equation is not exact.

9. Let $M = y^3 - y^2\sin x - x$ and $N = 3xy^2 + 2y\cos x$ so that $M_y = 3y^2 - 2y\sin x = N_x$. From $f_x = y^3 - y^2\sin x - x$ we obtain $f = xy^3 + y^2\cos x - \frac{1}{2}x^2 + h(y)$, $h'(y) = 0$, and $h(y) = 0$. A solution is $xy^3 + y^2\cos x - \frac{1}{2}x^2 = c$.

12. Let $M = 3x^2y + e^y$ and $N = x^3 + xe^y - 2y$ so that $M_y = 3x^2 + e^y = N_x$. From $f_x = 3x^2y + e^y$ we obtain $f = x^3y + xe^y + h(y)$, $h'(y) = -2y$, and $h(y) = -y^2$. A solution is $x^3y + xe^y - y^2 = c$.

15. Let $M = x^2y^3 - 1/(1+9x^2)$ and $N = x^3y^2$ so that $M_y = 3x^2y^2 = N_x$. From $f_x = x^2y^3 - 1/(1+9x^2)$ we obtain $f = \frac{1}{3}x^3y^3 - \frac{1}{3}\arctan(3x) + h(y)$, $h'(y) = 0$, and $h(y) = 0$. A solution is $x^3y^3 - \arctan(3x) = c$.

18. Let $M = 2y\sin x\cos x - y + 2y^2e^{xy^2}$ and $N = -x + \sin^2 x + 4xye^{xy^2}$ so that

$$M_y = 2\sin x\cos x - 1 + 4xy^3e^{xy^2} + 4ye^{xy^2} = N_x.$$

From $f_x = 2y\sin x\cos x - y + 2y^2e^{xy^2}$ we obtain $f = y\sin^2 x - xy + 2e^{xy^2} + h(y)$, $h'(y) = 0$, and $h(y) = 0$. A solution is $y\sin^2 x - xy + 2e^{xy^2} = c$.

21. Let $M = x^2 + 2xy + y^2$ and $N = 2xy + x^2 - 1$ so that $M_y = 2(x+y) = N_x$. From $f_x = x^2 + 2xy + y^2$ we obtain $f = \frac{1}{3}x^3 + x^2y + xy^2 + h(y)$, $h'(y) = -1$, and $h(y) = -y$. The solution is $\frac{1}{3}x^3 + x^2y + xy^2 - y = c$. If $y(1) = 1$ then $c = 4/3$ and a solution of the initial-value problem is $\frac{1}{3}x^3 + x^2y + xy^2 - y = \frac{4}{3}$.

24. Let $M = t/2y^4$ and $N = (3y^2 - t^2)/y^5$ so that $M_y = -2t/y^5 = N_t$. From $f_t = t/2y^4$ we obtain $f = \dfrac{t^2}{4y^4} + h(y)$, $h'(y) = \dfrac{3}{y^3}$, and $h(y) = -\dfrac{3}{2y^2}$. The solution is $\dfrac{t^2}{4y^4} - \dfrac{3}{2y^2} = c$. If $y(1) = 1$ then $c = -5/4$ and a solution of the initial-value problem is $\dfrac{t^2}{4y^4} - \dfrac{3}{2y^2} = -\dfrac{5}{4}$.

27. Equating $M_y = 3y^2 + 4kxy^3$ and $N_x = 3y^2 + 40xy^3$ we obtain $k = 10$.

30. Let $M = (x^2 + 2xy - y^2)/(x^2 + 2xy + y^2)$ and $N = (y^2 + 2xy - x^2)/(y^2 + 2xy + x^2)$ so that $M_y = -4xy/(x+y)^3 = N_x$. From $f_x = (x^2 + 2xy + y^2 - 2y^2)/(x+y)^2$ we obtain

$f = x + \dfrac{2y^2}{x+y} + h(y)$, $h'(y) = -1$, and $h(y) = -y$. A solution of the differential equation is $x^2 + y^2 = c(x+y)$.

33. We note that $(M_y - N_x)/N = 1/x$, so an integrating factor is $e^{\int dx/x} = x$. Let $M = 2xy^2 + 3x^2$ and $N = 2x^2y$ so that $M_y = 4xy = N_x$. From $f_x = 2xy^2 + 3x^2$ we obtain $f = x^2y^2 + x^3 + h(y)$, $h'(y) = 0$, and $h(y) = 0$. A solution of the differential equation is $x^2y^2 + x^3 = c$.

36. We note that $(M_y - N_x)/N = -\cot x$, so an integrating factor is $e^{-\int \cot x\, dx} = \csc x$. Let $M = \cos x \csc x = \cot x$ and $N = (1 + 2/y)\sin x \csc x = 1 + 2/y$, so that $M_y = 0 = N_x$. From $f_x = \cot x$ we obtain $f = \ln(\sin x) + h(y)$, $h'(y) = 1 + 2/y$, and $h(y) = y + \ln y^2$. A solution of the differential equation is $\ln(\sin x) + y + \ln y^2 = c$.

39. We note that $(M_y - N_x)/N = 2x/(4 + x^2)$, so an integrating factor is $e^{-2\int x\, dx/(4+x^2)} = 1/(4 + x^2)$. Let $M = x/(4 + x^2)$ and $N = (x^2y + 4y)/(4 + x^2) = y$, so that $M_y = 0 = N_x$. From $f_x = x(4 + x^2)$ we obtain $f = \frac{1}{2}\ln(4 + x^2) + h(y)$, $h'(y) = y$, and $h(y) = \frac{1}{2}y^2$. A solution of the differential equation is $\frac{1}{2}\ln(4 + x^2) + \frac{1}{2}y^2 = c$. Multiplying both sides by 2 the last equation can be written as $e^{y^2}(x^2 + 4) = c_1$. Using the initial condition $y(4) = 0$ we see that $c_1 = 20$. A solution of the initial-value problem is $e^{y^2}(x^2 + 4) = 20$.

2.5 | Solutions by Substitutions

3. Letting $x = vy$ we have

$$vy(v\,dy + y\,dv) + (y - 2vy)\,dy = 0$$

$$vy^2\,dv + y\left(v^2 - 2v + 1\right)dy = 0$$

$$\frac{v\,dv}{(v-1)^2} + \frac{dy}{y} = 0$$

$$\ln|v - 1| - \frac{1}{v-1} + \ln|y| = c$$

$$\ln\left|\frac{x}{y} - 1\right| - \frac{1}{x/y - 1} + \ln|y| = c$$

$$(x - y)\ln|x - y| - y = c(x - y).$$

6. Letting $y = ux$ and using partial fractions, we have

$$\left(u^2x^2 + ux^2\right)dx + x^2(u\,dx + x\,du) = 0$$

$$x^2\left(u^2 + 2u\right)dx + x^3\,du = 0$$

$$\frac{dx}{x} + \frac{du}{u(u+2)} = 0$$

$$\ln|x| + \frac{1}{2}\ln|u| - \frac{1}{2}\ln|u+2| = c$$

$$\frac{x^2 u}{u+2} = c_1$$

$$x^2 \frac{y}{x} = c_1\left(\frac{y}{x} + 2\right)$$

$$x^2 y = c_1(y + 2x).$$

9. Letting $y = ux$ we have

$$-ux\,dx + (x + \sqrt{u}\,x)(u\,dx + x\,du) = 0$$

$$(x^2 + x^2\sqrt{u}\,)\,du + xu^{3/2}\,dx = 0$$

$$\left(u^{-3/2} + \frac{1}{u}\right)du + \frac{dx}{x} = 0$$

$$-2u^{-1/2} + \ln|u| + \ln|x| = c$$

$$\ln|y/x| + \ln|x| = 2\sqrt{x/y} + c$$

$$y(\ln|y| - c)^2 = 4x.$$

12. Letting $y = ux$ we have

$$(x^2 + 2u^2 x^2)dx - ux^2(u\,dx + x\,du) = 0$$

$$x^2(1 + u^2)dx - ux^3\,du = 0$$

$$\frac{dx}{x} - \frac{u\,du}{1+u^2} = 0$$

$$\ln|x| - \frac{1}{2}\ln(1 + u^2) = c$$

$$\frac{x^2}{1+u^2} = c_1$$

$$x^4 = c_1(x^2 + y^2).$$

Using $y(-1) = 1$ we find $c_1 = 1/2$. The solution of the initial-value problem is $2x^4 = y^2 + x^2$.

15. From $y' + \frac{1}{x}y = \frac{1}{x}y^{-2}$ and $w = y^3$ we obtain $\frac{dw}{dx} + \frac{3}{x}w = \frac{3}{x}$. An integrating factor is x^3 so that $x^3 w = x^3 + c$ or $y^3 = 1 + cx^{-3}$.

18. From $y' - \left(1 + \frac{1}{x}\right)y = y^2$ and $w = y^{-1}$ we obtain $\frac{dw}{dx} + \left(1 + \frac{1}{x}\right)w = -1$. An integrating factor is xe^x so that $xe^x w = -xe^x + e^x + c$ or $y^{-1} = -1 + \frac{1}{x} + \frac{c}{x}e^{-x}$.

21. From $y' - \frac{2}{x}y = \frac{3}{x^2}y^4$ and $w = y^{-3}$ we obtain $\frac{dw}{dx} + \frac{6}{x}w = -\frac{9}{x^2}$. An integrating factor is x^6 so that $x^6 w = -\frac{9}{5}x^5 + c$ or $y^{-3} = -\frac{9}{5}x^{-1} + cx^{-6}$. If $y(1) = \frac{1}{2}$ then $c = \frac{49}{5}$ and $y^{-3} = -\frac{9}{5}x^{-1} + \frac{49}{5}x^{-6}$.

24. Let $u = x + y$ so that $du/dx = 1 + dy/dx$. Then $\dfrac{du}{dx} - 1 = \dfrac{1-u}{u}$ or $u\,du = dx$. Thus $\frac{1}{2}u^2 = x + c$ or $u^2 = 2x + c_1$, and $(x+y)^2 = 2x + c_1$.

27. Let $u = y - 2x + 3$ so that $du/dx = dy/dx - 2$. Then $\dfrac{du}{dx} + 2 = 2 + \sqrt{u}$ or $\dfrac{1}{\sqrt{u}}\,du = dx$. Thus $2\sqrt{u} = x + c$ and $2\sqrt{y - 2x + 3} = x + c$.

30. Let $u = 3x + 2y$ so that $du/dx = 3 + 2\,dy/dx$. Then $\dfrac{du}{dx} = 3 + \dfrac{2u}{u+2} = \dfrac{5u+6}{u+2}$ and $\dfrac{u+2}{5u+6}\,du = dx$. Now by long division

$$\frac{u+2}{5u+6} = \frac{1}{5} + \frac{4}{25u+30}$$

so we have

$$\int \left(\frac{1}{5} + \frac{4}{25u+30} \right) du = dx$$

and $\frac{1}{5}u + \frac{4}{25}\ln|25u + 30| = x + c$. Thus

$$\frac{1}{5}(3x + 2y) + \frac{4}{25}\ln|75x + 50y + 30| = x + c.$$

Setting $x = -1$ and $y = -1$ we obtain $c = \frac{4}{25}\ln 95$. The solution is

$$\frac{1}{5}(3x + 2y) + \frac{4}{25}\ln|75x + 50y + 30| = x + \frac{4}{25}\ln 95$$

or

$$5y - 5x + 2\ln|75x + 50y + 30| = 2\ln 95$$

39. Write the differential equation as $dP/dt - aP = -bP^2$ and let $u = P^{-1}$ or $P = u^{-1}$. Then

$$\frac{dp}{dt} = -u^{-2}\frac{du}{dt},$$

and substituting into the differential equation, we have

$$-u^{-2}\frac{du}{dt} - au^{-1} = -bu^{-2} \qquad \text{or} \qquad \frac{du}{dt} + au = b.$$

The latter differential equation is linear with integrating factor $e^{\int a\,dt} = e^{at}$, so

$$\frac{d}{dt}\left[e^{at}u\right] = be^{at}$$

and

$$e^{at}u = \frac{b}{a}e^{at} + c$$

$$e^{at}P^{-1} = \frac{b}{a}e^{at} + c$$

$$P^{-1} = \frac{b}{a} + ce^{-at}$$

$$P = \frac{1}{b/a + ce^{-at}} = \frac{a}{b + c_1 e^{-at}}.$$

2.6 ┃ A Numerical Method

3. Separating variables and integrating, we have

$$\frac{dy}{y} = dx \quad \text{and} \quad \ln|y| = x + c.$$

Thus $y = c_1 e^x$ and, using $y(0) = 1$, we find $c = 1$, so $y = e^x$ is the solution of the initial-value problem.

$h = 0.1$

x_n	y_n	Actual Value	Abs. Error	%Rel. Error
0.00	1.0000	1.0000	0.0000	0.00
1.10	1.1000	1.1052	0.0052	0.47
0.20	1.2100	1.2214	0.0114	0.93
0.30	1.3310	1.3499	0.0189	1.40
0.40	1.4641	1.4918	0.0277	1.86
0.50	1.6105	1.6487	0.0382	2.32
0.60	1.7716	1.8221	0.0506	2.77
0.70	1.9487	2.0138	0.0650	3.23
0.80	2.1436	2.2255	0.0820	3.68
0.90	2.3579	2.4596	0.1017	4.13
1.00	2.5937	2.7183	0.1245	4.58

$h = 0.05$

x_n	y_n	Actual Value	Abs. Error	%Rel. Error
0.00	1.0000	1.0000	0.0000	0.00
0.05	1.0500	1.0513	0.0013	0.12
0.10	1.1025	1.1052	0.0027	0.24
0.15	1.1576	1.1618	0.0042	0.36
0.20	1.2155	1.2214	0.0059	0.48
0.25	1.2763	1.2840	0.0077	0.60
0.30	1.3401	1.3499	0.0098	0.72
0.35	1.4071	1.4191	0.0120	0.84
0.40	1.4775	1.4918	0.0144	0.96
0.45	1.5513	1.5683	0.0170	1.08
0.50	1.6289	1.6487	0.0198	1.20
0.55	1.7103	1.7333	0.0229	1.32
0.60	1.7959	1.8221	0.0263	1.44
0.65	1.8856	1.9155	0.0299	1.56
0.70	1.9799	2.0138	0.0338	1.68
0.75	2.0789	2.1170	0.0381	1.80
0.80	2.1829	2.2255	0.0427	1.92
0.85	2.2920	2.3396	0.0476	2.04
0.90	2.4066	2.4596	0.0530	2.15
0.95	2.5270	2.5857	0.0588	2.27
1.00	2.6533	2.7183	0.0650	2.39

6.

$h = 0.1$

x_n	y_n
0.00	1.0000
0.10	1.1000
0.20	1.2220
0.30	1.3753
0.40	1.5735
0.50	1.8371

$h = 0.05$

x_n	y_n
0.00	1.0000
0.05	1.0500
0.10	1.1053
0.15	1.1668
0.20	1.2360
0.25	1.3144
0.30	1.4039
0.35	1.5070
0.40	1.6267
0.45	1.7670
0.50	1.9332

9.

$h = 0.1$

x_n	y_n
1.00	1.0000
1.10	1.0000
1.20	1.0191
1.30	1.0588
1.40	1.1231
1.50	1.2194

$h = 0.05$

x_n	y_n
1.00	1.0000
1.05	1.0000
1.10	1.0049
1.15	1.0147
1.20	1.0298
1.25	1.0506
1.30	1.0775
1.35	1.1115
1.40	1.1538
1.45	1.2057
1.50	1.2696

12. See the comments in Problem 11 above.

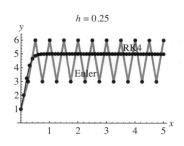
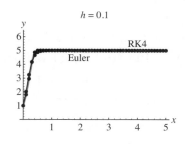
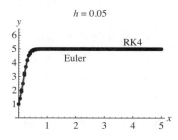

2.7 | Linear Models

3. Let $P = P(t)$ be the population at time t. Then $dP/dt = kP$ and $P = ce^{kt}$. From $P(0) = c = 500$ we see that $P = 500e^{kt}$. Since 15% of 500 is 75, we have $P(10) = 500e^{10k} = 575$. Solving for k, we get $k = \frac{1}{10} \ln \frac{575}{500} = \frac{1}{10} \ln 1.15$. When $t = 30$,

$$P(30) = 500e^{(1/10)(\ln 1.15)30} = 500e^{3 \ln 1.15} = 760 \text{ years}$$

and

$$P'(30) = kP(30) = \frac{1}{10}(\ln 1.15)760 = 10.62 \text{ persons/year.}$$

6. Let $A = A(t)$ be the amount present at time t. From $dA/dt = kA$ and $A(0) = 100$ we obtain $A = 100e^{kt}$. Using $A(6) = 97$ we find $k = \frac{1}{6} \ln 0.97$. Then $A(24) = 100e^{(1/6)(\ln 0.97)24} = 100(0.97)^4 \approx 88.5$ mg.

9. Using $A(t) = A_0 e^{kt}$ we know that

$$A_1 = A_0 e^{kt_1}$$
$$\frac{A_1}{A_0} = e^{kt_1} = \left(e^k\right)^{t_1}$$
$$e^k = \left(\frac{A_1}{A_0}\right)^{1/t_1}.$$

Therefore,

$$A(t) = A_0 e^{kt} = A_0 \left[\left(\frac{A_1}{A_0}\right)^{1/t_1}\right]^t = A_0 \left(\frac{A_1}{A_0}\right)^{t/t_1}.$$

12. From $dS/dt = rS$ we obtain $S = S_0 e^{rt}$ where $S(0) = S_0$.

    (a) If $S_0 = \$5000$ and $r = 5.75\%$ then $S(5) = \$6665.45$.

    (b) If $S(t) = \$10,000$ then $t = 12$ years.

    (c) $S \approx \$6651.82$

15. Assume that $dT/dt = k(T - 10)$ so that $T = 10 + ce^{kt}$. If $T(0) = 70°$ and $T(1/2) = 50°$ then $c = 60$ and $k = 2\ln(2/3)$ so that $T(1) = 36.67°$. If $T(t) = 15°$ then $t = 3.06$ minutes.

18. The differential equation for the first container is $dT_1/dt = k_1(T_1 - 0) = k_1 T_1$, whose solution is $T_1(t) = c_1 e^{k_1 t}$. Since $T_1(0) = 100$ (the initial temperature of the metal bar), we have $100 = c_1$ and $T_1(t) = 100 e^{k_1 t}$. After 1 minute, $T_1(1) = 100 e^{k_1} = 90°$C, so $k_1 = \ln 0.9$ and $T_1(t) = 100 e^{t \ln 0.9}$. After 2 minutes, $T_1(2) = 100 e^{2 \ln 0.9} = 100(0.9)^2 = 81°$C.

The differential equation for the second container is $dT_2/dt = k_2(T_2 - 100)$, whose solution is $T_2(t) = 100 + c_2 e^{k_2 t}$. When the metal bar is immersed in the second container, its initial temperature is $T_2(0) = 81$, so

$$T_2(0) = 100 + c_2 e^{k_2(0)} = 100 + c_2 = 81$$

and $c_2 = -19$. Thus $T_2(t) = 100 - 19 e^{k_2 t}$. After 1 minute in the second tank, the temperature of the metal bar is $91°$C, so

$$T_2(1) = 100 - 19 e^{k_2} = 91$$

$$e^{k_2} = \frac{9}{19}$$

$$k_2 = \ln \frac{9}{19}$$

and $T_2(t) = 100 - 19 e^{t \ln(9/19)}$. Setting $T_2(t) = 99.9$ we have

$$100 - 19 e^{t \ln(9/19)} = 99.9$$

$$e^{t \ln(9/19)} = \frac{0.1}{19}$$

$$t = \frac{\ln(0.1/19)}{\ln(9/19)} \approx 7.02.$$

Thus, from the start of the "double dipping" process, the total time until the bar reaches $99.9°$C in the second container is approximately 9.02 minutes.

21. According to Newton's Law of Cooling

$$\frac{dT}{dt} = k(T - T_m).$$

Separating variables we have

$$\frac{dT}{T - T_m} = k\,dt \quad \text{so} \quad \ln|T - T_m| = kt + c \quad \text{and} \quad T = T_m + c_1 e^{kt}.$$

Setting $T(0) = T_0$ we find $c_1 = T_0 - T_m$. Thus

$$T(t) = T_m + (T_0 - T_m)e^{kt}.$$

In this problem we use $T_0 = 98.6$ and $T_m = 70$. Now, let n denote the number of hours elapsed before the body was found. Then $T(n) = 85$ and $T(n+1) = 80$. Using this information, we have

$$70 + (98.6 - 70)e^{kn} = 85 \qquad \text{and} \qquad 70 + (98.6 - 70)e^{k(n+1)} = 80$$

or

$$28.6 e^{kn} = 15 \qquad \text{and} \qquad 28.6 e^{kn+k} = 28.6 e^{kn} e^k = 10.$$

The last equation is the same as $15 e^k = 10$. Solving for k, we have $k = \ln \frac{2}{3} \approx -0.4055$. Finally, solving $e^{-0.4055 n} = 15/28.6$ for n, we have

$$-0.4055 n = \ln\left(\frac{15}{28.6}\right)$$

$$n = \frac{1}{-0.4055} \ln\left(\frac{15}{28.6}\right) \approx 1.6.$$

Thus about 1.6 hours elapsed before the body was found.

24. From $dA/dt = 0 - A/50$ we obtain $A = ce^{-t/50}$. If $A(0) = 30$ then $c = 30$ and $A = 30 e^{-t/50}$.

27. From

$$\frac{dA}{dt} = 10 - \frac{10A}{500 - (10 - 5)t} = 10 - \frac{2A}{100 - t}$$

we obtain $A = 1000 - 10t + c(100 - t)^2$. If $A(0) = 0$ then $c = -\frac{1}{10}$. The tank is empty in 100 minutes.

30. (a) Initially the tank contains 300 gallons of solution. Since brine is pumped in at a rate of 3 gal/min and the mixture is pumped out at a rate of 2 gal/min, the net change is an increase of 1 gal/min. Thus in 100 minutes the tank will contain its capacity of 400 gallons.

(b) The differential equation describing the amount of salt in the tank is $A'(t) = 6 - 2A/(300 + t)$ with solution

$$A(t) = 600 + 2t - (4.95 \times 10^7)(300 + t)^{-2}, \qquad 0 \le t \le 100,$$

as noted in the discussion following Example 5 in the text. Thus the amount of salt in the tank when it overflows is

$$A(100) = 800 - (4.95 \times 10^7)(400)^{-2} = 490.625 \text{ lb}.$$

(c) When the tank is overflowing the amount of salt in the tank is governed by the differential equation

$$\frac{dA}{dt} = (3 \text{ gal/min})(2 \text{ lb/gal}) - \left(\frac{A}{400} \text{ lb/gal}\right)(3 \text{ gal/min})$$

$$= 6 - \frac{3A}{400}, \qquad A(100) = 490.625.$$

Solving the equation, we obtain $A(t) = 800 + ce^{-3t/400}$. The initial condition yields $c = -654.947$, so that

$$A(t) = 800 - 654.947e^{-3t/400}.$$

When $t = 150$, $A(150) = 587.37$ lb.

(d) As $t \to \infty$, the amount of salt is 800 lb, which is to be expected since (400 gal) (2 lb/gal)= 800 lb.

(e)

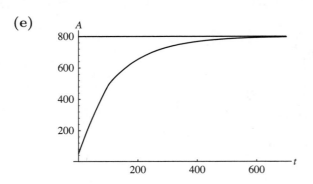

33. Assume $R\,dq/dt + (1/C)q = E(t)$, $R = 200$, $C = 10^{-4}$, and $E(t) = 100$ so that $q = 1/100 + ce^{-50t}$. If $q(0) = 0$ then $c = -1/100$ and $i = \frac{1}{2}e^{-50t}$.

36. We first solve $(1 - t/10)di/dt + 0.2i = 4$. Separating variables we obtain $di/(40 - 2i) = dt/(10 - t)$. Then

$$-\frac{1}{2}\ln|40 - 2i| = -\ln|10 - t| + c \quad \text{or} \quad \sqrt{40 - 2i} = c_1(10 - t).$$

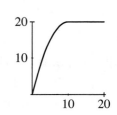

Since $i(0) = 0$ we must have $c_1 = 2/\sqrt{10}$. Solving for i we get $i(t) = 4t - \frac{1}{5}t^2$, $0 \le t < 10$. For $t \ge 10$ the equation for the current becomes $0.2i = 4$ or $i = 20$. Thus

$$i(t) = \begin{cases} 4t - \dfrac{1}{5}t^2, & 0 \le t < 10 \\[2mm] 20, & t \ge 10. \end{cases}$$

The graph of $i(t)$ is given in the figure.

39. When air resistance is proportional to velocity, the model for the velocity is $m\,dv/dt = -mg - kv$ (using the fact that the positive direction is upward.) Solving the differential equation using separation of variables we obtain $v(t) = -mg/k + ce^{-kt/m}$. From $v(0) = 300$ we get

$$v(t) = -\frac{mg}{k} + \left(300 + \frac{mg}{k}\right)e^{-kt/m}.$$

Integrating and using $s(0) = 0$ we find

$$s(t) = -\frac{mg}{k}t + \frac{m}{k}\left(300 + \frac{mg}{k}\right)(1 - e^{-kt/m}).$$

Setting $k = 0.0025$, $m = 16/32 = 0.5$, and $g = 32$ we have

$$s(t) = 1{,}340{,}000 - 6{,}400t - 1{,}340{,}000e^{-0.005t}$$

and

$$v(t) = -6{,}400 + 6{,}700e^{-0.005t}.$$

The maximum height is attained when $v = 0$, that is, at $t_a = 9.162$. The maximum height will be $s(9.162) = 1363.79\,\text{ft}$, which is less than the maximum height in Problem 38.

42. Separating variables, we obtain $dP/P = k\cos t\,dt$, so

$$\ln|P| = k\sin t + c \qquad \text{and} \qquad P = c_1 e^{k\sin t}.$$

If $P(0) = P_0$, then $c_1 = P_0$ and $P = P_0 e^{k\sin t}$.

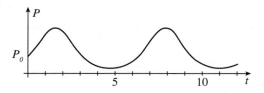

45. (a) Solving $r - kx = 0$ for x we find the equilibrium solution $x = r/k$. When $x < r/k$, $dx/dt > 0$ and when $x > r/k$, $dx/dt < 0$. From the phase portrait we see that $\displaystyle\lim_{t\to\infty} x(t) = r/k$.

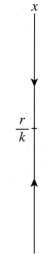

(b) From $dx/dt = r - kx$ and $x(0) = 0$ we obtain $x = r/k - (r/k)e^{-kt}$ so that $x \to r/k$ as $t \to \infty$. If $x(T) = r/2k$ then $T = (\ln 2)/k$.

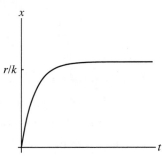

2.8   Nonlinear Models

3. From $dP/dt = P\left(10^{-1} - 10^{-7}P\right)$ and $P(0) = 5000$ we obtain $P = 500/(0.0005 + 0.0995e^{-0.1t})$ so that $P \to 1{,}000{,}000$ as $t \to \infty$. If $P(t) = 500{,}000$ then $t = 52.9\,\text{months}$.

6. The differential equation is

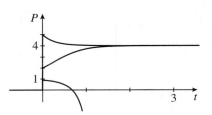

$$\frac{dP}{dt} = P(5-P) - 4 = -(P^2 - 5P + 4) = -(P-4)(P-1).$$

Separating variables and integrating, we obtain

$$\frac{dP}{(P-4)(P-1)} = -dt$$

$$\left(\frac{1/3}{P-4} - \frac{1/3}{P-1}\right) dP = -dt$$

$$\frac{1}{3}\ln\left|\frac{P-4}{P-1}\right| = -t + c$$

$$\frac{P-4}{P-1} = c_1 e^{-3t}.$$

Setting $t = 0$ and $P = P_0$ we find $c_1 = (P_0 - 4)/(P_0 - 1)$. Solving for P we obtain

$$P(t) = \frac{4(P_0 - 1) - (P_0 - 4)e^{-3t}}{(P_0 - 1) - (P_0 - 4)e^{-3t}}.$$

9. Solving $P(5-P) - 7 = 0$ for P we obtain complex roots, so there are no equilibrium solutions. Since $dP/dt < 0$ for all values of P, the population becomes extinct for any initial condition. Using separation of variables to solve the initial-value problem, we get

$$P(t) = \frac{5}{2} + \frac{\sqrt{3}}{2}\tan\left[\tan^{-1}\left(\frac{2P_0 - 5}{\sqrt{3}}\right) - \frac{\sqrt{3}}{2}t\right].$$

Solving $P(t) = 0$ for t we see that the time of extinction is

$$t = \frac{2}{3}\left(\sqrt{3}\tan^{-1}(5/\sqrt{3}) + \sqrt{3}\tan^{-1}\left[(2P_0 - 5)/\sqrt{3}\right]\right).$$

12. The number A is called the Allee threshold and $0 < A < K$. Without this value we have the autonomous equation $\frac{dP}{dt} = rP(1 - P/K)$ with critical numbers 0 and K. If we include the value for A into the model however, the corresponding one-dimensional phase portrait would look something like the figure on the right where K and 0 are attractors and A is a repeller. We expect the corresponding autonomous equation to have A as a critical number, thus a reasonable model would be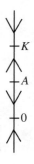

$$\frac{dP}{dt} = rP\left(1 - \frac{P}{K}\right)\left(\frac{P}{A} - 1\right)$$

For an initial value $P_0 < A$ the population decreases, that is, $P \to 0$ as $t \to \infty$.

15. (a) The initial-value problem is $dh/dt = -8A_h\sqrt{h}/A_w$, $h(0) = H$. Separating variables and integrating we have

$$\frac{dh}{\sqrt{h}} = -\frac{8A_h}{A_w}\,dt \quad \text{and} \quad 2\sqrt{h} = -\frac{8A_h}{A_w}t + c.$$

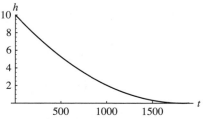

Using $h(0) = H$ we find $c = 2\sqrt{H}$, so the solution of the initial-value problem is $\sqrt{h(t)} = (A_w\sqrt{H} - 4A_h t)/A_w$, where $A_w\sqrt{H} - 4A_h t \geq 0$. Thus

$$h(t) = (A_w\sqrt{H} - 4A_h t)^2/A_w^2 \quad \text{for} \quad 0 \leq t \leq A_w H/4A_h.$$

(b) Identifying $H = 10$, $A_w = 4\pi$, and $A_h = \pi/576$ we have $h(t) = t^2/331{,}776 - (\sqrt{5/2}/144)t + 10$. Solving $h(t) = 0$ we see that the tank empties in $576\sqrt{10}$ seconds or 30.36 minutes.

18. When the height of the water is h, the radius of the top of the water is $\frac{2}{5}(20 - h)$ and $A_w = 4\pi(20 - h)^2/25$. The differential equation is

$$\frac{dh}{dt} = -c\frac{A_h}{A_w}\sqrt{2gh} = -0.6\frac{\pi(2/12)^2}{4\pi(20-h)^2/25}\sqrt{64h} = -\frac{5}{6}\frac{\sqrt{h}}{(20-h)^2}.$$

Separating variables and integrating we have

$$\frac{(20-h)^2}{\sqrt{h}}\,dh = -\frac{5}{6}\,dt \quad \text{and} \quad 800\sqrt{h} - \frac{80}{3}h^{3/2} + \frac{2}{5}h^{5/2} = -\frac{5}{6}t + c.$$

Using $h(0) = 20$ we find $c = 2560\sqrt{5}/3$, so an implicit solution of the initial-value problem is

$$800\sqrt{h} - \frac{80}{3}h^{3/2} + \frac{2}{5}h^{5/2} = -\frac{5}{6}t + \frac{2560\sqrt{5}}{3}.$$

To find the time it takes the tank to empty we set $h = 0$ and solve for t. The tank empties in $1024\sqrt{5}$ seconds or 38.16 minutes. Thus the tank empties more slowly when the base of the cone is on the bottom.

21. (a) Let ρ be the weight density of the water and V the volume of the object. Archimedes' principle states that the upward buoyant force has magnitude equal to the weight of the water displaced. Taking the positive direction to be down, the differential equation is

$$m\frac{dv}{dt} = mg - kv^2 - \rho V.$$

(b) Using separation of variables we have

$$\frac{m\,dv}{(mg - \rho V) - kv^2} = dt$$

$$\frac{m}{\sqrt{k}}\frac{\sqrt{k}\,dv}{(\sqrt{mg - \rho V})^2 - (\sqrt{k}\,v)^2} = dt$$

$$\frac{m}{\sqrt{k}}\frac{1}{\sqrt{mg - \rho V}}\tanh^{-1}\frac{\sqrt{k}\,v}{\sqrt{mg - \rho V}} = t + c.$$

Thus

$$v(t) = \sqrt{\frac{mg - \rho V}{k}} \tanh \left(\frac{\sqrt{kmg - k\rho V}}{m} t + c_1 \right).$$

(c) Since $\tanh t \to 1$ as $t \to \infty$, the terminal velocity is $\sqrt{(mg - \rho V)/k}$.

24. (a) Solving $r^2 + (10 - h)^2 = 10^2$ for r^2 we see that $r^2 = 20h - h^2$. Combining the rate of input of water, π, with the rate of output due to evaporation, $k\pi r^2 = k\pi(20h - h^2)$, we have $dV/dt = \pi - k\pi(20h - h^2)$. Using $V = 10\pi h^2 - \frac{1}{3}\pi h^3$, we see also that $dV/dt = (20\pi h - \pi h^2)dh/dt$. Thus

$$(20\pi h - \pi h^2)\frac{dh}{dt} = \pi - k\pi(20h - h^2) \quad \text{and} \quad \frac{dh}{dt} = \frac{1 - 20kh + kh^2}{20h - h^2}.$$

(b) Letting $k = 1/100$, separating variables and integrating (with the help of a CAS), we get

$$\frac{100h(h - 20)}{(h - 10)^2} dh = dt$$

and

$$\frac{100(h^2 - 10h + 100)}{10 - h} = t + c.$$

Using $h(0) = 0$ we find $c = 1000$, and solving for h we get $h(t) = 0.005 \left(\sqrt{t^2 + 4000t} - t \right)$, where the positive square root is chosen because $h \geq 0$.

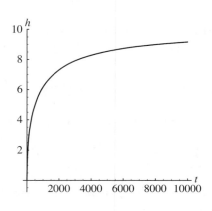

(c) The volume of the tank is $V = \frac{2}{3}\pi(10)^3$ feet, so at a rate of π cubic feet per minute, the tank will fill in $\frac{2}{3}(10)^3 \approx 666.67$ minutes ≈ 11.11 hours.

(d) At 666.67 minutes, the depth of the water is $h(666.67) = 5.486$ feet. From the graph in (b) we suspect that $\lim_{t \to \infty} h(t) = 10$, in which case the tank will never completely fill. To prove this we compute the limit of $h(t)$:

$$\lim_{t \to \infty} h(t) = 0.005 \lim_{t \to \infty} \left(\sqrt{t^2 + 4000t} - t \right) = 0.005 \lim_{t \to \infty} \frac{t^2 + 4000t - t^2}{\sqrt{t^2 + 4000t} + t}$$

$$= 0.005 \lim_{t \to \infty} \frac{4000t}{t\sqrt{1 + 4000/t} + t} = 0.005 \frac{4000}{1 + 1} = 0.005(2000) = 10.$$

27. The piecewise-defined function $w(x)$ is now

$$w(x) = \begin{cases} x, & 0 \leq x \leq \dfrac{\sqrt{2}}{2} \\[2mm] \sqrt{2} - x, & \dfrac{\sqrt{2}}{2} < x \leq \sqrt{2} \end{cases}$$

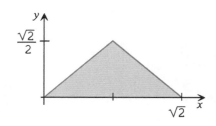

First, we solve

$$x\frac{dx}{dt} = 1, \qquad x(0) = 0$$

by separation of variables. This yields $x(t) = \sqrt{2t}$. The time interval corresponding to $0 \le x \le \frac{\sqrt{2}}{2}$ is defined by $0 \le t \le \frac{1}{4}$. Second, we solve

$$\left(\sqrt{2} - x\right)\frac{dx}{dt} = 1, \qquad x\left(\frac{1}{4}\right) = \frac{\sqrt{2}}{2}.$$

This gives $x^2 - 2\sqrt{2}\,x + 2t + 1 = 0$. Using the quadratic formula, we have $x(t) = \sqrt{2} - \sqrt{1 - 2t}$. The time interval corresponding to $\frac{\sqrt{2}}{2} < x \le \sqrt{2}$ is defined by $\frac{1}{4} \le t \le \frac{1}{2}$. Thus

$$x(t) = \begin{cases} \sqrt{2t}, & 0 \le t \le \frac{1}{4} \\ \sqrt{2} - \sqrt{1 - 2t}, & \frac{1}{4} < t \le \frac{1}{2}. \end{cases}$$

The time that it takes the saw to cut through the piece of wood is then $t = \frac{1}{2}$.

2.9   Modeling with Systems of First-Order DEs

3. The amounts x and y are the same at about $t = 5$ days. The amounts x and z are the same at about $t = 20$ days. The amounts y and z are the same at about $t = 147$ days. The time when y and z are the same makes sense because most of A and half of B are gone, so half of C should have been formed.

6. **(a)** From parts (a) and (c) from Problem 5:

$$\frac{A(t)}{P(t)} = \frac{\frac{\lambda_A}{\lambda_A + \lambda_C} P_0 \left(1 - e^{-(\lambda_A + \lambda_C)t}\right)}{P_0 e^{-(\lambda_A + \lambda_C)t}} = \frac{\lambda_A}{\lambda_A + \lambda_C}\left(e^{(\lambda_A + \lambda_C)t} - 1\right)$$

(b) Solving for t in part (a) we get

$$\frac{A(t)}{P(t)}\left(\frac{\lambda_A + \lambda_C}{\lambda_A}\right) = e^{(\lambda_A + \lambda_C)t} - 1$$

$$e^{(\lambda_A + \lambda_C)t} = \frac{A(t)}{P(t)}\left(\frac{\lambda_A + \lambda_C}{\lambda_A}\right) + 1$$

$$t = \frac{1}{\lambda_A + \lambda_C}\ln\left[\frac{A(t)}{P(t)}\left(\frac{\lambda_A + \lambda_C}{\lambda_A}\right) + 1\right]$$

(c) From part (b)

$$t = \frac{1}{5.543 \times 10^{-10}}\ln\left[\frac{8.6 \times 10^{-7}}{5.3 \times 10^{-10}}\left(\frac{5.543 \times 10^{-10}}{0.581 \times 10^{-10}}\right) + 1\right] \approx 1.69 \text{ billion years}$$

9. (a) A model is

$$\frac{dx_1}{dt} = 3 \cdot \frac{x_2}{100 - t} - 2 \cdot \frac{x_1}{100 + t}, \qquad x_1(0) = 100$$

$$\frac{dx_2}{dt} = 2 \cdot \frac{x_1}{100 + t} - 3 \cdot \frac{x_2}{100 - t}, \qquad x_2(0) = 50.$$

(b) Since the system is closed, no salt enters or leaves the system and $x_1(t) + x_2(t) = 100 + 50 = 150$ for all time. Thus $x_1 = 150 - x_2$ and the second equation in part (a) becomes

$$\frac{dx_2}{dt} = \frac{2(150 - x_2)}{100 + t} - \frac{3x_2}{100 - t} = \frac{300}{100 + t} - \frac{2x_2}{100 + t} - \frac{3x_2}{100 - t}$$

or

$$\frac{dx_2}{dt} + \left(\frac{2}{100 + t} + \frac{3}{100 - t}\right) x_2 = \frac{300}{100 + t},$$

which is linear in x_2. An integrating factor is

$$e^{2\ln(100+t) - 3\ln(100-t)} = (100 + t)^2 (100 - t)^{-3}$$

so

$$\frac{d}{dt}[(100 + t)^2 (100 - t)^{-3} x_2] = 300(100 + t)(100 - t)^{-3}.$$

Using integration by parts, we obtain

$$(100 + t)^2 (100 - t)^{-3} x_2 = 300 \left[\frac{1}{2}(100 + t)(100 - t)^{-2} - \frac{1}{2}(100 - t)^{-1} + c\right].$$

Thus

$$x_2 = \frac{300}{(100 + t)^2} \left[c(100 - t)^3 - \frac{1}{2}(100 - t)^2 + \frac{1}{2}(100 + t)(100 - t)\right]$$

$$= \frac{300}{(100 + t)^2}[c(100 - t)^3 + t(100 - t)].$$

Using $x_2(0) = 50$ we find $c = 5/3000$. At $t = 30$, $x_2 = (300/130^2)(70^3 c + 30 \cdot 70) \approx 47.4$ lb.

12. (a) The population $y(t)$ approaches 10,000, while the population $x(t)$ approaches extinction.

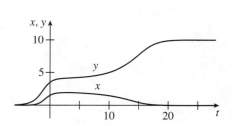

(b) The population $x(t)$ approaches 5,000, while the population $y(t)$ approaches extinction.

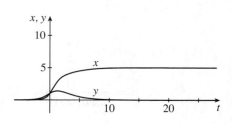

(c) The population $y(t)$ approaches 10,000, while the population $x(t)$ approaches extinction.

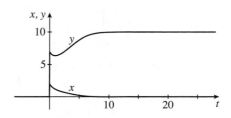

(d) The population $x(t)$ approaches 5,000, while the population $y(t)$ approaches extinction.

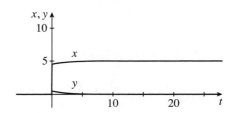

15. By Kirchhoff's first law we have $i_1 = i_2 + i_3$. Applying Kirchhoff's second law to each loop we obtain

$$E(t) = i_1 R_1 + L_1 \frac{di_2}{dt} + i_2 R_2$$

and

$$E(t) = i_1 R_1 + L_2 \frac{di_3}{dt} + i_3 R_3.$$

Combining the three equations, we obtain the system

$$L_1 \frac{di_2}{dt} + (R_1 + R_2)i_2 + R_1 i_3 = E$$

$$L_2 \frac{di_3}{dt} + R_1 i_2 + (R_1 + R_3)i_3 = E.$$

18. (a) If we know $s(t)$ and $i(t)$ then we can determine $r(t)$ from $s + i + r = n$.

(b) In this case the system is

$$\frac{ds}{dt} = -0.2si$$

$$\frac{di}{dt} = -0.7i + 0.2si.$$

We also note that when $i(0) = i_0$, $s(0) = 10 - i_0$ since $r(0) = 0$ and $i(t) + s(t) + r(t) = 0$ for all values of t. Now $k_2/k_1 = 0.7/0.2 = 3.5$, so we consider initial conditions $s(0) = 2$, $i(0) = 8$; $s(0) = 3.4$, $i(0) = 6.6$; $s(0) = 7$, $i(0) = 3$; and $s(0) = 9$, $i(0) = 1$.

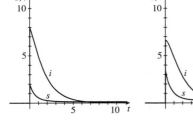 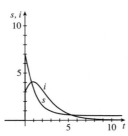 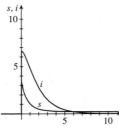 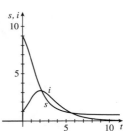

We see that an initial susceptible population greater than k_2/k_1 results in an epidemic in the sense that the number of infected persons increases to a maximum before decreasing to 0. On the other hand, when $s(0) < k_2/k_1$, the number of infected persons decreases from the start and there is no epidemic.

Chapter 2 in Review

3. By inspection, two solutions of the differential equation $y' + |y| = 5$ are $y = 5$ and $y = -5$.

6. The first-order differential equation, $\dfrac{d\theta}{dr} = r\theta + r + 1$ is not separable. False;
$\dfrac{d\theta}{dr} = (r + 1)(\theta + 1)$ is separable.

9. The differential equation $y' + y^2 = 1$ is a second-order equation. False.

12. $\dfrac{dy}{dx} = y(y - 2)^2(y - 4)$

15.

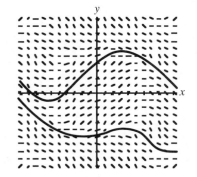

18. Write the differential equation in the form

$$y \ln \frac{x}{y}\, dx = \left(x \ln \frac{x}{y} - y\right) dy.$$

This is a homogeneous equation, so let $x = uy$. Then $dx = u\, dy + y\, du$ and the differential equation becomes

$$y \ln u(u\, dy + y\, du) = (uy \ln u - y)\, dy \quad \text{or} \quad y \ln u\, du = -dy.$$

Separating variables, we obtain

$$\ln u\, du = -\frac{dy}{y}$$

$$u \ln|u| - u = -\ln|y| + c$$

$$\frac{x}{y} \ln\left|\frac{x}{y}\right| - \frac{x}{y} = -\ln|y| + c$$

$$x(\ln x - \ln y) - x = -y \ln|y| + cy.$$

21. Write the equation in the form

$$\frac{dQ}{dt} + \frac{1}{t}Q = t^3 \ln t.$$

An integrating factor is $e^{\ln t} = t$, so

$$\frac{d}{dt}[tQ] = t^4 \ln t$$

$$tQ = -\frac{1}{25}t^5 + \frac{1}{5}t^5 \ln t + c$$

and

$$Q = -\frac{1}{25}t^4 + \frac{1}{5}t^4 \ln t + \frac{c}{t}.$$

24. Letting $M = 2r^2 \cos\theta \sin\theta + r\cos\theta$ and $N = 4r + \sin\theta - 2r\cos^2\theta$ we see that $M_r = 4r\cos\theta\sin\theta + \cos\theta = N_\theta$, so the differential equation is exact. From $f_\theta = 2r^2 \cos\theta\sin\theta + r\cos\theta$ we obtain $f = -r^2\cos^2\theta + r\sin\theta + h(r)$. Then $f_r = -2r\cos^2\theta + \sin\theta + h'(r) = 4r + \sin\theta - 2r\cos^2\theta$ and $h'(r) = 4r$ so $h(r) = 2r^2$. The solution is

$$-r^2\cos^2\theta + r\sin\theta + 2r^2 = c.$$

27. We put the equation $x\frac{dy}{dx} + 2y = xe^{x^2}$ in the standard form $\frac{dy}{dx} + \frac{2}{x}y = e^{x^2}$. Then the integrating factor is $e^{\int \frac{2}{x}\,dx} = e^{\ln x^2} = x^2$. Therefore

$$x^2\frac{dy}{dx} + 2xy = x^2e^{x^2}$$

$$\frac{d}{dx}[x^2y] = x^2e^{x^2}$$

$$\int_1^x \frac{d}{dt}[t^2 y(t)]\,dt = \int_1^x t^2 e^{t^2}\,dt$$

$$x^2y(x) - \overbrace{y(1)}^{3} = \int_1^x t^2 e^{t^2}\,dt$$

$$y(x) = \frac{3}{x^2} + \frac{1}{x^2}\int_1^x t^2 e^{t^2}\,dt$$

30. $\dfrac{dy}{dx} + P(x)y = e^x, \quad y(0) = -1, \qquad \text{where} \quad P(x) = \begin{cases} 1, & 0 \le x < 1 \\ -1, & x \ge 1 \end{cases}$

For $0 \le x < 1$,

$$\frac{d}{dx}[e^x y] = e^{2x}$$

$$e^x y = \frac{1}{2}e^{2x} + c_1$$

$$y = \frac{1}{2}e^x + c_1 e^{-x}$$

Using $y(0) = -1$, we have $c_1 = -\frac{3}{2}$. Therefore $y = \frac{1}{2}e^x - \frac{3}{2}e^{-x}$. Then for $x \geq 1$,

$$\frac{d}{dx}\left[e^{-x}y\right] = 1$$

$$e^{-x}y = x + c_2$$

$$y = xe^x + c_2 e^x$$

Requiring that $y(x)$ be continuous at $x = 1$ yields

$$e + c_2 e = \frac{1}{2}e - \frac{3}{2}e^{-1}$$

$$c_2 = -\frac{1}{2} - \frac{3}{2}e^{-2}$$

Therefore

$$y(x) = \begin{cases} \dfrac{1}{2}e^x - \dfrac{3}{2}e^{-x}, & 0 \leq x < 1 \\ xe^x - \dfrac{1}{2}e^x - \dfrac{3}{2}e^{x-2}, & x \geq 1 \end{cases}$$

33. (a) For $y < 0$, $\sqrt{y}$ is not a real number.

(b) Separating variables and integrating we have

$$\frac{dy}{\sqrt{y}} = dx \quad \text{and} \quad 2\sqrt{y} = x + c.$$

Letting $y(x_0) = y_0$ we get $c = 2\sqrt{y_0} - x_0$, so that

$$2\sqrt{y} = x + 2\sqrt{y_0} - x_0 \quad \text{and} \quad y = \frac{1}{4}(x + 2\sqrt{y_0} - x_0)^2.$$

Since $\sqrt{y} > 0$ for $y \neq 0$, we see that $dy/dx = \frac{1}{2}(x + 2\sqrt{y_0} - x_0)$ must be positive. Thus the interval on which the solution is defined is $(x_0 - 2\sqrt{y_0}, \infty)$.

36. The first step of Euler's method gives $y(1.1) \approx 9 + 0.1(1 + 3) = 9.4$. Applying Euler's method one more time gives $y(1.2) \approx 9.4 + 0.1(1 + 1.1\sqrt{9.4}) \approx 9.8373$.

39. (a) Since the amount of C-14 remaining is 53%, or a little more than half of the amount in a living person, and since the half-life of C-14 is 5730, the age of the iceman must be less than 5730 years – say 5400 years. Then the date of death of the iceman in 1991 was about $1991 - 5400 = -3409$ or 3409 BC.

(b) From the solution of Problem 13 in Section 2.7 we use $A(t) = A_0 e^{-0.0001204t}$. With $A(t) = 0.53A_0$ we have

$$-0.00012097t = \ln 0.53 \quad \text{or} \quad t = \frac{\ln 0.53}{-0.00012097} \approx 5248 \text{ years.}$$

This represents the iceman's age in 1991, so the approximate date of his death would be

$$1991 - 5248 = -3257 \quad \text{or} \quad 3257 \text{ BC.}$$

42. From $V\,dC/dt = kA(C_s - C)$ and $C(0) = C_0$ we obtain $C = C_s + (C_0 - C_s)e^{-kAt/V}$.

45. Separating variables, we obtain

$$\frac{dq}{E_0 - q/C} = \frac{dt}{k_1 + k_2 t}$$

$$-C \ln\left|E_0 - \frac{q}{C}\right| = \frac{1}{k_2} \ln|k_1 + k_2 t| + c_1$$

$$\frac{(E_0 - q/C)^{-C}}{(k_1 + k_2 t)^{1/k_2}} = c_2.$$

Setting $q(0) = q_0$ we find $c_2 = (E_0 - q_0/C)^{-C}/k_1^{1/k_2}$, so

$$\frac{(E_0 - q/C)^{-C}}{(k_1 + k_2 t)^{1/k_2}} = \frac{(E_0 - q_0/C)^{-C}}{k_1^{1/k_2}}$$

$$\left(E_0 - \frac{q}{C}\right)^{-C} = \left(E_0 - \frac{q_0}{C}\right)^{-C}\left(\frac{k_1}{k_1 + k_2 t}\right)^{-1/k_2}$$

$$E_0 - \frac{q}{C} = \left(E_0 - \frac{q_0}{C}\right)\left(\frac{k_1}{k_1 + k_2 t}\right)^{1/Ck_2}$$

$$q = E_0 C + (q_0 - E_0 C)\left(\frac{k_1}{k_1 + k_2 t}\right)^{1/Ck_2}.$$

48. One hour is 3,600 seconds, so the hour mark should be placed at

$$h(3600) = [\sqrt{2} - 0.00001628(3600)]^2 \approx 1.838\,\text{ft} \approx 22.0525\,\text{in.}$$

up from the bottom of the tank. The remaining marks corresponding to the passage of 2, 3, 4, ..., 12 hours are placed at the values shown in the table. The marks are not evenly spaced because the water is not draining out at a uniform rate; that is, $h(t)$ is not a linear function of time.

Time (seconds)	Height (inches)
0	24.0000
1	22.0520
2	20.1864
3	18.4033
4	16.7026
5	15.0844
6	13.5485
7	12.0952
8	10.7242
9	9.4357
10	8.2297
11	7.1060
12	6.0648

51. From $dx/dt = k_1 x(\alpha - x)$ we obtain

$$\left(\frac{1/\alpha}{x} + \frac{1/\alpha}{\alpha - x}\right) dx = k_1\,dt$$

so that $x = \alpha c_1 e^{\alpha k_1 t}/(1 + c_1 e^{\alpha k_1 t})$. From $dy/dt = k_2 xy$ we obtain

$$\ln|y| = \frac{k_2}{k_1} \ln\left|1 + c_1 e^{\alpha k_1 t}\right| + c \quad \text{or} \quad y = c_2\left(1 + c_1 e^{\alpha k_1 t}\right)^{k_2/k_1}.$$

54. **(a)** The graph of $P(t)$ in Problem 53 is given at the right.

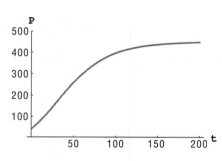

(b) Since $e^{-0.02948t} \to 0$ as $t \to \infty$, we have $P(t) \to 450$ as $t \to \infty$.

(c) By the properties of logarithms, the equation $dPdt = kP \ln \dfrac{450}{P}$ is the same as

$$P\left(k \ln 450 - k \ln P\right).$$

This is Gompertz's differential equation $dP/dt = P\left(a - b \ln P\right)$ with the identifications $a = k \ln 450$ and $b = k$.

57. From $y = -x - 1 + c_1 e^x$ we obtain $y' = y + x$ so that the differential equation of the orthogonal family is

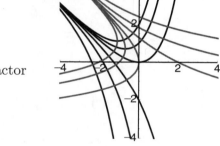

$$\frac{dy}{dx} = -\frac{1}{y + x} \qquad \text{or} \qquad \frac{dx}{dy} + x = -y.$$

This is a linear differential equation and has integrating factor $e^{\int dy} = e^y$, so

$$\frac{d}{dy}[e^y x] = -y e^y$$

$$e^y x = -y e^y + e^y + c_2$$

$$x = -y + 1 + c_2 e^{-y}.$$

60. Rewrite the equation $y^2 = 4c\left(x + c\right)$ as $4c^2 + 4cx - y^2 = 0$. Then by the quadratic formula we can solve for c:

$$c = \frac{-4x \pm 4\sqrt{x^2 + y^2}}{8} = -\frac{1}{2}x \pm \frac{1}{2}\sqrt{x^2 + y^2}$$

Differentiating the last equation gives

$$0 = -\frac{1}{2} \pm \frac{1}{4}\frac{1}{\sqrt{x^2 + y^2}}\left(2x + 2y\frac{dy}{dx}\right)$$

$$\frac{1}{2} = \pm\frac{1}{2}\frac{1}{\sqrt{x^2 + y^2}}\left(x + y\frac{dy}{dx}\right)$$

$$\pm\sqrt{x^2 + y^2} = x + y\frac{dy}{dx}$$

Squaring and simplifying then yields a differential equation of the family of parabolas:

$$x^2 + y^2 = x^2 + 2xy\frac{dy}{dx} + y^2\left(\frac{dy}{dx}\right)^2$$

$$0 = y^2\left(\frac{dy}{dx}\right)^2 + 2xy\frac{dy}{dx} - y^2$$

Replacing $\dfrac{dy}{dx}$ with $-\dfrac{dx}{dy}$, we obtain

$$y^2\left(-\frac{dx}{dy}\right)^2 + 2xy\left(-\frac{dx}{dy}\right) - y^2 = 0.$$

Dividing by $\left(-\dfrac{dx}{dy}\right)^2$ gives

$$y^2 - 2xy\frac{dy}{dx} - y^2\left(\frac{dy}{dx}\right)^2 = 0 \quad\text{or}\quad y^2\left(\frac{dy}{dx}\right)^2 + 2xy\frac{dy}{dx} - y^2 = 0.$$

We have shown that the differential equation of the family of orthogonal trajectories is the same as the differential equation of the given family. Hence the families are the same and so, by definition, are self orthogonal.

Chapter 3

Higher-Order Differential Equations

3.1 | Theory of Linear Equations

3. From $y = c_1 x + c_2 x \ln x$ we find $y' = c_1 + c_2(1 + \ln x)$. Then $y(1) = c_1 = 3$, $y'(1) = c_1 + c_2 = -1$ so that $c_1 = 3$ and $c_2 = -4$. The solution is $y = 3x - 4x \ln x$.

6. In this case we have $y(0) = c_1 = 0$, $y'(0) = 2c_2 \cdot 0 = 0$ so $c_1 = 0$ and c_2 is arbitrary. Two solutions are $y = x^2$ and $y = 2x^2$.

9. Since $a_2(x) = x - 2$ and $x_0 = 0$ the problem has a unique solution for $-\infty < x < 2$.

12. In this case we have $y(0) = c_1 = 1$, $y'(1) = 2c_2 = 6$ so that $c_1 = 1$ and $c_2 = 3$. The solution is $y = 1 + 3x^2$.

15. Since $(-4)x + (3)x^2 + (1)(4x - 3x^2) = 0$ the set of functions is linearly dependent.

18. Since $(1)\cos 2x + (1)1 + (-2)\cos^2 x = 0$ the set of functions is linearly dependent.

21. Suppose $c_1(1 + x) + c_2 x + c_3 x^2 = 0$. Then $c_1 + (c_1 + c_2)x + c_3 x^2 = 0$ and so $c_1 = 0$, $c_1 + c_2 = 0$, and $c_3 = 0$. Since $c_1 = 0$ we also have $c_2 = 0$. Thus the set of functions is linearly independent.

24. The functions satisfy the differential equation and are linearly independent since

$$W(\cosh 2x, \sinh 2x) = 2$$

for $-\infty < x < \infty$. The general solution is

$$y = c_1 \cosh 2x + c_2 \sinh 2x.$$

27. The functions satisfy the differential equation and are linearly independent since

$$W\left(x^3, x^4\right) = x^6 \neq 0$$

for $0 < x < \infty$. The general solution is

$$y = c_1 x^3 + c_2 x^4.$$

30. The functions satisfy the differential equation and are linearly independent since

$$W(1, x, \cos x, \sin x) = 1$$

for $-\infty < x < \infty$. The general solution is

$$y = c_1 + c_2 x + c_3 \cos x + c_4 \sin x.$$

33. The functions $y_1 = e^{2x}$ and $y_2 = xe^{2x}$ form a fundamental set of solutions of the associated homogeneous equation, and $y_p = x^2 e^{2x} + x - 2$ is a particular solution of the nonhomogeneous equation.

36. (a) $y_{p_1} = 5$

   (b) $y_{p_2} = -2x$

   (c) $y_p = y_{p_1} + y_{p_2} = 5 - 2x$

   (d) $y_p = \frac{1}{2} y_{p_1} - 2 y_{p_2} = \frac{5}{2} + 4x$

3.2 │ Reduction of Order

In Problems 1–8 we use reduction of order to find a second solution. In Problems 9–18 we use formula (5) from the text.

3. Define $y = u(x) \cos 4x$ so

$$y' = -4u \sin 4x + u' \cos 4x, \quad y'' = u'' \cos 4x - 8u' \sin 4x - 16u \cos 4x$$

and

$$y'' + 16y = (\cos 4x)u'' - 8(\sin 4x)u' = 0 \quad \text{or} \quad u'' - 8(\tan 4x)u' = 0.$$

If $w = u'$ we obtain the linear first-order equation $w' - 8(\tan 4x)w = 0$ which has the integrating factor $e^{-8 \int \tan 4x \, dx} = \cos^2 4x$. Now

$$\frac{d}{dx}[(\cos^2 4x)w] = 0 \quad \text{gives} \quad (\cos^2 4x)w = c.$$

Therefore $w = u' = c \sec^2 4x$ and $u = c_1 \tan 4x$. A second solution is $y_2 = \tan 4x \cos 4x = \sin 4x$.

6. Define $y = u(x)e^{5x}$ so

$$y' = 5e^{5x}u + e^{5x}u', \quad y'' = e^{5x}u'' + 10e^{5x}u' + 25e^{5x}u$$

and

$$y'' - 25y = e^{5x}(u'' + 10u') = 0 \quad \text{or} \quad u'' + 10u' = 0.$$

If $w = u'$ we obtain the linear first-order equation $w' + 10w = 0$ which has the integrating factor $e^{10 \int dx} = e^{10x}$. Now

$$\frac{d}{dx}\left[e^{10x}w\right] = 0 \quad \text{gives} \quad e^{10x}w = c.$$

Therefore $w = u' = ce^{-10x}$ and $u = c_1 e^{-10x}$. A second solution is $y_2 = e^{-10x}e^{5x} = e^{-5x}$.

9. Identifying $P(x) = -7/x$ we have

$$y_2 = x^4 \int \frac{e^{-\int (-7/x)\,dx}}{x^8}\,dx = x^4 \int \frac{1}{x}\,dx = x^4 \ln|x|.$$

A second solution is $y_2 = x^4 \ln|x|$.

12. Identifying $P(x) = 0$ we have

$$y_2 = x^{1/2}\ln x \int \frac{e^{-\int 0\,dx}}{x(\ln x)^2}\,dx = x^{1/2}\ln x \left(-\frac{1}{\ln x}\right) = -x^{1/2}.$$

A second solution is $y_2 = x^{1/2}$.

15. Identifying $P(x) = 2(1+x)/\left(1 - 2x - x^2\right)$ we have

$$y_2 = (x+1) \int \frac{e^{-\int 2(1+x)\,dx/\left(1-2x-x^2\right)}}{(x+1)^2}\,dx = (x+1) \int \frac{e^{\ln\left(1-2x-x^2\right)}}{(x+1)^2}\,dx$$

$$= (x+1) \int \frac{1 - 2x - x^2}{(x+1)^2}\,dx = (x+1) \int \left[\frac{2}{(x+1)^2} - 1\right]\,dx$$

$$= (x+1) \left[-\frac{2}{x+1} - x\right] = -2 - x^2 - x.$$

A second solution is $y_2 = x^2 + x + 2$.

18. Identifying $P(x) = 2\tan x$ we have

$$y_2 = \sin x \int \frac{e^{-\int 2\tan x\,dx}}{\sin^2 x}\,dx = \sin x \int \frac{e^{2\ln(\cos x)}}{\sin^2 x}\,dx = \sin x \int \frac{\cos^2 x}{\sin^2 x}\,dx$$

$$= \sin x \int \frac{1 - \sin^2 x}{\sin^2 x}\,dx = \sin x\,(-\cot x - x) = -\cos x - x\sin x.$$

A second solution is $y_2 = \cos x + x\sin x$.

21. Define $y = u(x)e^x$ so

$$y' = ue^x + u'e^x, \quad y'' = u''e^x + 2u'e^x + ue^x$$

and

$$y'' - 3y' + 2y = e^x u'' - e^x u' = 5e^{3x}.$$

If $w = u'$ we obtain the linear first-order equation $w' - w = 5e^{2x}$ which has the integrating factor $e^{-\int dx} = e^{-x}$. Now

$$\frac{d}{dx}[e^{-x}w] = 5e^x \quad \text{gives} \quad e^{-x}w = 5e^x + c_1.$$

Therefore $w = u' = 5e^{2x} + c_1 e^x$ and

$$u = \frac{5}{2}e^{2x} + c_1 e^x + c_2$$

$$y = \frac{5}{2}e^{3x} + c_1 e^{2x} + c_2 e^x$$

From the last equation we see that a second solution is $y_2 = e^{2x}$ and $y_p = \frac{5}{2}e^{3x}$.

24. Dividing by $2x$ we have

$$2xy'' - (2x + 1)y' + y = 0$$

$$y'' + \left(-1 - \frac{1}{2x}\right)y' + \frac{1}{2x} = 0$$

Using $P(x) = -1 - \dfrac{1}{2x}$ and formula (5) in the text we have

$$y_2(x) = e^x \int_{x_0}^x \frac{e^{-(1+1/2t)\,dt}}{e^{2t}}\,dt = e^x \int_{x_0}^x \frac{e^{t+\frac{1}{2}\ln t}}{e^{2t}}\,dt = e^x \int_{x_0}^x \frac{e^t \left(e^{\ln t}\right)^{1/2}}{e^{2t}}\,dt = e^x \int_{x_0}^x \frac{e^t \sqrt{t}}{e^{2t}}\,dt$$

Therefore $y_2(x) = e^x \displaystyle\int_{x_0}^x \sqrt{t}\,e^{-t}\,dt,\ x_0 \geq 0.$

3.3 | **Linear Equations with Constant Coefficients**

3. From $m^2 - m - 6 = 0$ we obtain $m_1 = 3$ and $m_2 = -2$ so that $y = c_1 e^{3x} + c_2 e^{-2x}$.

6. From $m^2 - 10m + 25 = 0$ we obtain $m_1 = 5$ and $m_2 = 5$ so that $y = c_1 e^{5x} + c_2 x e^{5x}$.

9. From $m^2 + 9 = 0$ we obtain $m_1 = 3i$ and $m_2 = -3i$ so that $y = c_1 \cos 3x + c_2 \sin 3x$.

12. From $2m^2 + 2m + 1 = 0$ we obtain $m = -1/2 \pm i/2$ so that

$$y = e^{-x/2}[c_1 \cos(x/2) + c_2 \sin(x/2)].$$

15. From $m^3 - 4m^2 - 5m = 0$ we obtain $m_1 = 0$, $m_2 = 5$, and $m_3 = -1$ so that

$$y = c_1 + c_2 e^{5x} + c_3 e^{-x}.$$

18. From $m^3 + 3m^2 - 4m - 12 = 0$ we obtain $m_1 = -2$, $m_2 = 2$, and $m_3 = -3$ so that

$$y = c_1 e^{-2x} + c_2 e^{2x} + c_3 e^{-3x}.$$

21. From $m^3 + 3m^2 + 3m + 1 = 0$ we obtain $m_1 = -1$, $m_2 = -1$, and $m_3 = -1$ so that

$$y = c_1 e^{-x} + c_2 x e^{-x} + c_3 x^2 e^{-x}.$$

24. From $m^4 - 2m^2 + 1 = 0$ we obtain $m_1 = 1$, $m_2 = 1$, $m_3 = -1$, and $m_4 = -1$ so that

$$y = c_1 e^x + c_2 x e^x + c_3 e^{-x} + c_4 x e^{-x}.$$

27. From $m^5 + 5m^4 - 2m^3 - 10m^2 + m + 5 = 0$ we obtain $m_1 = -1$, $m_2 = -1$, $m_3 = 1$, and $m_4 = 1$, and $m_5 = -5$ so that

$$u = c_1 e^{-r} + c_2 r e^{-r} + c_3 e^r + c_4 r e^r + c_5 e^{-5r}.$$

30. From $m^6 - 3m^4 + 3m^2 = 0$ we obtain $m_1 = 1$, $m_2 = 1$, $m_3 = 1$, $m_4 = -1$, $m_5 = -1$, and $m_6 = -1$ so that

$$y = c_1 e^x + c_2 x e^x + c_3 x^2 e^x + c_4 e^{-x} + c_5 x e^{-x} + c_6 x^2 e^{-x}.$$

33. From $m^2 - 4m - 5 = 0$ we obtain $m_1 = -1$ and $m_1 = 5$, so that $y = c_1 e^{-x} + c_2 e^{5x}$. If $y(1) = 0$ and $y'(1) = 2$, then $c_1 e^{-1} + c_2 e^5 = 0$, $-c_1 e^{-1} + 5c_2 e^5 = 2$, so $c_1 = -e/3$, $c_2 = e^{-5}/3$, and $y = -\frac{1}{3} e^{1-x} + \frac{1}{3} e^{5x-5}$.

36. From $m^2 - 2m + 1 = 0$ we obtain $m_1 = 1$ and $m_2 = 1$ so that $y = c_1 e^x + c_2 x e^x$. If $y(0) = 5$ and $y'(0) = 10$ then $c_1 = 5$, $c_1 + c_2 = 10$ so $c_1 = 5$, $c_2 = 5$, and $y = 5e^x + 5xe^x$.

39. From $m^2 - 10m + 25 = 0$ we obtain $m_1 = 5$ and $m_2 = 5$ so that $y = c_1 e^{5x} + c_2 x e^{5x}$. If $y(0) = 1$ and $y(1) = 0$ then $c_1 = 1$, $c_1 e^5 + c_2 e^5 = 0$, so $c_1 = 1$, $c_2 = -1$, and $y = e^{5x} - xe^{5x}$.

42. From $m^2 - 2m + 2 = 0$ we obtain $m = 1 \pm i$ so that $y = e^x(c_1 \cos x + c_2 \sin x)$. If $y(0) = 1$ and $y(\pi) = 1$ then $c_1 = 1$ and $y(\pi) = e^\pi \cos \pi = -e^\pi$. Since $-e^\pi \neq 1$, the boundary-value problem has no solution.

45. For $\alpha > 0$ the general solution of $y'' + \alpha^2 y = 0$ is $y = c_1 \cos \alpha x + c_2 \sin \alpha x$. Using $y(0) = 0$ results in

$$0 = c_1 \cos 0 + c_2 \sin 0 = c_1.$$

Hence $y = c_2 \sin \alpha x$. Since we want nontrivial solutions, using $y(2) = 0$ results in

$$0 = c_2 \sin 2\alpha, \ c_2 > 0.$$

So $\sin 2\alpha = 0$. Thus

$$2\alpha = \pi, 2\pi, 3\pi, \ldots, \quad \text{or} \quad 2\alpha = n\pi, \ n = 1, 2, 3, \ldots.$$

So $\alpha = n\pi/2$ for $n = 1, 2, 3, \ldots$.

48. The auxiliary equation should have one positive and one negative root, so that the solution has the form $y = c_1 e^{k_1 x} + c_2 e^{-k_2 x}$. Thus the differential equation is (a).

51. The differential equation should have the form $y'' + k^2 y = 0$ where $k = 1$ so that the period of the solution is 2π. Thus the differential equation is (d).

54. The equation $y = c_1 e^{-5x} + c_2 e^{-4x}$ suggests that the auxiliary equation has the roots $m_1 = -5$ and $m_2 = 4$ therefore the auxiliary equation itself has the form

$$(m+5)(m+4) = 0$$
$$m^2 + 9m + 20 = 0$$

Therefore the differential equation is $y'' + 9y' + 20y = 0$.

57. The equation $y = c_1 \cos 8x + c_2 \sin 8x$ suggests that the auxiliary equation has the roots $m_1 = 8i$ and $m_2 = -8i$ therefore the auxiliary equation itself has the form

$$(m - 8i)(m + 8i) = 0$$
$$m^2 + 64 = 0$$

Therefore the differential equation is $y'' + 64y = 0$.

60. The equation $y = c_1 + c_2 e^{-2x} \cos 5x + c_3 e^{-2x} \sin 5x$ suggests that the auxiliary equation has the roots $m_1 = 0$, $m_2 = -2 + 5i$, and $m_3 = -2 - 5i$ therefore the auxiliary equation itself has the form

$$(m)(m - (-2 + 5i))(m - (-2 - 5i)) = 0$$
$$m^3 + 4m^2 + 29m = 0$$

Therefore the differential equation is $y''' + 4y'' + 29y' = 0$.

3.4 | Undetermined Coefficients

3. From $m^2 - 10m + 25 = 0$ we find $m_1 = m_2 = 5$. Then $y_c = c_1 e^{5x} + c_2 x e^{5x}$ and we assume $y_p = Ax + B$. Substituting into the differential equation we obtain $25A = 30$ and $-10A + 25B = 3$. Then $A = \frac{6}{5}$, $B = \frac{6}{5}$, $y_p = \frac{6}{5}x + \frac{6}{5}$, and

$$y = c_1 e^{5x} + c_2 x e^{5x} + \frac{6}{5}x + \frac{6}{5}.$$

6. From $m^2 - 8m + 20 = 0$ we find $m_1 = 4 + 2i$ and $m_2 = 4 - 2i$. Then $y_c = e^{4x}(c_1 \cos 2x + c_2 \sin 2x)$ and we assume $y_p = Ax^2 + Bx + C + (Dx + E)e^x$. Substituting into the differential equation we obtain

$$2A - 8B + 20C = 0$$

$$-6D + 13E = 0$$

$$-16A + 20B = 0$$

$$13D = -26$$

$$20A = 100.$$

Then $A = 5$, $B = 4$, $C = \frac{11}{10}$, $D = -2$, $E = -\frac{12}{13}$, $y_p = 5x^2 + 4x + \frac{11}{10} + \left(-2x - \frac{12}{13}\right) e^x$ and

$$y = e^{4x}(c_1 \cos 2x + c_2 \sin 2x) + 5x^2 + 4x + \frac{11}{10} + \left(-2x - \frac{12}{13}\right) e^x.$$

9. From $m^2 - m = 0$ we find $m_1 = 1$ and $m_2 = 0$. Then $y_c = c_1 e^x + c_2$ and we assume $y_p = Ax$. Substituting into the differential equation we obtain $-A = -3$. Then $A = 3$, $y_p = 3x$ and $y = c_1 e^x + c_2 + 3x$.

12. From $m^2 - 16 = 0$ we find $m_1 = 4$ and $m_2 = -4$. Then $y_c = c_1 e^{4x} + c_2 e^{-4x}$ and we assume $y_p = Axe^{4x}$. Substituting into the differential equation we obtain $8A = 2$. Then $A = \frac{1}{4}$, $y_p = \frac{1}{4}xe^{4x}$ and

$$y = c_1 e^{4x} + c_2 e^{-4x} + \frac{1}{4}xe^{4x}.$$

15. From $m^2 + 1 = 0$ we find $m_1 = i$ and $m_2 = -i$. Then $y_c = c_1 \cos x + c_2 \sin x$ and we assume $y_p = (Ax^2 + Bx) \cos x + (Cx^2 + Dx) \sin x$. Substituting into the differential equation we obtain $4C = 0$, $2A + 2D = 0$, $-4A = 2$, and $-2B + 2C = 0$. Then $A = -\frac{1}{2}$, $B = 0$, $C = 0$, $D = \frac{1}{2}$, $y_p = -\frac{1}{2}x^2 \cos x + \frac{1}{2}x \sin x$, and

$$y = c_1 \cos x + c_2 \sin x - \frac{1}{2}x^2 \cos x + \frac{1}{2}x \sin x.$$

18. From $m^2 - 2m + 2 = 0$ we find $m_1 = 1 + i$ and $m_2 = 1 - i$. Then $y_c = e^x(c_1 \cos x + c_2 \sin x)$ and we assume $y_p = Ae^{2x} \cos x + Be^{2x} \sin x$. Substituting into the differential equation we obtain $A + 2B = 1$ and $-2A + B = -3$. Then $A = \frac{7}{5}$, $B = -\frac{1}{5}$, $y_p = \frac{7}{5}e^{2x} \cos x - \frac{1}{5}e^{2x} \sin x$ and

$$y = e^x(c_1 \cos x + c_2 \sin x) + \frac{7}{5}e^{2x} \cos x - \frac{1}{5}e^{2x} \sin x.$$

21. From $m^3 - 6m^2 = 0$ we find $m_1 = m_2 = 0$ and $m_3 = 6$. Then $y_c = c_1 + c_2 x + c_3 e^{6x}$ and we assume $y_p = Ax^2 + B \cos x + C \sin x$. Substituting into the differential equation we obtain $-12A = 3$, $6B - C = -1$, and $B + 6C = 0$. Then $A = -\frac{1}{4}$, $B = -\frac{6}{37}$, $C = \frac{1}{37}$, $y_p = -\frac{1}{4}x^2 - \frac{6}{37} \cos x + \frac{1}{37} \sin x$, and

$$y = c_1 + c_2 x + c_3 e^{6x} - \frac{1}{4}x^2 - \frac{6}{37} \cos x + \frac{1}{37} \sin x.$$

24. From $m^3 - m^2 - 4m + 4 = 0$ we find $m_1 = 1$, $m_2 = 2$, and $m_3 = -2$. Then $y_c = c_1 e^x + c_2 e^{2x} + c_3 e^{-2x}$ and we assume $y_p = A + Bxe^x + Cxe^{2x}$. Substituting into the differential equation we obtain $4A = 5$, $-3B = -1$, and $4C = 1$. Then $A = \frac{5}{4}$, $B = \frac{1}{3}$, $C = \frac{1}{4}$, $y_p = \frac{5}{4} + \frac{1}{3}xe^x + \frac{1}{4}xe^{2x}$, and

$$y = c_1 e^x + c_2 e^{2x} + c_3 e^{-2x} + \frac{5}{4} + \frac{1}{3}xe^x + \frac{1}{4}xe^{2x}.$$

27. We have $y_c = c_1 \cos 2x + c_2 \sin 2x$ and we assume $y_p = A$. Substituting into the differential equation we find $A = -\frac{1}{2}$. Thus $y = c_1 \cos 2x + c_2 \sin 2x - \frac{1}{2}$. From the initial conditions we obtain $c_1 = 0$ and $c_2 = \sqrt{2}$, so

$$y = \sqrt{2} \sin 2x - \frac{1}{2}.$$

30. We have $y_c = c_1 e^{-2x} + c_2 x e^{-2x}$ and we assume $y_p = (Ax^3 + Bx^2)e^{-2x}$. Substituting into the differential equation we find $A = \frac{1}{6}$ and $B = \frac{3}{2}$. Thus $y = c_1 e^{-2x} + c_2 x e^{-2x} + \left(\frac{1}{6}x^3 + \frac{3}{2}x^2\right)e^{-2x}$. From the initial conditions we obtain $c_1 = 2$ and $c_2 = 9$, so

$$y = 2e^{-2x} + 9xe^{-2x} + \left(\frac{1}{6}x^3 + \frac{3}{2}x^2\right)e^{-2x}.$$

33. We have $x_c = c_1 \cos \omega t + c_2 \sin \omega t$ and we assume $x_p = At \cos \omega t + Bt \sin \omega t$. Substituting into the differential equation we find $A = -F_0/2\omega$ and $B = 0$. Thus $x = c_1 \cos \omega t + c_2 \sin \omega t - (F_0/2\omega)t \cos \omega t$. From the initial conditions we obtain $c_1 = 0$ and $c_2 = F_0/2\omega^2$, so

$$x = (F_0/2\omega^2) \sin \omega t - (F_0/2\omega)t \cos \omega t.$$

36. We have $y_c = c_1 e^{-2x} + e^x(c_2 \cos \sqrt{3}\,x + c_3 \sin \sqrt{3}\,x)$ and we assume $y_p = Ax + B + Cxe^{-2x}$. Substituting into the differential equation we find $A = \frac{1}{4}$, $B = -\frac{5}{8}$, and $C = \frac{2}{3}$. Thus

$$y = c_1 e^{-2x} + e^x(c_2 \cos \sqrt{3}\,x + c_3 \sin \sqrt{3}\,x) + \frac{1}{4}x - \frac{5}{8} + \frac{2}{3}xe^{-2x}.$$

From the initial conditions we obtain $c_1 = -\frac{23}{12}$, $c_2 = -\frac{59}{24}$, and $c_3 = \frac{17}{72}\sqrt{3}$, so

$$y = -\frac{23}{12}e^{-2x} + e^x\left(-\frac{59}{24}\cos \sqrt{3}\,x + \frac{17}{72}\sqrt{3}\sin \sqrt{3}\,x\right) + \frac{1}{4}x - \frac{5}{8} + \frac{2}{3}xe^{-2x}.$$

39. The general solution of the differential equation $y'' + 3y = 6x$ is $y = c_1 \cos \sqrt{3}x + c_2 \sin \sqrt{3}x + 2x$. The condition $y(0) = 0$ implies $c_1 = 0$ and so $y = c_2 \sin \sqrt{3}x + 2x$. The condition $y(1) + y'(1) = 0$ implies $c_2 \sin \sqrt{3} + 2 + c_2\sqrt{3}\cos\sqrt{3} + 2 = 0$ so $c_2 = -4/(\sin\sqrt{3} + \sqrt{3}\cos\sqrt{3})$. The solution is

$$y = \frac{-4\sin\sqrt{3}x}{\sin\sqrt{3} + \sqrt{3}\cos\sqrt{3}} + 2x.$$

42. We have $y_c = e^x(c_1 \cos 3x + c_2 \sin 3x)$ and we assume $y_p = A$ on $[0, \pi]$. Substituting into the differential equation we find $A = 2$. Thus, $y = e^x(c_1 \cos 3x + c_2 \sin 3x) + 2$ on $[0, \pi]$. On (π, ∞) we have $y = e^x(c_3 \cos 3x + c_4 \sin 3x)$. From $y(0) = 0$ and $y'(0) = 0$ we obtain

$$c_1 = -2, \qquad c_1 + 3c_2 = 0.$$

Solving this system, we find $c_1 = -2$ and $c_2 = \frac{2}{3}$. Thus $y = e^x(-2\cos 3x + \frac{2}{3}\sin 3x) + 2$ on $[0, \pi]$. Now, continuity of y at $x = \pi$ implies

$$e^\pi\left(-2\cos 3\pi + \frac{2}{3}\sin 3\pi\right) + 2 = e^\pi(c_3\cos 3\pi + c_4\sin 3\pi)$$

or $2 + 2e^\pi = -c_3 e^\pi$ or $c_3 = -2e^{-\pi}(1 + e^\pi)$. Continuity of y' at π implies

$$\frac{20}{3}e^\pi\sin 3\pi = e^\pi[(c_3 + 3c_4)\cos 3\pi + (-3c_3 + c_4)\sin 3\pi]$$

or $-c_3 e^\pi - 3c_4 e^\pi = 0$. Since $c_3 = -2e^{-\pi}(1 + e^\pi)$ we have $c_4 = \frac{2}{3}e^{-\pi}(1 + e^\pi)$. Therefore

$$y(x) = \begin{cases} e^x(-2\cos 3x + \dfrac{2}{3}\sin 3x) + 2, & 0 \le x \le \pi \\[2mm] (1 + e^\pi)e^{x-\pi}(-2\cos 3x + \dfrac{2}{3}\sin 3x), & x > \pi. \end{cases}$$

3.5   Variation of Parameters

The particular solution, $y_p = u_1 y_1 + u_2 y_2$, in the following problems can take on a variety of forms, especially where trigonometric functions are involved. The validity of a particular form can best be checked by substituting it back into the differential equation.

3. The auxiliary equation is $m^2 + 1 = 0$, so $y_c = c_1\cos x + c_2\sin x$ and

$$W = \begin{vmatrix} \cos x & \sin x \\ -\sin x & \cos x \end{vmatrix} = 1$$

Identifying $f(x) = \sin x$ we obtain

$$u_1' = -\sin^2 x$$
$$u_2' = \cos x\sin x.$$

Then

$$u_1 = \frac{1}{4}\sin 2x - \frac{1}{2}x = \frac{1}{2}\sin x\cos x - \frac{1}{2}x$$

$$u_2 = -\frac{1}{2}\cos^2 x.$$

and

$$y = c_1\cos x + c_2\sin x + \frac{1}{2}\sin x\cos^2 x - \frac{1}{2}x\cos x - \frac{1}{2}\cos^2 x\sin x$$

$$= c_1\cos x + c_2\sin x - \frac{1}{2}x\cos x.$$

6. The auxiliary equation is $m^2 + 1 = 0$, so $y_c = c_1 \cos x + c_2 \sin x$ and

$$W = \begin{vmatrix} \cos x & \sin x \\ -\sin x & \cos x \end{vmatrix} = 1$$

Identifying $f(x) = \sec^2 x$ we obtain

$$u_1' = -\frac{\sin x}{\cos^2 x}$$

$$u_2' = \sec x.$$

Then

$$u_1 = -\frac{1}{\cos x} = -\sec x$$

$$u_2 = \ln|\sec x + \tan x|$$

and

$$y = c_1 \cos x + c_2 \sin x - \cos x \sec x + \sin x \ln|\sec x + \tan x|$$

$$= c_1 \cos x + c_2 \sin x - 1 + \sin x \ln|\sec x + \tan x|.$$

9. The auxiliary equation is $m^2 - 9 = 0$, so $y_c = c_1 e^{3x} + c_2 e^{-3x}$ and

$$W = \begin{vmatrix} e^{3x} & e^{-3x} \\ 3e^{3x} & -3e^{-3x} \end{vmatrix} = -6$$

Identifying $f(x) = 9x/e^{3x}$ we obtain $u_1' = \frac{3}{2}xe^{-6x}$ and $u_2' = -\frac{3}{2}x$. Then

$$u_1 = -\frac{1}{24}e^{-6x} - \frac{1}{4}xe^{-6x},$$

$$u_2 = -\frac{3}{4}x^2$$

and

$$y = c_1 e^{3x} + c_2 e^{-3x} - \frac{1}{24}e^{-3x} - \frac{1}{4}xe^{-3x} - \frac{3}{4}x^2 e^{-3x}$$

$$= c_1 e^{3x} + c_3 e^{-3x} - \frac{1}{4}xe^{-3x}(1 + 3x).$$

12. The auxiliary equation is $m^2 - 4m + 3 = 0$, so $y_c = c_1 e^x + c_2 e^{3x}$ and

$$W = \begin{vmatrix} e^x & e^{3x} \\ e^x & 3e^{3x} \end{vmatrix} = 2e^{4x}.$$

Identifying $f(x) = e^x$ we obtain $u_1' = -\dfrac{1}{2}$ and $u_2' = \dfrac{1}{2}e^{-2x}$. Then

$$u_1 = -\frac{1}{2}x,$$

$$u_2 = -\frac{1}{4}e^{-2x}$$

and

$$y = c_1 e^x + c_2 e^{3x} - \frac{1}{2}xe^x - \frac{1}{4}e^x = c_3 e^x + c_2 e^{3x} - \frac{1}{2}xe^x.$$

15. The auxiliary equation is $m^2 + 3m + 2 = (m+1)(m+2) = 0$, so $y_c = c_1 e^{-x} + c_2 e^{-2x}$ and

$$W = \begin{vmatrix} e^{-x} & e^{-2x} \\ -e^{-x} & = 2e^{-2} \end{vmatrix} = -e^{-3x}$$

Identifying $f(x) = \sin e^x$ we obtain

$$u_1' = \frac{e^{-2x}\sin e^x}{e^{-3x}} = e^x \sin e^x$$

$$u_2' = \frac{e^{-x}\sin e^x}{-e^{-3x}} = -e^{2x}\sin e^x.$$

Then $u_1 = -\cos e^x$, $u_2 = e^x \cos x - \sin e^x$, and

$$y = c_1 e^{-x} + c_2 e^{-2x} - e^{-x}\cos e^x + e^{-x}\cos e^x - e^{-2x}\sin e^x$$
$$= c_1 e^{-x} + c_2 e^{-2x} - e^{-2x}\sin e^x.$$

18. The auxiliary equation is $2m^2 + m = m(2m+1) = 0$, so $y_c = c_1 + c_2 e^{-x/2}$ and

$$W = \begin{vmatrix} 1 & e^{-x/2} \\ 0 & -\frac{1}{2}e^{-x/2} \end{vmatrix} = -\frac{1}{2}e^{-x/2}$$

Dividing by 2 we identify $f(t) = 3x$ and obtain

$$u_1' = 6x$$
$$u_2' = -6xe^{x/2}.$$

Then

$$u_1 = 3x^2$$
$$u_2 = -12xe^{x/2} + 24e^{x/2}.$$

and

$$y = c_1 + c_2 e^{-x/2} + 3x^2 - 12x + 24$$
$$= c_3 + c_2 e^{-x/2} + 3x^2 - 12x$$

21. The auxiliary equation is $4m^2 - 1 = (2m - 1)(2m + 1) = 0$, so $y_c = c_1 e^{x/2} + c_2 e^{-x/2}$ and

$$W = \begin{vmatrix} e^{x/2} & e^{-x/2} \\ \frac{1}{2}e^{x/2} & -\frac{1}{2}e^{-x/2} \end{vmatrix} = -1$$

Identifying $f(x) = xe^{x/2}/4$ we obtain $u_1' = x/4$ and $u_2' = -xe^x/4$. Then $u_1 = x^2/8$ and $u_2 = -xe^x/4 + e^x/4$. Thus

$$y = c_1 e^{x/2} + c_2 e^{-x/2} + \frac{1}{8}x^2 e^{x/2} - \frac{1}{4}xe^{x/2} + \frac{1}{4}e^{x/2}$$

$$= c_3 e^{x/2} + c_2 e^{-x/2} + \frac{1}{8}x^2 e^{x/2} - \frac{1}{4}xe^{x/2}$$

and

$$y' = \frac{1}{2}c_3 e^{x/2} - \frac{1}{2}c_2 e^{-x/2} + \frac{1}{16}x^2 e^{x/2} + \frac{1}{8}xe^{x/2} - \frac{1}{4}e^{x/2}.$$

The initial conditions imply

$$c_3 + c_2 = 0,$$

$$\frac{1}{2}c_3 - \frac{1}{2}c_2 - \frac{1}{4} = 0.$$

Thus $c_3 = 3/4$ and $c_2 = 1/4$, and

$$y = \frac{3}{4}e^{x/2} + \frac{1}{4}e^{-x/2} + \frac{1}{8}x^2 e^{x/2} - \frac{1}{4}xe^{x/2}.$$

24. The auxiliary equation is $2m^2 + m - 1 = (2m - 1)(m + 1) = 0$, so $y_c = c_1 e^{x/2} + c_2 e^{-x}$ and

$$W = \begin{vmatrix} e^{x/2} & e^{-x} \\ \frac{1}{2}e^{x/2} & -e^{-x} \end{vmatrix} = -\frac{3}{2}e^{-x/2}$$

Identifying $f(x) = (x + 1)/2$ we obtain

$$u_1' = \frac{1}{3}e^{-x/2}(x + 1)$$

$$u_2' = -\frac{1}{3}e^x(x + 1).$$

Then

$$u_1 = -e^{-x/2}\left(\frac{2}{3}x - 2\right)$$

$$u_2 = -\frac{1}{3}xe^x.$$

Thus

$$y = c_1 e^{x/2} + c_2 e^{-x} - x - 2$$

and

$$y' = \frac{1}{2}c_1 e^{x/2} - c_2 e^{-x} - 1.$$

The initial conditions imply

$$c_1 - c_2 - 2 = 1$$

$$\frac{1}{2}c_1 - c_2 - 1 = 0.$$

Thus $c_1 = 8/3$ and $c_2 = 1/3$, and

$$y = \frac{8}{3}e^{x/2} + \frac{1}{3}e^{-x} - x - 2.$$

27. The auxiliary equation is $m^2 + 1 = 0$, so $y_c = c_1 \cos x + c_2 \sin x$ and

$$W = \begin{vmatrix} \cos x & \sin x \\ -\sin x & \cos x \end{vmatrix} = 1$$

Identifying $f(x) = e^{x^2}$ we obtain $u_1' = -e^{x^2} \sin x$ and $u_2' = e^{x^2} \cos x$. Then

$$u_1 = -\int_{x_0}^{x} e^{t^2} \sin t \, dt$$

$$u_2 = \int_{x_0}^{x} e^{t^2} \cos t \, dt$$

and

$$y = c_1 \cos x + c_2 \sin x - \cos x \int_{x_0}^{x} e^{t^2} \sin t \, dt + \sin x \int_{x_0}^{x} e^{t^2} \cos t \, dt.$$

30. The auxiliary equation is $2m^2 + 2m + 1 = 0$, so $y_c = e^{-x/2}[c_1 \cos(x/2) + c_2 \sin(x/2)]$ and

$$W = \begin{vmatrix} e^{-x/2}\cos\dfrac{x}{2} & e^{-x/2}\sin\dfrac{x}{2} \\ \dfrac{1}{2}e^{-x/2}\cos\dfrac{x}{2} - \dfrac{1}{2}e^{-x/2}\sin\dfrac{x}{2} & \dfrac{1}{2}e^{-x/2}\cos\dfrac{x}{2} - \dfrac{1}{2}e^{x/2}\sin\dfrac{x}{2} \end{vmatrix} = \frac{1}{2}e^{-x}$$

Identifying $f(x) = 2\sqrt{x}$ we obtain

$$u_1' = -\frac{e^{-x/2}\sin(x/2)2\sqrt{x}}{\frac{1}{2}e^{-x/2}} = -4e^{x/2}\sqrt{x}\sin\frac{x}{2}$$

$$u_2' = -\frac{e^{-x/2}\cos(x/2)2\sqrt{x}}{\frac{1}{2}e^{-x/2}} = 4e^{x/2}\sqrt{x}\cos\frac{x}{2}.$$

Then

$$u_1 = -4\int_{x_0}^{x} e^{t/2}\sqrt{t}\sin\frac{t}{2}\,dt$$

$$u_2 = 4\int_{x_0}^{x} e^{t/2}\sqrt{t}\cos\frac{t}{2}\,dt$$

and

$$y = e^{-x/2} \left(c_1 \cos \frac{x}{2} + c_2 \sin \frac{x}{2} \right) - 4e^{-x/2} \cos \frac{x}{2} \int_{x_0}^{x} e^{t/2} \sqrt{t} \sin \frac{t}{2} \, dt$$

$$+ 4e^{-x/2} \sin \frac{x}{2} \int_{x_0}^{x} e^{t/2} \sqrt{t} \cos \frac{t}{2} \, dt.$$

33. The auxiliary equation is $m^3 + m = m(m^2 + 1) = 0$, so $y_c = c_1 + c_2 \cos x + c_3 \sin x$ and

$$W = \begin{vmatrix} 1 & \cos x & \sin x \\ 0 & -\sin x & \cos x \\ 0 & -\cos x & -\sin x \end{vmatrix} = 1$$

Identifying $f(x) = \tan x$ we obtain

$$u_1' = W_1 = \begin{vmatrix} 0 & \cos x & \sin x \\ 0 & -\sin x & \cos x \\ \tan x & -\cos x & -\sin x \end{vmatrix} = \tan x$$

$$u_2' = W_2 = \begin{vmatrix} 1 & 0 & \sin x \\ 0 & 0 & \cos x \\ 0 & \tan x & -\sin x \end{vmatrix} = -\sin x$$

$$u_3' = W_3 = \begin{vmatrix} 1 & \cos x & 0 \\ 0 & -\sin x & 0 \\ 0 & -\cos x & \tan x \end{vmatrix} = -\sin x \tan x = \frac{\cos^2 x - 1}{\cos x} = \cos x - \sec x$$

Then

$$u_1 = -\ln|\cos x|$$
$$u_2 = \cos x$$
$$u_3 = \sin x - \ln|\sec x + \tan x|$$

and

$$y = c_1 + c_2 \cos x + c_3 \sin x - \ln|\cos x| + \cos^2 x$$
$$+ \sin^2 x - \sin x \ln|\sec x + \tan x|$$
$$= c_4 + c_2 \cos x + c_3 \sin x - \ln|\cos x| - \sin x \ln|\sec x + \tan x|$$

for $-\pi/2 < x < \pi/2$.

36. The auxiliary equation is $m^3 - 3m^2 + 2 = m(m-1)(m-2) = 0$, so $y_c = c_1 + c_2 e^x + c_3 e^{2x}$ and

$$W = \begin{vmatrix} 1 & e^x & e^{2x} \\ 0 & e^x & 2e^{2x} \\ 0 & e^x & 4e^{2x} \end{vmatrix} = 2e^{3x}.$$

Identifying $f(x) = e^{2x}/(1 + e^x)$ we obtain

$$u_1' = \frac{1}{2e^{3x}} W_1 = \frac{1}{2e^{3x}} \begin{vmatrix} 0 & e^x & e^{2x} \\ 0 & e^x & 2e^{2x} \\ \dfrac{e^{2x}}{1+e^x} & e^x & 4e^{2x} \end{vmatrix} = \frac{1}{2} \cdot \frac{e^{2x}}{1+e^x},$$

$$u_2' = \frac{1}{2e^{3x}} W_2 = \frac{1}{2e^{3x}} \begin{vmatrix} 1 & 0 & e^{2x} \\ 0 & 0 & 2e^{2x} \\ 0 & \dfrac{e^{2x}}{1+e^x} & 4e^{2x} \end{vmatrix} = -\frac{e^x}{1+e^x},$$

$$u_3' = \frac{1}{2e^{3x}} W_3 = \frac{1}{2e^{3x}} \begin{vmatrix} 1 & e^x & 0 \\ 0 & e^x & 0 \\ 0 & e^x & \dfrac{e^{2x}}{1+e^x} \end{vmatrix} = \frac{1}{2} \cdot \frac{1}{1+e^x} = \frac{1}{2} \cdot \frac{e^{-x}}{1+e^{-x}}.$$

Then

$$u_1 = \frac{1}{2} e^x - \frac{1}{2} \ln\left(1 + e^x\right),$$

$$u_2 = -\ln\left(1 + e^x\right),$$

$$u_3 = -\frac{1}{2} \ln\left(1 + e^{-x}\right).$$

and

$$y = c_1 + c_2 e^x + c_3 e^{2x} + \frac{1}{2} e^x - \frac{1}{2} \ln\left(1 + e^x\right) - e^x \ln\left(1 + e^x\right) - \frac{1}{2} e^{2x} \ln\left(1 + e^{-x}\right)$$

$$= c_1 + \left(c_2 + \frac{1}{2}\right) e^x + c_3 e^{2x} - \left(\frac{1}{2} + e^x\right) \ln\left(1 + e^x\right) - \frac{1}{2} e^{2x} \ln\left(1 + e^{-x}\right)$$

$$= c_1 + \left(c_2 + \frac{1}{2}\right) e^x + c_3 e^{2x} - \left(\frac{1}{2} + e^x\right) \ln\left(1 + e^x\right) - \frac{1}{2} e^{2x} \ln\left(\frac{1 + e^x}{e^x}\right)$$

$$= c_1 + \left(c_2 + \frac{1}{2}\right) e^x + c_3 e^{2x} - \left(\frac{1}{2} + e^x\right) \ln\left(1 + e^x\right) - \frac{1}{2} e^{2x} \left(\ln\left(1 + e^x\right) - x\right)$$

$$= c_1 + \left(c_2 + \frac{1}{2}\right) e^x + c_3 e^{2x} - \left(\frac{1}{2} + e^x + \frac{1}{2} e^{2x}\right) \ln\left(1 + e^x\right)$$

Combining $c_2 + 1/2$ into a single constant, c_2 gives

$$y = c_1 + c_2 e^x + c_3 e^{2x} - \left(\frac{1}{2} + e^x + \frac{1}{2} e^{2x}\right) \ln\left(1 + e^x\right)$$

3.6 | Cauchy–Euler Equations

3. The auxiliary equation is $m^2 = 0$ so that $y = c_1 + c_2 \ln x$.

6. The auxiliary equation is $m^2 + 4m + 3 = (m+1)(m+3) = 0$ so that $y = c_1 x^{-1} + c_2 x^{-3}$.

9. The auxiliary equation is $25m^2 + 1 = 0$ so that $y = c_1 \cos\left(\frac{1}{5}\ln x\right) + c_2 \sin\left(\frac{1}{5}\ln x\right)$.

12. The auxiliary equation is $m^2 + 7m + 6 = (m+1)(m+6) = 0$ so that $y = c_1 x^{-1} + c_2 x^{-6}$.

15. Assuming that $y = x^m$ and substituting into the differential equation we obtain

$$m(m-1)(m-2) - 6 = m^3 - 3m^2 + 2m - 6 = (m-3)(m^2+2) = 0.$$

Thus

$$y = c_1 x^3 + c_2 \cos\left(\sqrt{2}\ln x\right) + c_3 \sin\left(\sqrt{2}\ln x\right).$$

18. Assuming that $y = x^m$ and substituting into the differential equation we obtain

$$m(m-1)(m-2)(m-3) + 6m(m-1)(m-2) + 9m(m-1) + 3m + 1 = m^4 + 2m^2 + 1 = (m^2+1)^2 = 0.$$

Thus

$$y = c_1 \cos\left(\ln x\right) + c_2 \sin\left(\ln x\right) + c_3(\ln x)\cos\left(\ln x\right) + c_4(\ln x)\sin\left(\ln x\right).$$

21. The auxiliary equation is $m^2 - 2m + 1 = (m-1)^2 = 0$ so that $y_c = c_1 x + c_2 x \ln x$ and

$$W\left(x, x^2\right) = \begin{vmatrix} x & x\ln x \\ 1 & 1 + \ln x \end{vmatrix} = x$$

Identifying $f(x) = 2/x$ we obtain $u_1' = -2(\ln x)/x$ and $u_2' = 2/x$. Then $u_1 = -(\ln x)^2$, $u_2 = 2\ln x$, and

$$y = c_1 x + c_2 x \ln x - x(\ln x)^2 + 2x(\ln x)^2$$
$$= c_1 x + c_2 x \ln x + x(\ln x)^2, \qquad x > 0.$$

24. The auxiliary equation $m(m-1) + m - 1 = m^2 - 1 = 0$ has roots $m_1 = -1$, $m_2 = 1$, so $y_c = c_1 x^{-1} + c_2 x$. With $y_1 = x^{-1}$, $y_2 = x$, and the identification $f(x) = 1/x^2(x+1)$, we get

$$W = 2x^{-1}, \qquad W_1 = -1/x(x+1), \qquad \text{and} \qquad W_2 = 1/x^3(x+1).$$

Then $u_1' = W_1/W = -1/2(x+1)$, $u_2' = W_2/W = 1/2x^2(x+1)$, and integration (by partial fractions for u_2') gives

$$u_1 = -\frac{1}{2}\ln(x+1)$$

$$u_2 = -\frac{1}{2}x^{-1} - \frac{1}{2}\ln x + \frac{1}{2}\ln(x+1),$$

so

$$y_p = u_1 y_1 + u_2 y_2 = \left[-\frac{1}{2} \ln (x+1) \right] x^{-1} + \left[-\frac{1}{2} x^{-1} - \frac{1}{2} \ln x + \frac{1}{2} \ln (x+1) \right] x$$

$$= -\frac{1}{2} - \frac{1}{2} x \ln x + \frac{1}{2} x \ln (x+1) - \frac{\ln (x+1)}{2x} = -\frac{1}{2} + \frac{1}{2} x \ln \left(1 + \frac{1}{x} \right) - \frac{\ln (x+1)}{2x}$$

and

$$y = y_c + y_p = c_1 x^{-1} + c_2 x - \frac{1}{2} + \frac{1}{2} x \ln \left(1 + \frac{1}{x} \right) - \frac{\ln (x+1)}{2x}, \qquad x > 0.$$

27. The auxiliary equation is $m^2 + 1 = 0$, so that

$$y = c_1 \cos (\ln x) + c_2 \sin (\ln x)$$

and

$$y' = -c_1 \frac{1}{x} \sin (\ln x) + c_2 \frac{1}{x} \cos (\ln x).$$

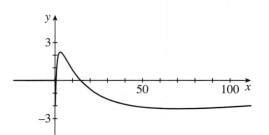

The initial conditions imply $c_1 = 1$ and $c_2 = 2$. Thus $y = \cos (\ln x) + 2 \sin (\ln x)$. The graph is given to the right.

30. The auxiliary equation is $m^2 - 6m + 8 = (m-2)(m-4) = 0$, so that $y_c = c_1 x^2 + c_2 x^4$ and

$$W \left(x^2, x^4 \right) = \begin{vmatrix} x^2 & x^4 \\ 2x & 4x^3 \end{vmatrix} = 2x^5$$

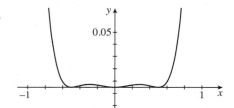

Identifying $f(x) = 8x^4$ we obtain $u_1' = -4x^3$ and $u_2' = 4x$. Then $u_1 = -x^4$, $u_2 = 2x^2$, and $y = c_1 x^2 + c_2 x^4 + x^6$. The initial conditions imply

$$\frac{1}{4} c_1 + \frac{1}{16} c_2 = -\frac{1}{64}$$

$$c_1 + \frac{1}{2} c_2 = -\frac{3}{16} \, .$$

Thus $c_1 = \frac{1}{16}$, $c_2 = -\frac{1}{2}$, and $y = \frac{1}{16} x^2 - \frac{1}{2} x^4 + x^6$. The graph is given to the right.

33. The solution $y = c_1 x^4 + c_2 x^{-2}$ suggests that the auxiliary equation has the roots $m = 4$ and $m = -2$ therefore the auxiliary equation itself has the form

$$am^2 + (b-a)m + c = 0$$
$$(m-4)(m+2) = 0$$
$$m^2 - 2m - 8 = 0$$

Therefore $a = 1$, $b - a = -2$, and $c = -8$. Now if $a = 1$ then $b - a = -2$ means that $b = -1$ and therefore the Cauchy–Euler equation is

$$ax^2 \frac{d^2y}{dx^2} + bx \frac{dy}{dx} + cy = 0$$

$$x^2 \frac{d^2y}{dx^2} - x \frac{dy}{dx} - 8y = 0$$

36. The solution $y = c_1 + c_2 x + c_3 x \ln x$ suggests that the auxiliary equation has the roots $m_1 = 0$ and $m_2 = m_3 = 1$ therefore the auxiliary equation itself has the form

$$a_3 m^3 + (a_2 - 3a_3)m^2 + (2a_3 - a_2 + a_1)m + a_0 = 0$$
$$m(m - 1)(m - 1) = 0$$
$$m^3 - 2m^2 + m = 0$$

Therefore $a_3 = 1$, $a_2 - 3a_3 = -2$, $2a_3 - a_2 + a_1 = 1$, $a_0 = 0$. Now if $a_3 = 1$ then $a_2 - 3a_3 = -2$ means that $a_2 = 1$ and so $2a_3 - a_2 + a_1 = 1$ means that $a_1 = 0$ therefore the Cauchy–Euler equation is

$$a_3 x^3 \frac{d^3y}{dx^3} + a_2 x^2 \frac{d^2y}{dx^2} + a_1 x \frac{dy}{dx} + a_0 y = 0$$

$$x^3 \frac{d^3y}{dx^3} + x^2 \frac{d^2y}{dx^2} = 0$$

39. If we force the function $y = (x - (-3))^m = (x + 3)^m$ into the equation we get

$$(x + 3)^2 y'' - 8(x + 3)y' + 14y = 0$$
$$(x + 3)^2 m(m - 1)(x + 3)^{m-2} - 8(x + 3)m(x + 3)^{m-1} + 14(x + 3)^m = 0$$
$$(x + 3)^m[m^2 - 9m + 14] = 0$$
$$(x + 3)^m(m - 2)(m - 7) = 0$$

The solutions to the auxiliary equation are therefore $m = 2$ and $m = 7$ from which we get the general solution $y = c_1(x + 3)^2 + c_2(x + 3)^7$.

42. If we force the function $y = (x - 4)^m$ into the equation we get

$$(x - 4)^2 y'' - 5(x - 4)y' + 9y = 0$$
$$(x - 4)^2 m(m - 1)(x - 4)^{m-2} - 5(x - 4)m(x - 4)^{m-1} + 9(x - 4)^m = 0$$
$$(x - 4)^m(m^2 - 6m + 9) = 0$$

The solutions to the auxiliary equation are repeated roots $m_1 = m_2 = 3$ from which we get the general solution $y = c_1(x - 4)^3 + c_2(x - 4)^3 \ln(x - 4)$.

45. Substituting $x = e^t$ into the differential equation we obtain

$$\frac{d^2y}{dt^2} + 9\frac{dy}{dt} + 8y = e^{2t}.$$

The auxiliary equation is $m^2 + 9m + 8 = (m+1)(m+8) = 0$ so that $y_c = c_1 e^{-t} + c_2 e^{-8t}$. Using undetermined coefficients we try $y_p = Ae^{2t}$. This leads to $30Ae^{2t} = e^{2t}$, so that $A = 1/30$ and

$$y = c_1 e^{-t} + c_2 e^{-8t} + \frac{1}{30}e^{2t} \quad \text{or} \quad y = c_1 x^{-1} + c_2 x^{-8} + \frac{1}{30}x^2.$$

48. From

$$\frac{d^2y}{dx^2} = \frac{1}{x^2}\left(\frac{d^2y}{dt^2} - \frac{dy}{dt}\right)$$

it follows that

$$\frac{d^3y}{dx^3} = \frac{1}{x^2}\frac{d}{dx}\left(\frac{d^2y}{dt^2} - \frac{dy}{dt}\right) - \frac{2}{x^3}\left(\frac{d^2y}{dt^2} - \frac{dy}{dt}\right)$$

$$= \frac{1}{x^2}\frac{d}{dx}\left(\frac{d^2y}{dt^2}\right) - \frac{1}{x^2}\frac{d}{dx}\left(\frac{dy}{dt}\right) - \frac{2}{x^3}\frac{d^2y}{dt^2} + \frac{2}{x^3}\frac{dy}{dt}$$

$$= \frac{1}{x^2}\frac{d^3y}{dt^3}\left(\frac{1}{x}\right) - \frac{1}{x^2}\frac{d^2y}{dt^2}\left(\frac{1}{x}\right) - \frac{2}{x^3}\frac{d^2y}{dt^2} + \frac{2}{x^3}\frac{dy}{dt}$$

$$= \frac{1}{x^3}\left(\frac{d^3y}{dt^3} - 3\frac{d^2y}{dt^2} + 2\frac{dy}{dt}\right).$$

Substituting into the differential equation we obtain

$$\frac{d^3y}{dt^3} - 3\frac{d^2y}{dt^2} + 2\frac{dy}{dt} - 3\left(\frac{d^2y}{dt^2} - \frac{dy}{dt}\right) + 6\frac{dy}{dt} - 6y = 3 + 3t$$

or

$$\frac{d^3y}{dt^3} - 6\frac{d^2y}{dt^2} + 11\frac{dy}{dt} - 6y = 3 + 3t.$$

The auxiliary equation is $m^3 - 6m^2 + 11m - 6 = (m-1)(m-2)(m-3) = 0$ so that $y_c = c_1 e^t + c_2 e^{2t} + c_3 e^{3t}$. Using undetermined coefficients we try $y_p = A + Bt$. This leads to $(11B - 6A) - 6Bt = 3 + 3t$, so that $A = -17/12$, $B = -1/2$, and

$$y = c_1 e^t + c_2 e^{2t} + c_3 e^{3t} - \frac{17}{12} - \frac{1}{2}t \quad \text{or} \quad y = c_1 x + c_2 x^2 + c_3 x^3 - \frac{17}{12} - \frac{1}{2}\ln x.$$

51. Use the chain rule to expand the left side of the equation and get

$$\frac{d^2T}{dr^2} + \frac{1}{r}\frac{dT}{dr} = \frac{T}{r^2}$$

$$\frac{d^2T}{dr^2} + \frac{1}{r}\frac{dT}{dr} - \frac{T}{r^2} = 0$$

$$r^2\frac{d^2T}{dr^2} + r\frac{dT}{dr} - T = 0$$

The result is a Cauchy–Euler equation whose auxiliary equation is $m(m-1) + m - 1 = 0$ or $m^2 - 1 = 0$. The solutions $m = \pm 1$ yield the general solution $T(r) = c_1 r + c_2 r^{-1}$. Next we apply the given conditions $T(2) = T_0$ and $T'(1) = 0$, which leads to the system of equations

$$2c_1 + \frac{1}{2}c_2 = T_0$$

$$c_1 - c_2 = 0$$

Solving gives $c_1 = c_2 = 2T_0/5$. Therefore $T(r) = \frac{2T_0}{5}\left(r + r^{-1}\right)$.

54. (a) Using the product rule twice, the left-hand side becomes

$$\frac{d}{dr}\left[\frac{1}{r}\frac{d}{dr}\left(r\frac{dw}{dr}\right)\right] = \frac{d}{dr}\left[\frac{1}{r}\left(r\frac{d^2w}{dr^2} + \frac{dw}{dr}\right)\right] = \frac{d}{dr}\left[\frac{d^2w}{dr^2} + \frac{1}{r}\frac{dw}{dr}\right]$$

$$= \frac{d^3w}{dr^3} + \frac{1}{r}\frac{d^2w}{dr^2} - \frac{1}{r^2}\frac{dw}{dr}$$

This is the left-hand side of the first equation in Problem 53.

(b) Integrating both sides with respect to r gives

$$\frac{d}{dr}\left[\frac{1}{r}\frac{d}{dr}\left(r\frac{dw}{dr}\right)\right] = \frac{q}{2D}r$$

$$\frac{1}{r}\frac{d}{dr}\left(r\frac{dw}{dr}\right) = \frac{q}{4D}r^2 + C_1$$

$$\frac{d}{dr}\left(r\frac{dw}{dr}\right) = \frac{1}{4D}r^3 + C_1 r$$

$$r\frac{dw}{dr} = \frac{q}{16D}r^4 + \frac{C_1}{2}r^2 + C_2$$

$$\frac{dw}{dr} = \frac{1}{16D}r^3 + \frac{C_1}{2}r + \frac{C_2}{r}$$

$$w(r) = \frac{q}{64D}r^2 + \frac{C_1}{4}r^2 + C_2\ln r + C_3\,.$$

Relabeling $c_1 = C_3$, $c_2 = C_2$, and $c_3 = C_1/4$ shows this is the same as the general solution obtained in part (a) of Problem 53.

3.7 | Nonlinear Equations

3. Let $u = y'$ so that $u' = y''$. The equation becomes $u' = -u - 1$ which is separable. Thus

$$\frac{du}{u^2 + 1} = -dx$$

$$\tan^{-1} u = -x + c_1$$

$$y' = \tan(c_1 - x)$$

$$y = \ln|\cos(c_1 - x)| + c_2.$$

6. Let $u = y'$ so that $u' = y''$. The equation becomes $e^{-x} u' = u^2$. Separating variables we obtain

$$\frac{du}{u^2} = e^x\, dx$$

$$-\frac{1}{u} = e^x + c_1$$

$$u = -\frac{1}{e^x + c_1} = -\frac{e^{-x}}{1 + c_1 e^{-x}} = \frac{1}{c_1}\frac{-c_1 e^{-x}}{1 + c_1 e^{-x}}$$

$$y = \frac{1}{c_1}\ln\left|1 + c_1 e^{-x}\right| + c_2$$

9. Let $u = y'$ so that $y'' = u\, du/dy$. The equation becomes $u\, du/dy + 2yu^3 = 0$. Separating variables we obtain

$$\frac{du}{u^2} + 2y\, dy = 0$$

$$-\frac{1}{u} + y^2 = c$$

$$u = \frac{1}{y^2 + c_1}$$

$$\left(y^2 + c_1\right) dy = dx$$

$$\frac{1}{3}y^3 + c_1 y = x + c_2.$$

12. Let $u = y'$ so that $u' = y''$. The equation becomes $u' + xu^2 = 0$ or $du/dx = -xu^2$. Separating variables we obtain

$$\frac{du}{u^2} = -x\, dx$$

$$-\frac{1}{u} = -\frac{1}{2}x^2 + c_1$$

$$u = \frac{2}{x^2 + c_2}$$

$$\frac{dy}{dx} = \frac{2}{x^2 + c_2}$$

Using $y'(1) = 2$ results in $\dfrac{2}{1 + c_2} = 2$ or $c_2 = 0$. So

$$\frac{dy}{dx} = \frac{2}{x^2} \quad \text{or} \quad y = -\frac{2}{x} + c_3$$

Using $y(1) = 4$ results in $-2 + c_3 = 4$ or $c_3 = 6$.

Therefore $y = -\dfrac{2}{x} + 6$.

15. Let $u = y'$ so that $u' = y''$. The equation becomes $u' - (1/x)u = (1/x)u^3$, which is Bernoulli. Using $w = u^{-2}$ we obtain $dw/dx + (2/x)w = -2/x$. An integrating factor is x^2, so

$$\frac{d}{dx}[x^2 w] = -2x$$

$$x^2 w = -x^2 + c_1$$

$$w = -1 + \frac{c_1}{x^2}$$

$$u^{-2} = -1 + \frac{c_1}{x^2}$$

$$u = \frac{x}{\sqrt{c_1 - x^2}}$$

$$\frac{dy}{dx} = \frac{x}{\sqrt{c_1 - x^2}}$$

$$y = -\sqrt{c_1 - x^2} + c_2$$

$$c_1 - x^2 = (c_2 - y)^2$$

$$x^2 + (c_2 - y)^2 = c_1.$$

In Problems 17–20 the thinner curve is obtained using a numerical solver, while the thicker curve is the graph of the Taylor polynomial.

18. We look for a solution of the form

$$y(x) = y(0) + y'(0)x + \frac{1}{2!}y''(0)x^2 + \frac{1}{3!}y'''(0)x^3 + \frac{1}{4!}y^{(4)}(0)x^4 + \frac{1}{5!}y^{(5)}(0)x^5.$$

From $y''(x) = 1 - y^2$ we compute

$$y'''(x) = -2yy'$$
$$y^{(4)}(x) = -2yy'' - 2(y')^2$$
$$y^{(5)}(x) = -2yy''' - 6y'y''.$$

Using $y(0) = 2$ and $y'(0) = 3$ we find

$$y''(0) = -3, \quad y'''(0) = -12, \quad y^{(4)}(0) = -6, \quad y^{(5)}(0) = 102.$$

An approximate solution is

$$y(x) = 2 + 3x - \frac{3}{2}x^2 - 2x^3 - \frac{1}{4}x^4 + \frac{17}{20}x^5.$$

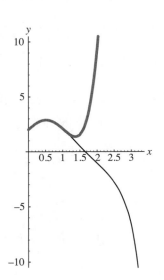

21. We need to solve $[1 + (y')^2]^{3/2} = y''$. Let $u = y'$ so that $u' = y''$. The equation becomes $(1 + u^2)^{3/2} = u'$ or $(1 + u^2)^{3/2} = du/dx$. Separating variables and using the substitution $u = \tan\theta$ we have

$$\frac{du}{(1 + u^2)^{3/2}} = dx$$

$$\int \frac{\sec^2\theta}{(1 + \tan^2\theta)^{3/2}}\, d\theta = x$$

$$\int \frac{\sec^2\theta}{\sec^3\theta}\, d\theta = x$$

$$\int \cos\theta\, d\theta = x$$

$$\sin\theta = x$$

$$\frac{u}{\sqrt{1 + u^2}} = x$$

$$\frac{y'}{\sqrt{1 + (y')^2}} = x$$

$$(y')^2 = x^2\left[1 + (y')^2\right] = x^2 + x^2(y')^2$$

$$(y')^2 = \frac{x^2}{1 - x^2}$$

$$y' = \frac{x}{\sqrt{1 - x^2}} \quad \text{(for } x > 0\text{)}$$

$$y = -\sqrt{1 - x^2}\,.$$

3.8   Linear Models: Initial-Value Problems

3. From $\frac{3}{4}x'' + 72x = 0$, $x(0) = -1/4$, and $x'(0) = 0$ we obtain $x = -\frac{1}{4}\cos 4\sqrt{6}\,t$.

6. From $50x'' + 200x = 0$, $x(0) = 0$, and $x'(0) = -10$ we obtain $x = -5\sin 2t$ and $x' = -10\cos 2t$.

9. (a) From $x'' + 16x = 0$, $x(0) = 1/2$, and $x'(0) = 3/2$, we get

$$x(t) = \frac{1}{2}\cos 2t + \frac{3}{4}\sin 2t$$

(b) We use $x(t) = A\sin(\omega t + \phi)$ where $A = \sqrt{c_1^2 + c_2^2} = \sqrt{(1/2)^2 + (3/4)^2} = \sqrt{13}/4$. From this we get

$$\tan\phi = \frac{c_1}{c_2} = \frac{2}{3}$$

$$\phi = \tan^{-1}\left(\frac{2}{3}\right) = 0.588$$

Therefore $x(t) = \frac{\sqrt{13}}{4}\sin(2t + 0.588)$.

(c) We use $x(t) = A\cos(\omega t - \phi)$ where $A = \sqrt{c_1^2 + c_2^2} = \sqrt{(1/2)^2 + (3/4)^2} = \sqrt{13}/4$. This time we get

$$\tan\phi = \frac{c_2}{c_1} = \frac{3}{2}$$

$$\phi = \tan^{-1}\left(\frac{3}{2}\right) = 0.983$$

Therefore $x(t) = \frac{\sqrt{13}}{4}\cos(2t - 0.983)$.

12. From $x'' + 9x = 0$, $x(0) = -1$, and $x'(0) = -\sqrt{3}$ we obtain

$$x = -\cos 3t - \frac{\sqrt{3}}{3}\sin 3t = \frac{2}{\sqrt{3}}\sin\left(3t + \frac{4\pi}{3}\right)$$

and $x' = 2\sqrt{3}\cos(3t + 4\pi/3)$. If $x' = 3$ then $t = -7\pi/18 + 2n\pi/3$ and $t = -\pi/2 + 2n\pi/3$ for $n = 1, 2, 3, \ldots$.

15. Using $k_1 = 40$ and $k_2 = 120$ we have

$$k_{\text{eff}} = \frac{40 \cdot 120}{40 + 120} = \frac{40 \cdot 120}{160} = 30 \text{ lb/ft}$$

$$\frac{20}{32}x'' + 30x = 0$$

$$x'' + 48x = 0$$

$$x(t) = c_1\cos 4\sqrt{3}\,t + c_2\sin 4\sqrt{3}\,t$$

Using the initial condition $x(0) = 0$ we have $c_1 = 0$ and therefore $x(t) = c_2\sin 4\sqrt{3}\,t$. Then $x'(t) = 4\sqrt{3}\,c_2\cos 4\sqrt{3}\,t$. Using the condition $x'(0) = 2$ we have $c_2 = \sqrt{3}/6$. Thus

$$x(t) = \frac{\sqrt{3}}{6}\sin 4\sqrt{3}\,t$$

18. For springs attached in series the effective spring constant is

$$k_{\text{eff}} = \frac{k_1 k_2}{k_1 + k_2}.$$

When $k = k_1 = k_2$, the effective spring constant is

$$k_{\text{eff}} = \frac{k^2}{k + k} = \frac{1}{2}k$$

Compared to a sing spring with spring constant k, the series-spring system is less stiff.

21. (a) above   **(b)** heading upward

24. (a) above   **(b)** heading downward

27. (a) From $x'' + 10x' + 16x = 0$, $x(0) = 1$, and $x'(0) = 0$ we obtain $x = \frac{4}{3}e^{-2t} - \frac{1}{3}e^{-8t}$.

(b) From $x'' + x' + 16x = 0$, $x(0) = 1$, and $x'(0) = -12$ then $x = -\frac{2}{3}e^{-2t} + \frac{5}{3}e^{-8t}$.

30. (a) From $\frac{1}{4}x'' + x' + 5x = 0$, $x(0) = 1/2$, and $x'(0) = 1$ we obtain $x = e^{-2t}\left(\frac{1}{2}\cos 4t + \frac{1}{2}\sin 4t\right)$.

(b) $x = \dfrac{1}{\sqrt{2}}e^{-2t}\sin\left(4t + \dfrac{\pi}{4}\right)$.

(c) If $x = 0$ then $4t + \pi/4 = \pi$, 2π, 3π, ... so that the times heading downward are $t = (7 + 8n)\pi/16$ for $n = 0, 1, 2, \ldots$.

(d)

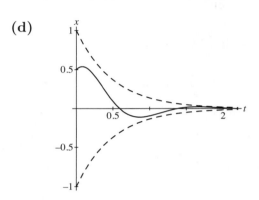

33. If $\frac{1}{2}x'' + \frac{1}{2}x' + 6x = 10\cos 3t$, $x(0) = -2$, and $x'(0) = 0$ then

$$x_c = e^{-t/2}\left(c_1\cos\left(\frac{\sqrt{47}}{2}t\right) + c_2\sin\left(\frac{\sqrt{47}}{2}t\right)\right)$$

and $x_p = \frac{10}{3}(\cos 3t + \sin 3t)$ so that the equation of motion is

$$x = e^{-t/2}\left(-\frac{4}{3}\cos\left(\frac{\sqrt{47}}{2}t\right) - \frac{64}{3\sqrt{47}}\sin\left(\frac{\sqrt{47}}{2}t\right)\right) + \frac{10}{3}(\cos 3t + \sin 3t).$$

36. From $x'' + 8x' + 16x = e^{-t}\sin 4t$, $x(0) = 0$, and $x'(0) = 0$ we obtain $x_c = c_1 e^{-4t} + c_2 t e^{-4t}$ and $x_p = -\frac{24}{625}e^{-t}\cos 4t - \frac{7}{625}e^{-t}\sin 4t$ so that

$$x = \frac{1}{625}e^{-4t}(24 + 100t) - \frac{1}{625}e^{-t}(24\cos 4t + 7\sin 4t).$$

As $t \to \infty$ the displacement $x \to 0$.

39. (a) By Hooke's law the external force is $F(t) = kh(t)$ so that $mx'' + \beta x' + kx = kh(t)$.

(b) From $\frac{1}{2}x'' + 2x' + 4x = 20\cos t$, $x(0) = 0$, and $x'(0) = 0$ we obtain $x_c = e^{-2t}(c_1\cos 2t + c_2\sin 2t)$ and $x_p = \frac{56}{13}\cos t + \frac{32}{13}\sin t$ so that

$$x = e^{-2t}\left(-\frac{56}{13}\cos 2t - \frac{72}{13}\sin 2t\right) + \frac{56}{13}\cos t + \frac{32}{13}\sin t.$$

42. The differential equation describes an aging spring/mass system that is damped and periodically driven. In this case, the variable stiffness is $\left(k_1 + (k - k_1)\, e^{-\alpha t}\right)$. We see the initial stiffness of the spring is k when $t = 0$ and decreases over time, that is, $\left(k_1 + (k - k_1)\, e^{-\alpha t}\right) \to k_1$ as $t \to \infty$.

45. (a) From $x'' + \omega^2 x = F_0 \cos \gamma t$, $x(0) = 0$, and $x'(0) = 0$ we obtain $x_c = c_1 \cos \omega t + c_2 \sin \omega t$ and $x_p = (F_0 \cos \gamma t)/\left(\omega^2 - \gamma^2\right)$ so that

$$x = -\frac{F_0}{\omega^2 - \gamma^2} \cos \omega t + \frac{F_0}{\omega^2 - \gamma^2} \cos \gamma t.$$

(b) $\displaystyle \lim_{\gamma \to \omega} \frac{F_0}{\omega^2 - \gamma^2}(\cos \gamma t - \cos \omega t) = \lim_{\gamma \to \omega} \frac{-F_0 t \sin \gamma t}{-2\gamma} = \frac{F_0}{2\omega} t \sin \omega t.$

51. Solving $\frac{1}{20} q'' + 2q' + 100q = 0$ we obtain $q(t) = e^{-20t}(c_1 \cos 40t + c_2 \sin 40t)$. The initial conditions $q(0) = 5$ and $q'(0) = 0$ imply $c_1 = 5$ and $c_2 = 5/2$. Thus

$$q(t) = e^{-20t} \left(5 \cos 40t + \frac{5}{2} \sin 40t \right) = \frac{5\sqrt{5}}{2} e^{-20t} \sin\left(40t + 1.1071\right)$$

and $q(0.01) \approx 4.5676$ coulombs. The charge is zero for the first time when $40t + 1.1071 = \pi$ or $t \approx 0.0509$ second.

54. Solving $q'' + 100q' + 2500q = 30$ we obtain $q(t) = c_1 e^{-50t} + c_2 t e^{-50t} + 0.012$. The initial conditions $q(0) = 0$ and $q'(0) = 2$ imply $c_1 = -0.012$ and $c_2 = 1.4$. Thus, using $i(t) = q'(t)$ we get

$$q(t) = -0.012 e^{-50t} + 1.4 t e^{-50t} + 0.012 \quad \text{and} \quad i(t) = 2 e^{-50t} - 70 t e^{-50t}.$$

Solving $i(t) = 0$ we see that the maximum charge occurs when $t = 1/35$ second and $q(1/35) \approx 0.01871$ coulomb.

57. The differential equation is $\frac{1}{2} q'' + 20q' + 1000q = 100 \sin 60t$. To use Example 10 in the text we identify $E_0 = 100$ and $\gamma = 60$. Then

$$X = L\gamma - \frac{1}{c\gamma} = \frac{1}{2}(60) - \frac{1}{0.001(60)} \approx 13.3333,$$

$$Z = \sqrt{X^2 + R^2} = \sqrt{X^2 + 400} \approx 24.0370,$$

and

$$\frac{E_0}{Z} = \frac{100}{Z} \approx 4.1603.$$

From Problem 56, then

$$i_p(t) \approx 4.1603 \sin\left(60t + \phi\right)$$

where $\sin \phi = -X/Z$ and $\cos \phi = R/Z$. Thus $\tan \phi = -X/R \approx -0.6667$ and ϕ is a fourth quadrant angle. Now $\phi \approx -0.5880$ and

$$i_p(t) = 4.1603 \sin\left(60t - 0.5880\right).$$

60. In Problem 56 it is shown that the amplitude of the steady-state current is E_0/Z, where $Z = \sqrt{X^2 + R^2}$ and $X = L\gamma - 1/C\gamma$. Since E_0 is constant the amplitude will be a maximum when Z is a minimum. Since R is constant, Z will be a minimum when $X = 0$. Solving $L\gamma - 1/C\gamma = 0$ for γ we obtain $\gamma = 1/\sqrt{LC}$. The maximum amplitude will be E_0/R.

63. In an LC-series circuit there is no resistor, so the differential equation is

$$L\frac{d^2q}{dt^2} + \frac{1}{C}q = E(t).$$

Then $q(t) = c_1 \cos\left(t/\sqrt{LC}\right) + c_2 \sin\left(t/\sqrt{LC}\right) + q_p(t)$ where $q_p(t) = A\sin\gamma t + B\cos\gamma t$. Substituting $q_p(t)$ into the differential equation we find

$$\left(\frac{1}{C} - L\gamma^2\right) A\sin\gamma t + \left(\frac{1}{C} - L\gamma^2\right) B\cos\gamma t = E_0\cos\gamma t.$$

Equating coefficients we obtain $A = 0$ and $B = E_0 C/(1 - LC\gamma^2)$. Thus, the charge is

$$q(t) = c_1 \cos\frac{1}{\sqrt{LC}}t + c_2 \sin\frac{1}{\sqrt{LC}}t + \frac{E_0 C}{1 - LC\gamma^2}\cos\gamma t.$$

The initial conditions $q(0) = q_0$ and $q'(0) = i_0$ imply $c_1 = q_0 - E_0 C/(1 - LC\gamma^2)$ and $c_2 = i_0\sqrt{LC}$. The current is $i(t) = q'(t)$ or

$$i(t) = -\frac{c_1}{\sqrt{LC}}\sin\frac{1}{\sqrt{LC}}t + \frac{c_2}{\sqrt{LC}}\cos\frac{1}{\sqrt{LC}}t - \frac{E_0 C\gamma}{1 - LC\gamma^2}\sin\gamma t$$

$$= i_0\cos\frac{1}{\sqrt{LC}}t - \frac{1}{\sqrt{LC}}\left(q_0 - \frac{E_0 C}{1 - LC\gamma^2}\right)\sin\frac{1}{\sqrt{LC}}t - \frac{E_0 C\gamma}{1 - LC\gamma^2}\sin\gamma t.$$

3.9   Linear Models: Boundary-Value Problems

3. (a) The general solution is

$$y(x) = c_1 + c_2 x + c_3 x^2 + c_4 x^3 + \frac{w_0}{24EI}x^4.$$

The boundary conditions are $y(0) = 0$, $y'(0) = 0$, $y(L) = 0$, $y''(L) = 0$. The first two conditions give $c_1 = 0$ and $c_2 = 0$. The conditions at $x = L$ give the system

$$c_3 L^2 + c_4 L^3 + \frac{w_0}{24EI}L^4 = 0$$

$$2c_3 + 6c_4 L + \frac{w_0}{2EI}L^2 = 0.$$

Solving, we obtain $c_3 = w_0 L^2/16EI$ and $c_4 = -5w_0 L/48EI$. The deflection is

$$y(x) = \frac{w_0}{48EI}(3L^2 x^2 - 5Lx^3 + 2x^4).$$

(b)

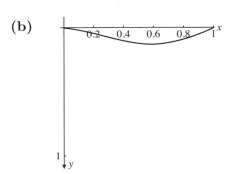

6. (a) $y_{\max} = y(L) = w_0 L^4 / 8EI$

(b) Replacing both L and x by $L/2$ in $y(x)$ we obtain $w_0 L^4/128EI$, which is $1/16$ of the maximum deflection when the length of the beam is L.

(c) $y_{\max} = y(L/2) = 5w_0 L^4 / 384EI$

(d) The maximum deflection in Example 1 is $y(L/2) = (w_0/24EI)L^4/16 = w_0 L^4/384EI$, which is $1/5$ of the maximum displacement of the beam in part (c).

9. (a) We solve $EIy^{(4)} = w(x)$ under the conditions $y'''(L) = y''(L) = y'(0) = y(0) = 0$. Integrating gives us

$$EIy''' = \begin{cases} c_1, & 0 \le x < L/2 \\ w_0 x + c_2, & L/2 \le x \le L. \end{cases}$$

Therefore $y'''(L) = w_0 L + c_2 = 0$ so that $c_2 = -w_0 L$. Assuming continuity at $x = L/2$ we get

$$w_0 \left(L/2 \right) - w_0 L = c_1$$
$$c_1 = -w_0 \left(L/2 \right)$$

So now we have

$$EIy''' = \begin{cases} -\dfrac{w_0 L}{2}, & 0 \le x < L/2 \\ w_0 \left(x - L \right), & L/2 \le x \le L. \end{cases}$$

(b) Following the same procedure as in part (a) we integrate again to get

$$EIy'' = \begin{cases} -\dfrac{w_0 L}{2} x + c_3, & 0 \le x < L/2 \\ w_0 \left(\dfrac{1}{2}x^2 - Lx \right) + c_4, & L/2 \le x \le L. \end{cases}$$

Therefore $y''(L) = w_0 \left(\frac{1}{2}L^2 - L^2\right) + c_4 = 0$ so that $c_4 = w_0 L^2/2$. Assuming continuity at $x = L/2$

$$w_0 \left(\frac{1}{2}\left(\frac{L}{2}\right)^2 L\left(\frac{L}{2}\right)\right) + \frac{w_0 L^2}{2} = -\frac{w_0 L}{2}\left(\frac{L}{2}\right) + c_3$$

$$c_3 = \frac{3w_0 L^2}{8}$$

So now we have

$$EIy'' = \begin{cases} \dfrac{w_0}{8}\left(-4Lx + 3L^2\right), & 0 \le x < L/2 \\[2mm] \dfrac{w_0}{2}\left(x^2 - 2Lx + L^2\right), & L/2 \le x \le L. \end{cases}$$

Repeat the same procedure to get

$$EIy' = \begin{cases} \dfrac{w_0}{8}\left(-2Lx^2 + 3L^2 x\right) + c_5, & 0 \le x < L/2 \\[2mm] \dfrac{w_0}{2}\left(\dfrac{1}{3}x^3 - Lx^2 + L^2 x\right) + c_6, & L/2 \le x \le L. \end{cases}$$

Therefore $y'(0) = \frac{w_0}{8}\left(-2L(0)^2 + 3L^2(0)\right) + c_5 = 0$ so that $c_5 = 0$. Assuming continuity at $xL/2$,

$$\frac{w_0}{2}\left(\frac{1}{3}\left(\frac{L}{2}\right)^3 - L\left(\frac{L}{2}\right)^2 + L^2\left(\frac{L}{2}\right)\right) + c_6 = \frac{w_0}{8}\left(-2L\left(\frac{L}{2}\right)^2 + 3L^2\left(\frac{L}{2}\right)\right)$$

$$c_6 = -\frac{w_0 L^3}{8}$$

So now we have

$$EIy' = \begin{cases} \dfrac{w_0}{8}\left(-2Lx^2 + 3L^2 x\right), & 0 \le x < L/2 \\[2mm] \dfrac{w_0}{48}\left(8x^3 - 24Lx^2 + 24L^2 x - L63\right), & L/2 \le x \le L. \end{cases}$$

Integrating one more time leads to

$$EIy = \begin{cases} \dfrac{w_0}{8}\left(-\dfrac{2L}{3}x^3 + \dfrac{3L^2}{2}x^2\right) + c_7, & 0 \le x < L/2 \\[2mm] \dfrac{w_0}{48}\left(2x^4 - 8Lx^3 + 12L^2 x^2 - L^3 x\right) + c_8, & L/2 \le x \le L. \end{cases}$$

Therefore $y(0) = \frac{w_0}{8}\left(-\frac{2L}{3}(0)^3 + \frac{3L^2}{2}(0)^2\right) + c_7 = 0$ so that $c_7 = 0$. Assuming continuity at $x = L/2$, we find $c_8 = w_0 L^4/384$. Therefore we have

$$y = \begin{cases} \dfrac{w_0}{48EI}\left(-4Lx^3 + 9L^2 x^2\right), & 0 \le x < L/2 \\[2mm] \dfrac{w_0}{384EI}\left(16x^4 - 64Lx^3 + 96L^2 x^2 - 8L^3 x + L^4\right), & L/2 \le x \le L. \end{cases}$$

(c) Substitute $x = L$ into the solution in part (b) to get $y(L) = 41w_0 L^4/384EI$.

12. This is Example 2 in the text with $L = \pi/4$. The eigenvalues are $\lambda_n = n^2\pi^2/(\pi/4)^2 = 16n^2$, $n = 1, 2, 3, \ldots$ and the eigenfunctions are $y_n = \sin(n\pi x/(\pi/4)) = \sin 4nx$, $n = 1, 2, 3, \ldots$.

15. For $\lambda = -\alpha^2 < 0$ the only solution of the boundary-value problem is $y = 0$. For $\lambda = 0$ we have $y = c_1 x + c_2$. Now $y' = c_1$ and $y'(0) = 0$ implies $c_1 = 0$. Then $y = c_2$ and $y'(\pi) = 0$. Thus, $\lambda = 0$ is an eigenvalue with corresponding eigenfunction $y = 1$. For $\lambda = \alpha^2 > 0$ we have

$$y = c_1 \cos \alpha x + c_2 \sin \alpha x.$$

Now

$$y'(x) = -c_1\alpha \sin \alpha x + c_2\alpha \cos \alpha x$$

and $y'(0) = 0$ implies $c_2 = 0$, so

$$y'(\pi) = -c_1\alpha \sin \alpha\pi = 0$$

gives

$$\alpha\pi = n\pi \quad \text{or} \quad \lambda = \alpha^2 = n^2, \ n = 1, 2, 3, \ldots.$$

The eigenvalues n^2 correspond to the eigenfunctions $\cos nx$ for $n = 0, 1, 2, \ldots$.

18. For $\lambda < -1$ the only solution of the boundary-value problem is $y = 0$. For $\lambda = -1$ we have $y = c_1 x + c_2$. Now $y' = c_1$ and $y'(0) = 0$ implies $c_1 = 0$. Then $y = c_2$ and $y'(1) = 0$. Thus, $\lambda = -1$ is an eigenvalue with corresponding eigenfunction $y = 1$. For $\lambda > -1$ or $\lambda + 1 = \alpha^2 > 0$ we have

$$y = c_1 \cos \alpha x + c_2 \sin \alpha x.$$

Now

$$y' = -c_1\alpha \sin \alpha x + c_2\alpha \cos \alpha x$$

and $y'(0) = 0$ implies $c_2 = 0$, so

$$y'(1) = -c_1\alpha \sin \alpha = 0$$

gives

$$\alpha = n\pi, \quad \lambda + 1 = \alpha^2 = n^2\pi^2, \quad \text{or} \quad \lambda_n = n^2\pi^2 - 1, \ n = 1, 2, 3, \ldots.$$

The eigenvalues $n^2\pi^2 - 1$ correspond to the eigenfunctions $y_n = \cos n\pi x$ for $n = 0, 1, 2, \ldots$.

21. For $\lambda = \alpha^4$, $\alpha > 0$, the general solution of the boundary-value problem

$$y^{(4)} - \lambda y = 0, \quad y(0) = 0, \ y''(0) = 0, \ y(1) = 0, \ y''(1) = 0$$

is

$$y = c_1 \cos \alpha x + c_2 \sin \alpha x + c_3 \cosh \alpha x + c_4 \sinh \alpha x.$$

The boundary conditions $y(0) = 0$, $y''(0) = 0$ give $c_1 + c_3 = 0$ and $-c_1\alpha^2 + c_3\alpha^2 = 0$, from which we conclude $c_1 = c_3 = 0$. Thus, $y = c_2 \sin \alpha x + c_4 \sinh \alpha x$. The boundary conditions $y(1) = 0$, $y''(1) = 0$ then give

$$c_2 \sin \alpha + c_4 \sinh \alpha = 0$$
$$-c_2\alpha^2 \sin \alpha + c_4\alpha^2 \sinh \alpha = 0.$$

In order to have nonzero solutions of this system, we must have the determinant of the coefficients equal zero, that is,

$$\begin{vmatrix} \sin \alpha & \sinh \alpha \\ -\alpha^2 \sin \alpha & \alpha^2 \sinh \alpha \end{vmatrix} = 0 \quad \text{or} \quad 2\alpha^2 \sinh \alpha \sin \alpha = 0$$

But since $\alpha > 0$, the only way that this is satisfied is to have $\sin \alpha = 0$ or $\alpha = n\pi$. The system is then satisfied by choosing $c_2 \neq 0$, $c_4 = 0$, and $\alpha = n\pi$. The eigenvalues and corresponding eigenfunctions are then

$$\lambda_n = \alpha^4 = (n\pi)^4, \ n = 1, 2, 3, \ldots \quad \text{and} \quad y_n = \sin n\pi x.$$

24. (a) The general solution of the differential equation is

$$y = c_1 \cos \sqrt{\frac{P}{EI}}\, x + c_2 \sin \sqrt{\frac{P}{EI}}\, x + \delta.$$

Since the column is embedded at $x = 0$, the boundary conditions are $y(0) = y'(0) = 0$. If $\delta = 0$ this implies that $c_1 = c_2 = 0$ and $y(x) = 0$. That is, there is no deflection.

(b) If $\delta \neq 0$, the boundary conditions give, in turn, $c_1 = -\delta$ and $c_2 = 0$. Then

$$y = \delta\left(1 - \cos \sqrt{\frac{P}{EI}}\, x\right).$$

In order to satisfy the boundary condition $y(L) = \delta$ we must have

$$\delta = \delta\left(1 - \cos \sqrt{\frac{P}{EI}}\, L\right) \quad \text{or} \quad \cos \sqrt{\frac{P}{EI}}\, L = 0.$$

This gives $\sqrt{P/EI}\, L = (2n-1)\pi/2$ for $n = 1, 2, 3, \ldots$. The smallest value of P_n, the Euler load, is then

$$\sqrt{\frac{P_1}{EI}}\, L = \frac{\pi}{2} \quad \text{or} \quad P_1 = \frac{1}{4}\left(\frac{\pi^2 EI}{L^2}\right).$$

27. The general solution is

$$y = c_1 \cos \sqrt{\frac{\rho}{T}}\, \omega x + c_2 \sin \sqrt{\frac{\rho}{T}}\, \omega x.$$

From $y(0) = 0$ we obtain $c_1 = 0$. Setting $y(L) = 0$ we find $\sqrt{\rho/T}\,\omega L = n\pi$, $n = 1, 2, 3, \ldots$. Thus, critical speeds are $\omega_n = n\pi\sqrt{T}/L\sqrt{\rho}$, $n = 1, 2, 3, \ldots$. The corresponding deflection curves are

$$y(x) = c_2 \sin \frac{n\pi}{L}\, x, \quad n = 1, 2, 3, \ldots,$$

where $c_2 \neq 0$.

30. The auxiliary equation is $m^2 = 0$ so that $u(r) = c_1 + c_2 \ln r$. The boundary conditions $u(a) = u_0$ and $u(b) = u_1$ yield the system $c_1 + c_2 \ln a = u_0$, $c_1 + c_2 \ln b = u_1$. Solving gives

$$c_1 = \frac{u_1 \ln a - u_0 \ln b}{\ln (a/b)} \quad \text{and} \quad c_2 = \frac{u_0 - u_1}{\ln (a/b)}.$$

Thus

$$u(r) = \frac{u_1 \ln a - u_0 \ln b}{\ln (a/b)} + \frac{u_0 - u_1}{\ln (a/b)} \ln r = \frac{u_0 \ln (r/b) - u_1 \ln (r/a)}{\ln(a/b)}.$$

3.10 Green's Functions

3.

$$y'' + 2y' + y = f(x)$$
$$y'' + 2y' + y = 0$$

$$y_1 = e^{-x}, \quad y_2 = xe^{-x}$$

$$W\left(e^{-x}, xe^{-x}\right) = \begin{vmatrix} e^{-x} & xe^{-x} \\ -e^{-x} & -xe^{-x} + e^{-x} \end{vmatrix} = e^{-2x}$$

$$G(x, t) = \frac{e^{-t}xe^{-x} - e^{-x}te^{-t}}{e^{-2t}} = (x - t)e^{-(x-t)}$$

$$y_p(x) = \int_{x_0}^{x} (x - t)e^{-(x-t)} f(t)\, dt$$

6.

$$y'' - 2y' + 2y = f(x)$$
$$y'' - 2y' + 2y = 0$$
$$y_1 = e^x \cos x, \quad y_2 = e^x \sin x$$

$$W\left(e^x \cos x, e^x \sin x\right) = \begin{vmatrix} e^x \cos x & e^x \sin x \\ -e^x \sin x + e^x \cos x & e^x \cos x + e^x \sin x \end{vmatrix} = e^{2x}$$

$$G(x, t) = \frac{e^t \cos t e^x \sin x - e^x \cos x e^t \sin t}{e^{2t}} = e^{x-t} \sin (x - t)$$

$$y_p(x) = \int_{x_0}^{x} e^{x-t} \sin (x - t) f(t)\, dt$$

9.
$$y'' + 2y' + y = e^{-x}$$
$$y = y_c + y_p$$
$$y = c_1 e^{-x} + c_2 x e^{-x} + y_p$$
$$y = c_1 e^{-x} + c_2 x e^{-x} + \int_{x_0}^{x} (x - t) e^{-(x-t)} e^{-t} \, dt$$

12.
$$y'' - 2y' + 2y = \cos^2 x$$
$$y = y_c + y_p$$
$$y = c_1 e^x \cos x + c_2 e^x \sin x + y_p$$
$$y = c_1 e^x \cos x + c_2 e^x \sin x = \int_{x_0}^{x} e^{(x-t)} \sin(x - t) \cos^2 t \, dt$$

15. The initial-value problem is $y'' - 10y' + 25y = e^{5x}, \quad y(0) = 0, \; y'(0) = 0$. Then we find that

$$y_1 = e^{5x}, \quad y_2 = x e^{5x}$$

$$W\left(e^{5x}, x e^{5x}\right) = \begin{vmatrix} e^{5x} & x e^{5x} \\ 5e^{5x} & 5x e^{5x} + e^{5x} \end{vmatrix} = e^{10x}.$$

Then

$$G(x, t) = \frac{e^{5t} x e^{5x} - e^{5x} t e^{5t}}{e^{10t}} = (x - t) e^{5(x-t)}$$

and the solution of the initial-value problem is

$$y_p(x) = \int_0^x (x - t) e^{5(x-t)} e^{5t} \, dt$$

$$= x e^{5x} \int_0^x dt - e^{5x} \int_0^x t \, dt$$

$$= x^2 e^{5x} - \frac{1}{2} x^2 e^{5x}$$

$$= \frac{1}{2} x^2 e^{5x}.$$

18. The initial-value problem is $y'' + y = \sec^2 x, \quad y(\pi) = 0, \; y'(\pi) = 0$. Then we find that

$$y_1 = \cos x, \quad y_2 = \sin x$$

$$W(\cos x, \sin x) = \begin{vmatrix} \cos x & \sin x \\ -\sin x & \cos x \end{vmatrix} = 1.$$

Then

$$G(x, t) = \cos t \sin x - \cos x \sin t = \sin(x - t)$$

and the solution of the initial-value problem is

$$y_p(x) = \int_\pi^x (\cos t \sin x - \cos x \sin t) \sec^2 t \, dt$$

$$= \sin x \int_\pi^x \sec t \, dt - \cos x \int_\pi^x \sec t \tan t \, dt$$

$$= -\cos x - 1 + \sin x \ln|\sec x + \tan x|.$$

21. The initial-value problem is $y'' - 10y' + 25y = e^{5x}$, $y(0) = -1$, $y'(0) = 1$, so $y(x) = c_1 e^{5x} + c_2 x e^{5x}$. The initial conditions give

$$c_1 = -1$$
$$5c_1 + c_2 = 1,$$

so $c_1 = -1$ and $c_2 = 6$, which implies that $y_h = -e^{5x} + 6xe^{5x}$. Now, y_p found in the solution of Problem 15 in this section gives

$$y = y_h + y_p = -e^{5x} + 6xe^{5x} + \frac{1}{2}x^2 e^{5x}.$$

24. The initial-value problem is $y'' + y = \sec^2 x$, $y(\pi) = \frac{1}{2}$, $y'(\pi) = -1$, so $y(x) = c_1 \cos x + c_2 \sin x$. The initial conditions give

$$-c_1 = \frac{1}{2}$$
$$-c_2 = -1,$$

so $c_1 = -\frac{1}{2}$ and $c_2 = 1$, which implies that $y_h = -\frac{1}{2}\cos x + \sin x$. Now, y_p found in the solution of Problem 18 in this section gives

$$y = y_h + y_p = -\frac{1}{2}\cos x + \sin x + (-\cos x - 1 + \sin x \ln|\sec x + \tan x|)$$

$$= -\frac{3}{2}\cos x + \sin x - 1 + \sin x \ln|\sec x + \tan x|.$$

27. The Cauchy–Euler initial-value problem $x^2 y'' - 2xy' + 2y = x$, $y(1) = 2$, $y'(1) = -1$, has auxiliary equation $m(m-1) - 2m + 2 = m^2 - 3m + 2 = (m-1)(m-2) = 0$ so $m_1 = 1$, $m_2 = 2$, $y(x) = c_1 x + c_2 x^2$, and $y' = c_1 + 2c_2 x$. The initial conditions give

$$c_1 + c_2 = 2$$
$$c_1 + 2c_2 = -1,$$

so $c_1 = 5$ and $c_2 = -3$, which implies that $y_h = 5x - 3x^2$. The Wronskian is

$$W(x, x^2) = \begin{vmatrix} x & x^2 \\ 1 & 2x \end{vmatrix} = x^2.$$

Then $G(x,t) = \dfrac{tx^2 - xt^2}{t^2} = \dfrac{x(x-t)}{t}$. From the standard form of the differential equation we identify the forcing function $f(t) = \dfrac{1}{t}$. Then, for $x > 1$,

$$
\begin{aligned}
y_p(x) &= \int_1^x \frac{x(x-t)}{t} \frac{1}{t}\, dt \\[2mm]
&= x^2 \int_1^x \frac{1}{t^2}\, dt - x \int_1^x \frac{1}{t}\, dt \\[2mm]
&= x^2 \left(-\frac{1}{x} + 1 \right) - x(\ln x - \ln 1) \\[2mm]
&= x^2 - x - x \ln x.
\end{aligned}
$$

The solution of the initial-value problem is

$$
y = y_h + y_p = (5x - 3x^2) + (x^2 - x - x \ln x) = 4x - 2x^2 - x \ln x.
$$

30. The Cauchy–Euler initial-value problem $x^2 y'' - xy' + y = x^2$, $y(1) = 4$, $y'(1) = 3$, has auxiliary equation $m(m-1) - m + 1 = m^2 - 2m + 1 = (m-1)^2 = 0$ so $m_1 = m_2 = 1$, $y(x) = c_1 x + c_2 x \ln x$, and $y' = c_1 + c_2(1 + \ln x)$. The initial conditions give

$$
c_1 = 4
$$
$$
c_1 + c_2 = 3,
$$

so $c_1 = 4$ and $c_2 = -1$, which implies that $y_h = 4x - x \ln x$. The Wronskian is

$$
W(x, x \ln x) = \begin{vmatrix} x & x \ln x \\ 1 & 1 + \ln x \end{vmatrix} = x.
$$

Then $G(x,t) = \dfrac{tx \ln x - xt \ln t}{t} = x \ln x - x \ln t$. From the standard form of the differential equation we identify the forcing function $f(t) = 1$. Then, for $x > 1$,

$$
\begin{aligned}
y_p(x) &= \int_1^x (x \ln x - x \ln t)\, dt = x \ln x \int_1^x dt - x \int_1^x \ln t\, dt \\[2mm]
&= x \ln x (x - 1) - x(x \ln x - x + 1) = x^2 - x - x \ln x.
\end{aligned}
$$

The solution of the initial-value problem is

$$
y = y_h + y_p = 4x - x \ln x + \left(x^2 - x - x \ln x \right) = x^2 + 3x - 2x \ln x.
$$

33. The initial-value problem is $y'' + y = f(x)$, $y(0) = 1$, $y'(0) = -1$, where

$$
f(x) = \begin{cases} 0, & x < 0 \\ 10, & 0 \le x \le 3\pi \\ 0, & x > 3\pi. \end{cases}
$$

We first find

$$y_h(x) = \cos x - \sin x \qquad \text{and} \qquad y_p(x) = \int_0^x \sin(x-t)f(t)\,dt.$$

Then for $x < 0$

$$y_p(x) = \int_0^x \sin(x-t)0\,dt = 0,$$

for $0 \le x \le 3\pi$

$$y_p(x) = 10\int_0^x \sin(x-t)\,dt = 10 - 10\cos x,$$

and for $x > 3\pi$

$$y_p(x) = 10\int_0^{3\pi} \sin(x-t)\,dt + \int_{3\pi}^x \sin(x-t)0\,dt = -20\cos x.$$

The solution is

$$y(x) = y_h(x) + y_p(x) = \cos x - \sin x + y_p(x),$$

where

$$y_p(x) = \begin{cases} 0, & x < 0 \\ 10 - 10\cos x, & 0 \le x \le 3\pi \\ -20\cos x, & x > 3\pi. \end{cases}$$

36. The boundary-value problem is $y'' = f(x)$, $\quad y(0) = 0$, $y(1) + y'(1) = 0$. The solution of the associated homogeneous equation is $y = c_1 + c_2 x$.

(a) To satisfy $y(0) = 0$ we take $y_1(x) = x$ and to satisfy $y(1) + y'(1) = 0$ we note that $y(1) + y'(1) = (c_1 + c_2) + c_2 = 0$ which implies that $c_1 = -2c_2$. Taking $c_2 = 1$, we find that $c_1 = -2$, so we have $y_2(x) = -2 + x$. The Wronskian of y_1 and y_2 is

$$W(y_1, y_2) = \begin{vmatrix} x & -2+x \\ 1 & 1 \end{vmatrix} = 2,$$

so

$$G(x,t) = \begin{cases} \dfrac{1}{2}t(x-2), & 0 \le t \le x \\[2mm] \dfrac{1}{2}x(t-2), & x \le t \le 1 \end{cases}$$

Therefore

$$y_p(x) = \int_0^1 G(x,t)f(t)\,dt = \frac{1}{2}(x-2)\int_0^x tf(t)\,dt + \frac{1}{2}x\int_x^1 (t-2)f(t)\,dt.$$

(b) By the Product Rule and the Fundamental Theorem of Calculus, the first two derivatives of $y_p(x)$ are

$$y_p'(x) = \frac{1}{2}(x-2)xf(x) + \frac{1}{2}\int_0^x tf(t)\,dt + \frac{1}{2}x[-(x-2)f(x)] + \frac{1}{2}\int_x^1 (t-2)f(t)\,dt$$

$$y_p''(x) = \frac{1}{2}(x-2)[xf'(x) + f(x)] + \frac{1}{2}xf(x) + \frac{1}{2}xf(x)$$

$$- \frac{1}{2}(x^2 - 2x)f'(x) - \frac{1}{2}(2x - 2)f(x) - \frac{1}{2}(x-2)f(x)$$

$$= \frac{1}{2}\left[x^2 f'(x) + xf(x) - 2xf'(x) - 2f(x) + xf(x) + xf(x) - x^2 f'(x)\right.$$

$$\left. + 2xf'(x) - 2xf(x) + 2f(x) - xf(x) + 2f(x)\right] = f(x)$$

Thus, $y_p(x)$ satisfies the differential equation. To see that the boundary conditions are satisfied we first compute

$$y_p(0) = \frac{1}{2}(0-2)\int_0^0 tf(t)\,dt + \frac{1}{2}(0)\int_0^1 (t-2)f(t)\,dt = 0.$$

Next we use $y_p'(x)$ found at the beginning of this part of the solution to compute

$$y_p(1) + y_p'(1) = \frac{1}{2}(1-2)\int_0^1 tf(t)\,dt + \frac{1}{2}(1)\int_1^1 (t-2)f(t)\,dt + \frac{1}{2}(1-2)f(x)$$

$$+ \frac{1}{2}\int_0^1 tf(t)\,dt + \frac{1}{2}(1)[-(1-2)f(x)] + \frac{1}{2}(1)\int_1^1 (1-2)f(t)\,dt$$

$$= \frac{1}{2}(-1)\int_0^1 tf(t)\,dt + \frac{1}{2}(-1)f(1) + \frac{1}{2}\int_0^1 tf(t)\,dt + \frac{1}{2}f(1) = 0.$$

39. The boundary-value problem is $y'' + y = 1$, $y(0) = 0$, $y(1) = 0$. The solution of the associated homogeneous equation is $y = c_1 \cos x + c_2 \sin x$. Since $y(0) = c_1 \cos 0 + c_2 \sin 0 = c_1 = 0$, we take $y_1(x) = \sin x$. To satisfy $y(1) = 0$ we note that $y(1) = c_1 \cos 1 + c_2 \sin 1 = 0$ which implies that $c_1 = -c_2 \sin 1 / \cos 1$ so

$$y(x) = -c_2 \frac{\sin 1}{\cos 1}\cos x + c_2 \sin x = -\frac{c_2}{\cos 1}(\sin x \cos 1 - \cos x \sin 1).$$

Taking $c_2 = -\cos 1$, we have

$$y_2(x) = \sin x \cos 1 - \cos x \sin 1 = \sin(x-1).$$

The Wronskian of y_1 and y_2 is

$$W(y_1, y_2) = \begin{vmatrix} \sin x & \sin(x-1) \\ \cos x & \cos(x-1) \end{vmatrix} = \sin x \cos(x-1) - \cos x \sin(x-1)$$

$$= \sin[x - (x-1)] = \sin 1,$$

so

$$G(x,t) = \begin{cases} \dfrac{\sin t \sin (x - 1)}{\sin 1}, & 0 \le t \le x \\[4mm] \dfrac{\sin x \sin (t - 1)}{\sin 1}, & x \le t \le 1. \end{cases}$$

Therefore, taking $f(t) = 1$,

$$y_p(x) = \int_0^1 G(x,t)\,dt = \frac{\sin (x-1)}{\sin 1} \int_0^x \sin t\,dt + \frac{\sin x}{\sin 1} \int_x^1 \sin (t-1)\,dt$$

$$= \frac{\sin (x-1)}{\sin 1}(-\cos x + 1) + \frac{\sin x}{\sin 1}[-1 + \cos (x-1)] = \frac{\sin (x-1)}{\sin 1} - \frac{\sin x}{\sin 1} + 1.$$

42. The boundary-value problem is $y'' - y' = e^{2x}$, $y(0) = 0$, $y(1) = 0$. The solution of the associated homogeneous equation is $y = c_1 + c_2 e^x$. Since $y(0) = c_1 + c_2 = 0$, we see that $c_1 = -c_2$ and we take $y_1(x) = 1 - e^x$. To satisfy $y(1) = 0$ we note that $y(1) = c_1 + c_2 e = 0$ which implies that $c_1 = -c_2 e$ so

$$y(x) = -c_2 e + c_2 e^x = -c_2 e(1 - e^{x-1}).$$

Taking $c_2 = -1/e$, we have $y_2(x) = 1 - e^{x-1}$. The Wronskian of y_1 and y_2 is

$$W(y_1, y_2) = \begin{vmatrix} 1 - e^x & 1 - e^{x-1} \\ -e^x & -e^{x-1} \end{vmatrix} = -e^{x-1} + e^{2x-1} + e^x - e^{2x-1} = e^x - e^{x-1} = e^x(1 - e^{-1}),$$

so

$$G(x,t) = \begin{cases} \dfrac{(1 - e^t)(1 - e^{x-1})}{e^t(1 - e^{-1})}, & 0 \le t \le x \\[4mm] \dfrac{(1 - e^x)(1 - e^{t-1})}{e^t(1 - e^{-1})}, & x \le t \le 1 \end{cases}$$

Therefore, taking $f(t) = e^{2t}$,

$$y_p(x) = \int_0^1 G(x,t) e^{2t}\,dt = \frac{1 - e^{x-1}}{1 - e^{-1}} \int_0^x (e^t - e^{2t})\,dt + \frac{1 - e^x}{1 - e^{-1}} \int_x^1 (e^t - e^{2t-1})\,dt$$

$$= \frac{1 - e^{x-1}}{1 - e^{-1}} \left(e^x - \frac{1}{2}e^{2x} - \frac{1}{2} \right) + \frac{1 - e^x}{1 - e^{-1}} \left(\frac{1}{2}e - e^x + \frac{1}{2}e^{2x-1} \right)$$

$$= \frac{1}{2}e^{2x} - \frac{1}{2}e^x = \frac{1}{2}e^{x+1} + \frac{1}{2}e.$$

3.11 Nonlinear Models

3. The period corresponding to $x(0) = 1$, $x'(0) = 1$ is approximately 5.8. The second initial-value problem does not have a periodic solution.

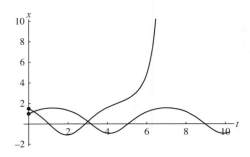

6. From the graphs we see that the interval is approximately $(-0.8, 1.1)$.

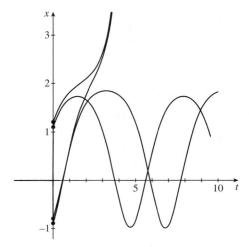

9. This is a damped hard spring, so x will approach 0 as t approaches ∞.

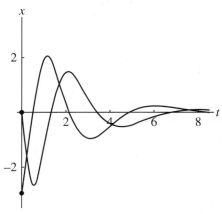

12. (a)

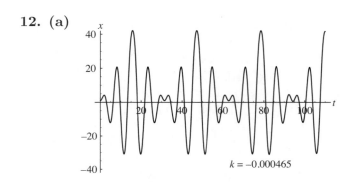
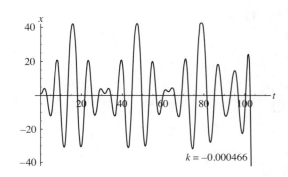

The system appears to be oscillatory for $-0.000465 \leq k_1 < 0$ and nonoscillatory for $k_1 \leq -0.000466$.

(b)

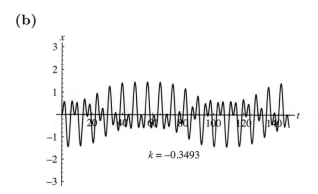

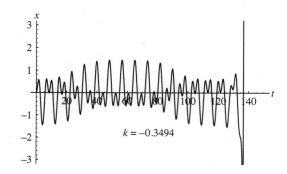

The system appears to be oscillatory for $-0.3493 \leq k_1 < 0$ and nonoscillatory for $k_1 \leq -0.3494$.

15. (a) Substitute $x(t) = k$ into equation (16) to get

$$(k) \cdot \frac{d^2}{dt^2}(k) + \left(\frac{d}{dt}(k)\right)^2 + 32(k) = 160$$

$$32k = 160$$

$$k = 5$$

(b) Intuitively one might expect a constant force of 5 lb to be able to lift only 5 feet of rope (weighing 5 lb).

(c) When $x = 0$, $\frac{dx}{dt} = \sqrt{160 - \frac{64}{3}x}$ yields $\frac{dx}{dt} = \sqrt{160} = 4\sqrt{10} \approx 12.65$ ft/s.

18. Using the results from Problem 17, when $m_b = 5\,\text{g}$, $m_w = 1\,\text{kg}$, and $h = 6$ cm, we have

$$v_b = \frac{1005}{5}\sqrt{2(980)(6)} \approx 21,797 \text{ cm/s}.$$

3.12   Solving Systems of Linear DEs

3. From $Dx = -y+t$ and $Dy = x-t$ we obtain $y = t-Dx$, $Dy = 1-D^2x$, and $(D^2+1)x = 1+t$. The solution is

$$x = c_1 \cos t + c_2 \sin t + 1 + t$$
$$y = c_1 \sin t - c_2 \cos t + t - 1.$$

6. From $(D+1)x + (D-1)y = 2$ and $3x + (D+2)y = -1$ we obtain $x = -\frac{1}{3} - \frac{1}{3}(D+2)y$, $Dx = -\frac{1}{3}(D^2 + 2D)y$, and $(D^2 + 5)y = -7$. The solution is

$$y = c_1 \cos \sqrt{5}\, t + c_2 \sin \sqrt{5}\, t - \frac{7}{5}$$

$$x = \left(-\frac{2}{3}c_1 - \frac{\sqrt{5}}{3}c_2\right) \cos \sqrt{5}\, t + \left(\frac{\sqrt{5}}{3}c_1 - \frac{2}{3}c_2\right) \sin \sqrt{5}\, t + \frac{3}{5}.$$

9. From $Dx + D^2 y = e^{3t}$ and $(D+1)x + (D-1)y = 4e^{3t}$ we obtain $D(D^2+1)x = 34e^{3t}$ and $D(D^2+1)y = -8e^{3t}$. The solution is

$$y = c_1 + c_2 \sin t + c_3 \cos t - \frac{4}{15}e^{3t}$$

$$x = c_4 + c_5 \sin t + c_6 \cos t + \frac{17}{15}e^{3t}.$$

Substituting into $(D+1)x + (D-1)y = 4e^{3t}$ gives

$$(c_4 - c_1) + (c_5 - c_6 - c_3 - c_2)\sin t + (c_6 + c_5 + c_2 - c_3)\cos t = 0$$

so that $c_4 = c_1$, $c_5 = c_3$, $c_6 = -c_2$, and

$$x = c_1 - c_2 \cos t + c_3 \sin t + \frac{17}{15}e^{3t}.$$

12. From $(2D^2 - D - 1)x - (2D+1)y = 1$ and $(D-1)x + Dy = -1$ we obtain $(2D+1)(D-1)(D+1)x = -1$ and $(2D+1)(D+1)y = -2$. The solution is

$$x = c_1 e^{-t/2} + c_2 e^{-t} + c_3 e^t + 1$$
$$y = c_4 e^{-t/2} + c_5 e^{-t} - 2.$$

Substituting into $(D-1)x + Dy = -1$ gives

$$\left(-\frac{3}{2}c_1 - \frac{1}{2}c_4\right)e^{-t/2} + (-2c_2 - c_5)e^{-t} = 0$$

so that $c_4 = -3c_1$, $c_5 = -2c_2$, and

$$y = -3c_1 e^{-t/2} - 2c_2 e^{-t} - 2.$$

15. Multiplying the first equation by $D+1$ and the second equation by $D^2 + 1$ and subtracting we obtain $(D^4 - D^2)x = 1$. Then

$$x = c_1 + c_2 t + c_3 e^t + c_4 e^{-t} - \frac{1}{2}t^2.$$

Multiplying the first equation by $D+1$ and subtracting we obtain $D^2(D+1)y = 1$. Then

$$y = c_5 + c_6 t + c_7 e^{-t} - \frac{1}{2}t^2.$$

Substituting into $(D-1)x + (D^2+1)y = 1$ gives

$$(-c_1 + c_2 + c_5 - 1) + (-2c_4 + 2c_7)e^{-t} + (-1 - c_2 + c_6)t = 1$$

so that $c_5 = c_1 - c_2 + 2$, $c_6 = c_2 + 1$, and $c_7 = c_4$. The solution of the system is

$$x = c_1 + c_2 t + c_3 e^t + c_4 e^{-t} - \frac{1}{2}t^2$$

$$y = (c_1 - c_2 + 2) + (c_2 + 1)t + c_4 e^{-t} - \frac{1}{2}t^2.$$

18. From $Dx + z = e^t$, $(D-1)x + Dy + Dz = 0$, and $x + 2y + Dz = e^t$ we obtain $z = -Dx + e^t$, $Dz = -D^2 x + e^t$, and the system $(-D^2 + D - 1)x + Dy = -e^t$ and $(-D^2 + 1)x + 2y = 0$. Then $y = \frac{1}{2}(D^2 - 1)x$, $Dy = \frac{1}{2}D(D^2 - 1)x$, and $(D-2)(D^2 + 1)x = -2e^t$ so that the solution is

$$x = c_1 e^{2t} + c_2 \cos t + c_3 \sin t + e^t$$

$$y = \frac{3}{2}c_1 e^{2t} - c_2 \cos t - c_3 \sin t$$

$$z = -2c_1 e^{2t} - c_3 \cos t + c_2 \sin t.$$

21. From $(D+5)x + y = 0$ and $4x - (D+1)y = 0$ we obtain $y = -(D+5)x$ so that $Dy = -(D^2 + 5D)x$. Then $4x + (D^2 + 5D)x + (D+5)x = 0$ and $(D+3)^2 x = 0$. Thus

$$x = c_1 e^{-3t} + c_2 t e^{-3t}$$

$$y = -(2c_1 + c_2)e^{-3t} - 2c_2 t e^{-3t}.$$

Using $x(1) = 0$ and $y(1) = 1$ we obtain

$$c_1 e^{-3} + c_2 e^{-3} = 0$$

$$-(2c_1 + c_2)e^{-3} - 2c_2 e^{-3} = 1$$

or

$$c_1 + c_2 = 0$$

$$2c_1 + 3c_2 = -e^3.$$

Thus $c_1 = e^3$ and $c_2 = -e^3$. The solution of the initial-value problem is

$$x = e^{-3t+3} - te^{-3t+3}$$

$$y = -e^{-3t+3} + 2te^{-3t+3}.$$

24. From Newton's second law in the x-direction we have

$$m\frac{d^2x}{dt^2} = -k\cos\theta = -k\frac{1}{v}\frac{dx}{dt} = -|c|\frac{dx}{dt}\,.$$

In the y-direction we have

$$m\frac{d^2y}{dt^2} = -mg - k\sin\theta = -mg - k\frac{1}{v}\frac{dy}{dt} = -mg - |c|\frac{dy}{dt}\,.$$

From $mD^2x + |c|Dx = 0$ we have $D(mD + |c|)x = 0$ so that $(mD + |c|)x = c_1$ or $(D + |c|/m)x = c_2$. This is a linear first-order differential equation. An integrating factor is $e^{\int |c|dt/m} = e^{|c|t/m}$ so that

$$\frac{d}{dt}\left[e^{|c|t/m}x\right] = c_2 e^{|c|t/m}$$

and $e^{|c|t/m}x = (c_2 m/|c|)e^{|c|t/m} + c_3$. The general solution of this equation is $x(t) = c_4 + c_3 e^{-|c|t/m}$. From $(mD^2 + |c|D)y = -mg$ we have $D(mD + |c|)y = -mg$ so that $(mD + |c|)y = -mgt + c_1$ or $(D + |c|/m)y = -gt + c_2$. This is a linear first-order differential equation with integrating factor $e^{\int |c|\,dt/m} = e^{|c|t/m}$. Thus

$$\frac{d}{dt}\left[e^{|c|t/m}y\right] = (-gt + c_2)e^{|c|t/m}$$

$$e^{|c|t/m}y = -\frac{mg}{|c|}te^{|c|t/m} + \frac{m^2g}{c^2}e^{|c|t/m} + c_3 e^{|c|t/m} + c_4$$

and

$$y(t) = -\frac{mg}{|c|}t + \frac{m^2g}{c^2} + c_3 + c_4 e^{-|c|t/m}.$$

Chapter 3 in Review

3. It is not true unless the differential equation is homogeneous. For example, $y_1 = x$ is a solution of $y'' + y = x$, but $y_2 = 5x$ is not.

6. $2\pi/5$, since $\frac{1}{4}x'' + 6.25x = 0$

9. The set is linearly independent over $(-\infty, \infty)$ and linearly dependent over $(0, \infty)$.

12. (a) The auxiliary equation is $am(m-1) + bm + c = am^2 + (b-a)m + c = 0$. If the roots are 3 and -1, then we want $(m-3)(m+1) = m^2 - 2m - 3 = 0$. Thus let $a = 1$, $b = -1$, and $c = -3$, so that the differential equation is $x^2y'' - xy' - 3y = 0$.

(b) In this case we want the auxiliary equation to be $m^2 + 1 = 0$, so let $a = 1$, $b = 1$, and $c = 1$. Then the differential equation is $x^2y'' + xy' + y = 0$.

15. From $m^2 - 2m - 2 = 0$ we obtain $m = 1 \pm \sqrt{3}$ so that

$$y = c_1 e^{(1+\sqrt{3})x} + c_2 e^{(1-\sqrt{3})x}.$$

18. From $2m^3 + 9m^2 + 12m + 5 = 0$ we obtain $m = -1$, $m = -1$, and $m = -5/2$ so that

$$y = c_1 e^{-5x/2} + c_2 e^{-x} + c_3 x e^{-x}.$$

21. Applying D^4 to the differential equation we obtain $D^4(D^2 - 3D + 5) = 0$. Then

$$y = \underbrace{e^{3x/2}\left(c_1 \cos \frac{\sqrt{11}}{2} x + c_2 \sin \frac{\sqrt{11}}{2} x\right)}_{y_c} + c_3 + c_4 x + c_5 x^2 + c_6 x^3.$$

and $y_p = A + Bx + Cx^2 + Dx^3$. Substituting y_p into the differential equation yields

$$(5A - 3B + 2C) + (5B - 6C + 6D)x + (5C - 9D)x^2 + 5Dx^3 = -2x + 4x^3.$$

Equating coefficients gives $A = -222/625$, $B = 46/125$, $C = 36/25$, and $D = 4/5$. The general solution is

$$y = e^{3x/2}\left(c_1 \cos \frac{\sqrt{11}}{2} x + c_2 \sin \frac{\sqrt{11}}{2} x\right) - \frac{222}{625} + \frac{46}{125}x + \frac{36}{25}x^2 + \frac{4}{5}x^3.$$

24. Applying D to the differential equation we obtain $D(D^3 - D^2) = D^3(D - 1) = 0$. Then

$$y = \underbrace{c_1 + c_2 x + c_3 e^x}_{y_c} + c_4 x^2$$

and $y_p = Ax^2$. Substituting y_p into the differential equation yields $-2A = 6$. Equating coefficients gives $A = -3$. The general solution is

$$y = c_1 + c_2 x + c_3 e^x - 3x^2.$$

27. The auxiliary equation is $6m^2 - m - 1 = 0$ so that

$$y = c_1 x^{1/2} + c_2 x^{-1/3}.$$

30. The auxiliary equation is $m^2 - 2m + 1 = (m - 1)^2 = 0$ and a particular solution is $y_p = \frac{1}{4}x^3$ so that

$$y = c_1 x + c_2 x \ln x + \frac{1}{4}x^3.$$

33. (a) The auxiliary equation is $m^4 - 2m^2 + 1 = (m^2 - 1)^2 = 0$, so the general solution of the differential equation is

$$y = c_1 \sinh x + c_2 \cosh x + c_3 x \sinh x + c_4 x \cosh x.$$

(b) Since both $\sinh x$ and $x \sinh x$ are solutions of the associated homogeneous differential equation, a particular solution of $y^{(4)} - 2y'' + y = \sinh x$ has the form
$y_p = Ax^2 \sinh x + Bx^2 \cosh x.$

36. The auxiliary equation is $m^2 + 2m + 1 = (m + 1)^2 = 0$, so that $y = c_1 e^{-x} + c_2 x e^{-x}$. Setting $y(-1) = 0$ and $y'(0) = 0$ we get $c_1 e - c_2 e = 0$ and $-c_1 + c_2 = 0$. Thus $c_1 = c_2$ and $y = c_1(e^{-x} + xe^{-x})$ is a solution of the boundary-value problem for any real number c_1.

39. Let $u = y'$ so that $u' = y''$. The equation becomes $u\,du/dx = 4x$. Separating variables we obtain

$$u\,du = 4x\,dx$$

$$\frac{1}{2}u^2 = 2x^2 + c_1$$

$$u^2 = 4x^2 + c_2.$$

When $x = 1$, $y' = u = 2$, so $4 = 4 + c_2$ and $c_2 = 0$. Then

$$u^2 = 4x^2$$

$$\frac{dy}{dx} = 2x \quad \text{or} \quad \frac{dy}{dx} = -2x$$

$$y = x^2 + c_3 \quad \text{or} \quad y = -x^2 + c_4.$$

When $x = 1$, $y = 5$, so $5 = 1 + c_3$ and $5 = -1 + c_4$. Thus $c_3 = 4$ and $c_4 = 6$. We have $y = x^2 + 4$ and $y = -x^2 + 6$. Note however that when $y = -x^2 + 6$, $y' = -2x$ and $y'(1) = -2 \neq 2$. Thus, the solution of the initial-value problem is $y = x^2 + 4$.

42. Consider $xy'' + y' = 0$ and look for a solution of the form $y = x^m$. Substituting into the differential equation we have

$$xy'' + y' = m(m-1)x^{m-1} + mx^{m-1} = m^2 x^{m-1}.$$

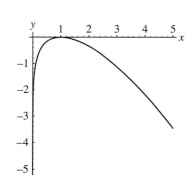

Thus, the general solution of $xy'' + y' = 0$ is $y_c = c_1 + c_2 \ln x$. To find a particular solution of $xy'' + y' = -\sqrt{x}$ we use variation of parameters.

The Wronskian is

$$W = \begin{vmatrix} 1 & \ln x \\ 0 & 1/x \end{vmatrix} = \frac{1}{x}.$$

Identifying $f(x) = -x^{-1/2}$ we obtain

$$u_1' = \frac{x^{-1/2} \ln x}{1/x} = \sqrt{x} \ln x \quad \text{and} \quad u_2' = \frac{-x^{-1/2}}{1/x} = -\sqrt{x},$$

so that

$$u_1 = x^{3/2}\left(\frac{2}{3}\ln x - \frac{4}{9}\right) \quad \text{and} \quad u_2 = -\frac{2}{3}x^{3/2}.$$

Then

$$y_p = x^{3/2} \left(\frac{2}{3} \ln x - \frac{4}{9} \right) - \frac{2}{3} x^{3/2} \ln x = -\frac{4}{9} x^{3/2}$$

and the general solution of the differential equation is

$$y = c_1 + c_2 \ln x - \frac{4}{9} x^{3/2}.$$

The initial conditions are $y(1) = 0$ and $y'(1) = 0$. These imply that $c_1 = \frac{4}{9}$ and $c_2 = \frac{2}{3}$. The solution of the initial-value problem is

$$y = \frac{4}{9} + \frac{2}{3} \ln x - \frac{4}{9} x^{3/2}.$$

The graph is shown above.

45. From $(D-2)x - y = -e^t$ and $-3x + (D-4)y = -7e^t$ we obtain $(D-1)(D-5)x = -4e^t$ so that

$$x = c_1 e^t + c_2 e^{5t} + t e^t.$$

Then

$$y = (D-2)x + e^t = -c_1 e^t + 3c_2 e^{5t} - t e^t + 2e^t.$$

48. (a) Solving $\frac{3}{8} x'' + 6x = 0$ subject to $x(0) = 1$ and $x'(0) = -4$ we obtain

$$x = \cos 4t - \sin 4t = \sqrt{2} \sin (4t + 3\pi/4).$$

(b) The amplitude is $\sqrt{2}$, period is $\pi/2$, and frequency is $2/\pi$.

(c) If $x = 1$ then $t = n\pi/2$ and $t = -\pi/8 + n\pi/2$ for $n = 1, 2, 3, \ldots$.

(d) If $x = 0$ then $t = \pi/16 + n\pi/4$ for $n = 0, 1, 2, \ldots$. The motion is upward for n even and downward for n odd.

(e) $x'(3\pi/16) = 0$

(f) If $x' = 0$ then $4t + 3\pi/4 = \pi/2 + n\pi$ or $t = 3\pi/16 + n\pi$.

51. From $q'' + 10^4 q = 100 \sin 50t$, $q(0) = 0$, and $q'(0) = 0$ we obtain $q_c = c_1 \cos 100t + c_2 \sin 100t$, $q_p = \frac{1}{75} \sin 50t$, and

(a) $q = -\frac{1}{150} \sin 100t + \frac{1}{75} \sin 50t$,

(b) $i = -\frac{2}{3} \cos 100t + \frac{2}{3} \cos 50t$, and

(c) $q = 0$ when $\sin 50t (1 - \cos 50t) = 0$ or $t = n\pi/50$ for $n = 0, 1, 2, \ldots$.

54. (a) The differential equation is $d^2r/dt^2 - \omega^2 r = -g\sin\omega t$. The auxiliary equation is $m^2 - \omega^2 = 0$, so $r_c = c_1 e^{\omega t} + c_2 e^{-\omega t}$. A particular solution has the form $r_p = A\sin\omega t + B\cos\omega t$. Substituting into the differential equation we find $-2A\omega^2\sin\omega t - 2B\omega^2\cos\omega t = -g\sin\omega t$. Thus, $B = 0$, $A = g/2\omega^2$, and $r_p = (g/2\omega^2)\sin\omega t$. The general solution of the differential equation is $r(t) = c_1 e^{\omega t} + c_2 e^{-\omega t} + (g/2\omega^2)\sin\omega t$. The initial conditions imply $c_1 + c_2 = r_0$ and $g/2\omega - \omega c_1 + \omega c_2 = v_0$. Solving for c_1 and c_2 we get

$$c_1 = (2\omega^2 r_0 + 2\omega v_0 - g)/4\omega^2 \quad\text{and}\quad c_2 = (2\omega^2 r_0 - 2\omega v_0 + g)/4\omega^2,$$

so that

$$r(t) = \frac{2\omega^2 r_0 + 2\omega v_0 - g}{4\omega^2}e^{\omega t} + \frac{2\omega^2 r_0 - 2\omega v_0 + g}{4\omega^2}e^{-\omega t} + \frac{g}{2\omega^2}\sin\omega t.$$

(b) The bead will exhibit simple harmonic motion when the exponential terms are missing. Solving $c_1 = 0$, $c_2 = 0$ for r_0 and v_0 we find $r_0 = 0$ and $v_0 = g/2\omega$.

To find the minimum length of rod that will accommodate simple harmonic motion we determine the amplitude of $r(t)$ and double it. Thus $L = g/\omega^2$.

(c) As t increases, $e^{\omega t}$ approaches infinity and $e^{-\omega t}$ approaches 0. Since $\sin\omega t$ is bounded, the distance, $r(t)$, of the bead from the pivot point increases without bound and the distance of the bead from P will eventually exceed $L/2$.

(d)

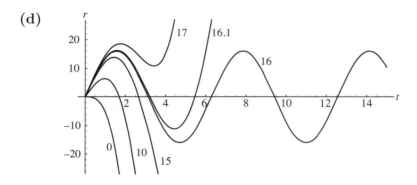

(e) For each v_0 we want to find the smallest value of t for which $r(t) = \pm 20$. Whether we look for $r(t) = -20$ or $r(t) = 20$ is determined by looking at the graphs in part (d). The total times that the bead stays on the rod is shown in the table below.

v_0	0	10	15	16.1	17
r	-20	-20	-20	20	20
t	1.55007	2.35494	3.43088	6.11627	4.22339

When $v_0 = 16$ the bead never leaves the rod.

57. The initial-value problem is $y'' + y = \tan x$, $y(0) = 2$, $y'(0) = -5$. We first solve the homogeneous version of this initial-value problem obtaining $y_h(x) = 2 \cos x - 5 \sin x$. In this case the Wronskian of $y_1(x) = \cos x$ and $y_2(x) = \sin x$ is 1, so the Green's function is

$$G(x, t) = \cos t \sin x - \cos x \sin t = \sin (x - t).$$

Therefore, taking $f(t) = \tan t$

$$y_p(x) = \int_0^x \sin (x - t) \tan t \, dt = \int_0^x (\sin x \cos t \tan t - \cos x \sin t \tan t) \, dt$$

$$= \sin x \int_0^x \sin t \, dt - \cos x \int_0^x \frac{\sin^2 t}{\cos t} \, dt = \sin x \int_0^x \sin t \, dt - \cos x \int_0^x \frac{1 - \cos^2 t}{\cos t} \, dt$$

$$= \sin x \int_0^x \sin t \, dt - \cos x \int_0^x (\sec t - \cos t) \, dt$$

$$= \sin x(- \cos x + 1) - \cos x(\ln | \sec x + \tan x| - \sin x) = \sin x - \cos x(\ln | \sec x + \tan x|).$$

Thus, the solution of the original initial-value problem is

$$y(x) = y_h(x) + y_p(x) = (2 \cos x - 5 \sin x) + (\sin x - \cos x \ln | \sec x + \tan x|)$$
$$= 2 \cos x - 4 \sin x - \cos x \ln | \sec x + \tan x|.$$

60. (a) The following graph shows the solutions (θ_1 and θ_2) to the linear initial-value problems.

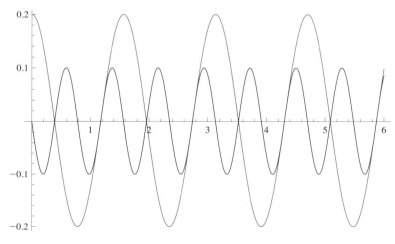

(b) The following graph shows the solutions (θ_1 and θ_2) to the nonlinear initial-value problems.

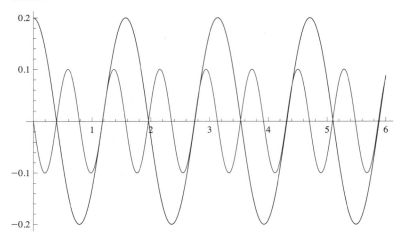

(c) The following three graphs compare the linear and nonlinear solutions for θ_1 on the same graphs for initial displacements of 0.2 radians, 0.5 radians, and 1.0 radians, respectively. The solutions begin to diverge beyond $\theta_0 = 0.2$.

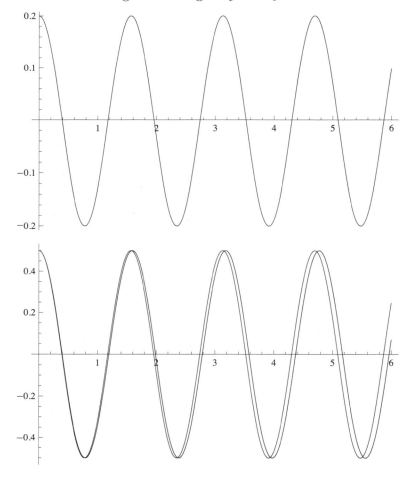

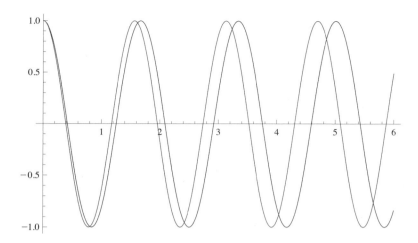

63. (a) Solve $x''(t) = 0$, subject to the conditions $x(0) = 0$ and $x'(0) = v_0 \cos t$ to get

$$x(t) = (v_0 \cos \theta)t$$

Similarly, solving for $y''(t) = -g$, subject to the conditions $y(0) = 0$ and $y'(0) = v_0 \sin \theta$ we get

$$y(t) = -\frac{1}{2}gt^2 + (v_0 \sin \theta)t$$

Substituting $t = x/(v_0 \cos \theta)$ into the equation for $y(t)$ yields

$$y(x) = -\frac{1}{2}g\left(\frac{x}{v_0 \cos \theta}\right)^2 + (v_0 \sin \theta)\left(\frac{x}{v_0 \cos \theta}\right) = -\frac{1}{2}\left(\frac{g}{v_0^2 \cos^2 \theta}\right)x^2 + (\tan \theta)\,x$$

(b) When the projectile hits the ground, $y(t) = -\frac{1}{2}gt^2 + (v_0 \sin \theta)t = 0$. The time the projectile hits the ground is found by solving this equation to get $t = (2v_0 \sin \theta)/g$. Substitute this value into the equation for $x(t)$ to get

$$x\left(\frac{2v_0 \sin \theta}{g}\right) = (v_0 \cos \theta)\left(\frac{2v_0 \sin \theta}{g}\right) = \frac{v_0^2 \cdot 2 \sin \theta \cos \theta}{g} = \frac{v_0^2 \sin 2\theta}{g}$$

The maximum value of the range equation occurs when $\sin 2\theta = 1$, that is, when $\theta = \pi/4$.

(c) At the angle $\theta = \pi/4$,

$$y(x) = -\frac{1}{2}\left(\frac{g}{v_0^2 \cos^2 (\pi/4)}\right)x^2 + \left(\tan \frac{\pi}{4}\right)x = -\frac{g}{v_0^2}x^2 + x.$$

The maximum height occurs at the vertex where

$$x = \frac{-(1)}{2\,(-g/v_0)} = \frac{v_0^2}{2g}$$

$$y\left(\frac{v_0^2}{2g}\right) = -\frac{g}{v_0^2}\left(\frac{v_0^2}{2g}\right)^2 + \left(\frac{v_0^2}{2g}\right) = \frac{v_0^2}{4g}$$

Chapter 4

The Laplace Transform

4.1 | **Definition of the Laplace Transform**

---

3. $\mathcal{L}\{f(t)\} = \int_0^1 te^{-st}\, dt + \int_1^\infty e^{-st}\, dt = \left(-\frac{1}{s}te^{-st} - \frac{1}{s^2}e^{-st}\right)\Big|_0^1 - \frac{1}{s}e^{-st}\Big|_1^\infty$

$$= \left(-\frac{1}{s}e^{-s} - \frac{1}{s^2}e^{-s}\right) - \left(0 - \frac{1}{s^2}\right) - \frac{1}{s}(0 - e^{-s}) = \frac{1}{s^2}(1 - e^{-s}), \quad s > 0$$

6. $\mathcal{L}\{f(t)\} = \int_{\pi/2}^\infty (\cos t)e^{-st}\, dt = \left(-\frac{s}{s^2+1}e^{-st}\cos t + \frac{1}{s^2+1}e^{-st}\sin t\right)\Big|_{\pi/2}^\infty$

$$= 0 - \left(0 + \frac{1}{s^2+1}e^{-\pi s/2}\right) = -\frac{1}{s^2+1}e^{-\pi s/2}, \quad s > 0$$

9. $f(t) = \begin{cases} 1 - t, & 0 < t < 1 \\ 0, & t > 1 \end{cases}$

$\mathcal{L}\{f(t)\} = \int_0^1 (1-t)e^{-st}\, dt + \int_1^\infty 0e^{-st}\, dt = \int_0^1 (1-t)e^{-st}\, dt$

$$= \left(-\frac{1}{s}(1-t)e^{-st} + \frac{1}{s^2}e^{-st}\right)\Big|_0^1 = \frac{1}{s^2}e^{-s} + \frac{1}{s} - \frac{1}{s^2}, \quad s > 0$$

12. $\mathcal{L}\{f(t)\} = \int_0^\infty e^{-2t-5}e^{-st}\, dt = e^{-5}\int_0^\infty e^{-(s+2)t}\, dt$

$$= -\frac{e^{-5}}{s+2}e^{-(s+2)t}\Big|_0^\infty = \frac{e^{-5}}{s+2}, \quad s > -2$$

15. $\mathscr{L}\{f(t)\} = \displaystyle\int_0^\infty e^{-t}(\sin t)e^{-st}\,dt = \int_0^\infty (\sin t)e^{-(s+1)t}\,dt$

$$= \left(\frac{-(s+1)}{(s+1)^2+1}e^{-(s+1)t}\sin t - \frac{1}{(s+1)^2+1}e^{-(s+1)t}\cos t \right)\Bigg|_0^\infty$$

$$= \frac{1}{(s+1)^2+1} = \frac{1}{s^2+2s+2}, \quad s > -1$$

18. $\mathscr{L}\{f(t)\} = \displaystyle\int_0^\infty t(\sin t)e^{-st}\,dt$

$$= \left[\left(-\frac{t}{s^2+1} - \frac{2s}{(s^2+1)^2} \right)(\cos t)e^{-st} - \left(\frac{st}{s^2+1} + \frac{s^2-1}{(s^2+1)^2} \right)(\sin t)e^{-st} \right]_0^\infty$$

$$= \frac{2s}{(s^2+1)^2}, \quad s > 0$$

21. $\mathscr{L}\{4t - 10\} = \dfrac{4}{s^2} - \dfrac{10}{s}$

24. $\mathscr{L}\{-4t^2 + 16t + 9\} = -\dfrac{8}{s^3} + \dfrac{16}{s^2} + \dfrac{9}{s}$

27. $\mathscr{L}\{1 + e^{4t}\} = \dfrac{1}{s} + \dfrac{1}{s-4}$

30. $\mathscr{L}\{e^{2t} - 2 + e^{-2t}\} = \dfrac{1}{s-2} - \dfrac{2}{s} + \dfrac{1}{s+2}$

33. $\mathscr{L}\{\sinh kt\} = \dfrac{1}{2}\mathscr{L}\{e^{kt} - e^{-kt}\} = \dfrac{1}{2}\left[\dfrac{1}{s-k} - \dfrac{1}{s+k} \right] = \dfrac{k}{s^2-k^2}$

36. $\mathscr{L}\{e^{-t}\cosh t\} = \mathscr{L}\left\{ e^{-t}\dfrac{e^t + e^{-t}}{2} \right\} = \mathscr{L}\left\{ \dfrac{1}{2} + \dfrac{1}{2}e^{-2t} \right\} = \dfrac{1}{2s} + \dfrac{1}{2(s+2)}$

39. Using the double-angle formula we have

$$\mathscr{L}\left\{ \cos^2 t - \sin^2 t \right\} = \mathscr{L}\left\{ \cos 2t \right\} = \frac{s}{s^2+4}$$

42. Expressing $\sin^4 t$ in terms of first-powers of cosine functions we have

$$\mathscr{L}\left\{ \sin^4 t \right\} = \mathscr{L}\left\{ (\sin^2 t)^2 \right\} = \mathscr{L}\left\{ \left[\frac{1}{2}(1 - \cos 2t) \right]^2 \right\} = \mathscr{L}\left\{ \frac{1}{4} - \frac{1}{2}\cos 2t + \frac{1}{4}\cos^2 2t \right\}$$

$$= \mathscr{L}\left\{ \frac{1}{4} - \frac{1}{2}\cos 2t + \frac{1}{4}\left[\frac{1}{2}(1 + \cos 2t) \right] \right\} = \mathscr{L}\left\{ \frac{3}{8} - \frac{1}{2}\cos 2t + \frac{1}{8}\cos 4t \right\}$$

$$= \frac{3}{8s} - \frac{s}{2(s^2+4)} + \frac{s}{8(s^2+16)}$$

45. Use integration by parts for $\alpha > 0$ to get

$$\Gamma(\alpha+1) = \int_0^\infty t^\alpha e^{-t}\,dt = -t^\alpha e^{-t}\Bigg|_0^\infty + \alpha \int_0^\infty t^{\alpha-1}e^{-t}\,dt = \alpha\Gamma(\alpha)$$

48. $\mathscr{L}\left\{ t^{1/2} \right\} = \dfrac{1}{s^{1/2+1}}\Gamma\left(\dfrac{1}{2} + 1 \right) = \dfrac{\Gamma\left(\frac{3}{2}\right)}{s^{3/2}} = \dfrac{\left(\frac{1}{2}\right)\Gamma\left(\frac{1}{2}\right)}{s^{3/2}} = \dfrac{\sqrt{\pi}}{2s^{3/2}}$

4.2 | Inverse Transforms and Transforms of Derivatives

3. $\mathscr{L}^{-1}\left\{\dfrac{1}{s^2} - \dfrac{48}{s^5}\right\} = \mathscr{L}^{-1}\left\{\dfrac{1}{s^2} - \dfrac{48}{24}\cdot\dfrac{4!}{s^5}\right\} = t - 2t^4$

6. $\mathscr{L}^{-1}\left\{\dfrac{(s+2)^2}{s^3}\right\} = \mathscr{L}^{-1}\left\{\dfrac{1}{s} + 4\cdot\dfrac{1}{s^2} + 2\cdot\dfrac{2}{s^3}\right\} = 1 + 4t + 2t^2$

9. $\mathscr{L}^{-1}\left\{\dfrac{1}{4s+1}\right\} = \dfrac{1}{4}\mathscr{L}^{-1}\left\{\dfrac{1}{s+1/4}\right\} = \dfrac{1}{4}e^{-t/4}$

12. $\mathscr{L}^{-1}\left\{\dfrac{10s}{s^2+16}\right\} = 10\cos 4t$

15. $\mathscr{L}^{-1}\left\{\dfrac{2s-6}{s^2+9}\right\} = \mathscr{L}^{-1}\left\{2\cdot\dfrac{s}{s^2+9} - 2\cdot\dfrac{3}{s^2+9}\right\} = 2\cos 3t - 2\sin 3t$

18. $\mathscr{L}^{-1}\left\{\dfrac{s+1}{s^2-4s}\right\} = \mathscr{L}^{-1}\left\{-\dfrac{1}{4}\cdot\dfrac{1}{s} + \dfrac{5}{4}\cdot\dfrac{1}{s-4}\right\} = -\dfrac{1}{4} + \dfrac{5}{4}e^{4t}$

21. $\mathscr{L}^{-1}\left\{\dfrac{0.9s}{(s-0.1)(s+0.2)}\right\} = \mathscr{L}^{-1}\left\{(0.3)\cdot\dfrac{1}{s-0.1} + (0.6)\cdot\dfrac{1}{s+0.2}\right\} = 0.3e^{0.1t} + 0.6e^{-0.2t}$

24. $\mathscr{L}^{-1}\left\{\dfrac{s^2+1}{s(s-1)(s+1)(s-2)}\right\} = \mathscr{L}^{-1}\left\{\dfrac{1}{2}\cdot\dfrac{1}{s} - \dfrac{1}{s-1} - \dfrac{1}{3}\cdot\dfrac{1}{s+1} + \dfrac{5}{6}\cdot\dfrac{1}{s-2}\right\}$

$$= \dfrac{1}{2} - e^t - \dfrac{1}{3}e^{-t} + \dfrac{5}{6}e^{2t}$$

27. $\mathscr{L}^{-1}\left\{\dfrac{2s-4}{(s^2+s)(s^2+1)}\right\} = \mathscr{L}^{-1}\left\{\dfrac{2s-4}{s(s+1)(s^2+1)}\right\}$

$$= \mathscr{L}^{-1}\left\{-\dfrac{4}{s} + \dfrac{3}{s+1} + \dfrac{s}{s^2+1} + \dfrac{3}{s^2+1}\right\}$$

$$= -4 + 3e^{-t} + \cos t + 3\sin t$$

30. $\mathscr{L}^{-1}\left\{\dfrac{6s+3}{(s^2+1)(s^2+4)}\right\} = \mathscr{L}^{-1}\left\{2\cdot\dfrac{s}{s^2+1} + \dfrac{1}{s^2+1} - 2\cdot\dfrac{s}{s^2+4} - \dfrac{1}{2}\cdot\dfrac{2}{s^2+4}\right\}$

$$= 2\cos t + \sin t - 2\cos 2t - \dfrac{1}{2}\sin 2t$$

33. The Laplace transform of the initial-value problem is

$$s\mathscr{L}\{y\} - y(0) - \mathscr{L}\{y\} = \dfrac{1}{s}.$$

Solving for $\mathscr{L}\{y\}$ we obtain

$$\mathscr{L}\{y\} = -\dfrac{1}{s} + \dfrac{1}{s-1}.$$

Thus

$$y = -1 + e^t.$$

36. The Laplace transform of the initial-value problem is

$$s\mathscr{L}\{y\} - \mathscr{L}\{y\} = \frac{2s}{s^2 + 25}.$$

Solving for $\mathscr{L}\{y\}$ we obtain

$$\mathscr{L}\{y\} = \frac{2s}{(s-1)(s^2+25)} = \frac{1}{13} \cdot \frac{1}{s-1} - \frac{1}{13}\frac{s}{s^2+25} + \frac{5}{13} \cdot \frac{5}{s^2+25}.$$

Thus

$$y = \frac{1}{13}e^t - \frac{1}{13}\cos 5t + \frac{5}{13}\sin 5t.$$

39. The Laplace transform of the initial-value problem is

$$s^2\mathscr{L}\{y\} - sy(0) + \mathscr{L}\{y\} = \frac{2}{s^2 + 2}.$$

Solving for $\mathscr{L}\{y\}$ we obtain

$$\mathscr{L}\{y\} = \frac{2}{(s^2+1)(s^2+2)} + \frac{10s}{s^2+1} = \frac{10s}{s^2+1} + \frac{2}{s^2+1} - \frac{2}{s^2+2}.$$

Thus

$$y = 10\cos t + 2\sin t - \sqrt{2}\sin\sqrt{2}\,t.$$

42. The Laplace transform of the initial-value problem is

$$s^3\mathscr{L}\{y\} - s^2(0) - sy'(0) - y''(0) + 2\left[s^2\mathscr{L}\{y\} - sy(0) - y'(0)\right] - \left[s\mathscr{L}\{y\} - y(0)\right] - 2\mathscr{L}\{y\}$$

$$= \frac{3}{s^2 + 9}.$$

Solving for $\mathscr{L}\{y\}$ we obtain

$$\mathscr{L}\{y\} = \frac{s^2 + 12}{(s-1)(s+1)(s+2)(s^2+9)}$$

$$= \frac{13}{60}\frac{1}{s-1} - \frac{13}{20}\frac{1}{s+1} + \frac{16}{39}\frac{1}{s+2} + \frac{3}{130}\frac{s}{s^2+9} - \frac{1}{65}\frac{3}{s^2+9}.$$

Thus

$$y = \frac{13}{60}e^t - \frac{13}{20}e^{-t} + \frac{16}{39}e^{-2t} + \frac{3}{130}\cos 3t - \frac{1}{65}\sin 3t.$$

45.

$$\mathscr{L}\left\{y^{(4)}\right\} - \mathscr{L}\{y\} = \mathscr{L}\{0\}$$

$$s^4 Y(s) - s^3 \overbrace{y(0)}^{1} - s^2 y'(0) - sy''(0) - y'''(0) - Y(s) = 0$$

$$s^4 Y(s) - s^3 - Y(s) = 0$$

$$\left(s^4 - 1\right)Y(s) = s^3$$

$$Y(s) = \frac{s^3}{s^2 - 1}$$

By partial fractions

$$Y(s) = \frac{1}{2}\frac{s}{s^2+1} + \frac{1}{4}\frac{1}{s-1} + \frac{1}{4}\frac{1}{s+1} = \frac{1}{2}\frac{s}{s^2+1} + \frac{1}{2}\frac{s}{s^2-1}.$$

Hence by (e) and (g) of Theorem 4.2.1,

$$y(t) = \frac{1}{2}\mathscr{L}^{-1}\left\{\frac{s}{s^2+1}\right\} + \frac{1}{2}\mathscr{L}^{-1}\left\{\frac{s}{s^2-1}\right\}$$

$$y(t) = \frac{1}{2}\cos t + \frac{1}{2}\cosh t.$$

48. The Laplace transform of the differential equation in the initial-value problem is

$$\left(s^2+2\right)Y(s) = \frac{1}{s^2+1}$$

$$Y(s) = \frac{1}{\left(s^2+1\right)\left(s^2+2\right)}$$

with the identification $a^2 = 1$ and $b^2 = 2$ or $a^2 - b^2 = -1$ we have from Problem 31:

$$\mathscr{L}^{-1}\left\{\frac{1}{\left(s^2+a^2\right)\left(s^2+b^2\right)}\right\} = \frac{a\sin bt - b\sin at}{ab\left(a^2-b^2\right)}$$

$$y(t) = \mathscr{L}^{-1}\left\{\frac{1}{\left(s^2+1\right)\left(s^2+2\right)}\right\} = \frac{\sin\sqrt{2}\,t - \sqrt{2}\sin t}{\sqrt{(-1)}}$$

$$y(t) = -\frac{1}{\sqrt{2}}\sin\sqrt{2}\,t + \sin t$$

51.

$$\mathscr{L}\left\{y^{(4)}\right\} + 16\mathscr{L}\left\{y\right\} = \mathscr{L}\left\{0\right\}$$

$$s^4 Y(s) - s^3 y(0) - s^2 y'(0) - s\overbrace{y''(0)}^{1} - y'''(0) + 16Y(s) = 0$$

$$s^4 Y(s) - s + 16Y(s) = 0$$

$$\left(s^4 + 16\right)Y(s) = s$$

$$Y(s) = \frac{s}{s^4+16}$$

Rewriting, we have

$$Y(s) = \frac{1}{4}\frac{4}{s^4+16} = \frac{1}{4}\frac{2\left(\sqrt{2}\right)^2}{s^4+2\left(\sqrt{2}\right)^2}$$

and identifying $k = \sqrt{2}$, it follows from formula 35 in Appendix C

$$y(t) = \frac{1}{4}\sin\left(\sqrt{2}\,t\right)\sinh\left(\sqrt{2}\,t\right).$$

4.3   Translation Theorems

3. $\mathscr{L}\left\{t^3 e^{-2t}\right\} = \dfrac{3!}{(s+2)^4}$

6. $\mathscr{L}\left\{e^{2t}(t-1)^2\right\} = \mathscr{L}\left\{t^2 e^{2t} - 2t e^{2t} + e^{2t}\right\} = \dfrac{2}{(s-2)^3} - \dfrac{2}{(s-2)^2} + \dfrac{1}{s-2}$

9. $\mathscr{L}\left\{(1 - e^t + 3e^{-4t})\cos 5t\right\} = \mathscr{L}\left\{\cos 5t - e^t \cos 5t + 3e^{-4t}\cos 5t\right\}$

$$= \frac{s}{s^2 + 25} - \frac{s-1}{(s-1)^2 + 25} + \frac{3(s+4)}{(s+4)^2 + 25}$$

12. $\mathscr{L}^{-1}\left\{\dfrac{1}{(s-1)^4}\right\} = \dfrac{1}{6}\mathscr{L}^{-1}\left\{\dfrac{3!}{(s-1)^4}\right\} = \dfrac{1}{6}t^3 e^t$

15. $\mathscr{L}^{-1}\left\{\dfrac{s}{s^2 + 4s + 5}\right\} = \mathscr{L}^{-1}\left\{\dfrac{s+2}{(s+2)^2 + 1^2} - 2\dfrac{1}{(s+2)^2 + 1^2}\right\} = e^{-2t}\cos t - 2e^{-2t}\sin t$

18. $\mathscr{L}^{-1}\left\{\dfrac{5s}{(s-2)^2}\right\} = \mathscr{L}^{-1}\left\{\dfrac{5(s-2) + 10}{(s-2)^2}\right\} = \mathscr{L}^{-1}\left\{\dfrac{5}{s-2} + \dfrac{10}{(s-2)^2}\right\} = 5e^{2t} + 10t e^{2t}$

21. $\mathscr{L}\left\{\sin kt \sinh kt\right\} = \mathscr{L}\left\{\sin kt \cdot \dfrac{1}{2}\left(e^{kt} - e^{-kt}\right)\right\} = \dfrac{1}{2}\mathscr{L}\left\{e^{kt}\sin kt\right\} - \dfrac{1}{2}\mathscr{L}\left\{e^{-kt}\sin kt\right\}$

$$= \frac{1}{2}\cdot\frac{k}{(s-k)^2 + k^2} - \frac{1}{2}\cdot\frac{k}{(s+k)^2 + k^2}$$

$$= \frac{k}{2}\left[\frac{(s+k)^2 + k^2 - (s-k)^2 - k^2}{\left((s-k)^2 + k^2\right)\left((s+k)^2 + k^2\right)}\right]$$

$$= \frac{2k^2 s}{\left((s^2 + 2k^2) - 2ks\right)\left((s^2 + 2k^2) + 2ks\right)} = \frac{2k^2 s}{(s^2 + 2k^2)^2 - 4k^2 s^2} = \frac{2k^2 s}{s^4 + 4k^4}$$

24. $\mathscr{L}\left\{\cos kt \cosh kt\right\} = \mathscr{L}\left\{\cos kt \cdot \dfrac{1}{2}\left(e^{kt} + e^{-kt}\right)\right\} = \dfrac{1}{2}\mathscr{L}\left\{e^{kt}\cos kt\right\} + \dfrac{1}{2}\mathscr{L}\left\{e^{-kt}\cos kt\right\}$

$$= \frac{1}{2}\cdot\frac{s-k}{(s-k)^2 + k^2} + \frac{1}{2}\cdot\frac{s+k}{(s+k)^2 + k^2}$$

$$= \frac{1}{2}\left[\frac{(s-k)\left((s+k)^2 + k^2\right) + (s+k)\left((s-k)^2 + k^2\right)}{\left((s-k)^2 + k^2\right)\left((s+k)^2 + k^2\right)}\right]$$

$$= \frac{s^3}{\left((s^2 + 2k^2) - 2ks\right)\left((s^2 + 2k^2) + 2ks\right)} = \frac{s^3}{s^4 + 4k^4}$$

27. The Laplace transform of the differential equation is

$$s^2 \mathscr{L}\{y\} - s y(0) - y'(0) + 2\left[s\mathscr{L}\{y\} - y(0)\right] + \mathscr{L}\{y\} = 0.$$

Solving for $\mathscr{L}\{y\}$ we obtain

$$\mathscr{L}\{y\} = \frac{s+3}{(s+1)^2} = \frac{1}{s+1} + \frac{2}{(s+1)^2}.$$

Thus

$$y = e^{-t} + 2te^{-t}.$$

30. The Laplace transform of the differential equation is

$$s^2\mathscr{L}\{y\} - sy(0) - y'(0) - 4\left[s\mathscr{L}\{y\} - y(0)\right] + 4\mathscr{L}\{y\} = \frac{6}{s^4}.$$

Solving for $\mathscr{L}\{y\}$ we obtain

$$\mathscr{L}\{y\} = \frac{s^5 - 4s^4 + 6}{s^4(s-2)^2} = \frac{3}{4}\frac{1}{s} + \frac{9}{8}\frac{1}{s^2} + \frac{3}{4}\frac{2}{s^3} + \frac{1}{4}\frac{3!}{s^4} + \frac{1}{4}\frac{1}{s-2} - \frac{13}{8}\frac{1}{(s-2)^2}.$$

Thus

$$y = \frac{3}{4} + \frac{9}{8}t + \frac{3}{4}t^2 + \frac{1}{4}t^3 + \frac{1}{4}e^{2t} - \frac{13}{8}te^{2t}.$$

33. The Laplace transform of the differential equation is

$$s^2\mathscr{L}\{y\} - sy(0) - y'(0) - \left[s\mathscr{L}\{y\} - y(0)\right] = \frac{s-1}{(s-1)^2 + 1}.$$

Solving for $\mathscr{L}\{y\}$ we obtain

$$\mathscr{L}\{y\} = \frac{1}{s(s^2 - 2s + 2)} = \frac{1}{2}\frac{1}{s} - \frac{1}{2}\frac{s-1}{(s-1)^2 + 1} + \frac{1}{2}\frac{1}{(s-1)^2 + 1}.$$

Thus

$$y = \frac{1}{2} - \frac{1}{2}e^t \cos t + \frac{1}{2}e^t \sin t.$$

36. Taking the Laplace transform of both sides of the differential equation and letting $c = y'(0)$ we obtain

$$\mathscr{L}\{y''\} + \mathscr{L}\{8y'\} + \mathscr{L}\{20y\} = 0$$
$$s^2\mathscr{L}\{y\} - y'(0) + 8s\mathscr{L}\{y\} + 20\mathscr{L}\{y\} = 0$$
$$s^2\mathscr{L}\{y\} - c + 8s\mathscr{L}\{y\} + 20\mathscr{L}\{y\} = 0$$
$$(s^2 + 8s + 20)\mathscr{L}\{y\} = c$$
$$\mathscr{L}\{y\} = \frac{c}{s^2 + 8s + 20} = \frac{c}{(s+4)^2 + 4}.$$

Therefore,

$$y(t) = \mathscr{L}^{-1}\left\{\frac{c}{(s+4)^2 + 4}\right\} = \frac{c}{2}e^{-4t}\sin 2t = c_1 e^{-4t}\sin 2t.$$

To find c_1 we let $y'(\pi) = 0$. Then $0 = y'(\pi) = c_1 e^{-4\pi}$ and $c_1 = 0$. Thus $y(t) = 0$. (Since the differential equation is homogeneous and both boundary conditions are 0, we can see immediately that $y(t) = 0$ is a solution. We have shown that it is the only solution.)

39. The differential equation is

$$\frac{d^2q}{dt^2} + 2\lambda dqdt + \omega^2 q = \frac{E_0}{L}, \quad q(0) = q'(0) = 0.$$

The Laplace transform of this equation is

$$s^2 \mathscr{L}\{q\} + 2\lambda s\mathscr{L}\{q\} + \omega^2 \mathscr{L}\{q\} = \frac{E_0}{L}\frac{1}{s}$$

or

$$\left(s^2 + 2\lambda s + \omega^2\right)\mathscr{L}\{q\} = \frac{E_0}{L}\frac{1}{s}.$$

Solving for $\mathscr{L}\{q\}$ and using partial fractions we obtain

$$\mathscr{L}\{q\} = \frac{E_0}{L}\left(\frac{1/\omega^2}{s} - \frac{(1/\omega^2)s + 2\lambda/\omega^2}{s^2 + 2\lambda s + \omega^2}\right) = \frac{E_0}{L\omega^2}\left(\frac{1}{s} - \frac{s + 2\lambda}{s^2 + 2\lambda s + \omega^2}\right).$$

For $\lambda > \omega$ we write $s^2 + 2\lambda s + \omega^2 = (s+\lambda)^2 - (\lambda^2 - \omega^2)$, so (recalling that $\omega^2 = 1/LC$)

$$\mathscr{L}\{q\} = E_0 C\left(\frac{1}{s} - \frac{s+\lambda}{(s+\lambda)^2 - (\lambda^2 - \omega^2)} - \frac{\lambda}{(s+\lambda)^2 - (\lambda^2 - \omega^2)}\right).$$

Thus for $\lambda > \omega$,

$$q(t) = E_0 C\left[1 - e^{-\lambda t}\left(\cosh\sqrt{\lambda^2 - \omega^2}\,t - \frac{\lambda}{\sqrt{\lambda^2 - \omega^2}}\sinh\sqrt{\lambda^2 - \omega^2}\,t\right)\right].$$

For $\lambda < \omega$ we write $s^2 + 2\lambda s + \omega^2 = (s+\lambda)^2 + (\omega^2 - \lambda^2)$, so

$$\mathscr{L}\{q\} = E_0 C\left(\frac{1}{s} - \frac{s+\lambda}{(s+\lambda)^2 + (\omega^2 - \lambda^2)} - \frac{\lambda}{(s+\lambda)^2 + (\omega^2 - \lambda^2)}\right).$$

Thus for $\lambda < \omega$,

$$q(t) = E_0 C\left[1 - e^{-\lambda t}\left(\cos\sqrt{\omega^2 - \lambda^2}\,t - \frac{\lambda}{\sqrt{\omega^2 - \lambda^2}}\sin\sqrt{\omega^2 - \lambda^2}\,t\right)\right].$$

For $\lambda = \omega$, $s^2 + 2\lambda + \omega^2 = (s+\lambda)^2$ and

$$\mathscr{L}\{q\} = \frac{E_0}{L}\frac{1}{s(s+\lambda)^2} = \frac{E_0}{L}\left(\frac{1/\lambda^2}{s} - \frac{1/\lambda^2}{s+\lambda} - \frac{1/\lambda}{(s+\lambda)^2}\right) = \frac{E_0}{L\lambda^2}\left(\frac{1}{s} - \frac{1}{s+\lambda} - \frac{\lambda}{(s+\lambda)^2}\right).$$

Thus for $\lambda = \omega$,

$$q(t) = E_0 C\left(1 - e^{-\lambda t} - \lambda t e^{-\lambda t}\right).$$

42. $\mathscr{L}\left\{e^{2-t}\mathscr{U}(t-2)\right\} = \mathscr{L}\left\{e^{-(t-2)}\mathscr{U}(t-2)\right\} = \dfrac{e^{-2s}}{s+1}$

45. $\mathscr{L}\{\cos 2t\,\mathscr{U}(t-\pi)\} = \mathscr{L}\{\cos 2(t-\pi)\mathscr{U}(t-\pi)\} = \dfrac{se^{-\pi s}}{s^2+4}$

Alternatively, (16) of this section could be used:

$$\mathscr{L}\{\cos 2t\,\mathscr{U}(t-\pi)\} = e^{-\pi s}\mathscr{L}\{\cos 2(t+\pi)\} = e^{-\pi s}\mathscr{L}\{\cos 2t\} = e^{-\pi s}\dfrac{s}{s^2+4}.$$

48. $\mathscr{L}^{-1}\left\{\dfrac{(1+e^{-2s})^2}{s+2}\right\} = \mathscr{L}^{-1}\left\{\dfrac{1}{s+2} + \dfrac{2e^{-2s}}{s+2} + \dfrac{e^{-4s}}{s+2}\right\}$

$$= e^{-2t} + 2e^{-2(t-2)}\mathscr{U}(t-2) + e^{-2(t-4)}\mathscr{U}(t-4)$$

51. $\mathscr{L}^{-1}\left\{\dfrac{e^{-s}}{s(s+1)}\right\} = \mathscr{L}^{-1}\left\{\dfrac{e^{-s}}{s} - \dfrac{e^{-s}}{s+1}\right\} = \mathscr{U}(t-1) - e^{-(t-1)}\mathscr{U}(t-1)$

54. (e)   **57. (a)**

60. $\mathscr{L}\{1 - \mathscr{U}(t-4) + \mathscr{U}(t-5)\} = \dfrac{1}{s} - \dfrac{e^{-4s}}{s} + \dfrac{e^{-5s}}{s}$

63. $\mathscr{L}\{t - t\,\mathscr{U}(t-2)\} = \mathscr{L}\{t - (t-2)\mathscr{U}(t-2) - 2\mathscr{U}(t-2)\} = \dfrac{1}{s^2} - \dfrac{e^{-2s}}{s^2} - \dfrac{2e^{-2s}}{s}$

66. $\mathscr{L}\{\sin t + \sin t\,\mathscr{U}(t-2\pi) + \sin t\,\mathscr{U}(t-2pi) + \sin t\,\mathscr{U}(t-4\pi)\}$

$$= \mathscr{L}\{\sin t + \sin(t-2\pi)\mathscr{U}(t-2\pi) + \sin(t-4\pi)\mathscr{U}(t-4\pi)\}$$

$$= \dfrac{1}{s^2+1} + \dfrac{e^{-2\pi s}}{s^2+1} + \dfrac{e^{-4\pi s}}{s^2+1}$$

69. $f(t) = \begin{cases} -t+2, & 0 \le t < 1 \\ 1, & 1 \le t < 2 \\ 0, & t \ge 2 \end{cases}$

$\mathscr{L}\{f(t)\} = \mathscr{L}\{-t + 2 - (-t+2)\mathscr{U}(t-1) + \mathscr{U}(t-1) - \mathscr{U}(t-2)\}$

$$= \mathscr{L}\{-t + 2 + (t-1)\mathscr{U}(t-1) - \mathscr{U}(t-2)\} = -\dfrac{1}{s^2} + \dfrac{2}{s} + \dfrac{e^{-s}}{s^2} - \dfrac{e^{-2s}}{s}$$

72. The Laplace transform of the differential equation is

$$s\mathscr{L}\{y\} - y(0) + \mathscr{L}\{y\} = \dfrac{1}{s} - \dfrac{2}{s}e^{-s}.$$

Solving for $\mathscr{L}\{y\}$ we obtain

$$\mathscr{L}\{y\} = \dfrac{1}{s(s+1)} - \dfrac{2e^{-s}}{s(s+1)} = \dfrac{1}{s} - \dfrac{1}{s+1} - 2e^{-s}\left[\dfrac{1}{s} - \dfrac{1}{s+1}\right].$$

Thus

$$y = 1 - e^{-t} - 2\left[1 - e^{-(t-1)}\right]\mathscr{U}(t-1).$$

75. The Laplace transform of the differential equation is

$$s^2 \mathscr{L}\{y\} - sy(0) - y'(0) + 4\mathscr{L}\{y\} = e^{-2\pi s}\frac{1}{s^2+1}.$$

Solving for $\mathscr{L}\{y\}$ we obtain

$$\mathscr{L}\{y\} = \frac{s}{s^2+4} + e^{-2\pi s}\left[\frac{1}{3}\frac{1}{s^2+1} - \frac{1}{6}\frac{2}{s^2+4}\right].$$

Thus

$$y = \cos 2t + \left[\frac{1}{3}\sin(t-2\pi) - \frac{1}{6}\sin 2(t-2\pi)\right]\mathscr{U}(t-2\pi).$$

78. The Laplace transform of the differential equation is

$$s^2 \mathscr{L}\{y\} - sy(0) - y'(0) + 4\left[s\mathscr{L}\{y\} - y(0)\right] + 3\mathscr{L}\{y\} = \frac{1}{s} - \frac{e^{-2s}}{s} - \frac{e^{-4s}}{s} + \frac{e^{-6s}}{s}.$$

Solving for $\mathscr{L}\{y\}$ we obtain

$$\mathscr{L}\{y\} = \frac{1}{3}\frac{1}{s} - \frac{1}{2}\frac{1}{s+1} + \frac{1}{6}\frac{1}{s+3} - e^{-2s}\left[\frac{1}{3}\frac{1}{s} - \frac{1}{2}\frac{1}{s+1} + \frac{1}{6}\frac{1}{s+3}\right]$$

$$- e^{-4s}\left[\frac{1}{3}\frac{1}{s} - \frac{1}{2}\frac{1}{s+1} + \frac{1}{6}\frac{1}{s+3}\right] + e^{-6s}\left[\frac{1}{3}\frac{1}{s} - \frac{1}{2}\frac{1}{s+1} + \frac{1}{6}\frac{1}{s+3}\right].$$

Thus

$$y = \frac{1}{3} - \frac{1}{2}e^{-t} + \frac{1}{6}e^{-3t} - \left[\frac{1}{3} - \frac{1}{2}e^{-(t-2)} + \frac{1}{6}e^{-3(t-2)}\right]\mathscr{U}(t-2)$$

$$- \left[\frac{1}{3} - \frac{1}{2}e^{-(t-4)} + \frac{1}{6}e^{-3(t-4)}\right]\mathscr{U}(t-4) + \left[\frac{1}{3} - \frac{1}{2}e^{-(t-6)} + \frac{1}{6}e^{-3(t-6)}\right]\mathscr{U}(t-6).$$

81. The differential equation is

$$2.5\frac{dq}{dt} + 12.5q = 5\mathscr{U}(t-3).$$

The Laplace transform of this equation is

$$s\mathscr{L}\{q\} + 5\mathscr{L}\{q\} = \frac{2}{s}e^{-3s}.$$

Solving for $\mathscr{L}\{q\}$ we obtain

$$\mathscr{L}\{q\} = \frac{2}{s(s+5)}e^{-3s} = \left(\frac{2}{5}\cdot\frac{1}{s} - \frac{2}{5}\cdot\frac{1}{s+5}\right)e^{-3s}.$$

Thus

$$q(t) = \frac{2}{5}\mathscr{U}(t-3) - \frac{2}{5}e^{-5(t-3)}\mathscr{U}(t-3).$$

84. (a) The differential equation is

$$50\frac{dq}{dt} + \frac{1}{0.01}q = E_0[\mathscr{U}(t-1) - \mathscr{U}(t-3)], \quad q(0) = 0$$

or

$$50\frac{dq}{dt} + 100q = E_0[\mathscr{U}(t-1) - \mathscr{U}(t-3)], \quad q(0) = 0.$$

The Laplace transform of this equation is

$$50s\mathscr{L}\{q\} + 100\mathscr{L}\{q\} = E_0\left(\frac{1}{s}e^{-s} - \frac{1}{s}e^{-3s}\right).$$

Solving for $\mathscr{L}\{q\}$ we obtain

$$\mathscr{L}\{q\} = \frac{E_0}{50}\left[\frac{e^{-s}}{s(s+2)} - \frac{e^{-3s}}{s(s+2)}\right] = \frac{E_0}{50}\left[\frac{1}{2}\left(\frac{1}{s} - \frac{1}{s+2}\right)e^{-s} - \frac{1}{2}\left(\frac{1}{s} - \frac{1}{s+2}\right)e^{-3s}\right].$$

Thus

$$q(t) = \frac{E_0}{100}\left[\left(1 - e^{-2(t-1)}\right)\mathscr{U}(t-1) - \left(1 - e^{-2(t-3)}\right)\mathscr{U}(t-3)\right].$$

(b)

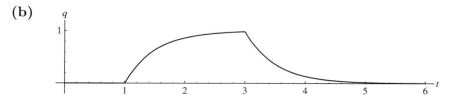

The maximum value of $q(t)$ is approximately 1 at $t = 3$.

87. The differential equation is

$$EI\frac{d^4y}{dx^4} = \frac{2w_0}{L}\left[\frac{L}{2} - x + \left(x - \frac{L}{2}\right)\mathscr{U}\left(x - \frac{L}{2}\right)\right].$$

Taking the Laplace transform of both sides and using $y(0) = y'(0) = 0$ we obtain

$$s^4\mathscr{L}\{y\} - sy''(0) - y'''(0) = \frac{2w_0}{EIL}\left[\frac{L}{2s} - \frac{1}{s^2} + \frac{1}{s^2}e^{-Ls/2}\right].$$

Letting $y''(0) = c_1$ and $y'''(0) = c_2$ we have

$$\mathscr{L}\{y\} = \frac{c_1}{s^3} + \frac{c_2}{s^4} + \frac{2w_0}{EIL}\left[\frac{L}{2s^5} - \frac{1}{s^6} + \frac{1}{s^6}e^{-Ls/2}\right]$$

so that

$$y(x) = \frac{1}{2}c_1x^2 + \frac{1}{6}c_2x^3 + \frac{2w_0}{EIL}\left[\frac{L}{48}x^4 - \frac{1}{120}x^5 + \frac{1}{120}\left(x - \frac{L}{2}\right)^5\mathscr{U}\left(x - \frac{L}{2}\right)\right]$$

$$= \frac{1}{2}c_1x^2 + \frac{1}{6}c_2x^3 + \frac{w_0}{60EIL}\left[\frac{5L}{2}x^4 - x^5 + \left(x - \frac{L}{2}\right)^5\mathscr{U}\left(x - \frac{L}{2}\right)\right].$$

To find c_1 and c_2 we compute

$$y''(x) = c_1 + c_2 x + \frac{w_0}{60EIL}\left[30Lx^2 - 20x^3 + 20\left(x - \frac{L}{2}\right)^3 \mathscr{U}\left(x - \frac{L}{2}\right)\right]$$

and

$$y'''(x) = c_2 + \frac{w_0}{60EIL}\left[60Lx - 60x^2 + 60\left(x - \frac{L}{2}\right)^2 \mathscr{U}\left(x - \frac{L}{2}\right)\right].$$

Then $y''(L) = y'''(L) = 0$ yields the system

$$c_1 + c_2 L + \frac{w_0}{60EIL}\left[30L^3 - 20L^3 + \frac{5}{2}L^3\right] = 0 \qquad\qquad c_1 + c_2 L + \frac{5w_0 L^2}{24EI} = 0$$

$$\text{or}$$

$$c_2 + \frac{w_0}{60EIL}[60L^2 - 60L^2 + 15L^2] = 0. \qquad\qquad c_2 + \frac{w_0 L}{4EI} = 0.$$

Solving for c_1 and c_2 we obtain $c_1 = w_0 L^2/24EI$ and $c_2 = -w_0 L/4EI$. Thus

$$y(x) = \frac{w_0 L^2}{48EI}x^2 - \frac{w_0 L}{24EI}x^3 + \frac{w_0}{60EIL}\left[\frac{5L}{2}x^4 - x^5 + \left(x - \frac{L}{2}\right)^5 \mathscr{U}\left(x - \frac{L}{2}\right)\right].$$

4.4 Additional Operational Properties

3. $\mathscr{L}\{t \cos 2t\} = -\dfrac{d}{ds}\left(\dfrac{s}{s^2 + 4}\right) = \dfrac{s^2 - 4}{(s^2 + 4)^2}$

6. $\mathscr{L}\{t^2 \cos t\} = \dfrac{d^2}{ds^2}\left(\dfrac{s}{s^2 + 1}\right) = \dfrac{d}{ds}\left(\dfrac{1 - s^2}{(s^2 + 1)^2}\right) = \dfrac{2s\left(s^2 - 3\right)}{(s^2 + 1)^3}$

9. The Laplace transform of the differential equation is

$$s\mathscr{L}\{y\} + \mathscr{L}\{y\} = \frac{2s}{(s^2 + 1)^2}.$$

Solving for $\mathscr{L}\{y\}$ we obtain

$$\mathscr{L}\{y\} = \frac{2s}{(s + 1)(s^2 + 1)^2} = -\frac{1}{2}\frac{1}{s + 1} - \frac{1}{2}\frac{1}{s^2 + 1} + \frac{1}{2}\frac{s}{s^2 + 1} + \frac{1}{(s^2 + 1)^2} + \frac{s}{(s^2 + 1)^2}.$$

Thus

$$y(t) = -\frac{1}{2}e^{-t} - \frac{1}{2}\sin t + \frac{1}{2}\cos t + \frac{1}{2}(\sin t - t\cos t) + \frac{1}{2}t\sin t$$

$$= -\frac{1}{2}e^{-t} + \frac{1}{2}\cos t - \frac{1}{2}t\cos t + \frac{1}{2}t\sin t.$$

12. The Laplace transform of the differential equation is

$$s^2 \mathscr{L}\{y\} - sy(0) - y'(0) + \mathscr{L}\{y\} = \frac{1}{s^2 + 1} \, .$$

Solving for $\mathscr{L}\{y\}$ we obtain

$$\mathscr{L}\{y\} = \frac{s^3 - s^2 + s}{(s^2 + 1)^2} = \frac{s}{s^2 + 1} - \frac{1}{s^2 + 1} + \frac{1}{(s^2 + 1)^2} \, .$$

Thus

$$y = \cos t - \sin t + \left(\frac{1}{2} \sin t - \frac{1}{2} t \cos t \right) = \cos t - \frac{1}{2} \sin t - \frac{1}{2} t \cos t.$$

15.

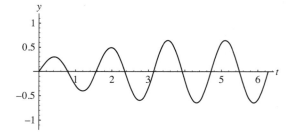

18. From (7) of Section 4.2 in the text along with Theorem 4.4.1,

$$\mathscr{L}\{ty''\} = -\frac{d}{ds} \mathscr{L}\{y''\} = -\frac{d}{ds} \left(s^2 Y(s) - sy(0) - y'(0) \right) = -s^2 Y'(s) - 2sY(s),$$

$$\mathscr{L}\{ty'\} = -\frac{d}{ds} \mathscr{L}\{y'\} = -\frac{d}{ds} \left(sY(s) - y(0) \right) = -sY'(s) - Y(s).$$

So that the transform of the given second-order differential equation is the linear first-order differential equation in $Y(s)$:

$$-s^2 Y'(s) - 2sY(s) + 4sY'(s) + 4Y(s) - 4Y(s) = 0$$

$$\left(s^2 - 4s \right) Y'(s) + 2sY(s) = 0$$

$$Y'(s) + \frac{2}{s - 4} Y(s) = 0.$$

The solution of the last equation is

$$Y(s) = \frac{c}{(s - 4)^2}$$

so,

$$y(t) = \mathscr{L}^{-1}\{Y(s)\} = cte^{4t}.$$

Using $y'(0) = 5$ results in $ce^0 = 5$ or $c = 5$. Hence $y(t) = 5te^{4t}$.

21. From (7) of Section 4.2 in the text along with Theorem 4.4.1,

$$\mathscr{L}\{y''\} = s^2 Y(s) - sy(0) - y'(0) = s^2 Y(s),$$

$$\mathscr{L}\{ty'\} = -\frac{d}{ds}\mathscr{L}\{y'\} = -\frac{d}{ds}\left(sY(s) - y(0)\right) = -sY'(s) - Y(s).$$

So that the transform of the given second-order differential equation is the linear first-order differential equation in $Y(s)$:

$$2s^2 Y(s) - sY'(s) - Y(s) - 2Y(s) = \frac{10}{s}$$

$$sY'(s) + \left(3 - 2s^2\right)Y(s) = -\frac{10}{s}$$

$$Y'(s) + \left(\frac{3}{s} - 2s\right)Y(s) = -\frac{10}{s^2}.$$

The solution of the last equation is

$$Y(s) = \frac{5}{s^3} + \frac{c}{s^3}e^{s^2}.$$

If $Y(s)$ is the Laplace transform of a piecewise-continuous function of exponential order, we must have, in view of Theorem 4.2.3, $\lim_{s \to \infty} Y(s) = 0$. In order to obtain this condition we require $c = 0$. Hence

$$y(t) = \mathscr{L}^{-1}\left\{\frac{5}{s^3}\right\} = \frac{5}{2}t^2.$$

24. Identify $f(\tau) = \tau$ and $g(t - \tau) = e^{-t+\tau}$. Therefore,

$$f * g = \int_0^t \tau e^{-t+\tau}\,d\tau = e^{-t}\int_0^t \tau e^\tau\,d\tau = e^{-t}\left(te^t - e^t + 1\right) = t - 2 + e^{-t}$$

Then

$$\mathscr{L}\{f * g\} = \mathscr{L}\left\{t - 1 + e^{-t}\right\} = \frac{1}{s^2} - \frac{1}{2} + \frac{1}{s+1} = \frac{1}{s^2(s+1)}$$

27. $\mathscr{L}\left\{1 * t^3\right\} = \frac{1}{s}\frac{3!s^4}{=}\frac{6}{s^5}$

30. $\mathscr{L}\left\{e^{2t} * \sin t\right\} = \dfrac{1}{(s-2)(s^2+1)}$

33. $\mathscr{L}\left\{\displaystyle\int_0^t e^{-\tau}\cos\tau\,d\tau\right\} = \frac{1}{s}\mathscr{L}\left\{e^{-t}\cos t\right\} = \frac{1}{s}\frac{s+1}{(s+1)^2+1} = \frac{s+1}{s(s^2+2s+2)}$

36. $\mathscr{L}\left\{\displaystyle\int_0^t \sin\tau\cos(t-\tau)\,d\tau\right\} = \mathscr{L}\{\sin t\}\mathscr{L}\{\cos t\} = \dfrac{s}{(s^2+1)^2}$

39. $\mathscr{L}^{-1}\left\{\dfrac{1}{s(s-1)}\right\} = \mathscr{L}^{-1}\left\{\dfrac{1/(s-1)}{s}\right\} = \displaystyle\int_0^t e^\tau\,d\tau = e^t - 1$

42. Using $\mathscr{L}^{-1}\left\{\dfrac{1}{(s-a)^2}\right\} = te^{at}$, (9) in the text gives

$$\mathscr{L}^{-1}\left\{\frac{1}{s(s-a)^2}\right\} = \int_0^t \tau e^{a\tau}\,d\tau = \frac{1}{a^2}(ate^{at} - e^{at} + 1).$$

45. (a) The result in (5) in the text is $\mathscr{L}^{-1}\{F(s)G(s)\} = f * g$, so identify

$$F(s) = \frac{2k^3}{(s^2+k^2)^2} \quad \text{and} \quad G(s) = \frac{4s}{s^2+k^2}.$$

Then

$$f(t) = \sin kt - kt\cos kt \quad \text{and} \quad g(t) = 4\cos kt$$

so

$$\mathscr{L}^{-1}\left\{\frac{8k^3 s}{(s^2+k^2)^3}\right\} = \mathscr{L}^{-1}\{F(s)G(s)\} = f * g = 4\int_0^t f(\tau)g(t-\tau)\,dt$$

$$= 4\int_0^t (\sin k\tau - k\tau\cos k\tau)\cos k(t-\tau)\,d\tau.$$

Using a CAS to evaluate the integral we get

$$\mathscr{L}^{-1}\left\{\frac{8k^3 s}{(s^2+k^2)^3}\right\} = t\sin kt - kt^2\cos kt.$$

(b) Observe from part (a) that

$$\mathscr{L}\{t(\sin kt - kt\cos kt)\} = \frac{8k^3 s}{(s^2+k^2)^3},$$

and from Theorem 4.4.1 that $\mathscr{L}\{tf(t)\} = -F'(s)$. We saw in (6) in the text that

$$\mathscr{L}\{\sin kt - kt\cos kt\} = 2k^3/(s^2+k^2)^2,$$

so

$$\mathscr{L}\{t(\sin kt - kt\cos kt)\} = -\frac{d}{ds}\frac{2k^3}{(s^2+k^2)^2} = \frac{8k^3 s}{(s^2+k^2)^3}.$$

48. The Laplace transform of the given equation is

$$\mathscr{L}\{f\} = \mathscr{L}\{2t\} - 4\mathscr{L}\{\sin t\}\mathscr{L}\{f\}.$$

Solving for $\mathscr{L}\{f\}$ we obtain

$$\mathscr{L}\{f\} = \frac{2s^2+2}{s^2(s^2+5)} = \frac{2}{5}\frac{1}{s^2} + \frac{8}{5\sqrt{5}}\frac{\sqrt{5}}{s^2+5}.$$

Thus

$$f(t) = \frac{2}{5}t + \frac{8}{5\sqrt{5}}\sin\sqrt{5}\,t.$$

51. The Laplace transform of the given equation is

$$\mathscr{L}\{f\} + \mathscr{L}\{1\}\mathscr{L}\{f\} = \mathscr{L}\{1\}.$$

Solving for $\mathscr{L}\{f\}$ we obtain $\mathscr{L}\{f\} = \dfrac{1}{s+1}$. Thus $f(t) = e^{-t}$.

54. The Laplace transform of the given equation is

$$\mathscr{L}\{t\} - 2\mathscr{L}\{f\} = \mathscr{L}\left\{e^t - e^{-t}\right\}\mathscr{L}\{f\}.$$

Solving for $\mathscr{L}\{f\}$ we obtain

$$\mathscr{L}\{f\} = \frac{s^2 - 1}{2s^4} = \frac{1}{2}\frac{1}{s^2} - \frac{1}{12}\frac{3!}{s^4}.$$

Thus

$$f(t) = \frac{1}{2}t - \frac{1}{12}t^3.$$

57. The Laplace transform of the given equation is

$$s\mathscr{L}\{y\} - y(0) = \mathscr{L}\{10\} - \mathscr{L}\{e^{-4t}\}\mathscr{L}\{y\}.$$

Solving for $\mathscr{L}\{y\}$ we obtain

$$\mathscr{L}\{y\} = \frac{5\,(s+2)\,(s+4)}{s\,(s+2)^2} = \frac{5s+20}{s\,(s+2)} = \frac{10}{s} - \frac{5}{s+2}.$$

Thus

$$y = 10 - 5e^{-2t}.$$

60. Using $L = 1$ h, $C = 0.04$ f, $i(0) = 0$, and $E(t) = 100t - 100(t-1)\mathscr{U}(t-1) - 100\mathscr{U}(t-4)$, equation (14) becomes

$$\frac{di}{dt} + 25 \int_0^t i(\tau)\,d\tau = 100t - 100(t-1)\mathscr{U}(t-1) - 100\mathscr{U}(t-4)$$

$$\mathscr{L}\left\{\frac{di}{dt}\right\} + 25\mathscr{L}\left\{\int_0^t i(\tau)\,d\tau\right\} = 100\mathscr{L}\{t\} - 100\mathscr{L}\{(t-1)\mathscr{U}(t-1)\} - 100\mathscr{L}\{\mathscr{U}(t-4)\}$$

$$sI(s) + 25\frac{I(s)}{s} = 100\frac{1}{s^2} - 100\frac{e^{-s}}{s^2} - 100\frac{e^{-4s}}{s}$$

$$(s^2 + 25)\,I(s) = 100\frac{1}{s} - 100\frac{e^{-s}}{s} - 100e^{-4s}$$

$$I(s) = 100\frac{1}{s\,(s^2+25)} - 100\frac{1}{s\,(s^2+25)}e^{-s} - 100\frac{1}{s^2+25}e^{-4s}$$

$$i(t) = 4\mathscr{L}^{-1}\left\{\frac{1}{s} - \frac{s}{s^2+25}\right\} - 4\mathscr{L}^{-1}\left\{\left(\frac{1}{s} - \frac{s}{s^2+25}\right)e^{-s}\right\} - 20\mathscr{L}^{-1}\left\{\frac{5}{s^2+25}e^{-4s}\right\}$$

$$i(t) = 4 - 4\cos 5t - 4\mathscr{U}(t-1) + 4\cos 5(t-1)\mathscr{U}(t-1) - 20\sin 5(t-4)\mathscr{U}(t-4).$$

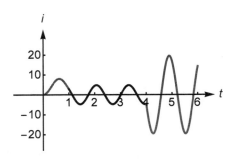

63. Let $\mathscr{L}\left\{e^{-t^2}\right\} = F(s)$ then

$$\mathscr{L}\left\{y''\right\} + 9\mathscr{L}\left\{y\right\} = 3\mathscr{L}\left\{e^{-t^2}\right\}$$

$$s^2 Y(s) + 9Y(s) = 3F(s)$$

$$Y(s) = \frac{3}{s^2 + 9} \cdot F(s)$$

Now take the inverse transform and use the inverse form of the convolution theorem to get

$$y(t) = \mathscr{L}^{-1}\left\{\frac{3}{s^2 + 9} \cdot F(s)\right\} = \sin 3t * e^{-t^2} = \int_0^t \sin 3\tau \, e^{-(t-\tau)^2} \, d\tau$$

or

$$y(t) = \mathscr{L}^{-1}\left\{F(s) \cdot \frac{3}{s^2 + 9}\right\} = e^{-t^2} * \sin 3t = \int_0^t e^{-\tau^2} \sin 3\,(t-\tau)\, d\tau$$

66. $\mathscr{L}\{f(t)\} = \dfrac{1}{1 - e^{-2as}} \displaystyle\int_a^{2a} e^{-st}\, dt = \dfrac{1}{1 - e^{-2as}} \cdot \dfrac{e^{-as} - e^{2as}}{s} = \dfrac{e^{-as}}{s\,(1 + e^{-as})}.$

69. $\mathscr{L}\{f(t)\} = \dfrac{1}{1 - e^{-\pi s}} \displaystyle\int_0^{\pi} e^{-st} \sin t\, dt = \dfrac{1}{s^2 + 1} \cdot \dfrac{e^{\pi s/2} + e^{-\pi s/2}}{e^{\pi s/2} - e^{-\pi s/2}} = \dfrac{1}{s^2 + 1} \coth \dfrac{\pi s}{2}$

72. The differential equation is $L\, di/dt + Ri = E(t)$, where $i(0) = 0$. The Laplace transform of the equation is

$$Ls\mathscr{L}\{i\} + R\mathscr{L}\{i\} = \mathscr{L}\{E(t)\}.$$

From Problem 67 we have

$$\mathscr{L}\{E(t)\} = \frac{1}{s}\left(\frac{1}{s} - \frac{1}{e^s - 1}\right) = \frac{1}{s^2} - \frac{1}{s}\frac{1}{e^s - 1}.$$

Thus

$$(Ls + R)\mathscr{L}\{i\} = \frac{1}{s^2} - \frac{1}{s}\frac{1}{e^s - 1}$$

and

$$\mathscr{L}\{i\} = \frac{1}{L}\frac{1}{s^2(s + R/L)} - \frac{1}{L}\frac{1}{s(s + R/L)}\frac{1}{e^s - 1}$$

$$= \frac{1}{R}\left(\frac{1}{s^2} - \frac{L}{R}\frac{1}{s} + \frac{L}{R}\frac{1}{s + R/L}\right) - \frac{1}{R}\left(\frac{1}{s} - \frac{1}{s + R/L}\right)\left(e^{-s} + e^{-2s} + e^{-3s} + \cdots\right).$$

Therefore

$$i(t) = \frac{1}{R}\left(t - \frac{L}{R} + \frac{L}{R}e^{-Rt/L}\right) - \frac{1}{R}\left(1 - e^{-R(t-1)/L}\right)\mathcal{U}(t-1)$$

$$- \frac{1}{R}\left(1 - e^{-R(t-2)/L}\right)\mathcal{U}(t-2) - \frac{1}{R}\left(1 - e^{-R(t-3)/L}\right)\mathcal{U}(t-3) - \cdots$$

$$= \frac{1}{R}\left(t - \frac{L}{R} + \frac{L}{R}e^{-Rt/L}\right) - \frac{1}{R}\sum_{n=1}^{\infty}\left(1 - e^{-R(t-n)/L}\right)\mathcal{U}(t-n).$$

The graph of $i(t)$ with $L = 1$ and $R = 1$ is shown below.

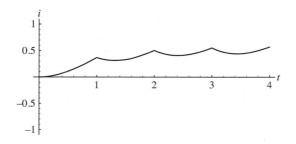

4.5   Dirac Delta Function

3. The Laplace transform of the differential equation yields

$$\mathscr{L}\{y\} = \frac{1}{s^2 + 1}\left(1 + e^{-2\pi s}\right)$$

so that

$$y = \sin t + \sin t\,\mathcal{U}(t - 2\pi).$$

6. The Laplace transform of the differential equation yields

$$\mathscr{L}\{y\} = \frac{s}{s^2 + 1} + \frac{1}{s^2 + 1}(e^{-2\pi s} + e^{-4\pi s})$$

so that

$$y = \cos t + \sin t[\mathcal{U}(t - 2\pi) + \mathcal{U}(t - 4\pi)].$$

9. The Laplace transform of the differential equation yields

$$\mathscr{L}\{y\} = \frac{1}{(s + 2)^2 + 1}e^{-2\pi s}$$

so that

$$y = e^{-2(t-2\pi)}\sin t\,\mathcal{U}(t - 2\pi).$$

12. The Laplace transform of the differential equation yields

$$\mathscr{L}\{y\} = \frac{1}{(s-1)^2(s-6)} + \frac{e^{-2s} + e^{-4s}}{(s-1)(s-6)}$$

$$= -\frac{1}{25}\frac{1}{s-1} - \frac{1}{5}\frac{1}{(s-1)^2} + \frac{1}{25}\frac{1}{s-6} + \left[-\frac{1}{5}\frac{1}{s-1} + \frac{1}{5}\frac{1}{s-6}\right](e^{-2s} + e^{-4s})$$

so that

$$y = -\frac{1}{25}e^t - \frac{1}{5}te^t + \frac{1}{25}e^{6t} + \left[-\frac{1}{5}e^{t-2} + \frac{1}{5}e^{6(t-2)}\right]\mathscr{U}(t-2) + \left[-\frac{1}{5}e^{t-4} + \frac{1}{5}e^{6(t-4)}\right]\mathscr{U}(t-4).$$

15. The Laplace transform of the differential equation yields

$$\mathscr{L}\{y''\} + \mathscr{L}\{y\} = \sum_{k=1}^{\infty}\mathscr{L}\{\delta(t-k\pi)\}$$

$$(s^2+1)\,Y(s) = 1 + \sum k = 1^{\infty}e^{-k\pi s}$$

$$Y(s) = \frac{1}{s^2+1} + \sum_{k=1}^{\infty}\frac{e^{-k\pi s}}{s^2+1}$$

so that

$$y(t) = \sin t + \sum_{k=1}^{\infty}\sin(t-k\pi)\mathscr{U}(t-k\pi)$$

$$= \sin t + \sin t\sum_{k=1}^{\infty}(-1)^k\,\mathscr{U}(t-k\pi) \qquad \longleftarrow \qquad \sin(t-k\pi) = (-1)^k\sin t$$

$$= \sin t - \sin t\,\mathscr{U}(t-\pi) + \sin t\,\mathscr{U}(t-2\pi) - \sin t\,\mathscr{U}(t-3\pi) + \sin t\,\mathscr{U}(t-4\pi) - \cdots$$

$$y = \begin{cases} \sin t, & 0 \le t < \pi \\ 0, & \pi \le t < 2\pi \\ \sin t, & 2\pi \le t < 3\pi \\ 0, & 3\pi \le t < 4\pi \\ \sin t, & 4\pi \le t < 5\pi \\ 0, & 5\pi \le t < 6\pi \\ \sin t, & 6\pi \le t < 7\pi \\ 0, & 7\pi \le t < 8\pi \\ \vdots \end{cases}$$

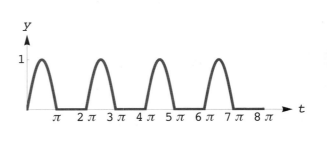

The graph of $y(t)$ given on the right is the half-wave rectification of $\sin t$. Also, see Figure 4.4.11 in Exercises 4.4.

18. From Problem 17 we have $y(x) = \dfrac{1}{2}\,y''(0)x^2 + \dfrac{1}{6}\,y'''(0)x^3 + \dfrac{w_0}{6EI}\left(x - \dfrac{L}{2}\right)^3\mathscr{U}\left(x - \dfrac{L}{2}\right)$. Using the conditions $y(L) = y'(L) = 0$ to find $y''(0)$ and $y'''(0)$ we get the system of equations

$$\begin{cases} (24L^2)y''(0) + (8L^3)y'''(0) = -\dfrac{w_0 L^3}{EI} \\[2mm] (8L)y''(0) + (4L^2)y'''(0) = -\dfrac{w_0 L^2}{EI} \end{cases}$$

Solving the system leads to the solution

$$y(x) = \frac{1}{2}\left(\frac{w_0 L}{8EI}\right)x^2 + \frac{1}{6}\left(-\frac{w_0}{2EI}\right)x^3 + \frac{w_0}{6EI}\left(x - \frac{L}{2}\right)^3\mathscr{U}\left(x - \frac{L}{2}\right)$$

We could also write the solution as

$$y(x) = \begin{cases} \left(\dfrac{w_0 L}{16EI}\right)x^2 - \left(\dfrac{w_0}{12EI}\right)x^3, & 0 \le x < \dfrac{L}{2} \\[4mm] \left(\dfrac{w_0 L}{16EI}\right)x^2 - \left(\dfrac{w_0}{12EI}\right)x^3 + \dfrac{w_0}{6EI}\left(x - \dfrac{L}{2}\right)^3, & \dfrac{L}{2} \le x \le L \end{cases}$$

4.6   Systems of Linear Differential Equations

3. Taking the Laplace transform of the system gives

$$s\mathscr{L}\{x\} + 1 = \mathscr{L}\{x\} - 2\mathscr{L}\{y\}$$
$$s\mathscr{L}\{y\} - 2 = 5\mathscr{L}\{x\} - \mathscr{L}\{y\}$$

so that

$$\mathscr{L}\{x\} = \frac{-s - 5}{s^2 + 9} = -\frac{s}{s^2 + 9} - \frac{5}{3}\frac{3}{s^2 + 9}$$

and

$$x = -\cos 3t - \frac{5}{3}\sin 3t.$$

Then

$$y = \frac{1}{2}x - \frac{1}{2}x' = 2\cos 3t - \frac{7}{3}\sin 3t.$$

6. Taking the Laplace transform of the system gives

$$(s + 1)\mathscr{L}\{x\} - (s - 1)\mathscr{L}\{y\} = -1$$
$$s\mathscr{L}\{x\} + (s + 2)\mathscr{L}\{y\} = 1$$

so that

$$\mathscr{L}\{y\} = \frac{s + 1/2}{s^2 + s + 1} = \frac{s + 1/2}{(s + 1/2)^2 + (\sqrt{3}/2)^2}$$

and

$$\mathscr{L}\{x\} = \frac{-3/2}{s^2 + s + 1} = -\sqrt{3}\,\frac{\sqrt{3}/2}{(s + 1/2)^2 + (\sqrt{3}/2)^2}\,.$$

Then

$$y = e^{-t/2}\cos\frac{\sqrt{3}}{2}t \qquad\text{and}\qquad x = -\sqrt{3}\,e^{-t/2}\sin\frac{\sqrt{3}}{2}t.$$

9. Adding the equations and then subtracting them gives

$$\frac{d^2x}{dt^2} = \frac{1}{2}t^2 + 2t$$

$$\frac{d^2y}{dt^2} = \frac{1}{2}t^2 - 2t.$$

Taking the Laplace transform of the system gives

$$\mathscr{L}\{x\} = 8\frac{1}{s} + \frac{1}{24}\frac{4!}{s^5} + \frac{1}{3}\frac{3!}{s^4}$$

and

$$\mathscr{L}\{y\} = \frac{1}{24}\frac{4!}{s^5} - \frac{1}{3}\frac{3!}{s^4}$$

so that

$$x = 8 + \frac{1}{24}t^4 + \frac{1}{3}t^3 \qquad\text{and}\qquad y = \frac{1}{24}t^4 - \frac{1}{3}t^3.$$

12. Taking the Laplace transform of the system gives

$$(s - 4)\mathscr{L}\{x\} + 2\mathscr{L}\{y\} = \frac{2e^{-s}}{s}$$

$$-3\mathscr{L}\{x\} + (s + 1)\mathscr{L}\{y\} = \frac{1}{2} + \frac{e^{-s}}{s}$$

so that

$$\mathscr{L}\{x\} = \frac{-1/2}{(s - 1)(s - 2)} + e^{-s}\frac{1}{(s - 1)(s - 2)}$$

$$= \frac{1}{2}\frac{1}{s - 1} - \frac{1}{2}\frac{1}{s - 2} + e^{-s}\left[-\frac{1}{s - 1} + \frac{1}{s - 2}\right]$$

and

$$\mathscr{L}\{y\} = \frac{e^{-s}}{s} + \frac{s/4 - 1}{(s - 1)(s - 2)} + e^{-s}\frac{-s/2 + 2}{(s - 1)(s - 2)}$$

$$= \frac{3}{4}\frac{1}{s - 1} - \frac{1}{2}\frac{1}{s - 2} + e^{-s}\left[\frac{1}{s} - \frac{3}{2}\frac{1}{s - 1} + \frac{1}{s - 2}\right].$$

Then

$$x = \frac{1}{2}e^t - \frac{1}{2}e^{2t} + \left[-e^{t-1} + e^{2(t-1)}\right]\mathscr{U}(t - 1)$$

and

$$y = \frac{3}{4}e^t - \frac{1}{2}e^{2t} + \left[1 - \frac{3}{2}e^{t-1} + e^{2(t-1)}\right]\mathscr{U}(t - 1).$$

15. (a) By Kirchhoff's first law we have $i_1 = i_2 + i_3$. By Kirchhoff's second law, on each loop we have $E(t) = Ri_1 + L_1 i_2'$ and $E(t) = Ri_1 + L_2 i_3'$ or $L_1 i_2' + Ri_2 + Ri_3 = E(t)$ and $L_2 i_3' + Ri_2 + Ri_3 = E(t)$.

(b) Taking the Laplace transform of the system

$$0.01 i_2' + 5i_2 + 5i_3 = 100$$
$$0.0125 i_3' + 5i_2 + 5i_3 = 100$$

gives

$$(s + 500)\mathscr{L}\{i_2\} + 500\mathscr{L}\{i_3\} = \frac{10{,}000}{s}$$

$$400\mathscr{L}\{i_2\} + (s + 400)\mathscr{L}\{i_3\} = \frac{8{,}000}{s}$$

so that

$$\mathscr{L}\{i_3\} = \frac{8{,}000}{s^2 + 900s} = \frac{80}{9}\frac{1}{s} - \frac{80}{9}\frac{1}{s + 900}.$$

Then

$$i_3 = \frac{80}{9} - \frac{80}{9}e^{-900t} \qquad \text{and} \qquad i_2 = 20 - 0.0025 i_3' - i_3 = \frac{100}{9} - \frac{100}{9}e^{-900t}.$$

(c) $i_1 = i_2 + i_3 = 20 - 20e^{-900t}$

18. Taking the Laplace transform of the system

$$0.5 i_1' + 50 i_2 = 60$$
$$0.005 i_2' + i_2 - i_1 = 0$$

gives

$$s\mathscr{L}\{i_1\} + 100\mathscr{L}\{i_2\} = \frac{120}{s}$$

$$-200\mathscr{L}\{i_1\} + (s + 200)\mathscr{L}\{i_2\} = 0$$

so that

$$\mathscr{L}\{i_2\} = \frac{24{,}000}{s(s^2 + 200s + 20{,}000)} = \frac{6}{5}\frac{1}{s} - \frac{6}{5}\frac{s + 100}{(s + 100)^2 + 100^2} - \frac{6}{5}\frac{100}{(s + 100)^2 + 100^2}.$$

Then

$$i_2 = \frac{6}{5} - \frac{6}{5}e^{-100t}\cos 100t - \frac{6}{5}e^{-100t}\sin 100t$$

and

$$i_1 = 0.005 i_2' + i_2 = \frac{6}{5} - \frac{6}{5}e^{-100t}\cos 100t.$$

21. (a) Solve the system by using the Laplace transform, subject to the conditions $x(0) = 0$, $x'(0) = v_0 \cos t$, $y(0) = 0$, and $y'(0) = v_0 \sin \theta$ to get

$$\begin{cases} x''(t) = 0 \\ y''(t) = -g \end{cases} \quad \text{or} \quad \begin{cases} s^2 X(s) = v_0 \cos \theta \\ s^2 Y(s) = v_0 \sin \theta - \dfrac{g}{s} \end{cases}$$

Then
$$\begin{cases} X(s) = (v_0 \cos \theta) \dfrac{1}{s^2} \\ Y(s) = (v_0 \sin \theta) \dfrac{1}{s^2} - \dfrac{g}{s^3} \end{cases} \quad \text{or} \quad \begin{cases} x(t) = (v_0 \cos \theta) t \\ y(t) = (v_0 \sin \theta) t - \dfrac{1}{2} g t^2 \end{cases}$$

(b) Substituting $t = x/(v_0 \cos \theta)$ into the equation for $y(t)$ yields

$$y(x) = -\frac{1}{2} g \left(\frac{x}{v_0 \cos \theta} \right)^2 + (v_0 \sin \theta) \left(\frac{x}{v_0 \cos \theta} \right) = -\frac{1}{2} \left(\frac{g}{v_0^2 \cos^2 \theta} \right) x^2 + (\tan \theta)$$

(c) Solve $y(x) = 0$ to get the horizontal range R

$$x \left(\frac{-g}{2 v_0^2 \cos^2 \theta} x + \tan \theta \right) = 0$$

$$x = \frac{(\tan \theta) \, 2 v_0^2 \cos^2 \theta}{g} = \frac{v_0^2 \sin 2\theta}{g}$$

Note that this result will be the same for the angle $\pi/2 - \theta$

$$\frac{v_0^2 \sin 2 \left(\frac{\pi}{2} - \theta \right)}{g} = \frac{v_0^2 \sin (\pi - 2\theta)}{g} = \frac{v_0^2 (\sin \pi \cos 2\theta - \cos \pi \sin 2\theta)}{g} = \frac{v_0^2 \sin 2\theta}{g}$$

(d) The maximum height occurs at the vertex of

$$y(x) = -\frac{g}{2 v_0^2 \cos^2 \theta} x^2 + (\tan \theta) x \quad \text{where} \quad x = \frac{v_0^2 \sin 2\theta}{2g}$$

so

$$y \left(\frac{v_0^2 \sin 2\theta}{2g} \right) = -\frac{g}{2 v_0^2 \cos^2 \theta} \left(\frac{v_0^2 \sin 2\theta}{2g} \right)^2 + (\tan \theta) \left(\frac{v_0^2 \sin 2\theta}{2g} \right) = \frac{v_0^2}{2g} \sin^2 \theta$$

(e) Under the given parameter values we get the following

$$\text{For } \theta = 38° \begin{cases} R = \dfrac{(300 \text{ ft/sec})^2}{32 \text{ ft/sec}^2} \cdot \sin 2\,(38°) = 2728.96 \text{ ft} \\[4mm] H = \dfrac{(300 \text{ ft/sec})^2}{2 \left(32 \text{ ft/sec}^2 \right)} \cdot (\sin 38°)^2 = 533.02 \text{ ft} \end{cases}$$

$$\text{For } \theta = 52° \begin{cases} R = \dfrac{(300 \text{ ft/sec})^2}{32 \text{ ft/sec}^2} \cdot \sin 2\,(52°) = 2728.96 \text{ ft} \\[4mm] H = \dfrac{(300 \text{ ft/sec})^2}{2 \left(52 \text{ ft/sec}^2 \right)} \cdot (\sin 52°)^2 = 873.23 \text{ ft} \end{cases}$$

(f) When the projectile hits the ground, $y(t) = -\frac{1}{2}gt^2 + (v_0 \sin\theta)\,t = 0$. The time the projectile hits the ground is found by solving this equation to get $t = (2v_0 \sin\theta)\,/g$. For $\theta = 38°$, this time is $t = (2 \cdot 300 \cdot \sin 38°)/32 = 11.5437$ s, therefore the range is $x(t) = (v_0 \cos\theta)t = (300 \cdot \cos 38°)(11.5437) = 2728.97$ ft. To find the maximum height, solve the equation $y'(t) = -gt + v_0 \sin\theta = 0$ to get $t = (v_0 \sin\theta)\,/g$. Now for $\theta = 38°$ we get $t = (300 \cdot \sin 38°)/32 = 5.7718$ s, therefore the maximum height is

$$y(5.7718) = (300 \sin 38°)(5.7718) - \frac{1}{2}(32)(5.7718)^2 = 533.02 \text{ ft}$$

Repeat this process for $\theta = 52°$ and find that the time to hit the ground is $t = 14.7752$ s, therefore the range is $x(14.7752) = 2728.96$ ft. The maximum height occurs when $t = 7.3876$ s and so the maximum height is $y(7.3876) = 873.23$ ft.

(g) The two trajectories or ballistic curves for the two launch angles (in radians) are given by the parametric equations

$$C_1 : \begin{cases} x(t) = \left(300 \cos \dfrac{38\pi}{180}\right) t \\[2mm] y(t) = \left(300 \sin \dfrac{38\pi}{180}\right) t - 16t^2 \end{cases} \quad \text{and} \quad C_2 : \begin{cases} x(t) = \left(300 \cos \dfrac{52\pi}{180}\right) t \\[2mm] y(t) = \left(300 \sin \dfrac{52\pi}{180}\right) t - 16t^2 \end{cases}$$

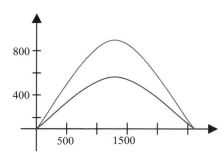

Chapter 4 in Review

3. False; consider $f(t) = t^{-1/2}$.

6. False; consider $f(t) = 1$ and $g(t) = 1$.

9. $\mathscr{L}\{\sin 2t\} = \dfrac{2}{s^2 + 4}$

12. $\mathscr{L}\{\sin 2t\,\mathscr{U}(t - \pi)\} = \mathscr{L}\{\sin 2(t - \pi)\mathscr{U}(t - \pi)\} = \dfrac{2}{s^2 + 4}e^{-\pi s}$

15. $\mathscr{L}\left\{\displaystyle\int_0^t e^\tau \sin(t - \tau)\,d\tau\right\} = \mathscr{L}\{e^t\}\mathscr{L}\{\sin t\} = \dfrac{1}{s - 1} \cdot \dfrac{1}{s^2 + 1} = \dfrac{1}{(s - 1)(s^2 + 1)}$

18. $\mathscr{L}\{1 - \delta(t - 5)\} = \mathscr{L}\{1\} - \mathscr{L}\{\delta(t - 5)\} = \dfrac{1}{s} - e^{-5s}$

21. $\mathscr{L}^{-1}\left\{\dfrac{1}{(s-5)^3}\right\} = \dfrac{1}{2}\mathscr{L}^{-1}\left\{\dfrac{2}{(s-5)^3}\right\} = \dfrac{1}{2}t^2 e^{5t}$

24. $\mathscr{L}^{-1}\left\{\dfrac{1}{s^2+s-2}\right\} = \dfrac{1}{3}\mathscr{L}^{-1}\left\{\dfrac{1}{s-1}-\dfrac{1}{s+2}\right\} = \dfrac{1}{3}\left(e^t - e^{-2t}\right)$

27. $\mathscr{L}^{-1}\left\{\dfrac{4s-86}{s^2+2s+101}\right\} = \mathscr{L}^{-1}\left\{\dfrac{4(s+1)-90}{(s+1)^2+10^2}\right\}$

$$= \mathscr{L}^{-1}\left\{4\,\dfrac{s+1}{(s+1)^2+10^2} - 9\,\dfrac{10}{(s+1)^2+10^2}\right\}$$

$$= 4e^{-t}\cos 10t - 9e^{-t}\sin 10t$$

30. $\mathscr{L}^{-1}\left\{\dfrac{1}{L^2 s^2 + n^2\pi^2}\right\} = \dfrac{1}{L^2}\,\dfrac{L}{n\pi}\,\mathscr{L}^{-1}\left\{\dfrac{n\pi/L}{s^2+(n^2\pi^2)/L^2}\right\} = \dfrac{1}{Ln\pi}\sin\dfrac{n\pi}{L}t$

33. $\mathscr{L}\{e^{at}f(t-k)\,\mathscr{U}(t-k)\} = e^{-ks}\mathscr{L}\{e^{a(t+k)}f(t)\} = e^{-ks}e^{ak}\mathscr{L}\{e^{at}f(t)\} = e^{-k(s-a)}F(s-a)$

36. $t^2 * t = \displaystyle\int_0^t \tau^2(1-\tau)\,d\tau = t\int_0^t \tau^2\,d\tau - \int_0^t \tau^3\,d\tau = t\,\dfrac{1}{3}\tau^3\Big|_0^t - \dfrac{1}{4}\tau^4\Big|_0^t = \dfrac{1}{3}t^4 - \dfrac{1}{4}t^4 = \dfrac{1}{12}t^4$

39. $f(t)\mathscr{U}(t-t_0)$

42. $f(t) - f(t)\mathscr{U}(t-t_0) + f(t)\mathscr{U}(t-t_1)$

45. $f(t) = 2 - 2\mathscr{U}(t-2) + [(t-2)+2]\mathscr{U}(t-2) = 2 + (t-2)\mathscr{U}(t-2)$

$\mathscr{L}\{f(t)\} = \dfrac{2}{s} + \dfrac{1}{s^2}e^{-2s}$

$\mathscr{L}\left\{e^t f(t)\right\} = \dfrac{2}{s-1} + \dfrac{1}{(s-1)^2}e^{-2(s-1)}$

48. The graph of

$$f(t) = \sum_{k=0}^{\infty}(2k+1-t)\left[\mathscr{U}(t-2k) - \mathscr{U}(t-2k-1)\right]$$

$$= (1-t)\left[\mathscr{U}(t) - \mathscr{U}(t-1)\right] + (3-t)\left[\mathscr{U}(t-2) - \mathscr{U}(t-3)\right] + \cdots$$

is

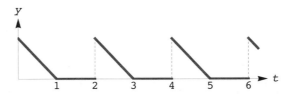

Since f is a periodic function it can also be defined by

$$f(t) = \begin{cases} 1-t, & 0 \le t < 1 \\ 0, & 1 \le t < 2 \end{cases} \quad \text{where } f(t+2) = f(t).$$

By Theorem 4.4.3 with $p = 2$ we get

$$\mathscr{L}\{f(t)\} = \frac{1}{1-e^{-2s}}\left(\int_0^1 (1-t)\,e^{-st}\,dt + \int_1^2 0\cdot e^{-st}\,dt\right) = \frac{1}{1-e^{-2s}}\left(\frac{1}{s} - \frac{1}{s^2}e^{-s} + \frac{1}{s^2}\right)$$

$$= \frac{s+1-e^{-s}}{s^2\left(1-e^{-2s}\right)}.$$

51. Taking the Laplace transform of the given differential equation we obtain

$$\mathscr{L}\{y\} = \frac{s^3+6s^2+1}{s^2(s+1)(s+5)} - \frac{1}{s^2(s+1)(s+5)}e^{-2s} - \frac{2}{s(s+1)(s+5)}e^{-2s}$$

$$= -\frac{6}{25}\cdot\frac{1}{s} + \frac{1}{5}\cdot\frac{1}{s^2} + \frac{3}{2}\cdot\frac{1}{s+1} - \frac{13}{50}\cdot\frac{1}{s+5}$$

$$- \left(-\frac{6}{25}\cdot\frac{1}{s} + \frac{1}{5}\cdot\frac{1}{s^2} + \frac{1}{4}\cdot\frac{1}{s+1} - \frac{1}{100}\cdot\frac{1}{s+5}\right)e^{-2s}$$

$$- \left(\frac{2}{5}\cdot\frac{1}{s} - \frac{1}{2}\cdot\frac{1}{s+1} + \frac{1}{10}\cdot\frac{1}{s+5}\right)e^{-2s}$$

so that

$$y = -\frac{6}{25} + \frac{1}{5}t + \frac{3}{2}e^{-t} - \frac{13}{50}e^{-5t} - \frac{4}{25}\mathscr{U}(t-2) - \frac{1}{5}(t-2)\mathscr{U}(t-2)$$

$$+ \frac{1}{4}e^{-(t-2)}\mathscr{U}(t-2) - \frac{9}{100}e^{-5(t-2)}\mathscr{U}(t-2).$$

54. Taking the Laplace transform of the differential equation we obtain

$$\mathscr{L}\{y\} = \frac{3}{s^2+5s+4} + \sum_{k=0}^{\infty}(-1)^k\,\frac{12}{s\left(s^2+5s+4\right)}e^{-ns}$$

$$= \left(\frac{1}{s+1} - \frac{1}{s+4}\right) + \sum_{k=0}^{\infty}(-1)^k\left(2\cdot\frac{1}{s} - 4\cdot\frac{1}{s+1} + \frac{1}{s+4}\right)e^{-ns}$$

so that

$$y = e^{-t} - e^{-4t} + \sum_{k=0}^{\infty}(-1)^k\left(3 - 4e^{-(t-n)} + e^{-4(t-n)}\right)\mathscr{U}(t-n).$$

57. Taking the Laplace transform of the integral equation we obtain

$$\mathscr{L}\{y\} = \frac{s^2+1}{s^2(s-2)} = -\frac{1}{4}\cdot\frac{1}{s} - \frac{1}{2}\cdot\frac{1}{s^2} + \frac{5}{4}\cdot\frac{1}{s-2}$$

so that

$$y = -\frac{1}{4} - \frac{1}{2}t + \frac{5}{4}e^{2t}.$$

60. Taking the Laplace transform of the system gives

$$s^2 \mathcal{L}\{x\} + s^2 \mathcal{L}\{y\} = \frac{1}{s-2}$$

$$2s \mathcal{L}\{x\} + s^2 \mathcal{L}\{y\} = -\frac{1}{s-2}$$

so that

$$\mathcal{L}\{x\} = \frac{2}{s(s-2)^2} = \frac{1}{2}\frac{1}{s} - \frac{1}{2}\frac{1}{s-2} + \frac{1}{(s-2)^2}$$

and

$$\mathcal{L}\{y\} = \frac{-s-2}{s^2(s-2)^2} = -\frac{3}{4}\frac{1}{s} - \frac{1}{2}\frac{1}{s^2} + \frac{3}{4}\frac{1}{s-2} - \frac{1}{(s-2)^2}.$$

Then

$$x = \frac{1}{2} - \frac{1}{2}e^{2t} + te^{2t} \quad \text{and} \quad y = -\frac{3}{4} - \frac{1}{2}t + \frac{3}{4}e^{2t} - te^{2t}.$$

63. Taking the Laplace transform of the given differential equation we obtain

$$\mathcal{L}\{y\} = \frac{2w_0}{EIL}\left(\frac{L}{48}\cdot\frac{4!}{s^5} - \frac{1}{120}\cdot\frac{5!}{s^6} + \frac{1}{120}\cdot\frac{5!}{s^6}e^{-sL/2}\right) + \frac{c_1}{2}\cdot\frac{2!}{s^3} + \frac{c_2}{6}\cdot\frac{3!}{s^4}$$

so that

$$y = \frac{2w_0}{EIL}\left[\frac{L}{48}x^4 - \frac{1}{120}x^5 + \frac{1}{120}\left(x - \frac{L}{2}\right)^5 \mathscr{U}\left(x - \frac{L}{2}\right) + \frac{c_1}{2}x^2 + \frac{c_2}{6}x^3\right]$$

where $y''(0) = c_1$ and $y'''(0) = c_2$. Using $y''(L) = 0$ and $y'''(L) = 0$ we find

$$c_1 = w_0 L^2/24EI, \qquad c_2 = -w_0 L/4EI.$$

Hence

$$y = \frac{w_0}{12EIL}\left[-\frac{1}{5}x^5 + \frac{L}{2}x^4 - \frac{L^2}{2}x^3 + \frac{L^3}{4}x^2 + \frac{1}{5}\left(x - \frac{L}{2}\right)^5 \mathscr{U}\left(x - \frac{L}{2}\right)\right].$$

Chapter 5

Series Solutions of Linear Equations

5.1 | Solutions about Ordinary Points |

3. By the ratio test,

$$\lim_{n \to \infty} \left| \frac{(x-5)^{n+1}/10^{n+1}}{(x-5)^n/10^n} \right| = \lim_{n \to \infty} \frac{1}{10}|x-5| = \frac{1}{10}|x-5|.$$

The series is absolutely convergent for $\frac{1}{10}|x-5| < 1$, $|x-5| < 10$, or on $(-5, 15)$. The radius of convergence is $R = 10$. At $x = -5$, the series $\sum_{n=1}^{\infty}(-1)^n(-10)^n/10^n = \sum_{n=1}^{\infty}1$ diverges by the nth term test. At $x = 15$, the series $\sum_{n=1}^{\infty}(-1)^n10^n/10^n = \sum_{n=1}^{\infty}(-1)^n$ diverges by the nth term test. Thus the series converges on $(-5, 15)$.

6. $e^{-x}\cos x = \left(1 - x + \dfrac{x^2}{2} - \dfrac{x^3}{6} + \dfrac{x^4}{24} - \cdots\right)\left(1 - \dfrac{x^2}{2} + \dfrac{x^4}{24} - \cdots\right) = 1 - x + \dfrac{x^3}{3} - \dfrac{x^4}{6} + \cdots$

9. Let $k = n + 2$ so that $n = k - 2$ and

$$\sum_{n=1}^{\infty} nc_n x^{n+2} = \sum_{k=3}^{\infty}(k-2)c_{k-2}x^k.$$

12. $\displaystyle\sum_{n=2}^{\infty} n(n-1)c_n x^n + 2\sum_{n=2}^{\infty} n(n-1)c_n x^{n-2} + 3\sum_{n=1}^{\infty} nc_n x^n$

$$= 2 \cdot 2 \cdot 1c_2 x^0 + 2 \cdot 3 \cdot 2c_3 x^1 + 3 \cdot 1 \cdot c_1 x^1$$

$$+ \underbrace{\sum_{n=2}^{\infty} n(n-1)c_n x^n}_{k=n} + 2\underbrace{\sum_{n=4}^{\infty} n(n-1)c_n x^{n-2}}_{k=n-2} + 3\underbrace{\sum_{n=2}^{\infty} nc_n x^n}_{k=n}$$

$$= 4c_2 + (3c_1 + 12c_3)x + \sum_{k=2}^{\infty} k(k-1)c_k x^k + 2\sum_{k=2}^{\infty}(k+2)(k+1)c_{k+2}x^k + 3\sum_{k=2}^{\infty} kc_k x^k$$

$$= 4c_2 + (3c_1 + 12c_3)x + \sum_{k=2}^{\infty}\left[(k(k-1) + 3k)c_k + 2(k+2)(k+1)c_{k+2}\right]x^k$$

$$= 4c_2 + (3c_1 + 12c_3)x + \sum_{k=2}^{\infty}\left[k(k+2)c_k + 2(k+1)(k+2)c_{k+2}\right]x^k$$

15. $y' = \sum_{n=1}^{\infty} (-1)^{n+1} x^{n-1}, \qquad y'' = \sum_{n=2}^{\infty} (-1)^{n+1}(n-1)x^{n-2}$

$$(x+1)y'' + y' = (x+1)\sum_{n=2}^{\infty}(-1)^{n+1}(n-1)x^{n-2} + \sum_{n=1}^{\infty}(-1)^{n+1}x^{n-1}$$

$$= \sum_{n=2}^{\infty}(-1)^{n+1}(n-1)x^{n-1} + \sum_{n=2}^{\infty}(-1)^{n+1}(n-1)x^{n-2} + \sum_{n=1}^{\infty}(-1)^{n+1}x^{n-1}$$

$$= -x^0 + x^0 + \underbrace{\sum_{n=2}^{\infty}(-1)^{n+1}(n-1)x^{n-1}}_{k=n-1} + \underbrace{\sum_{n=3}^{\infty}(-1)^{n+1}(n-1)x^{n-2}}_{k=n-2} + \underbrace{\sum_{n=2}^{\infty}(-1)^{n+1}x^{n-1}}_{k=n-1}$$

$$= \sum_{k=1}^{\infty}(-1)^{k+2}kx^k + \sum_{k=1}^{\infty}(-1)^{k+3}(k+1)x^k + \sum_{k=1}^{\infty}(-1)^{k+2}x^k$$

$$= \sum_{k=1}^{\infty}\left[(-1)^{k+2}k - (-1)^{k+2}k - (-1)^{k+2} + (-1)^{k+2}\right]x^k = 0$$

18. The singular points of $(x^2 - 2x + 10)y'' + xy' - 4y = 0$ are $1 + 3i$ and $1 - 3i$. The distance from 0 to either of these points is $\sqrt{10}$. The distance from 1 to either of these points is 3.

21. Substituting $y = \sum_{n=0}^{\infty} c_n x^n$ into the differential equation we have

$$y'' - 2xy' + y = \underbrace{\sum_{n=2}^{\infty} n(n-1)c_n x^{n-2}}_{k=n-2} - 2\underbrace{\sum_{n=1}^{\infty} nc_n x^n}_{k=n} + \underbrace{\sum_{n=0}^{\infty} c_n x^n}_{k=n}$$

$$= \sum_{k=0}^{\infty}(k+2)(k+1)c_{k+2}x^k - 2\sum_{k=1}^{\infty} kc_k x^k + \sum_{k=0}^{\infty} c_k x^k$$

$$= 2c_2 + c_0 + \sum_{k=1}^{\infty}[(k+2)(k+1)c_{k+2} - (2k-1)c_k]x^k = 0.$$

Thus

$$2c_2 + c_0 = 0$$

$$(k+2)(k+1)c_{k+2} - (2k-1)c_k = 0$$

and

$$c_2 = -\frac{1}{2}c_0$$

$$c_{k+2} = \frac{2k-1}{(k+2)(k+1)}c_k, \quad k = 1, 2, 3, \dots .$$

Choosing $c_0 = 1$ and $c_1 = 0$ we find

$$c_2 = -\frac{1}{2} \qquad c_3 = c_5 = c_7 = \cdots = 0 \qquad c_4 = -\frac{1}{8} \qquad c_6 = -\frac{7}{240}$$

and so on. For $c_0 = 0$ and $c_1 = 1$ we obtain

$$c_2 = c_4 = c_6 = \cdots = 0 \qquad c_3 = \frac{1}{6} \qquad c_5 = \frac{1}{24} \qquad c_7 = 1112$$

and so on. Thus two solutions are

$$y_1 = 1 - \frac{1}{2}x^2 - \frac{1}{8}x^4 - \frac{7}{240}x^6 - \cdots \qquad \text{and} \qquad y_2 = x + \frac{1}{6}x^3 + \frac{1}{24}x^5 + \frac{1}{112}x^7 + \cdots .$$

24. Substituting $y = \displaystyle\sum_{n=0}^{\infty} c_n x^n$ into the differential equation we have

$$y'' + 2xy' + 2y = \underbrace{\sum_{n=2}^{\infty} n(n-1)c_n x^{n-2}}_{k=n-2} + 2\underbrace{\sum_{n=1}^{\infty} nc_n x^n}_{k=n} + 2\underbrace{\sum_{n=0}^{\infty} c_n x^n}_{k=n}$$

$$= \sum_{k=0}^{\infty} (k+2)(k+1)c_{k+2}x^k + 2\sum_{k=1}^{\infty} kc_k x^k + 2\sum_{k=0}^{\infty} c_k x^k$$

$$= 2c_2 + 2c_0 + \sum_{k=1}^{\infty} [(k+2)(k+1)c_{k+2} + 2(k+1)c_k]x^k = 0.$$

Thus

$$2c_2 + 2c_0 = 0$$

$$(k+2)(k+1)c_{k+2} + 2(k+1)c_k = 0$$

and

$$c_2 = -c_0$$

$$c_{k+2} = -\frac{2}{k+2}\,c_k, \quad k = 1, 2, 3, \ldots .$$

Choosing $c_0 = 1$ and $c_1 = 0$ we find

$$c_2 = -1 \qquad c_3 = c_5 = c_7 = \cdots = 0 \qquad c_4 = \frac{1}{2} \qquad c_6 = -\frac{1}{6}$$

and so on. For $c_0 = 0$ and $c_1 = 1$ we obtain

$$c_2 = c_4 = c_6 = \cdots = 0 \qquad c_3 = -\frac{2}{3} \qquad c_5 = 415 \qquad c_7 = -8105$$

and so on. Thus two solutions are

$$y_1 = 1 - x^2 + \frac{1}{2}x^4 - \frac{1}{6}x^6 + \cdots \quad \text{and} \quad y_2 = x - \frac{2}{3}x^3 + \frac{4}{15}x^5 - \frac{8}{105}x^7 + \cdots .$$

27. Substituting $y = \displaystyle\sum_{n=0}^{\infty} c_n x^n$ into the differential equation we have

$$y'' - (x+1)y' - y = \underbrace{\sum_{n=2}^{\infty} n(n-1)c_n x^{n-2}}_{k=n-2} - \underbrace{\sum_{n=1}^{\infty} nc_n x^n}_{k=n} - \underbrace{\sum_{n=1}^{\infty} nc_n x^{n-1}}_{k=n-1} - \underbrace{\sum_{n=0}^{\infty} c_n x^n}_{k=n}$$

$$= \sum_{k=0}^{\infty} (k+2)(k+1)c_{k+2} x^k - \sum_{k=1}^{\infty} kc_k x^k - \sum_{k=0}^{\infty} (k+1)c_{k+1} x^k - \sum_{k=0}^{\infty} c_k x^k$$

$$= 2c_2 - c_1 - c_0 + \sum_{k=1}^{\infty} [(k+2)(k+1)c_{k+2} - (k+1)c_{k+1} - (k+1)c_k] x^k = 0.$$

Thus

$$2c_2 - c_1 - c_0 = 0$$

$$(k+2)(k+1)c_{k+2} - (k+1)(c_{k+1} + c_k) = 0$$

and

$$c_2 = \frac{c_1 + c_0}{2}$$

$$c_{k+2} = \frac{c_{k+1} + c_k}{k+2}, \quad k = 1, 2, 3, \ldots.$$

Choosing $c_0 = 1$ and $c_1 = 0$ we find

$$c_2 = \frac{1}{2}, \qquad c_3 = \frac{1}{6}, \qquad c_4 = \frac{1}{6},$$

and so on. For $c_0 = 0$ and $c_1 = 1$ we obtain

$$c_2 = \frac{1}{2}, \qquad c_3 = \frac{1}{2}, \qquad c_4 = \frac{1}{4},$$

and so on. Thus two solutions are

$$y_1 = 1 + \frac{1}{2}x^2 + \frac{1}{6}x^3 + \frac{1}{6}x^4 + \cdots \quad \text{and} \quad y_2 = x + \frac{1}{2}x^2 + \frac{1}{2}x^3 + \frac{1}{4}x^4 + \cdots.$$

30. Substituting $y = \displaystyle\sum_{n=0}^{\infty} c_n x^n$ into the differential equation we have

$$\left(x^2 - 1\right) y'' + xy' - y = \underbrace{\sum_{n=2}^{\infty} n(n-1)c_n x^n}_{k=n} - \underbrace{\sum_{n=2}^{\infty} n(n-1)c_n x^{n-2}}_{k=n-2} + \underbrace{\sum_{n=1}^{\infty} nc_n x^n}_{k=n} - \underbrace{\sum_{n=0}^{\infty} c_n x^n}_{k=n}$$

$$= \sum_{k=2}^{\infty} k(k-1)c_k x^k - \sum_{k=0}^{\infty} (k+2)(k+1)c_{k+2} x^k + \sum_{k=1}^{\infty} kc_k x^k - \sum_{k=0}^{\infty} c_k x^k$$

$$= (-2c_2 - c_0) - 6c_3 x + \sum_{k=2}^{\infty} \left[-(k+2)(k+1)c_{k+2} + \left(k^2 - 1\right)c_k\right] x^k = 0.$$

Thus

$$-2c_2 - c_0 = 0$$
$$-6c_3 = 0$$
$$-(k+2)(k+1)c_{k+2} + (k-1)(k+1)c_k = 0$$

and

$$c_2 = -\frac{1}{2}c_0 \qquad c_3 = 0 \qquad c_{k+2} = \frac{k-1}{k+2}c_k, \quad k = 2, 3, 4, \ldots .$$

Choosing $c_0 = 1$ and $c_1 = 0$ we find

$$c_2 = -\frac{1}{2} \qquad c_3 = c_5 = c_7 = \cdots = 0 \qquad c_4 = -\frac{1}{8}$$

and so on. For $c_0 = 0$ and $c_1 = 1$ we obtain

$$c_2 = c_4 = c_6 = \cdots = 0 \qquad c_3 = c_5 = c_7 = \cdots = 0.$$

Thus two solutions are

$$y_1 = 1 - \frac{1}{2}x^2 - \frac{1}{8}x^4 - \cdots \qquad \text{and} \qquad y_2 = x.$$

33. Substituting $y = \displaystyle\sum_{n=0}^{\infty} c_n x^n$ into the differential equation we have

$$y'' - 2xy' + 8y = \underbrace{\sum_{n=2}^{\infty} n(n-1)c_n x^{n-2}}_{k=n-2} - 2\underbrace{\sum_{n=1}^{\infty} nc_n x^n}_{k=n} + 8\underbrace{\sum_{n=0}^{\infty} c_n x^n}_{k=n}$$

$$= \sum_{k=0}^{\infty}(k+2)(k+1)c_{k+2}x^k - 2\sum_{k=1}^{\infty} kc_k x^k + 8\sum_{k=0}^{\infty} c_k x^k$$

$$= 2c_2 + 8c_0 + \sum_{k=1}^{\infty}[(k+2)(k+1)c_{k+2} + (8-2k)c_k]x^k = 0.$$

Thus

$$2c_2 + 8c_0 = 0$$
$$(k+2)(k+1)c_{k+2} + (8-2k)c_k = 0$$

and

$$c_2 = -4c_0 \qquad c_{k+2} = \frac{2(k-4)}{(k+2)(k+1)}c_k, \quad k = 1, 2, 3, \ldots .$$

Choosing $c_0 = 1$ and $c_1 = 0$ we find

$$c_2 = -4 \qquad c_3 = c_5 = c_7 = \cdots = 0 \qquad c_4 = \frac{4}{3} \qquad c_6 = c_8 = c_{10} = \cdots = 0.$$

For $c_0 = 0$ and $c_1 = 1$ we obtain

$$c_2 = c_4 = c_6 = \cdots = 0 \qquad\qquad c_3 = -1 \qquad\qquad c_5 = \frac{1}{10}$$

and so on. Thus

$$y = C_1 \left(1 - 4x^2 + \frac{4}{3}x^4 \right) + C_2 \left(x - x^3 + \frac{1}{10}x^5 + \cdots \right)$$

and

$$y' = C_1 \left(-8x + \frac{16}{3}x^3 \right) + C_2 \left(1 - 3x^2 + \frac{1}{2}x^4 + \cdots \right).$$

The initial conditions imply $C_1 = 3$ and $C_2 = 0$, so

$$y = 3 \left(1 - 4x^2 + \frac{4}{3}x^4 \right) = 3 - 12x^2 + 4x^4.$$

36. Substituting $y = \displaystyle\sum_{n=0}^{\infty} c_n x^n$ into the differential equation we have

$$y'' + e^x y' - y = \sum_{n=2}^{\infty} n(n-1)c_n x^{n-2}$$

$$+ \left(1 + x + \frac{1}{2}x^2 + \frac{1}{6}x^3 + \cdots \right) \left(c_1 + 2c_2 x + 3c_3 x^2 + 4c_4 x^3 + \cdots \right)$$

$$- \sum_{n=0}^{\infty} c_n x^n$$

$$= \left[2c_2 + 6c_3 x + 12c_4 x^2 + 20c_5 x^3 + \cdots \right]$$

$$+ \left[c_1 + (2c_2 + c_1)x + \left(3c_3 + 2c_2 + \frac{1}{2}c_1 \right) x^2 + \cdots \right]$$

$$- \left[c_0 + c_1 x + c_2 x^2 + \cdots \right]$$

$$= (2c_2 + c_1 - c_0) + (6c_3 + 2c_2)x + \left(12c_4 + 3c_3 + c_2 + \frac{1}{2}c_1 \right) x^2 + \cdots = 0.$$

Thus

$$2c_2 + c_1 - c_0 = 0$$

$$6c_3 + 2c_2 = 0$$

$$12c_4 + 3c_3 + c_2 + \frac{1}{2}c_1 = 0$$

and

$$c_2 = \frac{1}{2}c_0 - \frac{1}{2}c_1 \qquad c_3 = -\frac{1}{3}c_2 \qquad c_4 = -\frac{1}{4}c_3 + \frac{1}{12}c_2 - \frac{1}{24}c_1.$$

Choosing $c_0 = 1$ and $c_1 = 0$ we find

$$c_2 = \frac{1}{2}, \qquad c_3 = -\frac{1}{6}, \qquad c_4 = 0$$

and so on. For $c_0 = 0$ and $c_1 = 1$ we obtain

$$c_2 = -\frac{1}{2}, \qquad c_3 = \frac{1}{6}, \qquad c_4 = -\frac{1}{24}$$

and so on. Thus two solutions are

$$y_1 = 1 + \frac{1}{2}x^2 - \frac{1}{6}x^3 + \cdots \qquad \text{and} \qquad y_2 = x - \frac{1}{2}x^2 + \frac{1}{6}x^3 - \frac{1}{24}x^4 + \cdots .$$

5.2 | Solutions about Singular Points

3. Irregular singular point: $x = 3$; regular singular point: $x = -3$

6. Irregular singular point: $x = 5$; regular singular point: $x = 0$

9. Irregular singular point: $x = 0$; regular singular points: $x = 2, \pm 5$

12. Writing the differential equation in the form

$$y'' + \frac{x+3}{x}y' + 7xy = 0$$

we see that $x_0 = 0$ is a regular singular point. Multiplying by x^2, the differential equation can be put in the form

$$x^2 y'' + x(x+3)y' + 7x^3 y = 0.$$

We identify $p(x) = x + 3$ and $q(x) = 7x^3$.

15. Substituting $y = \sum_{n=0}^{\infty} c_n x^{n+r}$ into the differential equation and collecting terms, we obtain

$$2xy'' - y' + 2y = \left(2r^2 - 3r\right)c_0 x^{r-1} + \sum_{k=1}^{\infty} \left[2(k+r-1)(k+r)c_k - (k+r)c_k + 2c_{k-1}\right]x^{k+r-1}$$

$$= 0,$$

which implies

$$2r^2 - 3r = r(2r - 3) = 0$$

and

$$(k+r)(2k+2r-3)c_k + 2c_{k-1} = 0.$$

The indicial roots are $r = 0$ and $r = 3/2$. For $r = 0$ the recurrence relation is

$$c_k = -\frac{2c_{k-1}}{k(2k-3)}, \quad k = 1, 2, 3, \ldots,$$

and

$$c_1 = 2c_0, \qquad c_2 = -2c_0, \qquad c_3 = \frac{4}{9}c_0,$$

and so on. For $r = 3/2$ the recurrence relation is

$$c_k = -\frac{2c_{k-1}}{(2k+3)k}, \quad k = 1, 2, 3, \ldots,$$

and

$$c_1 = -\frac{2}{5}c_0, \qquad c_2 = \frac{2}{35}c_0, \qquad c_3 = -\frac{4}{945}c_0,$$

and so on. The general solution on $(0, \infty)$ is

$$y = C_1\left(1 + 2x - 2x^2 + \frac{4}{9}x^3 + \cdots\right) + C_2 x^{3/2}\left(1 - \frac{2}{5}x + \frac{2}{35}x^2 - \frac{4}{945}x^3 + \cdots\right).$$

18. Substituting $y = \displaystyle\sum_{n=0}^{\infty} c_n x^{n+r}$ into the differential equation and collecting terms, we obtain

$$2x^2 y'' - xy' + \left(x^2 + 1\right)y = \left(2r^2 - 3r + 1\right)c_0 x^r + \left(2r^2 + r\right)c_1 x^{r+1}$$

$$+ \sum_{k=2}^{\infty}[2(k+r)(k+r-1)c_k - (k+r)c_k + c_k + c_{k-2}]x^{k+r}$$

$$= 0,$$

which implies

$$2r^2 - 3r + 1 = (2r - 1)(r - 1) = 0,$$
$$\left(2r^2 + r\right)c_1 = 0,$$

and

$$[(k+r)(2k + 2r - 3) + 1]c_k + c_{k-2} = 0.$$

The indicial roots are $r = 1/2$ and $r = 1$, so $c_1 = 0$. For $r = 1/2$ the recurrence relation is

$$c_k = -\frac{c_{k-2}}{k(2k-1)}, \quad k = 2, 3, 4, \ldots,$$

and

$$c_2 = -\frac{1}{6}c_0, \qquad c_3 = 0, \qquad c_4 = \frac{1}{168}c_0,$$

and so on. For $r = 1$ the recurrence relation is

$$c_k = -\frac{c_{k-2}}{k(2k+1)}, \quad k = 2, 3, 4, \ldots,$$

and

$$c_2 = -\frac{1}{10}c_0, \qquad c_3 = 0, \qquad c_4 = \frac{1}{360}c_0,$$

and so on. The general solution on $(0, \infty)$ is

$$y = C_1 x^{1/2}\left(1 - \frac{1}{6}x^2 + \frac{1}{168}x^4 + \cdots\right) + C_2 x\left(1 - \frac{1}{10}x^2 + \frac{1}{360}x^4 + \cdots\right).$$

21. Substituting $y = \sum_{n=0}^{\infty} c_n x^{n+r}$ into the differential equation and collecting terms, we obtain

$$2xy'' - (3 + 2x)y' + y = \left(2r^2 - 5r\right)c_0 x^{r-1} + \sum_{k=1}^{\infty}[2(k+r)(k+r-1)c_k$$
$$- 3(k+r)c_k - 2(k+r-1)c_{k-1} + c_{k-1}]x^{k+r-1}$$
$$= 0,$$

which implies

$$2r^2 - 5r = r(2r - 5) = 0$$

and

$$(k+r)(2k+2r-5)c_k - (2k+2r-3)c_{k-1} = 0.$$

The indicial roots are $r = 0$ and $r = 5/2$. For $r = 0$ the recurrence relation is

$$c_k = \frac{(2k-3)c_{k-1}}{k(2k-5)}, \quad k = 1, 2, 3, \ldots,$$

and

$$c_1 = \frac{1}{3}c_0, \qquad c_2 = -\frac{1}{6}c_0, \qquad c_3 = -\frac{1}{6}c_0,$$

and so on. For $r = 5/2$ the recurrence relation is

$$c_k = \frac{2(k+1)c_{k-1}}{k(2k+5)}, \quad k = 1, 2, 3, \ldots,$$

and

$$c_1 = \frac{4}{7}c_0, \qquad c_2 = \frac{4}{21}c_0, \qquad c_3 = \frac{32}{693}c_0,$$

and so on. The general solution on $(0, \infty)$ is

$$y = C_1\left(1 + \frac{1}{3}x - \frac{1}{6}x^2 - \frac{1}{6}x^3 + \cdots\right) + C_2 x^{5/2}\left(1 + \frac{4}{7}x + \frac{4}{21}x^2 + \frac{32}{693}x^3 + \cdots\right).$$

24. Substituting $y = \sum_{n=0}^{\infty} c_n x^{n+r}$ into the differential equation and collecting terms, we obtain

$$2x^2 y'' + 3xy' + (2x - 1)y = \left(2r^2 + r - 1\right) c_0 x^r$$

$$+ \sum_{k=1}^{\infty} [2(k+r)(k+r-1)c_k + 3(k+r)c_k - c_k + 2c_{k-1}] x^{k+r}$$

$$= 0,$$

which implies

$$2r^2 + r - 1 = (2r - 1)(r + 1) = 0$$

and

$$[(k+r)(2k + 2r + 1) - 1]c_k + 2c_{k-1} = 0.$$

The indicial roots are $r = -1$ and $r = 1/2$. For $r = -1$ the recurrence relation is

$$c_k = -\frac{2c_{k-1}}{k(2k-3)}, \quad k = 1, 2, 3, \ldots,$$

and

$$c_1 = 2c_0, \qquad c_2 = -2c_0, \qquad c_3 = \frac{4}{9}c_0,$$

and so on. For $r = 1/2$ the recurrence relation is

$$c_k = -\frac{2c_{k-1}}{k(2k+3)}, \quad k = 1, 2, 3, \ldots,$$

and

$$c_1 = -\frac{2}{5}c_0, \qquad c_2 = \frac{2}{35}c_0, \qquad c_3 = -\frac{4}{945}c_0,$$

and so on. The general solution on $(0, \infty)$ is

$$y = C_1 x^{-1} \left(1 + 2x - 2x^2 + \frac{4}{9}x^3 + \cdots\right) + C_2 x^{1/2} \left(1 - \frac{2}{5}x + \frac{2}{35}x^2 - \frac{4}{945}x^3 + \cdots\right).$$

27. Substituting $y = \sum_{n=0}^{\infty} c_n x^{n+r}$ into the differential equation and collecting terms, we obtain

$$xy'' - xy' + y = \left(r^2 - r\right) c_0 x^{r-1} + \sum_{k=0}^{\infty} [(k+r+1)(k+r)c_{k+1} - (k+r)c_k + c_k] x^{k+r} = 0,$$

which implies

$$r^2 - r = r(r - 1) = 0$$

and

$$(k+r+1)(k+r)c_{k+1} - (k+r-1)c_k = 0.$$

The indicial roots are $r_1 = 1$ and $r_2 = 0$. For $r_1 = 1$ the recurrence relation is

$$c_{k+1} = \frac{kc_k}{(k+2)(k+1)}, \quad k = 0, 1, 2, \ldots,$$

and one solution is $y_1 = c_0 x$. A second solution is

$$y_2 = x \int \frac{e^{-\int -1\, dx}}{x^2}\, dx = x \int \frac{e^x}{x^2}\, dx = x \int \frac{1}{x^2}\left(1 + x + \frac{1}{2}x^2 + \frac{1}{3!}x^3 + \cdots\right) dx$$

$$= x \int \left(\frac{1}{x^2} + \frac{1}{x} + \frac{1}{2} + \frac{1}{3!}x + \frac{1}{4!}x^2 + \cdots\right) dx = x\left[-\frac{1}{x} + \ln x + \frac{1}{2}x + \frac{1}{12}x^2 + \frac{1}{72}x^3 + \cdots\right]$$

$$= x \ln x - 1 + \frac{1}{2}x^2 + \frac{1}{12}x^3 + \frac{1}{72}x^4 + \cdots.$$

The general solution on $(0, \infty)$ is

$$y = C_1 x + C_2 y_2(x).$$

30. Substituting $y = \displaystyle\sum_{n=0}^{\infty} c_n x^{n+r}$ into the differential equation and collecting terms, we obtain

$$xy'' + y' + y = r^2 c_0 x^{r-1} + \sum_{k=1}^{\infty}[(k+r)(k+r-1)c_k + (k+r)c_k + c_{k-1}]x^{k+r-1} = 0,$$

which implies $r^2 = 0$ and

$$(k+r)^2 c_k + c_{k-1} = 0.$$

The indicial roots are $r_1 = r_2 = 0$ and the recurrence relation is

$$c_k = -\frac{c_{k-1}}{k^2}, \quad k = 1, 2, 3, \ldots.$$

One solution is

$$y_1 = c_0\left(1 - x + \frac{1}{2^2}x^2 - \frac{1}{(3!)^2}x^3 + \frac{1}{(4!)^2}x^4 - \cdots\right) = c_0 \sum_{n=0}^{\infty} \frac{(-1)^n}{(n!)^2}x^n.$$

A second solution is

$$y_2 = y_1 \int \frac{e^{-\int (1/x)\, dx}}{y_1^2}\, dx = y_1 \int \frac{dx}{x\left(1 - x + \frac{1}{4}x^2 - \frac{1}{36}x^3 + \cdots\right)^2}$$

$$= y_1 \int \frac{dx}{x\left(1 - 2x + \frac{3}{2}x^2 - \frac{5}{9}x^3 + \frac{35}{288}x^4 - \cdots\right)}$$

$$= y_1 \int \frac{1}{x}\left(1 + 2x + \frac{5}{2}x^2 + \frac{23}{9}x^3 + \frac{677}{288}x^4 + \cdots\right) dx$$

$$= y_1 \int \left(\frac{1}{x} + 2 + \frac{5}{2}x + \frac{23}{9}x^2 + \frac{677}{288}x^3 + \cdots\right) dx$$

$$= y_1\left[\ln x + 2x + \frac{5}{4}x^2 + \frac{23}{27}x^3 + \frac{677}{1,152}x^4 + \cdots\right]$$

The general solution on $(0, \infty)$ is

$$y = C_1 y_1(x) + C_2 y_2(x).$$

33. (a) From $t = 1/x$ we have $dt/dx = -1/x^2 = -t^2$. Then

$$\frac{dy}{dx} = \frac{dy}{dt}\frac{dt}{dx} = -t^2\frac{dy}{dt}$$

and

$$\frac{d^2y}{dx^2} = \frac{d}{dx}\left(\frac{dy}{dx}\right) = \frac{d}{dx}\left(-t^2\frac{dy}{dt}\right) = -t^2\frac{d^2y}{dt^2}\frac{dt}{dx} - \frac{dy}{dt}\left(2t\frac{dt}{dx}\right) = t^4\frac{d^2y}{dt^2} + 2t^3\frac{dy}{dt}.$$

Now

$$x^4\frac{d^2y}{dx^2} + \lambda y = \frac{1}{t^4}\left(t^4\frac{d^2y}{dt^2} + 2t^3\frac{dy}{dt}\right) + \lambda y = \frac{d^2y}{dt^2} + \frac{2}{t}\frac{dy}{dt} + \lambda y = 0$$

becomes

$$t\frac{d^2y}{dt^2} + 2\frac{dy}{dt} + \lambda t y = 0.$$

(b) Substituting $y = \displaystyle\sum_{n=0}^{\infty} c_n t^{n+r}$ into the differential equation and collecting terms, we obtain

$$t\frac{d^2y}{dt^2} + 2\frac{dy}{dt} + \lambda t y = (r^2 + r)c_0 t^{r-1} + (r^2 + 3r + 2)c_1 t^r$$

$$+ \sum_{k=2}^{\infty}[(k+r)(k+r-1)c_k + 2(k+r)c_k + \lambda c_{k-2}]t^{k+r-1}$$

$$= 0,$$

which implies

$$r^2 + r = r(r+1) = 0,$$
$$\left(r^2 + 3r + 2\right)c_1 = 0,$$

and

$$(k+r)(k+r+1)c_k + \lambda c_{k-2} = 0.$$

The indicial roots are $r_1 = 0$ and $r_2 = -1$, so $c_1 = 0$. For $r_1 = 0$ the recurrence relation is

$$c_k = -\frac{\lambda c_{k-2}}{k(k+1)}, \quad k = 2, 3, 4, \ldots,$$

and

$$c_2 = -\frac{\lambda}{3!}c_0 \qquad c_3 = c_5 = c_7 = \cdots = 0 \qquad c_4 = \frac{\lambda^2}{5!}c_0 \qquad c_{2n} = (-1)^n\frac{\lambda^n}{(2n+1)!}c_0.$$

For $r_2 = -1$ the recurrence relation is

$$c_k = -\frac{\lambda c_{k-2}}{k(k-1)}, \quad k = 2, 3, 4, \ldots,$$

and

$$c_2 = -\frac{\lambda}{2!}c_0 \qquad c_3 = c_5 = c_7 = \cdots = 0 \qquad c_4 = \frac{\lambda^2}{4!}c_0 \qquad c_{2n} = (-1)^n \frac{\lambda^n}{(2n)!}c_0.$$

The general solution on $(0, \infty)$ is

$$y(t) = c_1 \sum_{n=0}^{\infty} \frac{(-1)^n}{(2n+1)!}(\sqrt{\lambda}\,t)^{2n} + c_2 t^{-1} \sum_{n=0}^{\infty} \frac{(-1)^n}{(2n)!}(\sqrt{\lambda}\,t)^{2n}$$

$$= \frac{1}{t}\left[C_1 \sum_{n=0}^{\infty} \frac{(-1)^n}{(2n+1)!}(\sqrt{\lambda}\,t)^{2n+1} + C_2 \sum_{n=0}^{\infty} \frac{(-1)^n}{(2n)!}(\sqrt{\lambda}\,t)^{2n}\right]$$

$$= \frac{1}{t}\left[C_1 \sin\sqrt{\lambda}\,t + C_2 \cos\sqrt{\lambda}\,t\right].$$

(c) Using $t = 1/x$, the solution of the original equation is

$$y(x) = C_1 x \sin\frac{\sqrt{\lambda}}{x} + C_2 x \cos\frac{\sqrt{\lambda}}{x}.$$

5.3 Special Functions

3. Since $\nu^2 = 25/4$ the general solution is $y = c_1 J_{5/2}(x) + c_2 J_{-5/2}(x)$.

6. Since $\nu^2 = 4$ the general solution is $y = c_1 J_2(x) + c_2 Y_2(x)$.

9. We identify $\alpha = 4$ and $\nu = 2/3$. Then the general solution is $y = c_1 I_{2/3}(4x) + c_2 K_{2/3}(4x)$

12. If $y = \sqrt{x}\,v(x)$ then

$$y' = x^{1/2}v'(x) + \frac{1}{2}x^{-1/2}v(x)$$

$$y'' = x^{1/2}v''(x) + x^{-1/2}v'(x) - \frac{1}{4}x^{-3/2}v(x)$$

and

$$x^2 y'' + \left(\alpha^2 x^2 - \nu^2 + \frac{1}{4}\right)y = x^{5/2}v''(x) + x^{3/2}v'(x) - \frac{1}{4}x^{1/2}v(x) + \left(\alpha^2 x^2 - \nu^2 + \frac{1}{4}\right)x^{1/2}v(x)$$

$$= x^{5/2}v''(x) + x^{3/2}v'(x) + (\alpha^2 x^{5/2} - \nu^2 x^{1/2})v(x) = 0.$$

Multiplying by $x^{-1/2}$ we obtain

$$x^2 v''(x) + xv'(x) + (\alpha^2 x^2 - \nu^2)v(x) = 0,$$

whose solution is $v(x) = c_1 J_\nu(\alpha x) + c_2 Y_\nu(\alpha x)$. Then $y = c_1 \sqrt{x}\, J_\nu(\alpha x) + c_2 \sqrt{x}\, Y_\nu(\alpha x)$.

15. Write the differential equation in the form $y'' - (1/x)y' + y = 0$. This is the form of (18) in the text with $a = 1$, $c = 1$, $b = 1$, and $p = 1$, so, by (19) in the text, the general solution is

$$y = x[c_1 J_1(x) + c_2 Y_1(x)].$$

18. Write the differential equation in the form $y'' + (4 + 1/4x^2)y = 0$. This is the form of (18) in the text with $a = \frac{1}{2}$, $c = 1$, $b = 2$, and $p = 0$, so, by (19) in the text, the general solution is

$$y = x^{1/2}[c_1 J_0(2x) + c_2 Y_0(2x)].$$

21. Using the fact that $i^2 = -1$, along with the definition of $J_\nu(x)$ in (7) in the text, we have

$$I_\nu(x) = i^{-\nu} J_\nu(ix) = i^{-\nu} \sum_{n=0}^{\infty} \frac{(-1)^n}{n!\Gamma(1 + \nu + n)} \left(\frac{ix}{2}\right)^{2n+\nu}$$

$$= \sum_{n=0}^{\infty} \frac{(-1)^n}{n!\Gamma(1 + \nu + n)} i^{2n+\nu-\nu} \left(\frac{x}{2}\right)^{2n+\nu}$$

$$= \sum_{n=0}^{\infty} \frac{(-1)^n}{n!\Gamma(1 + \nu + n)} (i^2)^n \left(\frac{x}{2}\right)^{2n+\nu}$$

$$= \sum_{n=0}^{\infty} \frac{(-1)^{2n}}{n!\Gamma(1 + \nu + n)} \left(\frac{x}{2}\right)^{2n+\nu}$$

$$= \sum_{n=0}^{\infty} \frac{1}{n!\Gamma(1 + \nu + n)} \left(\frac{x}{2}\right)^{2n+\nu},$$

which is a real function.

24. Write the differential equation in the form $y'' + (4/x)y' + (1 + 2/x^2)y = 0$. This is the form of (18) in the text with

$$1 - 2a = 4 \qquad\qquad 2c - 2 = 0 \qquad\qquad b^2 c^2 = 1 \qquad\qquad a^2 - p^2 c^2 = 2$$

$$a = -\frac{3}{2} \qquad\qquad c = 1 \qquad\qquad b = 1 \qquad\qquad p = \frac{1}{2}.$$

Then, by (19), (23), and (24) in the text,

$$y = x^{-3/2}[c_1 J_{1/2}(x) + c_2 J_{-1/2}(x)] = x^{-3/2}\left[c_1\sqrt{\frac{2}{\pi x}}\sin x + c_2\sqrt{\frac{2}{\pi x}}\cos x\right]$$

$$= C_1 \frac{1}{x^2}\sin x + C_2\frac{1}{x^2}\cos x.$$

27. (a) The recurrence relation follows from

$$-\nu J_\nu(x) + x J_{\nu-1}(x) = -\sum_{n=0}^{\infty} \frac{(-1)^n \nu}{n!\Gamma(1+\nu+n)} \left(\frac{x}{2}\right)^{2n+\nu} + x \sum_{n=0}^{\infty} \frac{(-1)^n}{n!\Gamma(\nu+n)} \left(\frac{x}{2}\right)^{2n+\nu-1}$$

$$= -\sum_{n=0}^{\infty} \frac{(-1)^n \nu}{n!\Gamma(1+\nu+n)} \left(\frac{x}{2}\right)^{2n+\nu}$$

$$+ \sum_{n=0}^{\infty} \frac{(-1)^n (\nu+n)}{n!\Gamma(1+\nu+n)} \cdot 2 \left(\frac{x}{2}\right) \left(\frac{x}{2}\right)^{2n+\nu-1}$$

$$= \sum_{n=0}^{\infty} \frac{(-1)^n (2n+\nu)}{n!\Gamma(1+\nu+n)} \left(\frac{x}{2}\right)^{2n+\nu} = x J_\nu'(x).$$

(b) The formula in part (a) is a linear first-order differential equation in $J_\nu(x)$. An integrating factor for this equation is x^ν, so

$$\frac{d}{dx}\left[x^\nu J_\nu(x)\right] = x^\nu J_{\nu-1}(x).$$

30. From (20) we obtain $J_0'(x) = -J_1(x)$, and from (21) we obtain $J_0'(x) = J_{-1}(x)$. Thus $J_0'(x) = J_{-1}(x) = -J_1(x)$.

33. Letting

$$s = \frac{2}{\alpha}\sqrt{\frac{k}{m}}\, e^{-\alpha t/2},$$

we have

$$\frac{dx}{dt} = \frac{dx}{ds}\frac{ds}{dt} = \frac{dx}{dt}\left[\frac{2}{\alpha}\sqrt{\frac{k}{m}}\left(-\frac{\alpha}{2}\right)e^{-\alpha t/2}\right] = \frac{dx}{ds}\left(-\sqrt{\frac{k}{m}}\,e^{-\alpha t/2}\right)$$

and

$$\frac{d^2x}{dt^2} = \frac{d}{dt}\left(\frac{dx}{dt}\right) = \frac{dx}{ds}\left(\frac{\alpha}{2}\sqrt{\frac{k}{m}}\,e^{-\alpha t/2}\right) + \frac{d}{dt}\left(\frac{dx}{ds}\right)\left(-\sqrt{\frac{k}{m}}\,e^{-\alpha t/2}\right)$$

$$= \frac{dx}{ds}\left(\frac{\alpha}{2}\sqrt{\frac{k}{m}}\,e^{-\alpha t/2}\right) + \frac{d^2x}{ds^2}\frac{ds}{dt}\left(-\sqrt{\frac{k}{m}}\,e^{-\alpha t/2}\right)$$

$$= \frac{dx}{ds}\left(\frac{\alpha}{2}\sqrt{\frac{k}{m}}\,e^{-\alpha t/2}\right) + \frac{d^2x}{ds^2}\left(\frac{k}{m}\,e^{-\alpha t}\right).$$

Then

$$m\frac{d^2x}{dt^2} + ke^{-\alpha t}x = ke^{-\alpha t}\frac{d^2x}{ds^2} + \frac{m\alpha}{2}\sqrt{\frac{k}{m}}\,e^{-\alpha t/2}\frac{dx}{ds} + ke^{-\alpha t}x = 0.$$

Multiplying by $2^2/\alpha^2 m$ we have

$$\frac{2^2}{\alpha^2}\frac{k}{m}e^{-\alpha t}\frac{d^2x}{ds^2} + \frac{2}{\alpha}\sqrt{\frac{k}{m}}\,e^{-\alpha t/2}\frac{dx}{ds} + \frac{2^2}{\alpha^2}\frac{k}{m}e^{-\alpha t}x = 0$$

or, since $s = (2/\alpha)\sqrt{k/m}\, e^{-\alpha t/2}$,

$$s^2 \frac{d^2 x}{ds^2} + s \frac{dx}{ds} + s^2 x = 0.$$

36. The general solution of the differential equation is

$$y(x) = c_1 J_0(\alpha x) + c_2 Y_0(\alpha x).$$

In order to satisfy the conditions that $\lim\limits_{x \to 0^+} y(x)$ and $\lim\limits_{x \to 0^+} y'(x)$ are finite we are forced to define $c_2 = 0$. Thus $y(x) = c_1 J_0(\alpha x)$. The second boundary condition, $y(2) = 0$, implies $c_1 = 0$ or $J_0(2\alpha) = 0$. In order to have a nontrivial solution we require that $J_0(2\alpha) = 0$. From Table 5.3.1, the first three positive zeros of J_0 are found to be

$$2\alpha_1 = 2.4048, \quad 2\alpha_2 = 5.5201, \quad 2\alpha_3 = 8.6537$$

and so $\alpha_1 = 1.2024$, $\alpha_2 = 2.7601$, $\alpha_3 = 4.3269$. The eigenfunctions corresponding to the eigenvalues $\lambda_1 = \alpha_1^2$, $\lambda_2 = \alpha_2^2$, $\lambda_3 = \alpha_3^2$ are $J_0(1.2024x)$, $J_0(2.7601x)$, and $J_0(4.3269x)$.

39. (a) Using the results of Problem 32 we get

$$j_1(x) = \sqrt{\frac{\pi}{2x}}\, J_{3/2}(x) = \sqrt{\frac{\pi}{2x}} \cdot \sqrt{\frac{2}{\pi x}} \left(\frac{\sin x}{x} - \cos x \right) = \frac{\sin x}{x^2} - \frac{\cos x}{x}$$

$$j_2(x) = \sqrt{\frac{\pi}{2x}}\, J_{5/2}(x) = \sqrt{\frac{\pi}{2x}} \cdot \sqrt{\frac{2}{\pi x}} \left[\left(\frac{3}{x^2} - 1 \right) \sin x - \frac{3 \cos x}{x} \right]$$

$$= \left(\frac{3}{x^3} - \frac{1}{x} \right) \sin x - \frac{3 \cos x}{x^2}$$

$$j_3(x) = \sqrt{\frac{\pi}{2x}}\, J_{7/2}(x) = \sqrt{\frac{\pi}{2x}} \cdot \sqrt{\frac{2}{\pi x}} \left[\left(\frac{15}{x^3} - \frac{6}{x} \right) \sin x - \left(\frac{15}{x^2} - 1 \right) \cos x \right]$$

$$= \left(\frac{15}{x^4} - \frac{6}{x^2} \right) \sin x - \left(\frac{15}{x^3} - \frac{1}{x} \right) \cos x$$

(b)

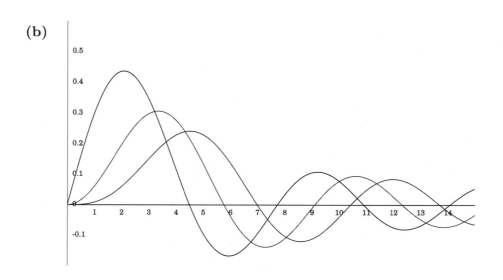

42. (a) Compare $x^2 Z''(x) + x Z'(x) + [\alpha^2 x^2 - (n + \frac{1}{2})^2]Z = 0$ to the parametric Bessel equation of order ν $x^2 y'' + xy' + (\alpha^2 x^2 - \nu^2)y = 0$ whose solution is $y = C_1 J_\nu(\alpha x) + C_2 Y_\nu(\alpha x)$. If we take $\nu = n + \frac{1}{2}$ then

$$Z(x) = C_1 J_{n+1/2}(\alpha x) + C_2 Y_{n+1/2}(\alpha x)$$

(b) Since $R(x) = (\alpha x)^{-1/2} Z(x)$ we get

$$R(x) = (ax)^{-1/2} Z(x) = C_1 (\alpha x)^{-1/2} J_{n+1/2}(\alpha x) + C_2 (\alpha x)^{-1/2} Y_{n+1/2}(\alpha x)$$

(c) Rename the constants C_1 and C_2 in part (b) to get

$$R(x) = \left(c_1 \sqrt{\frac{\pi}{2}}\right) \frac{J_{n+1/2}(\alpha x)}{\sqrt{\alpha x}} + \left(c_2 \sqrt{\frac{\pi}{2}}\right) \frac{Y_{n+1/2}(\alpha x)}{\sqrt{ax}}$$

$$= \left(c_1 \sqrt{\frac{\pi}{2\alpha x}}\right) J_{n+1/2}(\alpha x) + \left(c_2 \sqrt{\frac{\pi}{2\alpha x}}\right) Y_{n+1/2}(\alpha x)$$

$$= c_1 j_n(\alpha x) + c_2 y_n(\alpha x)$$

51. The recurrence relation can be written

$$P_{k+1}(x) = \frac{2k+1}{k+1} x P_k(x) - \frac{k}{k+1} P_{k-1}(x), \qquad k = 2,\ 3,\ 4,\ \dots\ .$$

$k = 1:$ $P_2(x) = \dfrac{3}{2}x^2 - \dfrac{1}{2}$

$k = 2:$ $P_3(x) = \dfrac{5}{3}x\left(\dfrac{3}{2}x^2 - \dfrac{1}{2}\right) - \dfrac{2}{3}x = \dfrac{5}{2}x^3 - \dfrac{3}{2}x$

$k = 3:$ $P_4(x) = \dfrac{7}{4}x\left(\dfrac{5}{2}x^3 - \dfrac{3}{2}x\right) - \dfrac{3}{4}\left(\dfrac{3}{2}x^2 - \dfrac{1}{2}\right) = \dfrac{35}{8}x^4 - \dfrac{30}{8}x^2 + \dfrac{3}{8}$

$k = 4:$ $P_5(x) = \dfrac{9}{5}x\left(\dfrac{35}{8}x^4 - \dfrac{30}{8}x^2 + \dfrac{3}{8}\right) - \dfrac{4}{5}\left(\dfrac{5}{2}x^3 - \dfrac{3}{2}x\right) = \dfrac{63}{8}x^5 - \dfrac{35}{4}x^3 + \dfrac{15}{8}x$

$k = 5:$ $P_6(x) = \dfrac{11}{6}x\left(\dfrac{63}{8}x^5 - \dfrac{35}{4}x^3 + \dfrac{15}{8}x\right) - \dfrac{5}{6}\left(\dfrac{35}{8}x^4 - \dfrac{30}{8}x^2 + \dfrac{3}{8}\right)$

$\qquad\qquad = \dfrac{231}{16}x^6 - \dfrac{315}{16}x^4 + \dfrac{105}{16}x^2 - \dfrac{5}{16}$

$k = 6:$ $P_7(x) = \dfrac{13}{7}x\left(\dfrac{231}{16}x^6 - \dfrac{315}{16}x^4 + \dfrac{105}{16}x^2 - \dfrac{5}{16}\right) - \dfrac{6}{7}\left(\dfrac{63}{8}x^5 - \dfrac{35}{4}x^3 + \dfrac{15}{8}x\right)$

$\qquad\qquad = \dfrac{429}{16}x^7 - \dfrac{693}{16}x^5 + \dfrac{315}{16}x^3 - \dfrac{35}{16}x$

54. (a)

$$P_0^0(x) = \left(1 - x^2\right)^0 \cdot \frac{d^0}{dx^0} P_0(x) = 1$$

$$P_1^0(x) = \left(1 - x^2\right)^0 \cdot \frac{d^0}{dx^0} P_1(x) = x$$

$$P_1^1(x) = \left(1 - x^2\right)^{1/2} \cdot \frac{d^1}{dx^1} P_1(x) = \left(1 - x^2\right)^{1/2}$$

$$P_2^1(x) = \left(1 - x^2\right)^{1/2} \cdot \frac{d^1}{dx^1} P_2(x) = \left(1 - x^2\right)^{1/2} \cdot (3x)$$

$$P_2^2(x) = \left(1 - x^2\right) \cdot \frac{d^2}{dx^2} P_2(x) = \left(1 - x^2\right) \cdot (3)$$

$$P_3^1(x) = \left(1 - x^2\right)^{1/2} \cdot \frac{d^1}{dx^1} P_3(x) = \left(1 - x^2\right)^{1/2} \cdot \frac{1}{2}\left(15x^2 - 3\right)$$

$$P_3^2(x) = \left(1 - x^2\right) \cdot \frac{d^2}{dx^2} P_3(x) = \left(1 - x^2\right) \cdot (15x)$$

$$P_3^3(x) = \left(1 - x^2\right)^{3/2} \cdot \frac{d^3}{dx^3} P_3(x) = \left(1 - x^2\right)^{3/2} \cdot (15)$$

(b) When m is an even nonnegative integer, $P_n^m(x)$ is a polynomial.

(c) Because $P_n(x)$ is a polynomial, $\frac{d^m}{dx^m} P_n(x) = 0$ for $m > n$. Therefore $P_n^m(x) = 0$ for $m > n$.

(d) When $n = m = 1$, the associated Legendre equation is $(1 - x^2)y'' - 2xy' + \left[2 - \frac{1}{1-x^2}\right]y = 0$ now force the function $y = P_1^1(x) = \left(1 - x^2\right)^{1/2}$ into the equation to get

$$\left(1 - x^2\right) \frac{d^2 \left(1 - x^2\right)^{1/2}}{dx^2} - 2x \frac{d \left(1 - x^2\right)^{1/2}}{dx} + \left[2 - \frac{1}{1 - x^2}\right]\left(1 - x^2\right)^{1/2} = 0$$

$$\left(1 - x^2\right)\left(-\frac{1}{(1 - x^2)^{3/2}}\right) - 2x\left(-\frac{x}{(1 - x^2)^{1/2}}\right) + \left[2 - \frac{1}{1 - x^2}\right]\left(1 - x^2\right)^{1/2} = 0$$

$$-\frac{1}{(1 - x^2)^{1/2}} + \frac{2x^2}{(1 - x^2)^{1/2}} + \left[2\left(1 - x^2\right)^{1/2} - \frac{1}{(1 - x^2)^{1/2}}\right] = 0$$

$$-\frac{1}{(1 - x^2)^{1/2}} + \frac{2x^2}{(1 - x^2)^{1/2}} + \left[\frac{1 - 2x^2}{(1 - x^2)^{1/2}}\right] = 0$$

$$0 = 0$$

Chapter 5 in Review

3. $x = -1$ is the nearest singular point to the ordinary point $x = 0$. Theorem 5.1.1 guarantees the existence of two power series solutions $y = \sum\limits_{n=1}^{\infty} c_n x^n$ of the differential equation that converge at least for $-1 < x < 1$. Since $-\frac{1}{2} \le x \le \frac{1}{2}$ is properly contained in $-1 < x < 1$, both power series must converge for all points contained in $-\frac{1}{2} \le x \le \frac{1}{2}$.

6. We have

$$f(x) = \frac{\sin x}{\cos x} = \frac{x - \dfrac{x^3}{6} + \dfrac{x^5}{120} - \cdots}{1 - \dfrac{x^2}{2} + \dfrac{x^4}{24} - \cdots} = x + \frac{x^3}{3} + \frac{2x^5}{15} + \cdots.$$

9. Substituting $y = \sum\limits_{n=0}^{\infty} c_n x^{n+r}$ into the differential equation we obtain

$$2xy'' + y' + y = \left(2r^2 - r\right) c_0 x^{r-1} + \sum_{k=1}^{\infty} [2(k+r)(k+r-1)c_k + (k+r)c_k + c_{k-1}]x^{k+r-1} = 0,$$

which implies

$$2r^2 - r = r(2r - 1) = 0$$

and

$$(k+r)(2k + 2r - 1)c_k + c_{k-1} = 0.$$

The indicial roots are $r = 0$ and $r = 1/2$. For $r = 0$ the recurrence relation is

$$c_k = -\frac{c_{k-1}}{k(2k-1)}, \quad k = 1, 2, 3, \ldots,$$

so

$$c_1 = -c_0, \qquad c_2 = \frac{1}{6}c_0, \qquad c_3 = -\frac{1}{90}c_0.$$

For $r = 1/2$ the recurrence relation is

$$c_k = -\frac{c_{k-1}}{k(2k+1)}, \quad k = 1, 2, 3, \ldots,$$

so

$$c_1 = -\frac{1}{3}c_0, \qquad c_2 = \frac{1}{30}c_0, \qquad c_3 = -\frac{1}{630}c_0.$$

Two linearly independent solutions are

$$y_1 = 1 - x + \frac{1}{6}x^2 - \frac{1}{90}x^3 + \cdots$$

and

$$y_2 = x^{1/2}\left(1 - \frac{1}{3}x + \frac{1}{30}x^2 - \frac{1}{630}x^3 + \cdots\right).$$

12. Substituting $y = \displaystyle\sum_{n=0}^{\infty} c_n x^n$ into the differential equation we obtain

$$y'' - x^2 y' + xy = 2c_2 + (6c_3 + c_0)x + \sum_{k=1}^{\infty}[(k+3)(k+2)c_{k+3} - (k-1)c_k]x^{k+1} = 0,$$

which implies $c_2 = 0$, $c_3 = -c_0/6$, and

$$c_{k+3} = \frac{k-1}{(k+3)(k+2)}c_k, \quad k = 1, 2, 3, \dots .$$

Choosing $c_0 = 1$ and $c_1 = 0$ we find

$$c_3 = -\frac{1}{6} \qquad c_4 = c_7 = c_{10} = \cdots = 0 \qquad c_5 = c_8 = c_{11} = \cdots = 0 \qquad c_6 = -\frac{1}{90}$$

and so on. For $c_0 = 0$ and $c_1 = 1$ we obtain

$$c_3 = c_6 = c_9 = \cdots = 0 \qquad c_4 = c_7 = c_{10} = \cdots = 0 \qquad c_5 = c_8 = c_{11} = \cdots = 0$$

and so on. Thus two solutions are

$$y_1 = 1 - \frac{1}{6}x^3 - \frac{1}{90}x^6 - \cdots \quad \text{and} \quad y_2 = x.$$

15. Substituting $y = \displaystyle\sum_{n=0}^{\infty} c_n x^n$ into the differential equation we have

$$y'' + xy' + 2y = \underbrace{\sum_{n=2}^{\infty} n(n-1)c_n x^{n-2}}_{k=n-2} + \underbrace{\sum_{n=1}^{\infty} nc_n x^n}_{k=n} + 2\underbrace{\sum_{n=0}^{\infty} c_n x^n}_{k=n}$$

$$= \sum_{k=0}^{\infty}(k+2)(k+1)c_{k+2}x^k + \sum_{k=1}^{\infty} kc_k x^k + 2\sum_{k=0}^{\infty} c_k x^k$$

$$= 2c_2 + 2c_0 + \sum_{k=1}^{\infty}[(k+2)(k+1)c_{k+2} + (k+2)c_k]x^k = 0.$$

Thus

$$2c_2 + 2c_0 = 0$$

$$(k+2)(k+1)c_{k+2} + (k+2)c_k = 0$$

and

$$c_2 = -c_0$$

$$c_{k+2} = -\frac{1}{k+1}c_k, \quad k = 1, 2, 3, \dots .$$

Choosing $c_0 = 1$ and $c_1 = 0$ we find

$$c_2 = -1 \qquad c_3 = c_5 = c_7 = \cdots = 0 \qquad c_4 = \frac{1}{3} \qquad c_6 = -\frac{1}{15}$$

and so on. For $c_0 = 0$ and $c_1 = 1$ we obtain

$$c_2 = c_4 = c_6 = \cdots = 0 \qquad c_3 = -\frac{1}{2} \qquad c_5 = \frac{1}{8} \qquad c_7 = -\frac{1}{48}$$

and so on. Thus the general solution is

$$y = C_0 \left(1 - x^2 + \frac{1}{3}x^4 - \frac{1}{15}x^6 + \cdots \right) + C_1 \left(x - \frac{1}{2}x^3 + \frac{1}{8}x^5 - \frac{1}{48}x^7 + \cdots \right)$$

and

$$y' = C_0 \left(-2x + \frac{4}{3}x^3 - \frac{2}{5}x^5 + \cdots \right) + C_1 \left(1 - \frac{3}{2}x^2 + \frac{5}{8}x^4 - \frac{7}{48}x^6 + \cdots \right).$$

Setting $y(0) = 3$ and $y'(0) = -2$ we find $c_0 = 3$ and $c_1 = -2$. Therefore, the solution of the initial-value problem is

$$y = 3 - 2x - 3x^2 + x^3 + x^4 - \frac{1}{4}x^5 - \frac{1}{5}x^6 + \frac{1}{24}x^7 + \cdots .$$

18. While we can find two solutions of the form

$$y_1 = c_0 [1 + \cdots] \quad \text{and} \quad y_2 = c_1 [x + \cdots],$$

the initial conditions at $x = 1$ give solutions for c_0 and c_1 in terms of infinite series. Letting $t = x - 1$ the initial-value problem becomes

$$\frac{d^2 y}{dt^2} + (t + 1)\frac{dy}{dt} + y = 0, \qquad y(0) = -6, \ y'(0) = 3.$$

Substituting $y = \sum_{n=0}^{\infty} c_n t^n$ into the differential equation, we have

$$\frac{d^2 y}{dt^2} + (t+1)\frac{dy}{dt} + y = \underbrace{\sum_{n=2}^{\infty} n(n-1)c_n t^{n-2}}_{k=n-2} + \underbrace{\sum_{n=1}^{\infty} nc_n t^n}_{k=n} + \underbrace{\sum_{n=1}^{\infty} nc_n t^{n-1}}_{k=n-1} + \underbrace{\sum_{n=0}^{\infty} c_n t^n}_{k=n}$$

$$= \sum_{k=0}^{\infty} (k+2)(k+1)c_{k+2} t^k + \sum_{k=1}^{\infty} kc_k t^k + \sum_{k=0}^{\infty} (k+1)c_{k+1} t^k + \sum_{k=0}^{\infty} c_k t^k$$

$$= 2c_2 + c_1 + c_0 + \sum_{k=1}^{\infty} \left[(k+2)(k+1)c_{k+2} + (k+1)c_{k+1} + (k+1)c_k \right] t^k$$

$$= 0.$$

Thus

$$2c_2 + c_1 + c_0 = 0$$

$$(k+2)(k+1)c_{k+2} + (k+1)c_{k+1} + (k+1)c_k = 0$$

and

$$c_2 = -\frac{c_1 + c_0}{2}$$

$$c_{k+2} = -\frac{c_{k+1} + c_k}{k+2}, \quad k = 1, 2, 3, \ldots .$$

Choosing $c_0 = 1$ and $c_1 = 0$ we find

$$c_2 = -\frac{1}{2}, \qquad c_3 = \frac{1}{6}, \qquad c_4 = \frac{1}{12},$$

and so on. For $c_0 = 0$ and $c_1 = 1$ we find

$$c_2 = -\frac{1}{2}, \qquad c_3 = -\frac{1}{6}, \qquad c_4 = \frac{1}{6},$$

and so on. Thus the general solution is

$$y = c_0 \left[1 - \frac{1}{2}t^2 + \frac{1}{6}t^3 + \frac{1}{12}t^4 + \cdots \right] + c_1 \left[t - \frac{1}{2}t^2 - \frac{1}{6}t^3 + \frac{1}{6}t^4 + \cdots \right].$$

The initial conditions then imply $c_0 = -6$ and $c_1 = 3$. Thus the solution of the initial-value problem is

$$y = -6 \left[1 - \frac{1}{2}(x-1)^2 + \frac{1}{6}(x-1)^3 + \frac{1}{12}(x-1)^4 + \cdots \right]$$

$$+ 3 \left[(x-1) - \frac{1}{2}(x-1)^2 - \frac{1}{6}(x-1)^3 + \frac{1}{6}(x-1)^4 + \cdots \right].$$

21. Substituting $y = \sum_{n=0}^{\infty} c_n x^n$ into the differential equation we have

$$y'' + x^2 y' + 2xy = \underbrace{\sum_{n=2}^{\infty} n(n-1)c_n x^{n-2}}_{k=n-2} + \underbrace{\sum_{n=1}^{\infty} nc_n x^{n+1}}_{k=n+1} + 2 \underbrace{\sum_{n=0}^{\infty} c_n x^{n+1}}_{k=n+1}$$

$$= \sum_{k=0}^{\infty} (k+2)(k+1)c_{k+2} x^k + \sum_{k=2}^{\infty} (k-1)c_{k-1} x^k + 2 \sum_{k=1}^{\infty} c_{k-1} x^k$$

$$= 2c_2 + (6c_3 + 2c_0)x + \sum_{k=2}^{\infty} [(k+2)(k+1)c_{k+2} + (k+1)c_{k-1}]x^k = 5 - 2x + 10x^3.$$

Thus equating coefficients of like powers of x gives

$$2c_2 = 5 \qquad 6c_3 + 2c_0 = -2 \qquad 12c_4 + 3c_1 = 0 \qquad 20c_5 + 4c_2 = 10$$

and

$$(k+2)(k+1)c_{k+2} + (k+1)c_{k-1} = 0, \quad k = 4, 5, 6, \dots.$$

Therefore

$$c_2 = \frac{5}{2} \qquad c_3 = -\frac{1}{3}c_0 - \frac{1}{3} \qquad c_4 = -\frac{1}{4}c_1 \qquad c_5 = \frac{1}{2} - \frac{1}{5}c_2 = \frac{1}{2} - \frac{1}{5}\left(\frac{5}{2}\right) = 0$$

and

$$c_{k+2} = -\frac{1}{k+2}\,c_{k-1}.$$

Using the recurrence relation, we find

$$c_6 = -\frac{1}{6}c_3 = \frac{1}{3 \cdot 6}(c_0 + 1) = \frac{1}{3^2 \cdot 2!}c_0 + \frac{1}{3^2 \cdot 2!}$$

$$c_7 = -\frac{1}{7}c_4 = \frac{1}{4 \cdot 7}c_1$$

$$c_8 = c_{11} = c_{14} = \cdots = 0$$

$$c_9 = -\frac{1}{9}c_6 = -\frac{1}{3^3 \cdot 3!}c_0 - \frac{1}{3^3 \cdot 3!}$$

$$c_{10} = -\frac{1}{10}c_7 = -\frac{1}{4 \cdot 7 \cdot 10}c_1$$

$$c_{12} = -\frac{1}{12}c_9 = \frac{1}{3^4 \cdot 4!}c_0 + \frac{1}{3^4 \cdot 4!}$$

$$c_{13} = -\frac{1}{13}c_0 = \frac{1}{4 \cdot 7 \cdot 10 \cdot 13}c_1$$

and so on. Thus

$$y = c_0\left[1 - \frac{1}{3}x^3 + \frac{1}{3^2 \cdot 2!}x^6 - \frac{1}{3^3 \cdot 3!}x^9 + \frac{1}{3^4 \cdot 4!}x^{12} - \cdots\right]$$

$$+ c_1\left[x - \frac{1}{4}x^4 + \frac{1}{4 \cdot 7}x^7 - \frac{1}{4 \cdot 7 \cdot 10}x^{10} + \frac{1}{4 \cdot 7 \cdot 10 \cdot 13}x^{13} - \cdots\right]$$

$$+ \left[\frac{5}{2}x^2 - \frac{1}{3}x^3 + \frac{1}{3^2 \cdot 2!}x^6 - \frac{1}{3^3 \cdot 3!}x^9 + \frac{1}{3^4 \cdot 4!}x^{12} - \cdots\right].$$

24. (a) Using formula (5) of Section 3.2 in the text, we find that a second solution of $(1 - x^2)y'' - 2xy' = 0$ is

$$y_2(x) = 1 \cdot \int \frac{e^{\int 2x\,dx/(1-x^2)}}{1^2}\,dx = \int e^{-\ln(1-x^2)}\,dx$$

$$= \int \frac{dx}{1 - x^2} = \frac{1}{2}\ln\left(\frac{1+x}{1-x}\right),$$

where partial fractions was used to obtain the last integral.

(b) Using formula (5) of Section 3.2 in the text, we find that a second solution of $(1 - x^2)y'' - 2xy' + 2y = 0$ is

$$y_2(x) = x \cdot \int \frac{e^{\int 2x\,dx/(1-x^2)}}{x^2}\,dx = x \int \frac{e^{-\ln(1-x^2)}}{x^2}\,dx$$

$$= x \int \frac{dx}{x^2(1-x^2)} = x \left[\frac{1}{2} \ln\left(\frac{1+x}{1-x}\right) - \frac{1}{x} \right] = \frac{x}{2} \ln\left(\frac{1+x}{1-x}\right) - 1,$$

where partial fractions was used to obtain the last integral.

(c)

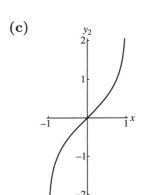

$$y_2(x) = \frac{1}{2} \ln\left(\frac{1+x}{1-x}\right)$$

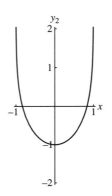

$$y_2 = \frac{x}{2} \ln\left(\frac{1+x}{1-x}\right) - 1$$

27. Force $y = \sum_{k=0}^{\infty} c_k x^k$ into the equation to get $\sum_{k=0}^{\infty} [(k+2)(k+1)c_{k+2} - 2kc_k + 2\alpha c_k]x^k = 0$
therefore

$$c_{k+2} = -\frac{(2\alpha - 2k)}{(k+2)(k+1)}c_k.$$

Now for $k = 0, 1, 2, 3, 4, 5, \ldots$ we get

$$c_2 = -\frac{(2\alpha)}{(2)(1)}c_0 = -\frac{2\alpha}{2!}c_0$$

$$c_3 = -\frac{(2\alpha - 2)}{(3)(2)}c_1 = -\frac{2(\alpha - 1)}{3!}c_1$$

$$c_4 = -\frac{(2\alpha - 4)}{(4)(3)}c_2 = -\frac{2(\alpha - 2)}{4 \cdot 3} \cdot \left(-\frac{2\alpha}{2!}c_0\right) = \frac{2^2\alpha(\alpha - 2)}{4!}c_0$$

$$c_5 = -\frac{(2\alpha - 6)}{(5)(4)}c_3 = -\frac{2(\alpha - 3)}{5 \cdot 4} \cdot \left(-\frac{2(\alpha - 1)}{3!}c_1\right) = \frac{2^2(\alpha - 1)(\alpha - 3)}{5!}c_1$$

$$c_6 = -\frac{(2\alpha - 8)}{(6)(5)}c_4 = -\frac{2(\alpha - 4)}{6 \cdot 5} \cdot \left(\frac{2^2\alpha(\alpha - 2)}{4!}c_0\right) = -\frac{2^3\alpha(\alpha - 2)(\alpha - 4)}{6!}c_0$$

$$c_7 = -\frac{(2\alpha - 10)}{(7)(6)}c_5 = -\frac{2(\alpha - 5)}{7 \cdot 6} \cdot \left(\frac{2^2(\alpha - 1)(\alpha - 3)}{5!}c_1\right) = -\frac{2^3(\alpha - 1)(\alpha - 3)(\alpha - 5)}{7!}c_1$$

and so on. Therefore we have

$$y = \sum_{k=0}^{\infty} c_k x^k = c_0 + c_1 x + c_2 x^2 + c_3 x^3 + c_4 x^4 + c_5 x^5 + c_6 x^6 + \dots$$

$$= c_0 + c_1 x - \frac{2\alpha}{2!} c_0 x^2 - \frac{2(\alpha-1)}{3!} c_1 x^3 + \frac{2^2\alpha(\alpha-2)}{4!} c_0 x^4$$

$$+ \frac{2^2(\alpha-1)(\alpha-3)}{5!} c_1 x^5 - \frac{2^3\alpha(\alpha-2)(\alpha-4)}{6!} c_0 x^6 + \dots$$

$$= c_0 \left[1 - \frac{2\alpha}{2!} x^2 + \frac{2^2\alpha(\alpha-2)}{4!} x^4 - \frac{2^3\alpha(\alpha-2)(\alpha-4)}{6!} x^6 + \dots \right]$$

$$+ c_1 \left[x - \frac{2(\alpha-1)}{3!} x^3 + \frac{2^2(\alpha-1)(\alpha-3)}{5!} x^5 + \dots \right]$$

$$= c_0 y_1(x) + c_1 y_2(x)$$

where we take

$$y_1(x) = 1 + \sum_{k=1}^{\infty} \frac{(-1)^k 2^k \alpha(\alpha-2)(\alpha-4)\dots(\alpha-(2k-2))}{(2k)!} x^{2k}$$

$$y_2(x) = x + \sum_{k=1}^{\infty} \frac{(-1)^k 2^k (\alpha-1)(\alpha-3)\dots(\alpha-(2k-1))}{(2k+1)!} x^{2k+1}$$

30. For $n = 0, 1, 2, 3, 4$, Rodrigues' formula gives

$$L_0(x) = \frac{e^x}{0!} \cdot \frac{d^0}{dx^0} x^0 e^{-x} = e^x \cdot e^{-x} = 1$$

$$L_1(x) = \frac{e^x}{1!} \cdot \frac{d}{dx} x e^{-x} = e^x \left(e^{-x} - x e^{-x} \right) = -x + 1$$

$$L_2(x) = \frac{e^x}{2!} \cdot \frac{d^2}{dx^2} x^2 e^{-x} = \frac{e^x}{2} \left(2e^{-x} - 4x e^{-x} + x^2 e^{-x} \right) = \frac{1}{2} \left(x^2 - 4x + 2 \right)$$

$$L_3(x) = \frac{e^x}{3!} \cdot \frac{d^3}{dx^3} x^3 e^{-x} = \frac{e^x}{6} \left(6e^{-x} - 18x e^{-x} + 9x^2 e^{-x} - x^3 e^{-x} \right) = \frac{1}{6} \left(-x^3 + 9x^2 - 18x + 6 \right)$$

$$L_4(x) = \frac{e^x}{4!} \cdot \frac{d^4}{dx^4} x^4 e^{-x} = \frac{e^x}{24} \left(24e^{-x} - 96x e^{-x} + 72x^2 e^{-x} - 16x^3 e^{-x} + x^4 e^{-x} \right)$$

$$= \frac{1}{24} \left(x^4 - 16x^3 + 72x^2 - 96x + 24 \right)$$

Chapter 6

Numerical Solutions of Ordinary Differential Equations

6.1 | Euler Methods and Error Analysis

3.

$h = 0.1$

x_n	y_n
0.00	0.0000
0.10	0.1005
0.20	0.2030
0.30	0.3098
0.40	0.4234
0.50	0.5470

$h = 0.05$

x_n	y_n
0.00	0.0000
0.05	0.0501
0.10	0.1004
0.15	0.1512
0.20	0.2028
0.25	0.2554
0.30	0.3095
0.35	0.3652
0.40	0.4230
0.45	0.4832
0.50	0.5465

6.

$h = 0.1$

x_n	y_n
0.00	0.0000
0.10	0.0050
0.20	0.0200
0.30	0.0451
0.40	0.0805
0.50	0.1266

$h = 0.05$

x_n	y_n
0.00	0.0000
0.05	0.0013
0.10	0.0050
0.15	0.0113
0.20	0.0200
0.25	0.0313
0.30	0.0451
0.35	0.0615
0.40	0.0805
0.45	0.1022
0.50	0.1266

9.

$h = 0.1$

x_n	y_n
1.00	1.0000
1.10	1.0095
1.20	1.0404
1.30	1.0967
1.40	1.1866
1.50	1.3260

$h = 0.05$

x_n	y_n
1.00	1.0000
1.05	1.0024
1.10	1.0100
1.15	1.0228
1.20	1.0414
1.25	1.0663
1.30	1.0984
1.35	1.1389
1.40	1.1895
1.45	1.2526
1.50	1.3315

12. (a)

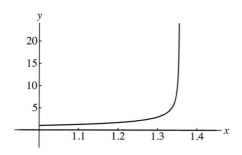

(b)

x_n	*Euler*	*Imp. Euler*
1.00	1.0000	1.0000
1.10	1.2000	1.2469
1.20	1.4938	1.6430
1.30	1.9711	2.4042
1.40	2.9060	4.5085

15. (a) Using Euler's method we obtain $y(0.1) \approx y_1 = 0.8$.

(b) Using $y'' = 5e^{-2x}$ we see that the local truncation error is

$$5e^{-2c}\frac{(0.1)^2}{2} = 0.025e^{-2c}.$$

Since e^{-2x} is a decreasing function, $e^{-2c} \le e^0 = 1$ for $0 \le c \le 0.1$. Thus an upper bound for the local truncation error is $0.025(1) = 0.025$.

(c) Since $y(0.1) = 0.8234$, the actual error is $y(0.1) - y_1 = 0.0234$, which is less than 0.025.

(d) Using Euler's method with $h = 0.05$ we obtain $y(0.1) \approx y_2 = 0.8125$.

(e) The error in (d) is $0.8234 - 0.8125 = 0.0109$. With global truncation error $O(h)$, when the step size is halved we expect the error for $h = 0.05$ to be one-half the error when $h = 0.1$. Comparing 0.0109 with 0.0234 we see that this is the case.

18. (a) Using $y''' = -114e^{-3(x-1)}$ we see that the local truncation error is

$$\left| y'''(c)\frac{h^3}{6} \right| = 114e^{-3(x-1)}\frac{h^3}{6} = 19h^3 e^{-3(c-1)}.$$

(b) Since $e^{-3(x-1)}$ is a decreasing function for $1 \le x \le 1.5$, $e^{-3(c-1)} \le e^{-3(1-1)} = 1$ for $1 \le c \le 1.5$ and

$$\left| y'''(c)\frac{h^3}{6} \right| \le 19(0.1)^3(1) = 0.019.$$

(c) Using the improved Euler's method with $h = 0.1$ we obtain $y(1.5) \approx 2.080108$. With $h = 0.05$ we obtain $y(1.5) \approx 2.059166$.

(d) Since $y(1.5) = 2.053216$, the error for $h = 0.1$ is $E_{0.1} = 0.026892$, while the error for $h = 0.05$ is $E_{0.05} = 0.005950$. With global truncation error $O(h^2)$ we expect $E_{0.1}/E_{0.05} \approx 4$. We actually have $E_{0.1}/E_{0.05} = 4.52$.

6.2 Runge–Kutta Methods

3.

x_n	y_n
1.00	5.0000
1.10	3.9724
1.20	3.2284
1.30	2.6945
1.40	2.3163
1.50	2.0533

6.

x_n	y_n
0.00	1.0000
0.10	1.1115
0.20	1.2530
0.30	1.4397
0.40	1.6961
0.50	2.0670

9.

x_n	y_n
0.00	0.5000
0.10	0.5213
0.20	0.5358
0.30	0.5443
0.40	0.5482
0.50	0.5493

12.

x_n	y_n
0.00	0.5000
0.10	0.5250
0.20	0.5498
0.30	0.5744
0.40	0.5987
0.50	0.6225

15. (a)

x_n	$h = 0.05$	$h = 0.1$
1.00	1.0000	1.0000
1.05	1.1112	
1.10	1.2511	1.2511
1.15	1.4348	
1.20	1.6934	1.6934
1.25	1.1047	
1.30	2.9560	2.9425
1.35	7.8981	
1.40	1.0608×10^{15}	903.0282

(b)

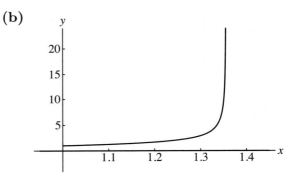

18. (a) Using $y^{(5)} = -1026e^{-3(x-1)}$ we see that the local truncation error is

$$\left| y^{(5)}(c)\, \frac{h^5}{120} \right| = 8.55 h^5 e^{-3(c-1)}.$$

(b) Since $e^{-3(x-1)}$ is a decreasing function for $1 \le x \le 1.5$, $e^{-3(c-1)} \le e^{-3(1-1)} = 1$ for $1 \le c \le 1.5$ and

$$y^{(5)}(c)\, \frac{h^5}{120} \le 8.55(0.1)^5(1) = 0.0000855.$$

(c) Using the RK4 method with $h = 0.1$ we obtain $y(1.5) \approx 2.053338827$. With $h = 0.05$ we obtain $y(1.5) \approx 2.053222989$.

6.3 Multistep Methods

In the tables in this section "ABM" stands for Adams-Bashforth-Moulton.

3. The first predictor is $y_4^* = 0.73318477$.

x_n	y_n	
0.0	1.00000000	init. cond.
0.2	0.73280000	RK4
0.4	0.64608032	RK4
0.6	1.65851653	RK4
0.8	1.72319464	ABM

6. The first predictor for $h = 0.2$ is $y_4^* = 3.34828434$.

x_n	$h = 0.2$		$h = 0.1$	
0.0	1.00000000	init. cond.	1.00000000	init. cond.
0.1			1.21017082	RK4
0.2	1.44139950	RK4	1.44140511	RK4
0.3			1.69487942	RK4
0.4	1.97190167	RK4	1.97191536	ABM
0.5			2.27400341	ABM
0.6	2.60280694	RK4	2.60283209	ABM
0.7			2.96031780	ABM
0.8	3.34860927	ABM	3.34863769	ABM
0.9			3.77026548	ABM
1.0	4.22797875	ABM	4.22801028	ABM

6.4 Higher-Order Equations and Systems

3. The substitution $y' = u$ leads to the system

$$y' = u, \qquad u' = 4u - 4y.$$

Using formula (4), we obtain the table shown.

x_n	$h = 0.2$ y_n	$h = 0.2$ u_n	$h = 0.1$ y_n	$h = 0.1$ u_n
0.0	−2.0000	1.0000	−2.0000	1.0000
0.1			−1.8321	2.4427
0.2	−1.4928	4.4731	−1.4919	4.4753

6. Using $h = 0.1$, the RK4 method for a system, and a numerical solver, we obtain

t_n	$h = 0.2$ i_{1n}	$h = 0.2$ i_{3n}
0.0	0.0000	0.0000
0.1	2.5000	3.7500
0.2	2.8125	5.7813
0.3	2.0703	7.4023
0.4	0.6104	9.1919
0.5	−1.5619	11.4877

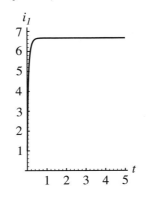

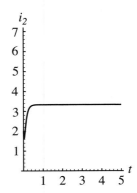

9.

t_n	$h = 0.2$ x_n	$h = 0.2$ y_n	$h = 0.1$ x_n	$h = 0.1$ y_n
0.0	−3.0000	5.0000	−3.0000	5.0000
0.1			−3.4790	4.6707
0.2	−3.9123	4.2857	−3.9123	4.2857

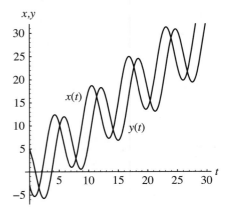

12. Solving for x' and y' we obtain the system

$$x' = \frac{1}{2}y - 3t^2 + 2t - 5$$

$$y' = -\frac{1}{2}y + 3t^2 + 2t + 5.$$

t_n	$h = 0.2$ x_n	$h = 0.2$ y_n	$h = 0.1$ x_n	$h = 0.1$ y_n
0.0	3.0000	−1.0000	3.0000	−1.0000
0.1			2.4727	−0.4527
0.2	1.9867	0.0933	1.9867	0.0933

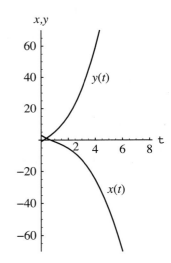

6.5 | Second-Order Boundary-Value Problems

3. We identify $P(x) = 2$, $Q(x) = 1$, $f(x) = 5x$, and $h = (1 - 0)/5 = 0.2$. Then the finite difference equation is

$$1.2y_{i+1} - 1.96y_i + 0.8y_{i-1} = 0.04(5x_i).$$

The solution of the corresponding linear system gives

x	0.0	0.2	0.4	0.6	0.8	1.0
y	0.0000	−0.2259	−0.3356	−0.3308	−0.2167	0.0000

6. We identify $P(x) = 5$, $Q(x) = 0$, $f(x) = 4\sqrt{x}$, and $h = (2-1)/6 = 0.1667$. Then the finite difference equation is

$$1.4167y_{i+1} - 2y_i + 0.5833y_{i-1} = 0.2778(4\sqrt{x_i}).$$

The solution of the corresponding linear system gives

x	1.0000	1.1667	1.3333	1.5000	1.6667	1.8333	2.0000
y	1.0000	−0.5918	−1.1626	−1.3070	−1.2704	−1.1541	−1.0000

9. We identify $P(x) = 1 - x$, $Q(x) = x$, $f(x) = x$, and $h = (1-0)/10 = 0.1$. Then the finite difference equation is

$$[1 + 0.05(1 - x_i)]y_{i+1} + [-2 + 0.01x_i]y_i + [1 - 0.05(1 - x_i)]y_{i-1} = 0.01x_i.$$

The solution of the corresponding linear system gives

x	0.0	0.1	0.2	0.3	0.4	0.5	0.6
y	0.0000	0.2660	0.5097	0.7357	0.9471	1.1465	1.3353

0.7	0.8	0.9	1.0
1.5149	1.6855	1.8474	2.0000

12. We identify $P(r) = 2/r$, $Q(r) = 0$, $f(r) = 0$, and $h = (4-1)/6 = 0.5$. Then the finite difference equation is

$$\left(1 + \frac{0.5}{r_i}\right)u_{i+1} - 2u_i + \left(1 - \frac{0.5}{r_i}\right)u_{i-1} = 0.$$

The solution of the corresponding linear system gives

r	1.0	1.5	2.0	2.5	3.0	3.5	4.0
u	50.0000	72.2222	83.3333	90.0000	94.4444	97.6190	100.0000

Chapter 6 in Review

3.

x_n	Euler $h = 0.1$	Euler $h = 0.05$	Imp. Euler $h = 0.1$	Imp. Euler $h = 0.05$	RK4 $h = 0.1$	RK4 $h = 0.05$
0.50	0.5000	0.5000	0.5000	0.5000	0.5000	0.5000
0.55		0.5500		0.5512		0.5512
0.60	0.6000	0.6024	0.6048	0.6049	0.6049	0.6049
0.65		0.6573		0.6609		0.6610
0.70	0.7090	0.7144	0.7191	0.7193	0.7194	0.7194
0.75		0.7739		0.7800		0.7801
0.80	0.8283	0.8356	0.8427	0.8430	0.8431	0.8431
0.85		0.8996		0.9082		0.9083
0.90	0.9559	0.9657	0.9752	0.9755	0.9757	0.9757
0.95		1.0340		1.0451		1.0452
1.00	1.0921	1.1044	1.1163	1.1168	1.1169	1.1169

6. The first predictor is $y_3^* = 1.14822731$.

x_n	y_n	
0.0	2.00000000	init. cond.
0.1	1.65620000	RK4
0.2	1.41097281	RK4
0.3	1.24645047	RK4
0.4	1.14796764	ABM

Chapter 7

Vectors

7.1 | Vectors in 2-Space

3. (a) $\langle 12, 0 \rangle$ **(b)** $\langle 4, -5 \rangle$ **(c)** $\langle 4, 5 \rangle$ **(d)** $\sqrt{41}$ **(e)** $\sqrt{41}$

6. (a) $\langle 3, 9 \rangle$ **(b)** $\langle -4, -12 \rangle$ **(c)** $\langle 6, 18 \rangle$ **(d)** $4\sqrt{10}$ **(e)** $6\sqrt{10}$

9. (a) $\langle 4, -12 \rangle - \langle -2, 2 \rangle = \langle 6, -14 \rangle$ **(b)** $\langle -3, 9 \rangle - \langle -5, 5 \rangle = \langle 2, 4 \rangle$

12. (a) $\langle 8, 0 \rangle - \langle 0, -6 \rangle = \langle 8, 6 \rangle$ **(b)** $\langle -6, 0 \rangle - \langle 0, -15 \rangle = \langle -6, 15 \rangle$

15.

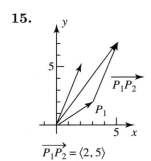

$$\overrightarrow{P_1P_2} = \langle 2, 5 \rangle$$

18.

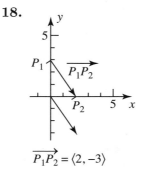

$$\overrightarrow{P_1P_2} = \langle 2, -3 \rangle$$

21. $a(= -\mathbf{a})$, $b(= -\frac{1}{4}\mathbf{a})$, $c(= \frac{5}{2}\mathbf{a})$, $e(= 2\mathbf{a})$, and $f(= -\frac{1}{2}\mathbf{a})$ are parallel to **(a)**.

24. $\langle 5, 2 \rangle$

27. $\|\mathbf{a}\| = 5$;   **(a)** $\mathbf{u} = \frac{1}{5}\langle 0, -5 \rangle = \langle 0, -1 \rangle$;   **(b)** $-\mathbf{u} = \langle 0, 1 \rangle$

30. $\|2\mathbf{a} - 3\mathbf{b}\| = \|\langle -5, 4 \rangle\| = \sqrt{25 + 16} = \sqrt{41}$; $\mathbf{u} = \frac{1}{\sqrt{41}}\langle -5, 4 \rangle = \left\langle -\frac{5}{\sqrt{41}}, \frac{4}{\sqrt{41}} \right\rangle$

33. $-\frac{3}{4}\mathbf{a} = \langle -3, -15/2 \rangle$

36.

39.

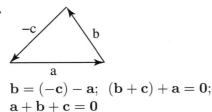

$\mathbf{b} = (-\mathbf{c}) - \mathbf{a};\quad (\mathbf{b} + \mathbf{c}) + \mathbf{a} = \mathbf{0};$
$\mathbf{a} + \mathbf{b} + \mathbf{c} = \mathbf{0}$

42. From $2\mathbf{i} + 3\mathbf{j} = k_1\mathbf{b} + k_2\mathbf{c} = k_1(-2\mathbf{i} + 4\mathbf{j}) + k_2(5\mathbf{i} + 7\mathbf{j}) = (-2k_1 + 5k_2)\mathbf{i} + (4k_1 + 7k_2)\mathbf{j}$ we obtain the system of equations $-2k_1 + 5k_2 = 2$, $4k_1 + 7k_2 = 3$. Solving, we find $k_1 = \frac{1}{34}$ and $k_2 = \frac{7}{17}$.

45. (a) Since $\mathbf{F}_f = -\mathbf{F}_g$, $\|\mathbf{F}_g\| = \|\mathbf{F}_f\| = \mu\|\mathbf{F}_n\|$ and $\tan\theta = \|\mathbf{F}_g\|/\|\mathbf{F}_n\| = \mu\|\mathbf{F}_n\|/\|\mathbf{F}_n\| = \mu$.

(b) $\theta = \tan^{-1} 0.6 \approx 31°$

48. Place one corner of the parallelogram at the origin and let two adjacent sides be $\overrightarrow{OP_1}$ and $\overrightarrow{OP_2}$. Let M be the midpoint of the diagonal connecting P_1 and P_2 and N be the midpoint of the other diagonal. Then

$\overrightarrow{OM} = \frac{1}{2}(\overrightarrow{OP_1} + \overrightarrow{OP_2})$. Since $\overrightarrow{OP_1} + \overrightarrow{OP_2}$ is the main diagonal of the parallelogram and N is its midpoint, $\overrightarrow{ON} = \frac{1}{2}(\overrightarrow{OP_1} + \overrightarrow{OP_2})$. Thus, $\overrightarrow{OM} = \overrightarrow{ON}$ and the diagonals bisect each other.

7.2   Vectors in 3-Space

3. – 6.

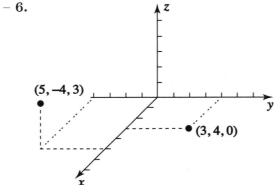

9. A line perpendicular to the xy-plane at $(2, 3, 0)$

12.

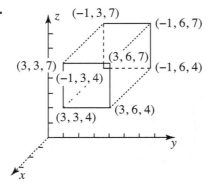

15. The union of the planes $x = 0$, $y = 0$, and $z = 0$

18. The union of the planes $x = 2$ and $z = 8$

21. $d = \sqrt{(3-6)^2 + (-1-4)^2 + (2-8)^2} = \sqrt{70}$

24. (a) 2;

(b) $d = \sqrt{(-6)^2 + 2^2 + (-3)^2} = 7$

27. $d(P_1, P_2) = \sqrt{(4-1)^2 + (1-2)^2 + (3-3)^2} = \sqrt{10}$;
$d(P_1, P_3) = \sqrt{(4-1)^2 + (6-2)^2 + (4-3)^2} = \sqrt{26}$
$d(P_2, P_3) = \sqrt{(4-4)^2 + (6-1)^2 + (4-3)^2} = \sqrt{26}$;
The triangle is an isosceles triangle.

30. $d(P_1, P_2) = \sqrt{(1-2)^2 + (4-3)^2 + (4-2)^2} = \sqrt{6}$
$d(P_1, P_3) = \sqrt{(5-2)^2 + (0-3)^2 + (-4-2)^2} = 3\sqrt{6}$
$d(P_2, P_3) = \sqrt{(5-1)^2 + (0-4)^2 + (-4-4)^2} = 4\sqrt{6}$
Since $d(P_1, P_2) + d(P_1, P_3) = d(P_2, P_3)$, the points P_1, P_2, and P_3 are collinear.

33. $\left(\dfrac{1+7}{2}, \dfrac{3+(-2)}{2}, \dfrac{1/2 + 5/2}{2} \right) = (4, 1/2, 3/2)$

36. $\qquad (-3 + (-5))/2 = x_3; \qquad (4+8)/2 = y_3 = 6; \qquad (1+3)/2 = z_3 = 2$

The coordinates of P_3 are $(-4, 6, 2)$.

(a) $\left(\dfrac{-3 + (-4)}{2}, \dfrac{4+6}{2}, \dfrac{1+2}{2} \right) = (-7/2, 5, 3/2)$

(b) $\left(\dfrac{-4 + (-5)}{2}, \dfrac{6+8}{2}, \dfrac{2+3}{2} \right) = (-9/2, 7, 5/2)$

39. $\overrightarrow{P_1 P_2} = \langle 2, 1, 1 \rangle$

42. $2\mathbf{a} - (\mathbf{b} - \mathbf{c}) = \langle 2, -6, 4 \rangle - \langle -3, -5, -8 \rangle = \langle 5, -1, 12 \rangle$

45. $\|\mathbf{a} + \mathbf{c}\| = \|\langle 3, 3, 11 \rangle\| = \sqrt{9 + 9 + 121} = \sqrt{139}$

48. $\|\mathbf{b}\|\mathbf{a} + \|\mathbf{a}\|\mathbf{b} = \sqrt{1+1+1}\,\langle 1, -3, 2 \rangle + \sqrt{1+9+4}\,\langle -1, 1, 1 \rangle$
$\qquad = \langle \sqrt{3}, -3\sqrt{3}, 2\sqrt{3} \rangle + \langle -\sqrt{14}, \sqrt{14}, \sqrt{14} \rangle$
$\qquad = \langle \sqrt{3} - \sqrt{14}, -3\sqrt{3} + \sqrt{14}, 2\sqrt{3} + \sqrt{14} \rangle$

51. $\mathbf{b} = 4\mathbf{a} = 4\mathbf{i} - 4\mathbf{j} + 4\mathbf{k}$

7.3 | Dot Product |

3. $\mathbf{a} \cdot \mathbf{c} = 2(3) + (-3)6 + 4(-1) = -16$ **6.** $\mathbf{b} \cdot (\mathbf{a} - \mathbf{c}) = (-1)(-1) + 2(-9) + 5(5) = 8$

9. $\mathbf{a} \cdot (\mathbf{a} + \mathbf{b} + \mathbf{c}) = 2(4) + (-3)5 + 4(8) = 25$

12. $(\mathbf{c} \cdot \mathbf{b})\mathbf{a} = [3(-1) + 6(2) + (-1)5]\langle 2, -3, 4 \rangle = 4\langle 2, -3, 4 \rangle = \langle 8, -12, 16 \rangle$

15. a and f, b and e, c and d

18. If $\mathbf{a}$ and $\mathbf{b}$ represent adjacent sides of the rhombus, then $\|\mathbf{a}\| = \|\mathbf{b}\|$, the diagonals of the rhombus are $\mathbf{a} + \mathbf{b}$ and $\mathbf{a} - \mathbf{b}$, and

$$(\mathbf{a} + \mathbf{b}) \cdot (\mathbf{a} - \mathbf{b}) = \mathbf{a} \cdot \mathbf{a} - \mathbf{a} \cdot \mathbf{b} + \mathbf{b} \cdot \mathbf{a} - \mathbf{b} \cdot \mathbf{b} = \mathbf{a} \cdot \mathbf{a} - \mathbf{b} \cdot \mathbf{b} = \|\mathbf{a}\|^2 - \|\mathbf{b}\|^2 = 0.$$

Thus the diagonals are perpendicular.

21. $\mathbf{a} \cdot \mathbf{b} = 3(2) + (-1)2 = 4; \quad \|\mathbf{a}\| = \sqrt{10}, \quad \|\mathbf{b}\| = 2\sqrt{2}$

$$\cos\theta = \frac{4}{(\sqrt{10})(2\sqrt{2})} = \frac{1}{\sqrt{5}}$$

$$\theta = \cos^{-1}\frac{1}{\sqrt{5}}$$

$$1.11 \text{ rad} \approx 63.43°$$

24. $\mathbf{a} \cdot \mathbf{b} = \frac{1}{2}(2) + \frac{1}{2}(-4) + \frac{3}{2}(6) = 8; \quad \|\mathbf{a}\| = \sqrt{11}/2, \quad \|\mathbf{b}\| = 2\sqrt{14}$

$$\cos\theta = \frac{8}{(\sqrt{11}/2)(2\sqrt{14})} = \frac{8}{\sqrt{154}}$$

$$\theta = \cos^{-1}(8/\sqrt{154}) \approx 0.87 \text{ rad} \approx 49.86°$$

27. $\|\mathbf{a}\| = 2; \quad \cos\alpha = 1/2, \alpha = 60°; \quad \cos\beta = 0, \beta = 90°; \quad \cos\gamma = -\sqrt{3}/2, \gamma = 150°$

30. If $\mathbf{a}$ and $\mathbf{b}$ are orthogonal, then $\mathbf{a} \cdot \mathbf{b} = 0$ and

$$\cos\alpha_1 \cos\alpha_2 + \cos\beta_1 \cos\beta_2 + \cos\gamma_1 \cos\gamma_2 = \frac{a_1}{\|\mathbf{a}\|}\frac{b_1}{\|\mathbf{b}\|} + \frac{a_2}{\|\mathbf{a}\|}\frac{b_2}{\|\mathbf{b}\|} + \frac{a_3}{\|\mathbf{a}\|}\frac{b_3}{\|\mathbf{b}\|}$$

$$= \frac{1}{\|\mathbf{a}\|\,\|\mathbf{b}\|}(a_1 b_1 + a_2 b_2 + a_3 b_3)$$

$$= \frac{1}{\|\mathbf{a}\|\,\|\mathbf{b}\|}(\mathbf{a} \cdot \mathbf{b}) = 0.$$

33. $\text{comp}_{\mathbf{b}}\mathbf{a} = \mathbf{a} \cdot \mathbf{b}/\|\mathbf{b}\| = \langle 1, -1, 3 \rangle \cdot \langle 2, 6, 3 \rangle/7 = 5/7$

36. $\mathbf{a} + \mathbf{b} = \langle 3, 5, 6 \rangle$; $2\mathbf{b} = \langle 4, 12, 6 \rangle$; $\text{comp}_{2\mathbf{b}}(\mathbf{a} + \mathbf{b}) \cdot 2\mathbf{b}/|2\mathbf{b}| = \langle 3, 5, 6 \rangle \cdot \langle 4, 12, 6 \rangle/14 = 54/7$

39. $\text{comp}_{\mathbf{b}}\mathbf{a} = \mathbf{a} \cdot \mathbf{b}/\|\mathbf{b}\| = (-5\mathbf{i} + 5\mathbf{j}) \cdot (-3\mathbf{i} + 4\mathbf{j})/5 = 7$

$\text{proj}_{\mathbf{b}}\mathbf{a} = (\text{comp}_{\mathbf{b}}\mathbf{a})\mathbf{b}/\|\mathbf{b}\| = 7(-3\mathbf{i} + 4\mathbf{j})/5 = -\dfrac{21}{5}\mathbf{i} + \dfrac{28}{5}\mathbf{j}$

42. $\text{comp}_{\mathbf{b}}\mathbf{a} = \mathbf{a} \cdot \mathbf{b}/\|\mathbf{b}\| = \langle 1, 1, 1 \rangle \cdot \langle -2, 2, -1 \rangle/3 = -1/3$

$\text{proj}_{\mathbf{b}}\mathbf{a} = (\text{comp}_{\mathbf{b}}\mathbf{a})\mathbf{b}/\|\mathbf{b}\| = -\dfrac{1}{3}\langle -2, 2, -1 \rangle/3 = \langle 2/9, -2/9, 1/9 \rangle$

45. We identify $\|\mathbf{F}\| = 20$, $\theta = 60°$ and $\|\mathbf{d}\| = 100$. Then $W = \|\mathbf{F}\|\,\|\mathbf{d}\|\cos\theta = 20(100)(\frac{1}{2}) = 1000$ ft-lb.

48. Using $\mathbf{d} = 6\mathbf{i} + 2\mathbf{j}$ and $\mathbf{F} = 3(\frac{3}{5}\mathbf{i} + \frac{4}{5}\mathbf{j})$, $W = \mathbf{F} \cdot \mathbf{d} = \langle \frac{9}{5}, \frac{12}{5} \rangle \cdot \langle 6, 2 \rangle = \frac{78}{5}$ ft-lb.

7.4 Cross Product

3. $\mathbf{a} \times \mathbf{b} = \begin{vmatrix} \mathbf{i} & \mathbf{j} & \mathbf{k} \\ 1 & -3 & 1 \\ 2 & 0 & 4 \end{vmatrix} = \begin{vmatrix} -3 & 1 \\ 0 & 4 \end{vmatrix}\mathbf{i} - \begin{vmatrix} 1 & 1 \\ 2 & 4 \end{vmatrix}\mathbf{j} + \begin{vmatrix} 1 & -3 \\ 2 & 0 \end{vmatrix}\mathbf{k} = \langle -12, -2, 6 \rangle$

6. $\mathbf{a} \times \mathbf{b} = \begin{vmatrix} \mathbf{i} & \mathbf{j} & \mathbf{k} \\ 4 & 1 & -5 \\ 2 & 3 & -1 \end{vmatrix} = \begin{vmatrix} 1 & -5 \\ 3 & -1 \end{vmatrix}\mathbf{i} - \begin{vmatrix} 4 & -5 \\ 2 & -1 \end{vmatrix}\mathbf{j} + \begin{vmatrix} 4 & 1 \\ 2 & 3 \end{vmatrix}\mathbf{k} = 14\mathbf{i} - 6\mathbf{j} + 10\mathbf{k}$

9. $\mathbf{a} \times \mathbf{b} = \begin{vmatrix} \mathbf{i} & \mathbf{j} & \mathbf{k} \\ 2 & 2 & -4 \\ -3 & -3 & 6 \end{vmatrix} = \begin{vmatrix} 2 & -4 \\ -3 & 6 \end{vmatrix}\mathbf{i} - \begin{vmatrix} 2 & -4 \\ -3 & 6 \end{vmatrix}\mathbf{j} + \begin{vmatrix} 2 & 2 \\ -3 & -3 \end{vmatrix}\mathbf{k} = \langle 0, 0, 0 \rangle$

12. $\overrightarrow{P_1 P_2} = (0, 1, 1)$; $\overrightarrow{P_1 P_3} = (1, 2, 2)$;

$\overrightarrow{P_1 P_2} \times \overrightarrow{P_1 P_3} = \begin{vmatrix} \mathbf{i} & \mathbf{j} & \mathbf{k} \\ 0 & 1 & 1 \\ 1 & 2 & 2 \end{vmatrix} = \begin{vmatrix} 1 & 1 \\ 2 & 2 \end{vmatrix}\mathbf{i} - \begin{vmatrix} 0 & 1 \\ 1 & 2 \end{vmatrix}\mathbf{j} + \begin{vmatrix} 0 & 1 \\ 1 & 2 \end{vmatrix}\mathbf{k} = \mathbf{j} - \mathbf{k}$

15. $\mathbf{a} \times \mathbf{b} = \begin{vmatrix} \mathbf{i} & \mathbf{j} & \mathbf{k} \\ 5 & -2 & 1 \\ 2 & 0 & -7 \end{vmatrix} = \begin{vmatrix} -2 & 1 \\ 0 & -7 \end{vmatrix}\mathbf{i} - \begin{vmatrix} 5 & 1 \\ 2 & -7 \end{vmatrix}\mathbf{j} + \begin{vmatrix} 5 & -2 \\ 2 & 0 \end{vmatrix}\mathbf{k} = \langle 14, 37, 4 \rangle$

$\mathbf{a} \cdot (\mathbf{a} \times \mathbf{b}) = \langle 5, -2, -1 \rangle \cdot \langle 14, 37, 4 \rangle = 70 - 74 + 4 = 0;$

$\mathbf{b} \cdot (\mathbf{a} \times \mathbf{b}) = \langle 2, 0, -7 \rangle \cdot \langle 14, 37, 4 \rangle = 28 + 0 - 28 = 0$

18. (a)
$$\mathbf{b} \times \mathbf{c} = \begin{vmatrix} \mathbf{i} & \mathbf{j} & \mathbf{k} \\ 1 & 2 & -1 \\ -1 & 5 & 8 \end{vmatrix} = \begin{vmatrix} 2 & -1 \\ 5 & 8 \end{vmatrix}\mathbf{i} - \begin{vmatrix} 1 & -1 \\ -1 & 8 \end{vmatrix}\mathbf{j} + \begin{vmatrix} 1 & 2 \\ -1 & 5 \end{vmatrix}\mathbf{k} = 21\mathbf{i} - 7\mathbf{j} + 7\mathbf{k}$$

$$\mathbf{a} \times (\mathbf{b} \times \mathbf{c}) = \begin{vmatrix} \mathbf{i} & \mathbf{j} & \mathbf{k} \\ 3 & 0 & -4 \\ 21 & -7 & 7 \end{vmatrix} = \begin{vmatrix} 0 & -4 \\ -7 & 7 \end{vmatrix}\mathbf{i} - \begin{vmatrix} 3 & -4 \\ 21 & 7 \end{vmatrix}\mathbf{j} + \begin{vmatrix} 3 & 0 \\ 21 & -7 \end{vmatrix}\mathbf{k}$$
$$= -28\mathbf{i} - 105\mathbf{j} - 21\mathbf{k}$$

(b) $\mathbf{a} \cdot \mathbf{c} = (3\mathbf{i} - 4\mathbf{k}) \cdot (-\mathbf{i} + 5\mathbf{j} + 8\mathbf{k}) = -35; \quad (\mathbf{a} \cdot \mathbf{c})\mathbf{b} = -35(\mathbf{i} + 2\mathbf{j} - \mathbf{k}) = -35\mathbf{i} - 70\mathbf{j} + 35\mathbf{k};$

$\mathbf{a} \cdot \mathbf{b} = (3\mathbf{i} - 4\mathbf{k}) \cdot (\mathbf{i} + 2\mathbf{j} - \mathbf{k}) = 7; \quad (\mathbf{a} \cdot \mathbf{b})\mathbf{c} = 7(-\mathbf{i} + 5\mathbf{j} + 8\mathbf{k}) = -7\mathbf{i} + 35\mathbf{j} + 56\mathbf{k};$

$\mathbf{a} \times (\mathbf{b} \times \mathbf{c}) = (\mathbf{a} \cdot \mathbf{c})\mathbf{b} - (\mathbf{a} \cdot \mathbf{b})\mathbf{c}(-35\mathbf{i} - 70\mathbf{j} + 35\mathbf{k}) - (-7\mathbf{i} + 35\mathbf{j} + 56\mathbf{k}) = -28\mathbf{i} - 105\mathbf{j} - 21\mathbf{k}$

21. $\mathbf{k} \times (2\mathbf{i} - \mathbf{j}) = \mathbf{k} \times (2\mathbf{i}) + \mathbf{k} \times (-\mathbf{j}) = 2(\mathbf{k} \times \mathbf{i}) - (\mathbf{k} \times \mathbf{j}) = 2\mathbf{j} - (-\mathbf{i}) = \mathbf{i} + 2\mathbf{j}$

24. $(2\mathbf{i} - \mathbf{j} + 5\mathbf{k}) \times \mathbf{i} = (2\mathbf{i} \times \mathbf{i}) + (-\mathbf{j} \times \mathbf{i}) + (5\mathbf{k} \times \mathbf{i}) = 2(\mathbf{i} \times \mathbf{i}) + (\mathbf{i} \times \mathbf{j}) + 5(\mathbf{k} \times \mathbf{i}) = 5\mathbf{j} + \mathbf{k}$

27. $\mathbf{k} \cdot (\mathbf{j} \times \mathbf{k}) = \mathbf{k} \cdot \mathbf{i} = 0$

30. $(\mathbf{i} \times \mathbf{j}) \cdot (3\mathbf{j} \times \mathbf{i}) = \mathbf{k} \cdot (-3\mathbf{k}) = -3(\mathbf{k} \cdot \mathbf{k}) = -3$

33. $(\mathbf{i} \times \mathbf{i}) \times \mathbf{j} = \mathbf{0} \times \mathbf{j} = \mathbf{0}$

36. $(\mathbf{i} \times \mathbf{k}) \times (\mathbf{j} \times \mathbf{i}) = (-\mathbf{j}) \times (-\mathbf{k}) = (-1)(-1)(\mathbf{j} \times \mathbf{k}) = \mathbf{j} \times \mathbf{k} = \mathbf{i}$

39. $(-\mathbf{a}) \times \mathbf{b} = -(\mathbf{a} \times \mathbf{b}) = -4\mathbf{i} + 3\mathbf{j} - 6\mathbf{k}$

42. $(\mathbf{a} \times \mathbf{b}) \cdot \mathbf{c} = 4(2) + (-3)4 + 6(-1) = -10$

45. (a) Let $A = (1, 3, 0)$, $B = (2, 0, 0)$, $C = (0, 0, 4)$, and $D = (1, -3, 4)$. Then $\overrightarrow{AB} = \mathbf{i} - 3\mathbf{j}$, $\overrightarrow{AC} = -\mathbf{i} - 3\mathbf{j} + 4\mathbf{k}$, $\overrightarrow{CD} = \mathbf{i} - 3\mathbf{j}$, and $\overrightarrow{BD} = -\mathbf{i} - 3\mathbf{j} + 4\mathbf{k}$. Since $\overrightarrow{AB} = \overrightarrow{CD}$ and $\overrightarrow{AC} = \overrightarrow{BD}$, the quadrilateral is a parallelogram.

(b) Computing

$$\overrightarrow{AB} \times \overrightarrow{AC} = \begin{vmatrix} \mathbf{i} & \mathbf{j} & \mathbf{k} \\ 1 & -3 & 0 \\ -1 & -3 & 4 \end{vmatrix} = -12\mathbf{i} - 4\mathbf{j} - 6\mathbf{k}$$

we find that the area is $\| -12\mathbf{i} - 4\mathbf{j} - 6\mathbf{k} \| = \sqrt{144 + 16 + 36} = 14$.

48. $\overrightarrow{P_1P_2} = \mathbf{j} + 2\mathbf{k};$ $\overrightarrow{P_2P_3} = 2\mathbf{i} + \mathbf{j} - 2\mathbf{k}$

$$\overrightarrow{P_1P_2} \times \overrightarrow{P_2P_3} = \begin{vmatrix} \mathbf{i} & \mathbf{j} & \mathbf{k} \\ 0 & 1 & 2 \\ 2 & 1 & -2 \end{vmatrix} = \begin{vmatrix} 1 & 2 \\ 1 & -2 \end{vmatrix}\mathbf{i} - \begin{vmatrix} 0 & 2 \\ 2 & -2 \end{vmatrix}\mathbf{j} + \begin{vmatrix} 0 & 1 \\ 2 & 1 \end{vmatrix}\mathbf{k} = -4\mathbf{i} + 4\mathbf{j} - 2\mathbf{k};$$

$A = \frac{1}{2}\| -4\mathbf{i} + 4\mathbf{j} - 2\mathbf{k} \| = 3$ sq. units

51. $\mathbf{b} \times \mathbf{c} = \begin{vmatrix} \mathbf{i} & \mathbf{j} & \mathbf{k} \\ -1 & 4 & 0 \\ 2 & 2 & 2 \end{vmatrix} = \begin{vmatrix} 4 & 0 \\ 2 & 2 \end{vmatrix} \mathbf{i} - \begin{vmatrix} -1 & 0 \\ 2 & 2 \end{vmatrix} \mathbf{j} + \begin{vmatrix} -1 & 4 \\ 2 & 2 \end{vmatrix} \mathbf{k} = 8\mathbf{i} + 2\mathbf{j} - 10\mathbf{k}$

$\mathbf{v} = |\mathbf{a} \cdot (\mathbf{b} \times \mathbf{c})| = |(\mathbf{i} + \mathbf{j}) \cdot (8\mathbf{i} + 2\mathbf{j} - 10\mathbf{k})| = |8 + 2 + 0| = 10$ cu. units

54. The four points will be coplanar if the three vectors $\overrightarrow{P_1 P_2} = \langle 3, -1, -1 \rangle$, $\overrightarrow{P_2 P_3} = \langle -3, -5, 13 \rangle$, and $\overrightarrow{P_3 P_4} = \langle -8, 7, -6 \rangle$ are coplanar.

$\overrightarrow{P_2 P_3} \times \overrightarrow{P_3 P_4} = \begin{vmatrix} \mathbf{i} & \mathbf{j} & \mathbf{k} \\ -3 & -5 & 12 \\ -8 & 7 & -6 \end{vmatrix} = \begin{vmatrix} -5 & 13 \\ 7 & -6 \end{vmatrix} \mathbf{i} - \begin{vmatrix} -3 & 13 \\ -8 & -6 \end{vmatrix} \mathbf{j} + \begin{vmatrix} -3 & -5 \\ -8 & 7 \end{vmatrix} \mathbf{k} = \langle -61, -122, -61 \rangle$

$\overrightarrow{P_1 P_2} \cdot (\overrightarrow{P_2 P_3} \times \overrightarrow{P_3 P_4}) = \langle 3, -1, -1 \rangle \cdot \langle -61, -122, -61 \rangle = -183 + 122 + 61 = 0$

The four points are coplanar.

57. (a) We note first that $\mathbf{a} \times \mathbf{b} = \mathbf{k}$, $\mathbf{b} \times \mathbf{c} = \frac{1}{2}(\mathbf{i} - \mathbf{k})$, $\mathbf{c} \times \mathbf{a} = \frac{1}{2}(\mathbf{j} - \mathbf{k})$, $\mathbf{a} \cdot (\mathbf{b} \times \mathbf{c}) = \frac{1}{2}$, $\mathbf{b} \cdot (\mathbf{c} \times \mathbf{a}) = \frac{1}{2}$, and $\mathbf{c} \cdot (\mathbf{a} \times \mathbf{b}) = \frac{1}{2}$. Then

$$A = \frac{\frac{1}{2}(\mathbf{i} - \mathbf{k})}{\frac{1}{2}} = \mathbf{i} - \mathbf{k}, \quad B = \frac{\frac{1}{2}(\mathbf{j} - \mathbf{k})}{\frac{1}{2}} = \mathbf{j} - \mathbf{k}, \quad \text{and} \quad C = \frac{\mathbf{k}}{\frac{1}{2}} = 2\mathbf{k}.$$

(b) We need to compute $\mathbf{A} \cdot (\mathbf{B} \times \mathbf{C})$. Using formula (10) in the text we have

$$\mathbf{B} \times \mathbf{C} = \frac{(\mathbf{c} \times \mathbf{a}) \times (\mathbf{a} \times \mathbf{b})}{[\mathbf{b} \cdot (\mathbf{c} \times \mathbf{a})][\mathbf{c} \cdot (\mathbf{a} \times \mathbf{b})]} = \frac{[(\mathbf{c} \times \mathbf{a}) \cdot \mathbf{b}]\mathbf{a} - [(\mathbf{c} \times \mathbf{a}) \cdot \mathbf{a}]\mathbf{b}}{[\mathbf{b} \cdot (\mathbf{c} \times \mathbf{a})][\mathbf{c} \cdot (\mathbf{a} \times \mathbf{b})]}$$

$$= \frac{\mathbf{a}}{\mathbf{c} \cdot (\mathbf{a} \times \mathbf{b})} \qquad \boxed{\text{since} (\mathbf{c} \times \mathbf{a}) \cdot \mathbf{a} = 0.}$$

Then

$$\mathbf{A} \cdot (\mathbf{B} \times \mathbf{C}) = \frac{\mathbf{b} \times \mathbf{c}}{\mathbf{a} \cdot (\mathbf{b} \times \mathbf{c})} \cdot \mathbf{a} \, c \cdot (\mathbf{a} \times \mathbf{b}) = \frac{1}{\mathbf{c} \cdot (\mathbf{a} \times \mathbf{b})}$$

and the volume of the unit cell of the reciprocal lattice is the reciprocal of the volume of the unit cell of the original lattice.

7.5 Lines and Planes in 3-Space

The equation of a line through P_1 and P_2 in 3-space with $\mathbf{r}_1 = \overrightarrow{OP_1}$ and $\mathbf{r}_2 = \overrightarrow{OP_2}$ can be expressed as $\mathbf{r} = \mathbf{r}_1 + t(k\mathbf{a})$ or $\mathbf{r} = \mathbf{r}_2 + t(k\mathbf{a})$ where $\mathbf{a} = \mathbf{r}_2 - \mathbf{r}_1$ and k is any non-zero scalar. Thus, the form of the equation of a line is not unique. (See the alternate solution to Problem 1.)

3. $\mathbf{a} = \langle 1/2 - (-3/2), -1/2 - 5/2, 1 - (-1/2) \rangle = \langle 2, -3, 3/2 \rangle$;

$\langle x, y, z \rangle = \langle 1/2, -1/2, 1 \rangle + t \langle 2, -3, 3/2 \rangle$

6. $\mathbf{a} = \langle 3 - 5/2, 2 - 1, 1 - (-2) \rangle = \langle 1/2, 1, 3 \rangle$; $\langle x, y, z \rangle = \langle 3, 2, 1 \rangle + t \langle 1/2, 1, 3 \rangle$

9. $\mathbf{a} = \langle 1 - 3, 0 - (-2), 0 - (-7) \rangle = \langle -2, 2, 7 \rangle$; $x = 1 - 2t$, $y = 2t$, $z = 7t$

12. $\mathbf{a} = \langle -3 - 4, 7 - (-8), 9 - (-1) \rangle = \langle -7, 15, 10 \rangle$; $x = -3 - 7t$, $y = 7 + 15t$, $z = 9 + 10t$

15. $a_1 = -7 - 4 = -11$, $a_2 = 2 - 2 = 0$, $a_3 = 5 - 1 = 4$; $\dfrac{x + 7}{-11} = \dfrac{z - 5}{4}$, $y = 2$

18. $a_1 = 5/6 - 1/3 = 1/2$; $a_2 = -1/4 - 3/8 = -5/8$; $a_3 = 1/5 - 1/10 = 1/10$

$$\frac{x - 5/6}{1/2} = \frac{y + 1/4}{-5/8} = \frac{z - 1/5}{1/10}$$

21. Parametric: $x = 5t$, $y = 9t$, $z = 4t$;   symmetric: $\dfrac{x}{5} = \dfrac{y}{9} = \dfrac{z}{4}$

24. A direction vector is $\langle 5, 1/3, -2 \rangle$. Symmetric equations for the line are
$(x - 4)/5 = (y + 11)/(1/3) = (z + 7)/(-2)$.

27. Both lines go through the points $(0, 0, 0)$ and $(6, 6, 6)$. Since two points determine a line, the lines are the same.

30. The parametric equations for the line are $x = 1 + 2t$, $y = -2 + 3t$, $z = 4 + 2t$. In the xy-plane, $z = 4 + 2t = 0$ and $t = -2$. Then $x = 1 + 2(-2) = -3$ and $y = -2 + 3(-2) = -8$. The point is $(-3, -8, 0)$. In the xz-plane, $y = -2 + 3t = 0$ and $t = 2/3$. Then $x = 1 + 2(2/3) = 7/3$ and $z = 4 + 2(2/3) = 16/3$. The point is $(7/3, 0, 16/3)$. In the yz-plane, $x = 1 + 2t = 0$ and $t = -1/2$. Then $y = -2 + 3(-1/2) = -7/2$ and $z = 4 + 2(-1/2) = 3$. The point is $(0, -7/2, 3)$.

33. The system of equations $2 - t = 4 + s$, $3 + t = 1 + s$, $1 + t = 1 - s$, or $t + s = -2$, $t - s = -2$, $t + s = 0$ has no solution since $-2 \neq 0$. Thus, the lines do not intersect.

36. $\mathbf{a} = \langle 2, 7, -1 \rangle$, $\mathbf{b} = \langle -2, 1, 4 \rangle$, $\mathbf{a} \cdot \mathbf{b} = -1$, $\|\mathbf{a}\| = 3\sqrt{6}$, $\|\mathbf{b}\| = \sqrt{21}$;

$$\cos\theta = \frac{\mathbf{a} \cdot \mathbf{b}}{\|\mathbf{a}\|\,\|\mathbf{b}\|} = \frac{-1}{(3\sqrt{6})(\sqrt{21})} = -\frac{1}{9\sqrt{14}}; \quad \theta = \cos^{-1}\left(-\frac{1}{9\sqrt{14}}\right) \approx 91.70°$$

39. $2(x - 5) - 3(y - 1) + 4(z - 3) = 0$; $2x - 3y + 4z = 19$

42. $6x - y + 3z = 0$

45. From the points $(3, 5, 2)$ and $(2, 3, 1)$ we obtain the vector $\mathbf{u} = \mathbf{i} + 2\mathbf{j} + \mathbf{k}$. From the points $(2, 3, 1)$ and $(-1, -1, 4)$ we obtain the vector $\mathbf{v} = 3\mathbf{i} + 4\mathbf{j} - 3\mathbf{k}$. From the points $(-1, -1, 4)$ and (x, y, z) we obtain the vector $\mathbf{w} = (x + 1)\mathbf{i} + (y + 1)\mathbf{j} + (z - 4)\mathbf{k}$. Then, a normal vector is

$$\mathbf{u} \times \mathbf{v} = \begin{vmatrix} \mathbf{i} & \mathbf{j} & \mathbf{k} \\ 1 & 2 & 1 \\ 3 & 4 & -3 \end{vmatrix} = -10\mathbf{i} + 6\mathbf{j} - 2\mathbf{k}.$$

A vector equation of the plane is $-10(x + 1) + 6(y + 1) - 2(z - 4) = 0$ or $5x - 3y + z = 2$.

48. The three points are not colinear and all satisfy $x = 0$, which is the equation of the plane.

51. A normal vector to $x + y - 4z = 1$ is $\langle 1, 1, -4 \rangle$. The equation of the parallel plane is $(x - 2) + (y - 3) - 4(z + 5) = 0$ or $x + y - 4z = 25$.

54. A normal vector is $\langle 0, 1, 0 \rangle$. The equation of the plane is $y + 5 = 0$ or $y = -5$.

57. A direction vector for the two lines is $\langle 1, 2, 1 \rangle$. Points on the lines are $(1, 1, 3)$ and $(3, 0, -2)$. Thus, another vector parallel to the plane is $\langle 1 - 3, 1 - 0, 3 + 2 \rangle = \langle -2, 1, 5 \rangle$. A normal vector to the plane is $\langle 1, 2, 1 \rangle \times \langle -2, 1, 5 \rangle = \langle 9, -7, 5 \rangle$. Using the point $(3, 0, -2)$ in the plane, the equation of the plane is $9(x - 3) - 7(y - 0) + 5(z + 2) = 0$ or $9x - 7y + 5z = 17$.

60. A normal vector to the plane is $\langle 2 - 1, 6 - 0, -3 + 2 \rangle = \langle 1, 6, -1 \rangle$. The equation of the plane is $(x - 1) + 6(y - 1) - (z - 1) = 0$ or $x + 6y - z = 6$.

63. A direction vector of the line is $\langle -6, 9, 3 \rangle$, and the normal vectors of the planes are **(a)** $\langle 4, 1, 2 \rangle$, **(b)** $\langle 2, -3, 1 \rangle$, **(c)** $\langle 10, -15, -5 \rangle$, **(d)** $\langle -4, 6, 2 \rangle$. Vectors **(c)** and **(d)** are multiples of the direction vector and hence the corresponding planes are perpendicular to the line.

66. Letting $y = t$ in both equations and solving $x - z = 2 - 2t$, $3x + 2z = 1 + t$, we obtain $x = 1 - \frac{3}{5}t$, $y = t$, $z = -1 + \frac{7}{5}t$ or, letting $t = 5s$, $x = 1 - 3s$, $y = 5s$, $z = -1 + 7s$.

69. Substituting the parametric equations into the equation of the plane, we obtain $2(1 + 2t) - 3(2 - t) + 2(-3t) = -7$ or $t = -3$. Letting $t = -3$ in the equation of the line, we obtain the point of intersection $(-5, 5, 9)$.

72. Substituting the parametric equations into the equation of the plane, we obtain $4 + t - 3(2 + t) + 2(1 + 5t) = 0$ or $t = 0$. Letting $t = 0$ in the equation of the line, we obtain the point of intersection $(4, 2, 1)$.

In Problems 75 and 76, the cross product of the direction vector of the line with the normal vector of the given plane will be a normal vector to the desired plane.

75. A direction vector of the line is $\langle 3, -1, 5 \rangle$ and a normal vector to the given plane is $\langle 1, 1, 1 \rangle$. A normal vector to the desired plane is $\langle 3, -1, 5 \rangle \times \langle 1, 1, 1 \rangle = \langle -6, 2, 4 \rangle$. A point on the line, and hence in the plane, is $(4, 0, 1)$. The equation of the plane is $-6(x - 4) + 2(y - 0) + 4(z - 1) = 0$ or $3x - y - 2z = 10$.

78.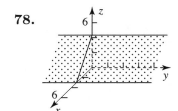

81.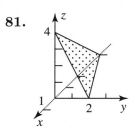

7.6 | Vector Spaces

3. Not a vector space. Axiom **(x)** is not satisfied.

6. A vector space

9. A vector space

12. Not a subspace. Axiom (**i**) is not satisfied.

15. A subspace

18. A subspace

21. Let (x_1, y_1, z_1) and (x_2, y_2, z_2) be in S. Then

$$(x_1, y_1, z_1) + (x_2, y_2, z_2) = (at_1, bt_1, ct_1) + (at_2, bt_2, ct_2) = (a(t_1 + t_2), b(t_1 + t_2), c(t_1 + t_2))$$

is in S. Also, for (x, y, z) in S then $k(x, y, z) = (kx, ky, kz) = (a(kt), b(kt), c(kt))$ is also in S.

24. (**a**) The assumption $c_1 p_1 + c_2 p_2 = 0$ is equivalent to $(c_1 + c_2)x + (c_1 - c_2) = 0$. Thus $c_1 + c_2 = 0$, $c_1 - c_2 = 0$. The only solution of this system is $c_1 = 0$, $c_2 = 0$.

    (**b**) Solving the system $c_1 + c_2 = 5$, $c_1 - c_2 = 2$ gives $c_1 = \frac{7}{2}$, $c_2 = \frac{3}{2}$. Thus $p(x) = \frac{7}{2}p_1(x) + \frac{3}{2}p_2(x)$

27. Linearly independent

30. $(x, \sin x) = \displaystyle\int_0^{2\pi} x \sin x \, dx = (-x \cos x + \sin x)\Big|_0^{2\pi} = -2\pi$

33. We need to show that $\text{Span}\{x_1, x_2, \ldots, x_n\}$ is closed under vector addition and scalar multiplication. Suppose $\mathbf{u}$ and $\mathbf{v}$ are in $\text{Span}\{x_1, x_2, \ldots, x_n\}$. Then $\mathbf{u} = a_1 x_1 + a_2 x_2 + \cdots + a_n x_n$ and $\mathbf{v} = b_1 x_1 + b_2 x_2 + \cdots + b_n x_n$, so that

$$\mathbf{u} + \mathbf{v} = (a_1 + b_1)x_1 + (a_2 + b_2)x_2 + \cdots + (a_n + b_n)x_n,$$

which is in $\text{Span}\{x_1, x_2, \ldots, x_n\}$. Also, for any real number k,

$$k\mathbf{u} = k(a_1 x_1 + a_2 x_2 + \cdots + a_n x_n) = ka_1 x_1 + ka_2 x_2 + \cdots + ka_n x_n,$$

which is in $\text{Span}\{x_1, x_2, \ldots, x_n\}$. Thus, $\text{Span}\{x_1, x_2, \ldots, x_n\}$ is a subspace of $\mathbf{V}$.

7.7   Gram–Schmidt Orthogonalization Process

Since the basis vectors in Problems 3 and 4 are orthogonal but not orthonormal, the result of Theorem 7.7.1 must be slightly modified to read

$$\mathbf{u} = \frac{\mathbf{u} \cdot \mathbf{w}_1}{||\mathbf{w}_1||^2} \mathbf{w}_1 + \frac{\mathbf{u} \cdot \mathbf{w}_2}{||\mathbf{w}_2||^2} \mathbf{w}_2 + \cdots + \frac{\mathbf{u} \cdot \mathbf{w}_n}{||\mathbf{w}_n||^2} \mathbf{w}_n.$$

The proof is very similar to that given in the text for Theorem 7.7.1.

3. Letting $\mathbf{w}_1 = \langle 1, 0, 1 \rangle$, $\mathbf{w}_2 = \langle 0, 1, 0 \rangle$, and $\mathbf{w}_3 = \langle -1, 0, 1 \rangle$ we have

$$\mathbf{w}_1 \cdot \mathbf{w}_2 = (1)(0) + (0)(1) + (1)(0) = 0$$

$$\mathbf{w}_1 \cdot \mathbf{w}_3 = (1)(-1) + (0)(0) + (1)(1) = 0$$

$$\mathbf{w}_2 \cdot \mathbf{w}_3 = (0)(-1) + (1)(0) + (0)(1) = 0$$

so the vectors are orthogonal. We also compute

$$\|\mathbf{w}_1\|^2 = 1^2 + 0^2 + 1^2 = 2$$

$$\|\mathbf{w}_2\|^2 = 0^2 + 1^2 + 0^2 = 1$$

$$\|\mathbf{w}_3\|^2 = (-1)^2 + 0^2 + 1^2 = 2$$

and, with $\mathbf{u} = \langle 10, 7, -13 \rangle$,

$$\mathbf{u} \cdot \mathbf{w}_1 = (10)(1) + (7)(0) + (-13)(1) = -3$$

$$\mathbf{u} \cdot \mathbf{w}_2 = (10)(0) + (7)(1) + (-13)(0) = 7$$

$$\mathbf{u} \cdot \mathbf{w}_3 = (10)(-1) + (7)(0) + (-13)(1) = -23.$$

Then, using the result given before the solution to this problem, we have

$$\mathbf{u} = -\frac{3}{2}\,\mathbf{w}_1 + 7\mathbf{w}_2 - \frac{23}{2}\,\mathbf{w}_3.$$

6. (a) We have $\mathbf{u}_1 = \langle -3, 4 \rangle$ and $\mathbf{u}_2 = \langle -1, 0 \rangle$. Taking $\mathbf{v}_1 = \mathbf{u}_1 = \langle -3, 4 \rangle$, and using $\mathbf{u}_2 \cdot \mathbf{v}_1 = 3$ and $\mathbf{v}_1 \cdot \mathbf{v}_1 = 25$ we obtain

$$\mathbf{v}_2 = \mathbf{u}_2 - \frac{\mathbf{u}_2 \cdot \mathbf{v}_1}{\mathbf{v}_1 \cdot \mathbf{v}_1}\,\mathbf{v}_1 = \langle -1, 0 \rangle - \frac{3}{25}\langle -3, 4 \rangle = \left\langle -\frac{16}{25}, -\frac{12}{25} \right\rangle.$$

Thus, an orthogonal basis is $\left\{ \langle -3, 4 \rangle, \left\langle -\frac{16}{25}, -\frac{12}{25} \right\rangle \right\}$ and an orthonormal basis is $\{\mathbf{w}_1', \mathbf{w}_2'\}$, where

$$\mathbf{w}_1' = \frac{1}{\|\langle -3, 4 \rangle\|}\,\langle -3, 4 \rangle = \frac{1}{5}\,\langle -3, 4 \rangle = \left\langle -\frac{3}{5}, \frac{4}{5} \right\rangle$$

and

$$\mathbf{w}_2' = \frac{1}{\|\langle -\frac{16}{25}, -\frac{12}{25} \rangle\|}\,\left\langle -\frac{16}{25}, -\frac{12}{25} \right\rangle = \frac{1}{4/5}\,\left\langle -\frac{16}{25}, -\frac{12}{25} \right\rangle = \left\langle -\frac{4}{5}, -\frac{3}{5} \right\rangle.$$

(b) We have $\mathbf{u}_1 = \langle -3, 4 \rangle$ and $\mathbf{u}_2 = \langle -1, 0 \rangle$. Taking $\mathbf{v}_1 = \mathbf{u}_2 = \langle -1, 0 \rangle$, and using $\mathbf{u}_1 \cdot \mathbf{v}_1 = 3$ and $\mathbf{v}_1 \cdot \mathbf{v}_1 = 1$ we obtain

$$\mathbf{v}_2 = \mathbf{u}_1 - \frac{\mathbf{u}_1 \cdot \mathbf{v}_1}{\mathbf{v}_1 \cdot \mathbf{v}_1}\,\mathbf{v}_1 = \langle -3, 4 \rangle - \frac{3}{1}\langle -1, 0 \rangle = \langle 0, 4 \rangle.$$

Thus, an orthogonal basis is $\{\langle -1, 0 \rangle, \langle 0, 4 \rangle\}$ and an orthonormal basis is $\{\mathbf{w}_3'', \mathbf{w}_4''\}$, where

$$\mathbf{w}_3'' = \frac{1}{\|\langle -1, 0 \rangle\|}\,\langle -1, 0 \rangle = \frac{1}{1}\,\langle -1, 0 \rangle = \langle -1, 0 \rangle$$

and

$$\mathbf{w}_4'' = \frac{1}{\|\langle 0, 4 \rangle\|}\,\langle 0, 4 \rangle = \frac{1}{4}\,\langle 0, 4 \rangle = \langle 0, 1 \rangle.$$

(c)

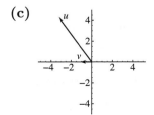 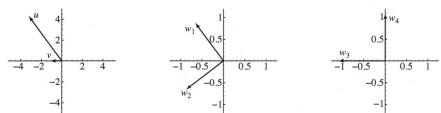 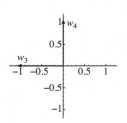

9. We have $\mathbf{u}_1 = \langle 1, 1, 0 \rangle$, $\mathbf{u}_2 = \langle 1, 2, 2 \rangle$, and $\mathbf{u}_3 = \langle 2, 2, 1 \rangle$. Taking $\mathbf{v}_1 = \mathbf{u}_1 = \langle 1, 1, 0 \rangle$ and using $\mathbf{u}_2 \cdot \mathbf{v}_1 = 3$ and $\mathbf{v}_1 \cdot \mathbf{v}_1 = 2$ we obtain

$$\mathbf{v}_2 = \mathbf{u}_2 - \frac{\mathbf{u}_2 \cdot \mathbf{v}_1}{\mathbf{v}_1 \cdot \mathbf{v}_1} \mathbf{v}_1 = \langle 1, 2, 2 \rangle - \frac{3}{2} \langle 1, 1, 0 \rangle = \left\langle -\frac{1}{2}, \frac{1}{2}, 2 \right\rangle.$$

Next, using $\mathbf{u}_3 \cdot \mathbf{v}_1 = 4$, $\mathbf{u}_3 \cdot \mathbf{v}_2 = 2$, and $\mathbf{v}_2 \cdot \mathbf{v}_2 = \frac{9}{2}$, we obtain

$$\mathbf{v}_3 = \mathbf{u}_3 - \frac{\mathbf{u}_3 \cdot \mathbf{v}_1}{\mathbf{v}_1 \cdot \mathbf{v}_1} \mathbf{v}_1 - \frac{\mathbf{u}_3 \cdot \mathbf{v}_2}{\mathbf{v}_2 \cdot \mathbf{v}_2} \mathbf{v}_2 = \langle 2, 2, 1 \rangle - \frac{4}{2} \langle 1, 1, 0 \rangle - \frac{2}{9/2} \left\langle -\frac{1}{2}, \frac{1}{2}, 2 \right\rangle = \left\langle \frac{2}{9}, -\frac{2}{9}, \frac{1}{9} \right\rangle.$$

Thus, an orthogonal basis is

$$B' = \left\{ \langle 1, 1, 0 \rangle, \left\langle -\frac{1}{2}, \frac{1}{2}, 2 \right\rangle, \left\langle \frac{2}{9}, -\frac{2}{9}, \frac{1}{9} \right\rangle \right\},$$

and an orthonormal basis is

$$B'' = \left\{ \left\langle \frac{1}{\sqrt{2}}, \frac{1}{\sqrt{2}}, 0 \right\rangle, \left\langle -\frac{1}{3\sqrt{2}}, \frac{1}{3\sqrt{2}}, \frac{4}{3\sqrt{2}} \right\rangle, \left\langle \frac{2}{3}, -\frac{2}{3}, \frac{1}{3} \right\rangle \right\}.$$

12. We have $\mathbf{u}_1 = \langle 1, 1, 1 \rangle$, $\mathbf{u}_2 = \langle 9, -1, 1 \rangle$, and $\mathbf{u}_3 = \langle -1, 4, -2 \rangle$. Taking $\mathbf{v}_1 = \mathbf{u}_1 = \langle 1, 1, 1 \rangle$ and using $\mathbf{u}_2 \cdot \mathbf{v}_1 = 9$ and $\mathbf{v}_1 \cdot \mathbf{v}_1 = 3$ we obtain

$$\mathbf{v}_2 = \mathbf{u}_2 - \frac{\mathbf{u}_2 \cdot \mathbf{v}_1}{\mathbf{v}_1 \cdot \mathbf{v}_1} \mathbf{v}_1 = \langle 9, -1, 1 \rangle - \frac{9}{3} \langle 1, 1, 1 \rangle = \langle 6, -4, -2 \rangle.$$

Next, using $\mathbf{u}_3 \cdot \mathbf{v}_1 = 1$, $\mathbf{u}_3 \cdot \mathbf{v}_2 = -18$, and $\mathbf{v}_2 \cdot \mathbf{v}_2 = 56$, we obtain

$$\mathbf{v}_3 = \mathbf{u}_3 - \frac{\mathbf{u}_3 \cdot \mathbf{v}_1}{\mathbf{v}_1 \cdot \mathbf{v}_1} \mathbf{v}_1 - \frac{\mathbf{u}_3 \cdot \mathbf{v}_2}{\mathbf{v}_2 \cdot \mathbf{v}_2} \mathbf{v}_2 = \langle -1, 4, -2 \rangle - \frac{1}{3} \langle 1, 1, 1 \rangle - \frac{-18}{56} \langle 6, -4, -2 \rangle$$

$$= \left\langle \frac{25}{42}, \frac{50}{21}, -\frac{125}{42} \right\rangle.$$

Thus, an orthogonal basis is

$$B' = \left\{ \langle 1, 1, 1 \rangle, \langle 6, -4, -2 \rangle, \left\langle \frac{25}{42}, \frac{50}{21}, -\frac{125}{42} \right\rangle \right\},$$

and an orthonormal basis is

$$B'' = \left\{ \left\langle \frac{1}{\sqrt{3}}, \frac{1}{\sqrt{3}}, \frac{1}{\sqrt{3}} \right\rangle, \left\langle \frac{3}{\sqrt{14}}, -\frac{2}{\sqrt{14}}, -\frac{1}{\sqrt{14}} \right\rangle, \left\langle \frac{1}{\sqrt{42}}, \frac{4}{\sqrt{42}}, -\frac{5}{\sqrt{42}} \right\rangle \right\}.$$

15. We have $\mathbf{u}_1 = \langle 1, -1, 1, -1 \rangle$, and $\mathbf{u}_2 = \langle 1, 3, 0, 1 \rangle$. Taking $\mathbf{v}_1 = \mathbf{u}_1 = \langle 1, -1, 1, -1 \rangle$ and using $\mathbf{u}_2 \cdot \mathbf{v}_1 = -3$ and $\mathbf{v}_1 \cdot \mathbf{v}_1 = 4$ we obtain

$$\mathbf{v}_2 = \mathbf{u}_2 - \frac{\mathbf{u}_2 \cdot \mathbf{v}_1}{\mathbf{v}_1 \cdot \mathbf{v}_1} \mathbf{v}_1 = \langle 1, 3, 0, 1 \rangle - \frac{-3}{4} \langle 1, -1, 1, -1 \rangle = \left\langle \frac{7}{4}, \frac{9}{4}, \frac{3}{4}, \frac{1}{4} \right\rangle.$$

Thus, an orthogonal basis is $B' = \left\{ \langle 1, -1, 1, -1 \rangle, \left\langle \frac{7}{4}, \frac{9}{4}, \frac{3}{4}, \frac{1}{4} \right\rangle \right\}$, and an orthonormal basis is

$$B'' = \left\{ \left\langle \frac{1}{2}, -\frac{1}{2}, \frac{1}{2}, -\frac{1}{2} \right\rangle, \left\langle \frac{7}{2\sqrt{35}}, \frac{9}{2\sqrt{35}}, \frac{3}{2\sqrt{35}}, \frac{1}{2\sqrt{35}} \right\rangle \right\}.$$

18. We have $u_1 = x^2 - x, u_2 = x^2 + 1$, and $u_3 = 1 - x^2$. Taking $v_1 = u_1 = x^2 - x$ and using

$$(u_2, v_1) = \int_{-1}^{1} (x^2 + 1)(x^2 - x) \, dx = \frac{16}{15} \quad \text{and} \quad (v_1, v_1) = \int_{-1}^{1} (x^2 - x)(x^2 - x) \, dx = \frac{16}{15}$$

we obtain

$$v_2 = u_2 - \frac{(u_2, v_1)}{(v_1, v_1)} v_1 = x^2 + 1 - \frac{16/15}{16/15}(x^2 - x) = x + 1.$$

Next, using

$$(u_3, v_1) = \int_{-1}^{1} (1 - x^2)(x^2 - x) \, dx = \frac{4}{15}, \quad (u_3, v_2) = \int_{-1}^{1} (1 - x^2)(x + 1) \, dx = \frac{4}{3},$$

and

$$(v_2, v_2) = \int_{-1}^{1} (x + 1)(x + 1) \, dx = \frac{8}{3},$$

we obtain

$$v_3 = u_3 - \frac{(u_3, v_1)}{(v_1, v_1)} v_1 - \frac{(u_3, v_2)}{(v_2, v_2)} v_2 = 1 - x^2 - \frac{4/15}{16/15}(x^2 - x) - \frac{4/3}{8/3}(x + 1) = -\frac{5}{4}x^3 - \frac{1}{4}x + \frac{1}{2}.$$

Thus, an orthogonal basis is $B' = \left\{ x^2 - x, x + 1, -\frac{5}{4}x^3 - \frac{1}{4}x + \frac{1}{2} \right\}$.

21. Using $w_1 = 1/\sqrt{2}, w_2 = 3x/\sqrt{6}$, and $w_3 = (15/2\sqrt{10})(x^2 - 1/3)$, and computing

$$(p, w_1) = \int_{-1}^{1} (9x^2 - 6x + 5) \frac{1}{\sqrt{2}} \, dx = 8\sqrt{2},$$

$$(p, w_2) = \int_{-1}^{1} (9x^2 - 6x + 5) \frac{3}{\sqrt{6}} x \, dx = -2\sqrt{6}$$

$$(p, w_3) = \int_{-1}^{1} (9x^2 - 6x + 5) \left[\frac{15}{2\sqrt{10}} \left(x^2 - \frac{1}{3} \right) \right] \, dx = \frac{12}{\sqrt{10}}$$

we find from Theorem 7.7.1

$$p(x) = 9x^2 - 6x + 5 = (p, w_1)w_1 + (p, w_2)w_2 + (p, w_3)w_3 = 8\sqrt{2}\, w_1 - 2\sqrt{6}\, w_2 + \frac{12}{\sqrt{10}}\, w_3.$$

Chapter 7 in Review

3. False; since a normal to the plane is $\langle 2, 3, -4 \rangle$ which is not a multiple of the direction vector $\langle 5, -2, 1 \rangle$ of the line.

6. True   **9.** True

12. Orthogonal

15. $\sqrt{(-12)^2 + 4^2 + 6^2} = 14$

18. The coordinates of $(1, -2, -10)$ satisfy the given equation.

21. $x_2 - 2 = 3, \ x_2 = 5; \ y_2 - 1 = 5, \ y_2 = 6; \ z_2 - 7 = -4, \ z_2 = 3; \ P_2 = (5, 6, 3)$

24. $2\mathbf{b} = \langle -2, 4, 2 \rangle; \ 4\mathbf{c} = \langle 0, -8, 8 \rangle; \ \mathbf{a} \cdot (2\mathbf{b} + 4\mathbf{c}) = \langle 3, 1, 0 \rangle \cdot \langle -2, -4, 10 \rangle = -10$

27. $A = \dfrac{1}{2}|5\mathbf{i} - 4\mathbf{j} - 7\mathbf{k}| = \dfrac{3\sqrt{10}}{2}$

30. Parallel: $-2c = 5, \ c = -5/2$; orthogonal: $1(-2) + 3(-6) + c(5) = 0, \ c = 4$

33. $\text{comp}_{\mathbf{b}}\mathbf{a} = \mathbf{a} \cdot \mathbf{b}/\|\mathbf{b}\| = \langle 1, 2, -2 \rangle \cdot \langle 4, 3, 0 \rangle / 5 = 2$

36. $\mathbf{a} - \mathbf{b} = \langle 1, 2, -2 \rangle - \langle 4, 3, 0 \rangle = \langle -3, -1, -2 \rangle$

$\text{comp}_{\mathbf{b}}(\mathbf{a} - \mathbf{b}) = (\mathbf{a} - \mathbf{b}) \cdot \mathbf{b}/\sqrt{16 + 9} = \frac{1}{5}(\mathbf{a} \cdot \mathbf{b} - \mathbf{b} \cdot \mathbf{b}) = \frac{1}{5}[(4 + 6 + 0) - (16 + 9)] = -3$

$\text{proj}_{\mathbf{b}}(\mathbf{a} - \mathbf{b}) = [\text{comp}_{\mathbf{b}}(\mathbf{a} - \mathbf{b})](\mathbf{b}/\|\mathbf{b}\|) = -3\langle \frac{4}{5}, \frac{3}{5}, 0 \rangle = \langle -\frac{12}{5}, -\frac{9}{5}, 0 \rangle$

39. A direction vector of the given line is $\langle 4, -2, 6 \rangle$. A parallel line containing $(7, 3, -5)$ is $(x - 7)/4 = (y - 3)/(-2) = (z + 5)/6$.

42. Vectors in the plane are $\langle 2, 3, 1 \rangle$ and $\langle 1, 0, 2 \rangle$. A normal vector is $\langle 2, 3, 1 \rangle \times \langle 1, 0, 2 \rangle = \langle 6, -3, -3 \rangle = 3\langle 2, -1, -1 \rangle$. An equation of the plane is $2x - y - z = 0$.

45. $\mathbf{F} = 10\dfrac{\mathbf{a}}{\|\mathbf{a}\|} = \dfrac{10}{\sqrt{2}}(\mathbf{i} + \mathbf{j}) = 5\sqrt{2}\,\mathbf{i} + 5\sqrt{2}\,\mathbf{j}; \ \mathbf{d} = \langle 7, 4, 0 \rangle - \langle 4, 1, 0 \rangle = 3\mathbf{i} + 3\mathbf{j}$

$W = \mathbf{F} \cdot \mathbf{d} = 15\sqrt{2} + 15\sqrt{2} = 30\sqrt{2}$ N-m

48. Let $\|\mathbf{F}_1\| = F_1$ and $\|\mathbf{F}_2\| = F_2$. Then $\mathbf{F}_1 = F_1[(\cos 45°)\mathbf{i} + (\sin 45°)\mathbf{j}]$ and $\mathbf{F}_2 = F_2[(\cos 120°)\mathbf{i} + (\sin 120°)\mathbf{j}]$, or $\mathbf{F}_1 = F_1(\frac{1}{\sqrt{2}}\mathbf{i} + \frac{1}{\sqrt{2}}\mathbf{j})$ and $\mathbf{F}_2 = F_2(-\frac{1}{2}\mathbf{i} + \frac{\sqrt{3}}{2}\mathbf{j})$. Since $\mathbf{w} + \mathbf{F}_1 + \mathbf{F}_2 = \mathbf{0}$,

$$F_1\left(\frac{1}{\sqrt{2}}\mathbf{i} + \frac{1}{\sqrt{2}}\mathbf{j}\right) + F_2\left(-\frac{1}{2}\mathbf{i} + \frac{\sqrt{3}}{2}\mathbf{j}\right) = 50\mathbf{j}, \quad \left(\frac{1}{\sqrt{2}}F_1 - \frac{1}{2}F_2\right)\mathbf{i} + \left(\frac{1}{\sqrt{2}}F_1 + \frac{\sqrt{3}}{2}F_2\right)\mathbf{j} = 50\mathbf{j}$$

and

$$\frac{1}{\sqrt{2}}F_1 - \frac{1}{2}F_2 = 0, \qquad \frac{1}{\sqrt{2}}F_1 + \frac{\sqrt{3}}{2}F_2 = 50.$$

Solving, we obtain $F_1 = 25(\sqrt{6} - \sqrt{2}) \approx 25.9$ lb and $F_2 = 50(\sqrt{3} - 1) \approx 36.6$ lb.

51. Let p_1 and p_2 be in P_n such that $\dfrac{d^2p_1}{dx^2} = 0$ and $\dfrac{d^2p_2}{dx^2} = 0$. Since

$$0 = \frac{d^2p_1}{dx^2} + \frac{d^2p_2}{dx^2} = \frac{d^2}{dx^2}(p_1 + p_2) \quad \text{and} \quad 0 = k\frac{d^2p_1}{dx^2} = \frac{d^2}{dx^2}(kp_1)$$

we conclude that the set of polynomials with the given property is a subspace of P_n. A basis for the subspace is $1, x$.

Chapter 8

Matrices

8.1 | Matrix Algebra

3. 3×3 **6.** 8×1 **9.** Not equal

12. Solving $x^2 = 9$, $y = 4x$ we obtain $x = 3$, $y = 12$ and $x = -3$, $t = -12$.

15. (a) $\mathbf{A} + \mathbf{B} = \begin{pmatrix} 4-2 & 5+6 \\ -6+8 & 9-10 \end{pmatrix} = \begin{pmatrix} 2 & 11 \\ 2 & -1 \end{pmatrix}$

(b) $\mathbf{B} - \mathbf{A} = \begin{pmatrix} -2-4 & 6-5 \\ 8+6 & -10-9 \end{pmatrix} = \begin{pmatrix} -6 & 1 \\ 14 & -19 \end{pmatrix}$

(c) $2\mathbf{A} + 3\mathbf{B} = \begin{pmatrix} 8 & 10 \\ -12 & 18 \end{pmatrix} + \begin{pmatrix} -6 & 18 \\ 24 & -30 \end{pmatrix} = \begin{pmatrix} 2 & 28 \\ 12 & -12 \end{pmatrix}$

18. (a) $\mathbf{AB} = \begin{pmatrix} -4+4 & 6-12 & -3+8 \\ -20+10 & 30-30 & -15+20 \\ -32+12 & 49-36 & -24+24 \end{pmatrix} = \begin{pmatrix} 0 & -6 & 5 \\ -10 & 0 & 5 \\ -20 & 12 & 0 \end{pmatrix}$

(b) $\mathbf{BA} = \begin{pmatrix} -4+30-24 & -16+60-36 \\ 1-15+16 & 4-30+24 \end{pmatrix} = \begin{pmatrix} 2 & 8 \\ 2 & -2 \end{pmatrix}$

21. (a) $\mathbf{A}^T\mathbf{A} = \begin{pmatrix} 4 & 8 & -10 \end{pmatrix} \begin{pmatrix} 4 \\ 8 \\ -10 \end{pmatrix} = \begin{pmatrix} 180 \end{pmatrix}$

(b) $\mathbf{B}^T\mathbf{B} = \begin{pmatrix} 2 \\ 4 \\ 5 \end{pmatrix} \begin{pmatrix} 2 & 4 & 5 \end{pmatrix} = \begin{pmatrix} 4 & 8 & 10 \\ 8 & 16 & 20 \\ 10 & 20 & 25 \end{pmatrix}$

(c) $\mathbf{A} + \mathbf{B}^T = \begin{pmatrix} 4 \\ 8 \\ -10 \end{pmatrix} + \begin{pmatrix} 2 \\ 4 \\ 5 \end{pmatrix} = \begin{pmatrix} 6 \\ 12 \\ -5 \end{pmatrix}$

24. (a) $\mathbf{A}^T + \mathbf{B} = \begin{pmatrix} 5 & -4 \\ 9 & 6 \end{pmatrix} + \begin{pmatrix} -3 & 11 \\ -7 & 2 \end{pmatrix} = \begin{pmatrix} 2 & 7 \\ 2 & 8 \end{pmatrix}$

(b) $2\mathbf{A} + \mathbf{B}^T = \begin{pmatrix} 10 & 18 \\ -8 & 12 \end{pmatrix} + \begin{pmatrix} -3 & -7 \\ 11 & 2 \end{pmatrix} = \begin{pmatrix} 7 & 11 \\ 3 & 14 \end{pmatrix}$

27. $\begin{pmatrix} -19 \\ 18 \end{pmatrix} - \begin{pmatrix} 19 \\ 20 \end{pmatrix} = \begin{pmatrix} -38 \\ -2 \end{pmatrix}$ **30.** 3×2

33. $(\mathbf{AB})^T = \begin{pmatrix} 16 & 40 \\ -8 & -20 \end{pmatrix}^T = \begin{pmatrix} 16 & -8 \\ 40 & -20 \end{pmatrix};$ $\mathbf{B}^T\mathbf{A}^T = \begin{pmatrix} 4 & 2 \\ 10 & 5 \end{pmatrix} \begin{pmatrix} 2 & -3 \\ 4 & 2 \end{pmatrix} = \begin{pmatrix} 16 & -8 \\ 40 & -20 \end{pmatrix}$

36. Using Problem 33 we have $(\mathbf{AA}^T)^T = (\mathbf{A}^T)^T\mathbf{A}^T = \mathbf{AA}^T$, so that $\mathbf{AA}^T$ is symmetric.

39. Since $(\mathbf{A} + \mathbf{B})^2 = (\mathbf{A} + \mathbf{B})(\mathbf{A} + \mathbf{B}) \neq \mathbf{A}^2 + \mathbf{AB} + \mathbf{BA} + \mathbf{B}^2$, and $\mathbf{AB} \neq \mathbf{BA}$ in general, $(\mathbf{A} + \mathbf{B})^2 \neq \mathbf{A}^2 + 2\mathbf{AB} + \mathbf{B}^2$.

42. $\begin{pmatrix} 2 & 6 & 1 \\ 1 & 2 & -1 \\ 5 & 7 & -4 \end{pmatrix} \begin{pmatrix} x_1 \\ x_2 \\ x_3 \end{pmatrix} = \begin{pmatrix} 7 \\ -1 \\ 9 \end{pmatrix}$

45. $\begin{pmatrix} x_1 \\ y_1 \end{pmatrix} = \begin{pmatrix} \cos\dfrac{\pi}{2} & -\sin\dfrac{\pi}{2} \\ \sin\dfrac{\pi}{2} & \cos\dfrac{\pi}{2} \end{pmatrix} \begin{pmatrix} 1 \\ 1 \end{pmatrix} = \begin{pmatrix} 0 & -1 \\ 1 & 0 \end{pmatrix} \begin{pmatrix} 1 \\ 1 \end{pmatrix} = \begin{pmatrix} -1 \\ 1 \end{pmatrix}$

Therefore $\mathbf{b} = \langle -1, 1 \rangle$.

48. $\begin{pmatrix} x_1 \\ y_1 \end{pmatrix} = \begin{pmatrix} \cos\dfrac{2\pi}{3} & -\sin\dfrac{2\pi}{3} \\ \sin\dfrac{2\pi}{3} & \cos\dfrac{2\pi}{3} \end{pmatrix} \begin{pmatrix} 1 \\ -1 \end{pmatrix} = \begin{pmatrix} -\dfrac{1}{2} & -\dfrac{\sqrt{3}}{2} \\ -\dfrac{\sqrt{3}}{2} & -\dfrac{1}{2} \end{pmatrix} \begin{pmatrix} 1 \\ -1 \end{pmatrix} = \begin{pmatrix} -\dfrac{1}{2} + \dfrac{\sqrt{3}}{2} \\ \dfrac{\sqrt{3}}{2} + \dfrac{1}{2} \end{pmatrix}$

Therefore $\mathbf{b} = \frac{1}{2}\langle -1 + \sqrt{3}, 1 + \sqrt{3} \rangle$.

51. (a) $M_Y \begin{pmatrix} x \\ y \\ z \end{pmatrix} = \begin{pmatrix} \cos\gamma & \sin\gamma & 0 \\ -\sin\gamma & \cos\gamma & 0 \\ 0 & 0 & 1 \end{pmatrix} \begin{pmatrix} x \\ y \\ z \end{pmatrix} = \begin{pmatrix} x\cos\gamma + y\sin\gamma \\ -x\sin\gamma + y\cos\gamma \\ z \end{pmatrix} = \begin{pmatrix} x_Y \\ y_Y \\ z_Y \end{pmatrix}$

(b) $M_R = \begin{pmatrix} \cos\beta & 0 & -\sin\beta \\ 0 & 1 & 0 \\ \sin\beta & 0 & \cos\beta \end{pmatrix};$ $M_P \begin{pmatrix} 1 & 0 & 0 \\ 0 & \cos\alpha & \sin\alpha \\ 0 & -\sin\alpha & \cos\alpha \end{pmatrix}$

(c) $M_P \begin{pmatrix} 1 \\ 1 \\ 1 \end{pmatrix} = \begin{pmatrix} 1 & 0 & 0 \\ 0 & \cos 30° & \sin 30° \\ 0 & -\sin 30° & \cos 30° \end{pmatrix} \begin{pmatrix} 1 \\ 1 \\ 1 \end{pmatrix} = \begin{pmatrix} 1 & 0 & 0 \\ 0 & \dfrac{\sqrt{3}}{2} & \dfrac{1}{2} \\ 0 & -\dfrac{1}{2} & \dfrac{\sqrt{3}}{2} \end{pmatrix} \begin{pmatrix} 1 \\ 1 \\ 1 \end{pmatrix} = \begin{pmatrix} 1 \\ \dfrac{1}{2}\left(\sqrt{3}+1\right) \\ \dfrac{1}{2}\left(\sqrt{3}-1\right) \end{pmatrix}$

$M_R M_P \begin{pmatrix} 1 \\ 1 \\ 1 \end{pmatrix} = \begin{pmatrix} \cos 45° & 0 & -\sin 45° \\ 0 & 1 & 0 \\ \sin 45° & 0 & \cos 45° \end{pmatrix} \begin{pmatrix} 1 \\ \dfrac{1}{2}\left(\sqrt{3}+1\right) \\ \dfrac{1}{2}\left(\sqrt{3}-1\right) \end{pmatrix}$

$= \begin{pmatrix} \dfrac{\sqrt{2}}{2} & 0 & -\dfrac{\sqrt{2}}{2} \\ 0 & 1 & 0 \\ \dfrac{\sqrt{2}}{2} & 0 & \dfrac{\sqrt{2}}{2} \end{pmatrix} \begin{pmatrix} 1 \\ \dfrac{1}{2}\left(\sqrt{3}+1\right) \\ \dfrac{1}{2}\left(\sqrt{3}-1\right) \end{pmatrix} = \begin{pmatrix} \dfrac{1}{4}\left(3\sqrt{2}-\sqrt{6}\right) \\ \dfrac{1}{2}\left(\sqrt{3}+1\right) \\ \dfrac{1}{4}\left(\sqrt{2}+\sqrt{6}\right) \end{pmatrix}$

$M_Y M_R M_P \begin{pmatrix} 1 \\ 1 \\ 1 \end{pmatrix} = \begin{pmatrix} \cos 60° & \sin 60° & 0 \\ -\sin 60° & \cos 45° & 0 \\ 0 & 0 & 1 \end{pmatrix} \begin{pmatrix} \dfrac{1}{4}\left(3\sqrt{2}-\sqrt{6}\right) \\ \dfrac{1}{2}\left(\sqrt{3}+1\right) \\ \dfrac{1}{4}\left(\sqrt{2}+\sqrt{6}\right) \end{pmatrix}$

$= \begin{pmatrix} \dfrac{1}{2} & \dfrac{\sqrt{3}}{2} & 0 \\ -\dfrac{\sqrt{3}}{2} & \dfrac{1}{2} & 0 \\ 0 & 0 & 1 \end{pmatrix} \begin{pmatrix} \dfrac{1}{4}\left(3\sqrt{2}-\sqrt{6}\right) \\ \dfrac{1}{2}\left(\sqrt{3}+1\right) \\ \dfrac{1}{4}\left(\sqrt{2}+\sqrt{6}\right) \end{pmatrix}$

$= \begin{pmatrix} \dfrac{1}{8}\left(3\sqrt{2}-\sqrt{6}+6+2\sqrt{3}\right) \\ \dfrac{1}{8}\left(-3\sqrt{6}+3\sqrt{2}+2\sqrt{3}+2\right) \\ \dfrac{1}{4}\left(\sqrt{2}+\sqrt{6}\right) \end{pmatrix}$

8.2 | Systems of Linear Equations

3. $\begin{pmatrix} 9 & 3 & | & -5 \\ 2 & -1 & | & -1 \end{pmatrix} \xrightarrow{\frac{1}{9}R_1} \begin{pmatrix} 1 & \frac{1}{3} & | & -\frac{5}{9} \\ 2 & 1 & | & -1 \end{pmatrix} \xrightarrow{-2R_1+R_2} \begin{pmatrix} 1 & \frac{1}{3} & | & -\frac{5}{9} \\ 0 & \frac{1}{3} & | & \frac{1}{9} \end{pmatrix} \xrightarrow{3R_2} \begin{pmatrix} 1 & \frac{1}{3} & | & -\frac{5}{9} \\ 0 & 1 & | & \frac{1}{3} \end{pmatrix}$

$\xrightarrow{-\frac{1}{3}R_2+R_1} \begin{pmatrix} 1 & 0 & | & -\frac{2}{3} \\ 0 & 1 & | & \frac{1}{3} \end{pmatrix}$

The solution is $x_1 = -\frac{2}{3}$, $x_2 = \frac{1}{3}$.

6. $\begin{pmatrix} 1 & 2 & -1 & | & 0 \\ 2 & 1 & 2 & | & 9 \\ 1 & -1 & 1 & | & 3 \end{pmatrix} \xrightarrow[-R_1+R_3]{-2R_1+R_2} \begin{pmatrix} 1 & 2 & -1 & | & 0 \\ 0 & -3 & 4 & | & 9 \\ 0 & -3 & 2 & | & 3 \end{pmatrix} \xrightarrow{-\frac{1}{3}R_2} \begin{pmatrix} 1 & 2 & -1 & | & 0 \\ 0 & 1 & -\frac{4}{3} & | & -3 \\ 0 & -3 & 2 & | & 3 \end{pmatrix}$

$\xrightarrow[3R_2+R_3]{-2R_2+R_1} \begin{pmatrix} 1 & 0 & \frac{5}{3} & | & 6 \\ 0 & 1 & -\frac{4}{3} & | & -3 \\ 0 & 0 & -2 & | & -6 \end{pmatrix} \xrightarrow{-\frac{1}{2}R_3} \begin{pmatrix} 1 & 0 & \frac{5}{3} & | & 6 \\ 0 & 1 & -\frac{4}{3} & | & -3 \\ 0 & 0 & 1 & | & 3 \end{pmatrix}$

$\xrightarrow[\frac{4}{3}R_3+R_2]{-\frac{5}{3}R_3+R_1} \begin{pmatrix} 1 & 0 & 0 & | & 1 \\ 0 & 1 & 0 & | & 1 \\ 0 & 0 & 1 & | & 3 \end{pmatrix}$

The solution is $x_1 = 1$, $x_2 = 1$, $x_3 = 3$.

9. $\begin{pmatrix} 1 & -1 & -1 & | & 8 \\ 1 & -1 & 1 & | & 3 \\ -1 & 1 & 1 & | & 4 \end{pmatrix} \xrightarrow[\text{operations}]{\text{row}} \begin{pmatrix} 1 & -1 & -1 & | & 8 \\ 0 & 0 & 2 & | & -5 \\ 0 & 0 & 0 & | & 12 \end{pmatrix}$

Since the bottom row implies $0 = 12$, the system is inconsistent.

12. $\begin{pmatrix} 1 & -1 & -2 & | & 0 \\ 2 & 4 & 5 & | & 0 \\ 6 & 0 & -3 & | & 0 \end{pmatrix} \xrightarrow[\text{operations}]{\text{row}} \begin{pmatrix} 1 & -1 & -2 & | & 0 \\ 0 & 1 & \frac{3}{2} & | & 0 \\ 0 & 0 & 0 & | & 0 \end{pmatrix}$

The solution is $x_1 = \frac{1}{2}t$, $x_2 = -\frac{3}{2}t$, $x_3 = t$.

15. $\begin{pmatrix} 1 & 1 & 1 & | & 3 \\ 1 & -1 & -1 & | & -1 \\ 3 & 1 & 1 & | & 5 \end{pmatrix} \xrightarrow[\text{operations}]{\text{row}} \begin{pmatrix} 1 & 1 & 1 & | & 3 \\ 0 & 1 & 1 & | & 2 \\ 0 & 0 & 0 & | & 0 \end{pmatrix}$

If $x_3 = t$ the solution is $x_1 = 1$, $x_2 = 2 - t$, $x_3 = t$.

18. $\begin{pmatrix} 2 & 1 & 1 & 0 & | & 3 \\ 3 & 1 & 1 & 1 & | & 4 \\ 1 & 2 & 2 & 3 & | & 3 \\ 4 & 5 & -2 & 1 & | & 16 \end{pmatrix} \xrightarrow[\text{operations}]{\text{row}} \begin{pmatrix} 1 & \frac{1}{2} & \frac{1}{2} & 0 & | & \frac{3}{2} \\ 0 & 1 & 1 & -2 & | & 1 \\ 0 & 0 & 1 & -1 & | & -1 \\ 0 & 0 & 0 & 1 & | & 0 \end{pmatrix}$

The solution is $x_1 = 1$, $x_2 = 2$, $x_3 = -1$, $x_4 = 0$.

21. $\begin{pmatrix} 1 & 1 & 1 & | & 4.280 \\ 0.2 & -0.1 & -0.5 & | & -1.978 \\ 4.1 & 0.3 & 0.12 & | & 1.686 \end{pmatrix} \xrightarrow[\text{operations}]{\text{row}} \begin{pmatrix} 1 & 1 & 1 & | & 4.28 \\ 0 & 1 & 2.333 & | & 9.447 \\ 0 & 0 & 1 & | & 4.1 \end{pmatrix}$

The solution is $x_1 = 0.3$, $x_2 = -0.12$, $x_3 = 4.1$.

24. From $x_1 KClO_3 \rightarrow x_2 KCl + x_3 O_2$ we obtain the system $x_1 = x_2$, $x_1 = x_2$, $3x_1 = 2x_3$. Letting $x_3 = t$ we see that a solution of the system is $x_1 = x_2 = \frac{2}{3}t$, $x_3 = t$. Taking $t = 3$ we obtain the balanced equation

$$2KClO_3 \rightarrow 2KCl + 3O_2.$$

27. From $x_1\text{Cu} + x_2\text{HNO}_3 \rightarrow x_3\text{Cu(NO}_3)_2 + x_4\text{H}_2\text{O} + x_5\text{NO}$ we obtain the system

$$x_1 = 3, \qquad x_2 = 2x_4, \qquad x_2 = 2x_3 + x_5, \qquad 3x_2 = 6x_3 + x_4 + x_5.$$

Letting $x_4 = t$ we see that $x_2 = 2t$ and

$$\begin{aligned} 2t &= 2x_3 + x_5 \\ 6t &= 6x_3 + t + x_5 \end{aligned} \qquad \text{or} \qquad \begin{aligned} 2x_3 + x_5 &= 2t \\ 6x_3 + x_5 &= 5t. \end{aligned}$$

Then $x_3 = \frac{3}{4}t$ and $x_5 = \frac{1}{2}t$. Finally, $x_1 = x_3 = \frac{3}{4}t$. Taking $t = 4$ we obtain the balanced equation

$$3\text{Cu} + 8\text{HNO}_3 \rightarrow 3\text{Cu(NO}_3)_2 + 4\text{H}_2\text{O} + 2\text{NO}.$$

30. The system of equations is

$$\begin{aligned} i_1 - i_2 - i_3 &= 0 \\ 52 - i_1 - 5i_2 &= 0 \\ -10i_3 + 5i_2 &= 0 \end{aligned} \qquad \text{or} \qquad \begin{aligned} i_1 - i_2 - i_3 &= 0 \\ i_1 + 5i_2 &= 52 \\ 5i_2 - 10i_3 &= 0 \end{aligned}$$

Gaussian elimination gives

$$\begin{pmatrix} 1 & -1 & -1 & 0 \\ 1 & 5 & 0 & 52 \\ 0 & 5 & -10 & 0 \end{pmatrix} \xrightarrow[\text{operations}]{\text{row}} \begin{pmatrix} 1 & -1 & -1 & 0 \\ 0 & 1 & 1/6 & 26/3 \\ 0 & 0 & 1 & 4 \end{pmatrix}$$

The solution is $i_1 = 12$, $i_2 = 8$, $i_3 = 4$.

33. AX = 0 is

$$\begin{pmatrix} 2 & -3 & 1 \\ 1 & 1 & -1 \\ 4 & -1 & -1 \end{pmatrix} \begin{pmatrix} x_1 \\ x_2 \\ x_3 \end{pmatrix} = \begin{pmatrix} -12 \\ 1 \\ -10 \end{pmatrix}$$

Now

$$\begin{pmatrix} 2 & -3 & 1 \\ 1 & 1 & -1 \\ 4 & -1 & -1 \end{pmatrix} \begin{pmatrix} -1 \\ 4 \\ 2 \end{pmatrix} + \begin{pmatrix} 2 & -3 & 1 \\ 1 & 1 & -1 \\ 4 & -1 & -1 \end{pmatrix} \begin{pmatrix} 2c_1 \\ 3c_1 \\ 5c_1 \end{pmatrix} = \begin{pmatrix} -2 - 12 + 2 \\ -1 + 4 - 2 \\ -4 - 4 - 2 \end{pmatrix} + \begin{pmatrix} 4c_1 - 9c_1 + 5c_1 \\ 2c_1 + 3c_1 - 5c_1 \\ 8c_1 - 3c_1 - 5c_1 \end{pmatrix}$$

$$= \begin{pmatrix} -12 \\ 1 \\ -10 \end{pmatrix} + \begin{pmatrix} 0 \\ 0 \\ 0 \end{pmatrix}$$

$$= \begin{pmatrix} -12 \\ 1 \\ -10 \end{pmatrix} \quad \text{for all values of } c_1.$$

36. Multiply row 3 by c in $\mathbf{I}_3$.

39. $\mathbf{EA} = \begin{pmatrix} a_{21} & a_{22} & a_{23} \\ a_{11} & a_{12} & a_{13} \\ a_{31} & a_{32} & a_{33} \end{pmatrix}$

42. $\mathbf{E}_1\mathbf{E}_2\mathbf{A}$

$$= \mathbf{E}_1 \begin{pmatrix} a_{11} & a_{12} & a_{13} \\ a_{21} & a_{22} & a_{23} \\ ca_{21}+a_{31} & ca_{22}+a_{32} & ca_{23}+a_{33} \end{pmatrix} = \begin{pmatrix} a_{21} & a_{22} & a_{23} \\ a_{11} & a_{12} & a_{13} \\ ca_{21}+a_{31} & ca_{22}+a_{32} & ca_{23}+a_{33} \end{pmatrix}$$

8.3 | Rank of a Matrix

3. $\begin{pmatrix} 2 & 1 & 3 \\ 6 & 3 & 9 \\ -1 & -\frac{1}{2} & -\frac{3}{2} \end{pmatrix} \xrightarrow[\text{operations}]{\text{row}} \begin{pmatrix} 1 & \frac{1}{2} & \frac{3}{2} \\ 0 & 0 & 0 \\ 0 & 0 & 0 \end{pmatrix};$ The rank is 1.

6. $\begin{pmatrix} 3 & -1 & 2 & 0 \\ 6 & 2 & 4 & 5 \end{pmatrix} \xrightarrow[\text{operations}]{\text{row}} \begin{pmatrix} 1 & -\frac{1}{3} & \frac{2}{3} & 0 \\ 0 & 1 & 0 & \frac{5}{4} \end{pmatrix};$ The rank is 2.

9. $\begin{pmatrix} 0 & 2 & 4 & 2 & 2 \\ 4 & 1 & 0 & 5 & 1 \\ 2 & 1 & \frac{2}{3} & 3 & \frac{1}{3} \\ 6 & 6 & 6 & 12 & 0 \end{pmatrix} \xrightarrow[\text{operations}]{\text{row}} \begin{pmatrix} 1 & \frac{1}{2} & \frac{1}{3} & \frac{3}{2} & \frac{1}{6} \\ 0 & 1 & \frac{4}{3} & 1 & -\frac{1}{3} \\ 0 & 0 & 1 & 0 & 2 \\ 0 & 0 & 0 & 0 & 0 \end{pmatrix};$ The rank is 3.

12. $\begin{pmatrix} 2 & 6 & 3 \\ 1 & -1 & 4 \\ 3 & 2 & 1 \\ 2 & 5 & 4 \end{pmatrix} \xrightarrow[\text{operations}]{\text{row}} \begin{pmatrix} 1 & -1 & 4 \\ 0 & 1 & -\frac{5}{8} \\ 0 & 0 & 1 \\ 0 & 0 & 0 & 1 \end{pmatrix};$

Since the rank of the matrix is 3 and there are 4 vectors, the vectors are linearly dependent.

15. Since the number of unknowns is $n = 8$ and the rank of the coefficient matrix is $r = 3$, the solution of the system has $n - r = 5$ parameters.

8.4 | Determinants

3. $C_{13} = (-1)^{1+3} \begin{vmatrix} 1 & -1 \\ -2 & 3 \end{vmatrix} = 1$

6. $M_{41} = \begin{vmatrix} 2 & 4 & 0 \\ 2 & -2 & 3 \\ 1 & 0 & -1 \end{vmatrix} = 24$

9. -7 **12.** $-1/2$

15. $\begin{vmatrix} 0 & 2 & 0 \\ 3 & 0 & 1 \\ 0 & 5 & 8 \end{vmatrix} = -3 \begin{vmatrix} 2 & 0 \\ 5 & 8 \end{vmatrix} = -48$

18. $\begin{vmatrix} 1 & -1 & -1 \\ 2 & 2 & -2 \\ 1 & 1 & 9 \end{vmatrix} = \begin{vmatrix} 2 & -2 \\ 1 & 9 \end{vmatrix} - 2 \begin{vmatrix} -1 & -1 \\ 1 & 9 \end{vmatrix} + \begin{vmatrix} -1 & -1 \\ 2 & -2 \end{vmatrix} = 20 - 2(-8) + 4 = 40$

21. $\begin{vmatrix} -2 & -1 & 4 \\ -3 & 6 & 1 \\ -3 & 4 & 8 \end{vmatrix} = -2 \begin{vmatrix} 6 & 1 \\ 4 & 8 \end{vmatrix} + 3 \begin{vmatrix} -1 & 4 \\ 4 & 8 \end{vmatrix} - 3 \begin{vmatrix} -1 & 4 \\ 6 & 1 \end{vmatrix} = -2(44) + 3(-24) - 3(-25) = -85$

24. $\begin{vmatrix} 1 & 1 & 1 \\ x & y & z \\ 2+x & 3+y & 4+z \end{vmatrix} = \begin{vmatrix} y & z \\ 3+y & 4+z \end{vmatrix} - \begin{vmatrix} y & z \\ 3+y & 4+z \end{vmatrix} + \begin{vmatrix} x & y \\ 2+x & 3+y \end{vmatrix}$

$$= (4y + yz - 3z - yz) - (4x + xz - 2z - xz) + (3x + xy - 2y - xy)$$

$$= -x + 2y - z$$

27. Expanding along the first column in the original matrix and each succeeding minor, we obtain $3(1)(2)(4)(2) = 48$.

30. Solving $-\lambda^3 + 3\lambda^2 - 2\lambda = -\lambda(\lambda - 2)(\lambda - 1) = 0$ we obtain $\lambda = 0, 1$, and 2.

8.5 Properties of Determinants

3. Theorem 8.5.7

6. Theorem 8.5.4 (twice)

9. Theorem 8.5.1

12. $\det \mathbf{B} = 2(3)(5) = 30$

15. By Theorem 8.5.7

$$\begin{vmatrix} a_1 & a_2 & a_3 \\ b_1 & b_2 & b_3 \\ c_1 & c_2 & c_3 \end{vmatrix} = \begin{vmatrix} a_1 - 2b_1 & a_2 - 2b_2 & a_3 - 2b_3 \\ b_1 & b_2 & b_3 \\ c_1 & c_2 & c_3 \end{vmatrix}$$

$$= \begin{vmatrix} a_1 - 2b_1 + 3c_1 & a_2 - 2b_2 + 3c_2 & a_3 - 2b_3 + 3c_3 \\ b_1 & b_2 & b_3 \\ c_1 & c_2 & c_3 \end{vmatrix} = 5$$

18. $\det \mathbf{B} = -a_{13}a_{22}a_{31}$

21. $\det \mathbf{A} = 14 = \det \mathbf{A}^T$

24. From Theorem 8.5.6, $(\det \mathbf{A})^2 = \det \mathbf{A}^2 = \det \mathbf{I} = 1$, so $\det \mathbf{A} = \pm 1$.

27. Using Theorems 8.5.7, 8.5.5, and 8.5.2,

$$\det \mathbf{A} = \begin{vmatrix} a & 1 & 2 \\ b & 1 & 2 \\ c & 1 & 2 \end{vmatrix} = 2 \begin{vmatrix} a & 1 & 1 \\ b & 1 & 1 \\ c & 1 & 1 \end{vmatrix} = 0.$$

30. $\begin{vmatrix} 2 & 4 & 5 \\ 4 & 2 & 0 \\ 8 & 7 & -2 \end{vmatrix} = \begin{vmatrix} 2 & 4 & 5 \\ 0 & -6 & -10 \\ 0 & -9 & -22 \end{vmatrix} = -2 \begin{vmatrix} 2 & 4 & 5 \\ 0 & 3 & 5 \\ 0 & -9 & -22 \end{vmatrix} = -2 \begin{vmatrix} 2 & 4 & 5 \\ 0 & 3 & 5 \\ 0 & 0 & -7 \end{vmatrix} = -2(2)(3)(-7) = 84$

33. $\begin{vmatrix} 1 & -2 & 2 & 1 \\ 2 & 1 & -2 & 3 \\ 3 & 4 & -8 & 1 \\ 3 & -11 & 12 & 2 \end{vmatrix} = \begin{vmatrix} 1 & -2 & 2 & 1 \\ 0 & 5 & -6 & 1 \\ 0 & 10 & -14 & -2 \\ 0 & -5 & 6 & -1 \end{vmatrix} = \begin{vmatrix} 1 & -2 & 2 & 1 \\ 0 & 5 & -6 & 1 \\ 0 & 0 & -2 & -4 \\ 0 & 0 & 0 & 0 \end{vmatrix} = 1(5)(-2)(0) = 0$

36. $\begin{vmatrix} 2 & 9 & 1 & 8 \\ 1 & 3 & 7 & 4 \\ 0 & 1 & 6 & 5 \\ 3 & 1 & 4 & 2 \end{vmatrix} = - \begin{vmatrix} 1 & 3 & 7 & 4 \\ 2 & 9 & 1 & 8 \\ 0 & 1 & 6 & 5 \\ 3 & 1 & 4 & 2 \end{vmatrix} = - \begin{vmatrix} 1 & 3 & 7 & 4 \\ 0 & 3 & -13 & 0 \\ 0 & 1 & 6 & 5 \\ 0 & -8 & -17 & -10 \end{vmatrix} = \begin{vmatrix} 1 & 3 & 7 & 4 \\ 0 & 1 & 6 & 5 \\ 0 & 3 & -13 & 0 \\ 0 & -8 & -17 & -10 \end{vmatrix}$

$= \begin{vmatrix} 1 & 3 & 7 & 4 \\ 0 & 1 & 6 & 5 \\ 0 & 3 & -31 & -15 \\ 0 & 0 & 31 & 30 \end{vmatrix} = \begin{vmatrix} 1 & 3 & 7 & 4 \\ 0 & 1 & 6 & 5 \\ 0 & 0 & -31 & -15 \\ 0 & 0 & 0 & 15 \end{vmatrix} = 1(1)(-31)(15) = -465$

39. Since $C_{11} = 4$, $C_{12} = 5$, and $C_{13} = -6$, we have

$$a_{21}C_{11} + a_{22}C_{12} + a_{23}C_{13} = (-1)(4) + 2(5) + 1(-6) = 0.$$

Since $C_{12} = 5$, $C_{22} = -7$, and $C_{23} = -3$, we have

$$a_{13}C_{12} + a_{23}C_{22} + a_{33}C_{32} = 2(5) + 1(-7) + 1(-3) = 0.$$

42. $\det(2\mathbf{A}) = 2^5 \det \mathbf{A} = 32(-7) = -224$

8.6 Inverse of a Matrix

3. $\det \mathbf{A} = 9$. $\mathbf{A}$ is nonsingular. $\mathbf{A}^{-1} = \dfrac{1}{9} \begin{pmatrix} 1 & 1 \\ -4 & 5 \end{pmatrix} = \begin{pmatrix} \frac{1}{9} & \frac{1}{9} \\ -\frac{4}{9} & \frac{5}{9} \end{pmatrix}$

6. $\det \mathbf{A} = -3\pi^2$. $\mathbf{A}$ is nonsingular. $\mathbf{A}^{-1} = -\dfrac{1}{3\pi^2} \begin{pmatrix} \pi & \pi \\ \pi & -2\pi \end{pmatrix} = \begin{pmatrix} -\frac{1}{3\pi} & -\frac{1}{3\pi} \\ -\frac{1}{3\pi} & \frac{2}{3\pi} \end{pmatrix}$

9. $\det \mathbf{A} = -30$. $\mathbf{A}$ is nonsingular. $\mathbf{A}^{-1} = -\dfrac{1}{30} \begin{pmatrix} -14 & 13 & 16 \\ -2 & 4 & -2 \\ -4 & -7 & -4 \end{pmatrix} = \begin{pmatrix} \frac{7}{15} & -\frac{13}{30} & -\frac{8}{15} \\ \frac{1}{15} & -\frac{2}{15} & \frac{1}{15} \\ \frac{2}{15} & \frac{7}{30} & \frac{2}{15} \end{pmatrix}$

12. $\det \mathbf{A} = 16$. $\mathbf{A}$ is nonsingular. $\mathbf{A}^{-1} = \dfrac{1}{16} \begin{pmatrix} 0 & 0 & 2 \\ 8 & 0 & 0 \\ 0 & 16 & 0 \end{pmatrix} = \begin{pmatrix} 0 & 0 & \frac{1}{8} \\ \frac{1}{2} & 0 & 0 \\ 0 & 1 & 0 \end{pmatrix}$

15. $\begin{pmatrix} 6 & -2 & | & 1 & 0 \\ 0 & 4 & | & 0 & 1 \end{pmatrix} \xrightarrow[\frac{1}{4}R_2]{\frac{1}{6}R_1} \begin{pmatrix} 1 & -\frac{1}{3} & | & \frac{1}{6} & 0 \\ 0 & 1 & | & 0 & \frac{1}{4} \end{pmatrix} \xrightarrow{\frac{1}{3}R_2 + R_1} \begin{pmatrix} 1 & - & | & \frac{1}{6} & \frac{1}{12} \\ 0 & 1 & | & 0 & \frac{1}{4} \end{pmatrix};$ $\mathbf{A}^{-1} = \begin{pmatrix} \frac{1}{6} & \frac{1}{12} \\ 0 & \frac{1}{4} \end{pmatrix}$

18. $\begin{pmatrix} 2 & -3 & | & 1 & 0 \\ -2 & 4 & | & 0 & 1 \end{pmatrix} \xrightarrow{\frac{1}{2}R_1} \begin{pmatrix} 1 & -\frac{3}{2} & | & \frac{1}{2} & 0 \\ -2 & 4 & | & 0 & 1 \end{pmatrix} \xrightarrow{2R_1+R_2} \begin{pmatrix} 1 & -\frac{3}{2} & | & \frac{1}{2} & 0 \\ 0 & 1 & | & 1 & 1 \end{pmatrix}$

$\xrightarrow{\frac{3}{2}R_2+R_1} \begin{pmatrix} 1 & 0 & | & 2 & \frac{3}{2} \\ 0 & 1 & | & 1 & 1 \end{pmatrix}; \quad \mathbf{A}^{-1} = \begin{pmatrix} 2 & \frac{3}{2} \\ 1 & 1 \end{pmatrix}$

21. $\begin{pmatrix} 4 & 2 & 3 & | & 1 & 0 & 0 \\ 2 & 1 & 0 & | & 0 & 1 & 0 \\ -1 & -2 & 0 & | & 0 & 0 & 1 \end{pmatrix} \xrightarrow{R_{13}} \begin{pmatrix} -1 & -2 & 0 & | & 0 & 0 & 1 \\ 2 & 1 & 0 & | & 0 & 1 & 0 \\ 4 & 2 & 3 & | & 1 & 0 & 0 \end{pmatrix}$

$\xrightarrow[\text{operations}]{\text{row}} \begin{pmatrix} 1 & 0 & 0 & | & 0 & \frac{2}{3} & \frac{1}{3} \\ 0 & 1 & 0 & | & 0 & -\frac{1}{3} & -\frac{2}{3} \\ 0 & 0 & 1 & | & \frac{1}{3} & -\frac{2}{3} & 0 \end{pmatrix}; \quad \mathbf{A}^{-1} = \begin{pmatrix} 0 & \frac{2}{3} & \frac{1}{3} \\ 0 & -\frac{1}{3} & -\frac{2}{3} \\ \frac{1}{3} & -\frac{2}{3} & 0 \end{pmatrix}$

24. $\begin{pmatrix} 1 & 2 & 3 & | & 1 & 0 & 0 \\ 0 & 1 & 4 & | & 0 & 1 & 0 \\ 0 & 0 & 8 & | & 0 & 0 & 1 \end{pmatrix} \xrightarrow[\text{operations}]{\text{row}} \begin{pmatrix} 1 & 0 & 0 & | & 1 & -2 & \frac{5}{8} \\ 0 & 1 & 0 & | & 0 & 1 & -\frac{1}{2} \\ 0 & 0 & 1 & | & 0 & 0 & \frac{1}{8} \end{pmatrix}; \quad \mathbf{A}^{-1} = \begin{pmatrix} 1 & -2 & \frac{5}{8} \\ 0 & 1 & -\frac{1}{2} \\ 0 & 0 & \frac{1}{8} \end{pmatrix}$

27. $(\mathbf{AB})^{-1} = \mathbf{B}^{-1}\mathbf{A}^{-1} = \begin{pmatrix} -\frac{1}{3} & \frac{1}{3} \\ -1 & \frac{10}{3} \end{pmatrix}$

30. $\mathbf{A}^T = \begin{pmatrix} 1 & 2 \\ 4 & 10 \end{pmatrix}; \quad (\mathbf{A}^T)^{-1} = \begin{pmatrix} 5 & -1 \\ -2 & \frac{1}{2} \end{pmatrix}; \quad \mathbf{A}^{-1} = \begin{pmatrix} 5 & -2 \\ -1 & \frac{1}{2} \end{pmatrix}; \quad (\mathbf{A}^{-1})^T = \begin{pmatrix} 5 & -1 \\ -2 & \frac{1}{2} \end{pmatrix}$

In Problems 45–48 in Exercises 8.1, the matrix $\boldsymbol{B} = MA$ gives the coordinates of the new rotated vector. So $\boldsymbol{A} = M^{-1}B$ gives the coordinates of the original vector (i.e., the vector that is to be rotated).

33. (a) $\mathbf{A}^T = \begin{pmatrix} \sin\theta & -\cos\theta \\ \cos\theta & \sin\theta \end{pmatrix} = \mathbf{A}^{-1}$ **(b)** $\mathbf{A}^T = \begin{pmatrix} \frac{1}{\sqrt{3}} & \frac{1}{\sqrt{3}} & \frac{1}{\sqrt{3}} \\ 0 & \frac{1}{\sqrt{2}} & -\frac{1}{\sqrt{2}} \\ -\frac{2}{\sqrt{6}} & \frac{1}{\sqrt{6}} & \frac{1}{\sqrt{6}} \end{pmatrix} = \mathbf{A}^{-1}$

36. Suppose $\mathbf{A}$ is singular. Then $\det \mathbf{A} = 0$, $\det \mathbf{AB} = \det \mathbf{A} \cdot \det \mathbf{B} = 0$, and $\mathbf{AB}$ is singular.

39. If $\mathbf{A}$ is nonsingular, then $\mathbf{A}^{-1}$ exists, and $\mathbf{AB} = \mathbf{0}$ implies $\mathbf{A}^{-1}\mathbf{AB} = \mathbf{A}^{-1}\mathbf{0}$, so $\mathbf{B} = \mathbf{0}$.

42. From Theorem 8.5.1, $\det \mathbf{A} = \det \mathbf{A}^T$. If $\mathbf{A}$ is a nonsingular matrix, then $\det \mathbf{A} \neq 0$ implies $\det \mathbf{A}^T \neq 0$ and therefore $\mathbf{A}^T$ is nonsingular.

45. $\mathbf{A}^{-1} = \begin{pmatrix} \frac{1}{3} & \frac{1}{3} \\ \frac{2}{3} & -\frac{1}{3} \end{pmatrix}; \quad \mathbf{A}^{-1}\begin{pmatrix} 4 \\ 14 \end{pmatrix} = \begin{pmatrix} 6 \\ -2 \end{pmatrix}; \quad x_1 = 6, \; x_2 = -2$

48. $\mathbf{A}^{-1} = \begin{pmatrix} -2 & 1 \\ \frac{3}{2} & -\frac{1}{2} \end{pmatrix}$; $\mathbf{A}^{-1} \begin{pmatrix} 4 \\ -3 \end{pmatrix} = \begin{pmatrix} -11 \\ \frac{15}{2} \end{pmatrix}$; $x_1 = -11, \; x_2 = \dfrac{15}{2}$

51. $\mathbf{A}^{-1} = \begin{pmatrix} -2 & -3 & 2 \\ \frac{1}{4} & -\frac{1}{4} & 0 \\ \frac{5}{4} & \frac{7}{4} & -1 \end{pmatrix}$; $\mathbf{A}^{-1} \begin{pmatrix} 1 \\ -3 \\ 7 \end{pmatrix} = \begin{pmatrix} 21 \\ 1 \\ -11 \end{pmatrix}$; $x_1 = 21, \; x_2 = 1, \; x_3 = -11$

54. $\begin{pmatrix} 1 & 2 & 5 \\ 2 & 3 & 8 \\ -1 & 1 & 2 \end{pmatrix} \begin{pmatrix} x_1 \\ x_2 \\ x_3 \end{pmatrix} = \begin{pmatrix} b_1 \\ b_2 \\ b_3 \end{pmatrix}$; $\mathbf{A}^{-1} = \begin{pmatrix} 2 & -1 & -1 \\ 12 & -7 & -2 \\ -5 & 3 & 1 \end{pmatrix}$; $\mathbf{X} = \mathbf{A}^{-1} \begin{pmatrix} -1 \\ 4 \\ 6 \end{pmatrix} = \begin{pmatrix} -12 \\ -52 \\ 23 \end{pmatrix}$;

$\mathbf{X} = \mathbf{A}^{-1} \begin{pmatrix} 3 \\ 3 \\ 3 \end{pmatrix} = \begin{pmatrix} 0 \\ 9 \\ -3 \end{pmatrix}$; $\mathbf{A}^{-1} \begin{pmatrix} 0 \\ -5 \\ 4 \end{pmatrix} = \begin{pmatrix} 1 \\ 27 \\ -11 \end{pmatrix}$

57. $\det \mathbf{A} = 0$, so the system has a nontrivial solution.

60. (a) We write the equations in the form

$$-4u_1 + u_2 + u_4 = -200$$
$$u_1 - 4u_2 + u_3 = -300$$
$$u_2 - 4u_3 + u_4 = -300$$
$$u_1 + u_3 - 4u_4 = -200$$

In matrix form this becomes $\begin{pmatrix} -4 & 1 & 0 & 1 \\ 1 & -4 & 1 & 0 \\ 0 & 1 & -4 & 1 \\ 1 & 0 & 1 & -4 \end{pmatrix} \begin{pmatrix} u_1 \\ u_2 \\ u_3 \\ u_4 \end{pmatrix} = \begin{pmatrix} -200 \\ -300 \\ -300 \\ -200 \end{pmatrix}$.

(b) $\mathbf{A}^{-1} = \begin{pmatrix} -\frac{7}{24} & -\frac{1}{12} & -\frac{1}{24} & -\frac{1}{12} \\ -\frac{1}{12} & -\frac{7}{24} & -\frac{1}{12} & -\frac{1}{24} \\ -\frac{1}{24} & -\frac{1}{12} & -\frac{7}{24} & -\frac{1}{12} \\ -\frac{1}{12} & -\frac{1}{24} & -\frac{1}{12} & -\frac{7}{24} \end{pmatrix}$; $\mathbf{A}^{-1} \begin{pmatrix} -200 \\ -300 \\ -300 \\ -200 \end{pmatrix} = \begin{pmatrix} \frac{225}{2} \\ \frac{275}{2} \\ \frac{275}{2} \\ \frac{225}{2} \end{pmatrix}$;

$u_1 = u_4 = \dfrac{225}{2}, \; u_2 = u_3 = \dfrac{275}{2}$

8.7 Cramer's Rule

3. $\det \mathbf{A} = 0.3$, $\det \mathbf{A}_1 = 0.03$, $\det \mathbf{A}_2 = -0.09$; $x_1 = \frac{0.03}{0.3} = 0.1$, $x_2 = \frac{-0.09}{0.3} = -0.3$

6. $\det \mathbf{A} = -70$, $\det \mathbf{A}_1 = -14$, $\det \mathbf{A}_2 = 35$; $r = \frac{-14}{-70} = \frac{1}{5}$, $s = \frac{35}{-70} = -\frac{1}{2}$

9. $\det \mathbf{A} = -12$, $\det \mathbf{A}_1 = -48$, $\det \mathbf{A}_2 = -18$, $\det \mathbf{A}_3 = -12$; $u = \frac{48}{12} = 4$, $v = \frac{18}{12} = \frac{3}{2}$,
$w = 1$

12. (a) $\det \mathbf{A} = \epsilon - 1$, $\det \mathbf{A}_1 = \epsilon - 2$, $\det \mathbf{A}_2 = 1$; $x_1 = \dfrac{\epsilon - 2}{\epsilon - 1} = \dfrac{\epsilon - 1 - 1}{\epsilon - 1} = 1 - \dfrac{1}{\epsilon - 1}$,

$x_2 = \dfrac{1}{\epsilon - 1}$

(b) When $\epsilon = 1.01$, $x_1 = -99$ and $x_2 = 100$. When $\epsilon = 0.99$, $x_1 = 101$ and $x_2 = -100$.

15. The system is

$$i_1 + i_2 - i_3 = 0$$
$$r_1 i_1 - r_2 i_2 = E_1 - E_2$$
$$r_2 i_2 + R i_3 = E_2$$

$\det \mathbf{A} = -r_1 R - r_2 R - r_1 r_2$, $\det A_3 = -r_1 E_2$, $-r_2 E_1$; $i_3 = \dfrac{r_1 E_2 + r_2 E_1}{r_1 R + r_2 R + r_1 r_2}$

8.8 Eigenvalue Problem

3. $\mathbf{K}_3$ since $\begin{pmatrix} 6 & 3 \\ 2 & 1 \end{pmatrix} \begin{pmatrix} -5 \\ 10 \end{pmatrix} = \begin{pmatrix} 0 \\ 0 \end{pmatrix} = 0 \begin{pmatrix} -5 \\ 10 \end{pmatrix}$; $\quad \lambda = 0$

6. $\mathbf{K}_2$ since $\begin{pmatrix} -1 & 1 & 0 \\ 1 & 2 & 1 \\ 0 & 3 & -1 \end{pmatrix} \begin{pmatrix} 1 \\ 4 \\ 3 \end{pmatrix} = \begin{pmatrix} 3 \\ 12 \\ 9 \end{pmatrix} = 3 \begin{pmatrix} 1 \\ 4 \\ 3 \end{pmatrix}$; $\quad \lambda = 3$

9. We solve $\det(\mathbf{A} - \lambda \mathbf{I}) = \begin{vmatrix} 8 - \lambda & -1 \\ 16 & -\lambda \end{vmatrix} = (\lambda + 4)^2 = 0$.

For $\lambda_1 = \lambda_2 = -4$ we have $\begin{pmatrix} -4 & -1 & | & 0 \\ 16 & 4 & | & 0 \end{pmatrix}$ or $\begin{pmatrix} 1 & 1/4 & | & 0 \\ 0 & 0 & | & 0 \end{pmatrix}$

so that $k_1 = -\frac{1}{4}k_2$. If $k_2 = 4$ then $\mathbf{K}_1 = \begin{pmatrix} -1 \\ 4 \end{pmatrix}$.

12. We solve $\det(\mathbf{A} - \lambda \mathbf{I}) = \begin{vmatrix} 1 - \lambda & -1 \\ 1 & 1 - \lambda \end{vmatrix} = \lambda^2 - 2\lambda + 2 = 0$.

For $\lambda_1 = 1 - i$ we have $\begin{pmatrix} i & -1 & | & 0 \\ 1 & i & | & 0 \end{pmatrix}$ or $\begin{pmatrix} i & -1 & | & 0 \\ 0 & 0 & | & 0 \end{pmatrix}$

so that $k_1 = -ik_2$. If $k_2 = 1$ then $\mathbf{K}_1 = \begin{pmatrix} -i \\ 1 \end{pmatrix}$ and $\mathbf{K}_2 = \overline{\mathbf{K}}_1 = \begin{pmatrix} i \\ 1 \end{pmatrix}$.

15. We solve $\det(\mathbf{A} - \lambda \mathbf{I}) = \begin{vmatrix} 5 - \lambda & -1 & 0 \\ 0 & -5 - \lambda & 9 \\ 5 & -1 & -\lambda \end{vmatrix} = \begin{vmatrix} 4 - \lambda & -1 & 0 \\ 4 - \lambda & -5 - \lambda & 9 \\ 4 - \lambda & -1 & -\lambda \end{vmatrix} = \lambda(4 - \lambda)(\lambda + 4) = 0$.

For $\lambda_1 = 0$ we have $\begin{pmatrix} 5 & -1 & 0 & 0 \\ 0 & -5 & 9 & 0 \\ 5 & -1 & 0 & 0 \end{pmatrix}$ or $\begin{pmatrix} 1 & 0 & -9/25 & 0 \\ 0 & 1 & -9/5 & 0 \\ 0 & 0 & 0 & 0 \end{pmatrix}$

so that $k_1 = \frac{9}{25}k_3$ and $k_2 = \frac{9}{5}k_3$. If $k_3 = 25$ then $\mathbf{K}_1 = \begin{pmatrix} 9 \\ 45 \\ 25 \end{pmatrix}$. For $\lambda_2 = 4$ we have

$$\begin{pmatrix} 1 & -1 & 0 & 0 \\ 0 & -9 & 9 & 0 \\ 5 & -1 & -4 & 0 \end{pmatrix} \quad \text{or} \quad \begin{pmatrix} 1 & 0 & -1 & 0 \\ 0 & 1 & -1 & 0 \\ 0 & 0 & 0 & 0 \end{pmatrix}$$

so that $k_1 = k_3$ and $k_2 = k_3$. If $k_3 = 1$ then $\mathbf{K}_2 = \begin{pmatrix} 1 \\ 1 \\ 1 \end{pmatrix}$. For $\lambda_3 = -4$ we have

$$\begin{pmatrix} 9 & -1 & 0 & 0 \\ 0 & -1 & 9 & 0 \\ 5 & -1 & 4 & 0 \end{pmatrix} \quad \text{or} \quad \begin{pmatrix} 1 & 0 & -1 & 0 \\ 0 & 1 & -9 & 0 \\ 0 & 0 & 0 & 0 \end{pmatrix}$$

so that $k_1 = k_3$ and $k_2 = 9k_3$. If $k_3 = 1$ then $\mathbf{K}_3 = \begin{pmatrix} 1 \\ 9 \\ 1 \end{pmatrix}$.

18. We solve $\det(\mathbf{A} - \lambda\mathbf{I}) = \begin{vmatrix} 1-\lambda & 6 & 0 \\ 0 & 2-\lambda & 1 \\ 0 & 1 & 2-\lambda \end{vmatrix} = \begin{vmatrix} 1-\lambda & 6 & 0 \\ 0 & 3-\lambda & 3-\lambda \\ 0 & 1 & 2-\lambda \end{vmatrix}$

$$= (3-\lambda)(1-\lambda)^2 = 0.$$

For $\lambda_1 = 3$ we have $\begin{pmatrix} -2 & 6 & 0 & 0 \\ 0 & 0 & 0 & 0 \\ 0 & 1 & -1 & 0 \end{pmatrix} \quad \text{or} \quad \begin{pmatrix} 1 & 0 & -3 & 0 \\ 0 & 1 & -1 & 0 \\ 0 & 0 & 0 & 0 \end{pmatrix}$

so that $k_1 = 3k_3$ and $k_2 = k_3$. If $k_3 = 1$ then $\mathbf{K}_1 = \begin{pmatrix} 3 \\ 1 \\ 1 \end{pmatrix}$. For $\lambda_2 = \lambda_3 = 1$ we have

$$\begin{pmatrix} 0 & 6 & 0 & 0 \\ 0 & 1 & 1 & 0 \\ 0 & 1 & 1 & 0 \end{pmatrix} \quad \text{or} \quad \begin{pmatrix} 0 & 1 & 0 & 0 \\ 0 & 0 & 1 & 0 \\ 0 & 0 & 0 & 0 \end{pmatrix}$$

so that $k_2 = 0$ and $k_3 = 0$. If $k_1 = 1$ then $\mathbf{K}_2 = \begin{pmatrix} 1 \\ 0 \\ 0 \end{pmatrix}$.

21. We solve $\det(\mathbf{A} - \lambda\mathbf{I}) = \begin{vmatrix} 1-\lambda & 2 & 3 \\ 0 & 5-\lambda & 6 \\ 0 & 0 & -7-\lambda \end{vmatrix} = -(\lambda-1)(\lambda-5)(\lambda+7) = 0.$

For $\lambda_1 = 1$ we have $\begin{pmatrix} 0 & 2 & 3 & 0 \\ 0 & 4 & 6 & 0 \\ 0 & 0 & -6 & 0 \end{pmatrix} \quad \text{or} \quad \begin{pmatrix} 0 & 1 & 0 & 0 \\ 0 & 0 & 1 & 0 \\ 0 & 0 & 0 & 0 \end{pmatrix}$

so that $k_2 = 0$ and $k_3 = 0$. If $k_1 = 1$ then $\mathbf{K}_1 = \begin{pmatrix} 1 \\ 0 \\ 0 \end{pmatrix}$. For $\lambda_2 = 5$ we have

$$\begin{pmatrix} -4 & 2 & 3 & 0 \\ 0 & 0 & 6 & 0 \\ 0 & 0 & -12 & 0 \end{pmatrix} \quad \text{or} \quad \begin{pmatrix} 1 & -1/2 & 0 & 0 \\ 0 & 0 & 1 & 0 \\ 0 & 0 & 0 & 0 \end{pmatrix}$$

so that $k_3 = 0$ and $k_2 = 2k_1$. If $k_1 = 1$ then $\mathbf{K}_2 = \begin{pmatrix} 1 \\ 2 \\ 0 \end{pmatrix}$. For $\lambda_3 = -7$ we have

$$\begin{pmatrix} 8 & 2 & 3 & 0 \\ 0 & 12 & 6 & 0 \\ 0 & 0 & 0 & 0 \end{pmatrix} \quad \text{or} \quad \begin{pmatrix} 1 & 0 & 1/4 & 0 \\ 0 & 1 & 1/2 & 0 \\ 0 & 0 & 0 & 0 \end{pmatrix}$$

so that $k_1 = -\frac{1}{4}k_3$ and $k_2 = -\frac{1}{2}k_3$. If $k_3 = 4$ then $\mathbf{K}_3 = \begin{pmatrix} -1 \\ -2 \\ 4 \end{pmatrix}$.

24. We solve $\det(\mathbf{A} - \lambda\mathbf{I}) = \begin{vmatrix} 4-\lambda & 2 \\ 7 & -1-\lambda \end{vmatrix} = (\lambda - 6)(\lambda + 3) = 0$.

For $\lambda_1 = 6$ we have $\begin{pmatrix} -2 & 2 & | & 0 \\ 7 & -7 & | & 0 \end{pmatrix} \quad \text{or} \quad \begin{pmatrix} 1 & -1 & | & 0 \\ 0 & 0 & | & 0 \end{pmatrix}$

so that $k_1 = k_2$. If $k_2 = 1$ then $\mathbf{K}_1 = \begin{pmatrix} 1 \\ 1 \end{pmatrix}$. For $\lambda_2 = -3$ we have

$$\begin{pmatrix} 7 & 2 & | & 0 \\ 7 & 2 & | & 0 \end{pmatrix} \quad \text{or} \quad \begin{pmatrix} 1 & 2/7 & | & 0 \\ 0 & 0 & | & 0 \end{pmatrix}$$

so that $k_1 = -\frac{2}{7}k_2$. If $k_2 = -7$ then $\mathbf{K}_2 = \begin{pmatrix} -2 \\ 7 \end{pmatrix}$.

By Theorem 8.8.3, $\mathbf{A}^{-1}$ has eigenvalues $\lambda_1 = \frac{1}{6}$ and $\lambda_2 = -\frac{1}{3}$ with corresponding eigenvectors $\mathbf{K}_1 = \begin{pmatrix} 1 \\ 1 \end{pmatrix}$ and $\mathbf{K}_2 = \begin{pmatrix} -2 \\ 7 \end{pmatrix}$, respectively.

8.9 Powers of Matrices

3. The characteristic equation is $\lambda^2 - 3\lambda - 10 = 0$, with eigenvalues -2 and 5. Substituting the eigenvalues into $\lambda^m = c_0 + c_1\lambda$ generates

$$(-2)^m = c_0 - 2c_1$$
$$5^m = c_0 + 5c_1.$$

Solving the system gives

$$c_0 = \frac{1}{7}[5(-2)^m + 2(5)^m], \qquad c_1 = \frac{1}{7}[-(-2)^m + 5^m].$$

Thus

$$\mathbf{A}^m = c_0\mathbf{I} + c_1\mathbf{A} = \begin{pmatrix} \frac{1}{7}\left[3(-1)^m 2^{m+1} + 5^m\right] & \frac{3}{7}\left[-(-2)^m + 5^m\right] \\ \frac{2}{7}\left[-(-2)^m + 5^m\right] & \frac{1}{7}\left[(-2)^m + 6(5)^m\right] \end{pmatrix}$$

and

$$\mathbf{A}^3 = \begin{pmatrix} 11 & 57 \\ 38 & 106 \end{pmatrix}.$$

6. The characteristic equation is $\lambda^2 + 4\lambda + 3 = 0$, with eigenvalues -3 and -1. Substituting the eigenvalues into $\lambda^m = c_0 + c_1\lambda$ generates

$$(-3)^m = c_0 - 3c_1$$
$$(-1)^m = c_0 - c_1.$$

Solving the system gives

$$c_0 = \frac{1}{2}[-(-3)^m + 3(-1)^m], \qquad c_1 = \frac{1}{2}[-(-3)^m + (-1)^m].$$

Thus

$$\mathbf{A}^m = c_0\mathbf{I} + c_1\mathbf{A} = \begin{pmatrix} (-1)^m & -(-3)^m + (-1)^m \\ 0 & (-3)^m \end{pmatrix}$$

and

$$\mathbf{A}^6 = \begin{pmatrix} 1 & -728 \\ 0 & 729 \end{pmatrix}.$$

9. The characteristic equation is $-\lambda^3 + 3\lambda^2 + 6\lambda - 8 = 0$, with eigenvalues -2, 1, and 4. Substituting the eigenvalues into $\lambda^m = c_0 + c_1\lambda + c_2\lambda^2$ generates

$$(-2)^m = c_0 - 2c_1 + 4c_2$$
$$1 = c_0 + c_1 + c_2$$
$$4^m = c_0 + 4c_1 + 16c_2.$$

Solving the system gives

$$c_0 = \frac{1}{9}[8 + (-1)^m 2^{m+1} - 4^m],$$

$$c_1 = \frac{1}{18}[4 - 5(-2)^m + 4^m],$$

$$c_2 = \frac{1}{18}[-2 + (-2)^m + 4^m].$$

Thus

$$\mathbf{A}^m = c_0\mathbf{I} + c_1\mathbf{A} + c_2\mathbf{A}^2 = \begin{pmatrix} \frac{1}{9}\left[(-2)^m + (-1)^m 2^{m+1} + 3\cdot 2^{2m+1}\right] & \frac{1}{3}\left[-(-2)^m + 4^m\right] & 0 \\ -\frac{2}{3}\left[(-2)^m - 4^m\right] & \frac{1}{3}\left[(-1)^m 2^{m+1} + 4^m\right] & 0 \\ \frac{1}{3}\left[-3 + (-2)^m + 2^{2m+1}\right] & \frac{1}{3}\left[-(-2)^m + 4^m\right] & 1 \end{pmatrix}$$

and

$$\mathbf{A}^{10} = \begin{pmatrix} 699392 & 349184 & 0 \\ 698368 & 350208 & 0 \\ 699391 & 349184 & 1 \end{pmatrix}$$

12. The characteristic equation is $-\lambda^3 - \lambda^2 + 21\lambda + 45 = 0$, with eigenvalues -3, -3, and 5. Substituting the eigenvalues into $\lambda^m = c_0 + c_1\lambda + c_2\lambda^2$ generates

$$(-3)^m = c_0 - 3c_1 + 9c_2$$
$$(-3)^{m-1}m = c_1 - 6c_2$$
$$5^m = c_0 + 5c_1 + 25c_2.$$

Solving the system gives

$$c_0 = \frac{1}{64}\left[73(-3)^m - 2(-1)^m 3^{m+2} + 9\cdot 5^m - 40(-3)^m m\right],$$

$$c_1 = \frac{1}{96}\left[-(-1)^m 3^{m+2} + 9\cdot 5^m - 8(-3)^m m\right],$$

$$c_2 = \frac{1}{64}\left[-(-3)^m + 5^m - 8(-3)^{m-1}m\right].$$

Thus $\mathbf{A}^m = c_0\mathbf{I} + c_1\mathbf{A} + c_2\mathbf{A}^2$

$$= \begin{pmatrix} \frac{1}{32}\left[31(-3)^m - (-1)^m 3^{m+1} + 4\cdot 5^m\right] & \frac{1}{16}\left[-(-3)^m - (-1)^m 3^{m+1} + 4\cdot 5^m\right] & \frac{1}{32}\left[(-3)^m + (-1)^m 3^{m+1} - 4\cdot 5^m\right] \\ \frac{1}{16}\left[-(-3)^m - (-1)^m 3^{m+1} + 4\cdot 5^m\right] & \frac{1}{8}\left[7(-3)^m - (-1)^m 3^{m+1} + 4\cdot 5^m\right] & \frac{1}{16}\left[(-3)^m + (-1)^m 3^{m+1} - 4\cdot 5^m\right] \\ \frac{3}{32}\left[(-3)^m + (-1)^m 3^{m+1} - 4\cdot 5^m\right] & \frac{3}{16}\left[(-3)^m + (-1)^m 3^{m+1} - 4\cdot 5^m\right] & \frac{1}{32}\left[29(-3)^m - (-1)^m 3^{m+2} + 12\cdot 5^m\right] \end{pmatrix}$$

and

$$\mathbf{A}^5 = \begin{pmatrix} 178 & 842 & -421 \\ 842 & 1441 & -842 \\ -1263 & -2526 & 1020 \end{pmatrix}$$

15. The characteristic equation of $\mathbf{A}$ is $\lambda^2 - 5\lambda + 10 = 0$, so $\mathbf{A}^2 - 5\mathbf{A} + 10\mathbf{I} = \mathbf{0}$ and $\mathbf{I} = -\frac{1}{10}\mathbf{A}^2 + \frac{1}{2}\mathbf{A}$. Multiplying by $\mathbf{A}^{-1}$ we find

$$\mathbf{A}^{-1} = -\frac{1}{10}\mathbf{A} + \frac{1}{2}\mathbf{I} = -\frac{1}{10}\begin{pmatrix} 2 & -4 \\ 1 & 3 \end{pmatrix} = \frac{1}{2}\begin{pmatrix} 1 & 0 \\ 0 & 1 \end{pmatrix} = \begin{pmatrix} \frac{3}{10} & \frac{2}{5} \\ -\frac{1}{10} & \frac{1}{5} \end{pmatrix}.$$

18. (a) If $\mathbf{A}^m = \mathbf{0}$ for some m, then $(\det\mathbf{A})^m = \det\mathbf{A}^m = \det\mathbf{0} = 0$, and $\mathbf{A}$ is a singular matrix.

(b) By (1) of Section 8.8 we have $\mathbf{A}\mathbf{K} = \lambda\mathbf{K}$, $\mathbf{A}^2\mathbf{K} = \lambda\mathbf{A}\mathbf{K} = \lambda^2\mathbf{K}$, $\mathbf{A}^3\mathbf{K} = \lambda^2\mathbf{A}\mathbf{K} = \lambda^3\mathbf{K}$, and, in general, $\mathbf{A}^m\mathbf{K} = \lambda^m\mathbf{K}$. If $\mathbf{A}$ is nilpotent with index m, then $\mathbf{A}^m = \mathbf{0}$ and $\lambda^m = 0$.

8.10   Orthogonal Matrices

3. (a)-(b)
$$\begin{pmatrix} 5 & 13 & 0 \\ 13 & 5 & 0 \\ 0 & 0 & -8 \end{pmatrix} \begin{pmatrix} \frac{\sqrt{2}}{2} \\ \frac{\sqrt{2}}{2} \\ 0 \end{pmatrix} = \begin{pmatrix} 9\sqrt{2} \\ 9\sqrt{2} \\ 0 \end{pmatrix} = 18 \begin{pmatrix} \frac{\sqrt{2}}{2} \\ \frac{\sqrt{2}}{2} \\ 0 \end{pmatrix}; \quad \lambda_1 = 18$$

$$\begin{pmatrix} 5 & 13 & 0 \\ 13 & 5 & 0 \\ 0 & 0 & -8 \end{pmatrix} \begin{pmatrix} \frac{\sqrt{3}}{3} \\ -\frac{\sqrt{3}}{3} \\ \frac{\sqrt{3}}{3} \end{pmatrix} = \begin{pmatrix} -\frac{8\sqrt{3}}{3} \\ \frac{8\sqrt{3}}{3} \\ -\frac{8\sqrt{3}}{3} \end{pmatrix} = (-8) \begin{pmatrix} \frac{\sqrt{3}}{3} \\ -\frac{\sqrt{3}}{3} \\ \frac{\sqrt{3}}{3} \end{pmatrix}; \quad \lambda_2 = -8$$

$$\begin{pmatrix} 5 & 13 & 0 \\ 13 & 5 & 0 \\ 0 & 0 & -8 \end{pmatrix} \begin{pmatrix} \frac{\sqrt{6}}{6} \\ -\frac{\sqrt{6}}{6} \\ -\frac{\sqrt{6}}{3} \end{pmatrix} = \begin{pmatrix} -\frac{8\sqrt{6}}{6} \\ \frac{8\sqrt{6}}{6} \\ \frac{8\sqrt{6}}{3} \end{pmatrix} = (-8) \begin{pmatrix} \frac{\sqrt{6}}{6} \\ -\frac{\sqrt{6}}{6} \\ -\frac{\sqrt{6}}{3} \end{pmatrix}; \quad \lambda_3 = -8$$

(c) $\mathbf{K}_1^T \mathbf{K}_2 = \begin{pmatrix} \frac{\sqrt{2}}{2} & \frac{\sqrt{2}}{2} & 0 \end{pmatrix} \begin{pmatrix} \frac{\sqrt{3}}{3} \\ -\frac{\sqrt{3}}{3} \\ \frac{\sqrt{3}}{3} \end{pmatrix} = \frac{\sqrt{6}}{6} - \frac{\sqrt{6}}{6} = 0;$

$\mathbf{K}_1^T \mathbf{K}_3 = \begin{pmatrix} \frac{\sqrt{2}}{2} & \frac{\sqrt{2}}{2} & 0 \end{pmatrix} \begin{pmatrix} \frac{\sqrt{6}}{6} \\ -\frac{\sqrt{6}}{6} \\ -\frac{\sqrt{6}}{3} \end{pmatrix} = \frac{\sqrt{12}}{12} - \frac{\sqrt{12}}{12} = 0;$

$\mathbf{K}_2^T \mathbf{K}_3 = \begin{pmatrix} \frac{\sqrt{3}}{3} & -\frac{\sqrt{3}}{3} & \frac{\sqrt{3}}{3} \end{pmatrix} \begin{pmatrix} \frac{\sqrt{6}}{6} \\ -\frac{\sqrt{6}}{6} \\ -\frac{\sqrt{6}}{3} \end{pmatrix} = \frac{\sqrt{18}}{18} + \frac{\sqrt{18}}{18} - \frac{\sqrt{18}}{9} = 0$

6. Not orthogonal. Columns one and three are not unit vectors.

9. Not orthogonal. Columns are not unit vectors.

12. $\lambda_1 = 7$, $\lambda_2 = 4$, $\mathbf{K}_1 = \begin{pmatrix} 1 \\ 0 \end{pmatrix}$, $\mathbf{K}_2 = \begin{pmatrix} 0 \\ 1 \end{pmatrix}$, $\mathbf{P} = \begin{pmatrix} 1 & 0 \\ 0 & 1 \end{pmatrix}$

15. $\lambda_1 = 0$, $\lambda_2 = 2$, $\lambda_3 = 1$, $\mathbf{K}_1 = \begin{pmatrix} -1 \\ 0 \\ 1 \end{pmatrix}$, $\mathbf{K}_2 = \begin{pmatrix} 1 \\ 0 \\ 1 \end{pmatrix}$, $\mathbf{K}_3 = \begin{pmatrix} 0 \\ 1 \\ 0 \end{pmatrix}$, $\mathbf{P} = \begin{pmatrix} -\frac{1}{\sqrt{2}} & \frac{1}{\sqrt{2}} & 0 \\ 0 & 0 & 1 \\ \frac{1}{\sqrt{2}} & \frac{1}{\sqrt{2}} & 0 \end{pmatrix}$

18. $\lambda_1 = -18$, $\lambda_2 = 0$, $\lambda_3 = 9$, $\mathbf{K}_1 = \begin{pmatrix} 1 \\ -2 \\ 2 \end{pmatrix}$, $\mathbf{K}_2 = \begin{pmatrix} -2 \\ 1 \\ 2 \end{pmatrix}$, $\mathbf{K}_3 = \begin{pmatrix} 2 \\ 2 \\ 1 \end{pmatrix}$,

$$\mathbf{P} = \begin{pmatrix} \frac{1}{3} & -\frac{2}{3} & \frac{2}{3} \\ -\frac{2}{3} & \frac{1}{3} & \frac{2}{3} \\ \frac{2}{3} & \frac{2}{3} & \frac{1}{3} \end{pmatrix}$$

21. (a)–(b) We compute

$$\mathbf{AK}_1 = \begin{pmatrix} 0 & 2 & 2 \\ 2 & 0 & 2 \\ 2 & 2 & 0 \end{pmatrix} \begin{pmatrix} 1 \\ -1 \\ 0 \end{pmatrix} = \begin{pmatrix} -2 \\ 2 \\ 0 \end{pmatrix} = -2 \begin{pmatrix} 1 \\ -1 \\ 0 \end{pmatrix} = -2\mathbf{K}_1$$

$$\mathbf{AK}_2 = \begin{pmatrix} 0 & 2 & 2 \\ 2 & 0 & 2 \\ 2 & 2 & 0 \end{pmatrix} \begin{pmatrix} 1 \\ 0 \\ -1 \end{pmatrix} = \begin{pmatrix} -2 \\ 0 \\ 2 \end{pmatrix} = -2 \begin{pmatrix} 1 \\ 0 \\ -1 \end{pmatrix} = -2\mathbf{K}_2$$

$$\mathbf{AK}_1 = \begin{pmatrix} 0 & 2 & 2 \\ 2 & 0 & 2 \\ 2 & 2 & 0 \end{pmatrix} \begin{pmatrix} 1 \\ 1 \\ 1 \end{pmatrix} = \begin{pmatrix} 4 \\ 4 \\ 4 \end{pmatrix} = 4 \begin{pmatrix} 1 \\ 1 \\ 1 \end{pmatrix} = 4\mathbf{K}_3$$

and observe that $\mathbf{K}_1$ is an eigenvector with corresponding eigenvalue -2, $\mathbf{K}_2$ is an eigenvector with corresponding eigenvalue -2, and $\mathbf{K}_3$ is an eigenvector with corresponding eigenvalue 4.

(c) Since $\mathbf{K}_1 \cdot \mathbf{K}_2 = 1 \neq 0$, $\mathbf{K}_1$ and $\mathbf{K}_2$ are not orthogonal, while $\mathbf{K}_1 \cdot \mathbf{K}_3 = 0$ and $\mathbf{K}_2 \cdot \mathbf{K}_3 = 0$ so $\mathbf{K}_3$ is orthogonal to both $\mathbf{K}_1$ and $\mathbf{K}_2$, To transform $\{\mathbf{K}_1, \mathbf{K}_2\}$ into an orthogonal set we let $\mathbf{V}_1 = \mathbf{K}_1$ and compute $\mathbf{K}_2 \cdot \mathbf{V}_1 = 1$ and $\mathbf{V}_1 \cdot \mathbf{V}_1 = 2$. Then

$$\mathbf{V}_2 = \mathbf{K}_2 - \frac{\mathbf{K}_2 \cdot \mathbf{V}_1}{\mathbf{V}_1 \cdot \mathbf{V}_1} \mathbf{V}_1 = \begin{pmatrix} 1 \\ 0 \\ -1 \end{pmatrix} - \frac{1}{2} \begin{pmatrix} 1 \\ -1 \\ 0 \end{pmatrix} = \begin{pmatrix} \frac{1}{2} \\ \frac{1}{2} \\ -1 \end{pmatrix}$$

Now, $\{\mathbf{V}_1, \mathbf{V}_2, \mathbf{K}_3\}$ is an orthogonal set of eigenvectors with

$$\|\mathbf{V}_1\| = \sqrt{2}, \quad \|\mathbf{V}_2\| = \frac{3}{\sqrt{6}}, \quad \text{and} \quad \|\mathbf{K}_3\| = \sqrt{3}.$$

An orthonormal set of vectors is

$$\begin{pmatrix} \frac{1}{\sqrt{2}} \\ -\frac{1}{\sqrt{2}} \\ 0 \end{pmatrix}, \quad \begin{pmatrix} \frac{1}{\sqrt{6}} \\ \frac{1}{\sqrt{6}} \\ -\frac{2}{\sqrt{6}} \end{pmatrix}, \quad \text{and} \quad \begin{pmatrix} \frac{1}{\sqrt{3}} \\ \frac{1}{\sqrt{3}} \\ \frac{1}{\sqrt{3}} \end{pmatrix}$$

and so the matrix

$$\mathbf{P} = \begin{pmatrix} \frac{1}{\sqrt{2}} & \frac{1}{\sqrt{6}} & \frac{1}{\sqrt{3}} \\ -\frac{1}{\sqrt{2}} & \frac{1}{\sqrt{6}} & \frac{1}{\sqrt{3}} \\ 0 & -\frac{2}{\sqrt{6}} & \frac{1}{\sqrt{3}} \end{pmatrix}$$

is orthogonal.

24. The eigenvalues and corresponding eigenvectors of $\mathbf{A}$ are

$$\lambda_1 = \lambda_2 = -1, \quad \lambda_3 = \lambda_4 = 3, \quad \text{and} \quad \mathbf{K}_1 = \begin{pmatrix} -1 \\ 1 \\ 0 \\ 0 \end{pmatrix}, \quad \mathbf{K}_2 = \begin{pmatrix} 0 \\ 0 \\ -1 \\ 1 \end{pmatrix}, \quad \mathbf{K}_3 = \begin{pmatrix} 0 \\ 0 \\ 1 \\ 1 \end{pmatrix},$$

$$\mathbf{K}_4 = \begin{pmatrix} 1 \\ 1 \\ 0 \\ 0 \end{pmatrix}.$$

Since $\mathbf{K}_1 \cdot \mathbf{K}_2 = \mathbf{K}_1 \cdot \mathbf{K}_3 = \mathbf{K}_1 \cdot \mathbf{K}_4 = \mathbf{K}_2 \cdot \mathbf{K}_3 = \mathbf{K}_2 \cdot \mathbf{K}_4 = \mathbf{K}_3 \cdot \mathbf{K}_4 = 0$, the vectors are orthogonal. Using $\|\mathbf{K}_1\| = \|\mathbf{K}_2\| = \|\mathbf{K}_3\| = \|\mathbf{K}_4\| = \sqrt{2}$, we construct the orthogonal matrix

$$\mathbf{P} = \begin{pmatrix} -\frac{1}{\sqrt{2}} & 0 & 0 & \frac{1}{\sqrt{2}} \\ \frac{1}{\sqrt{2}} & 0 & 0 & \frac{1}{\sqrt{2}} \\ 0 & -\frac{1}{\sqrt{2}} & \frac{1}{\sqrt{2}} & 0 \\ 0 & \frac{1}{\sqrt{2}} & \frac{1}{\sqrt{2}} & 0 \end{pmatrix}.$$

27. To show that $\mathbf{A}^{-1}$ is orthogonal we must demostrate that $\left(\mathbf{A}^{-1}\right)^{-1} = \left(\mathbf{A}^{-1}\right)^{T}$.

$$\left(\mathbf{A}^{-1}\right)^{-1} = \mathbf{A} = \left(\mathbf{A}^{T}\right)^{T}$$
$$\left(\mathbf{A}^{-1}\right)^{-1} = \left(\mathbf{A}^{-1}\right)^{T} \quad \leftarrow \quad \text{Since } \mathbf{A}^{T} = \mathbf{A}^{-1}$$

30. $\mathbf{A}\mathbf{A}^{T} = \begin{pmatrix} \cos\theta & -\sin\theta \\ \sin\theta & \cos\theta \end{pmatrix} \begin{pmatrix} \cos\theta & \sin\theta \\ -\sin\theta & \cos\theta \end{pmatrix} = \begin{pmatrix} \cos^2\theta + \sin^2\theta & 0 \\ 0 & \sin^2\theta + \cos^2\theta \end{pmatrix} = \begin{pmatrix} 1 & 0 \\ 0 & 1 \end{pmatrix}$

$= \mathbf{I}$.

Therefore $\mathbf{A}^{T} = \mathbf{A}^{-1}$.

8.11 | Approximation of Eigenvalues

3. Taking $\mathbf{X}_0 = \begin{pmatrix} 1 \\ 1 \end{pmatrix}$ and computing $\mathbf{A}\mathbf{X}_0 = \begin{pmatrix} 6 \\ 16 \end{pmatrix}$, we define $\mathbf{X}_1 = \frac{1}{16}\begin{pmatrix} 6 \\ 16 \end{pmatrix} = \begin{pmatrix} 0.375 \\ 1 \end{pmatrix}$.

Continuing in this manner we obtain

$$\mathbf{X}_2 = \begin{pmatrix} 0.3363 \\ 1 \end{pmatrix}, \quad \mathbf{X}_3 = \begin{pmatrix} 0.3335 \\ 1 \end{pmatrix}, \quad \mathbf{X}_4 = \begin{pmatrix} 0.3333 \\ 1 \end{pmatrix}.$$

We conclude that a dominant eigenvector is $\mathbf{K} = \begin{pmatrix} 0.3333 \\ 1 \end{pmatrix}$ with corresponding eigenvalue $\lambda = 14$.

6. Taking $\mathbf{X}_0 = \begin{pmatrix} 1 \\ 1 \\ 1 \end{pmatrix}$ and computing $\mathbf{A}\mathbf{X}_0 = \begin{pmatrix} 5 \\ 2 \\ 2 \end{pmatrix}$, we define $\mathbf{X}_1 = \frac{1}{5}\begin{pmatrix} 5 \\ 2 \\ 2 \end{pmatrix} = \begin{pmatrix} 1 \\ 0.4 \\ 0.4 \end{pmatrix}$. Continuing in this manner we obtain

$$\mathbf{X}_2 = \begin{pmatrix} 1 \\ 0.2105 \\ 0.2105 \end{pmatrix}, \quad \mathbf{X}_3 = \begin{pmatrix} 1 \\ 0.1231 \\ 0.1231 \end{pmatrix}, \quad \mathbf{X}_4 = \begin{pmatrix} 1 \\ 0.0758 \\ 0.0758 \end{pmatrix}, \quad \mathbf{X}_5 = \begin{pmatrix} 1 \\ 0.0481 \\ 0.0481 \end{pmatrix}.$$

At this point if we restart with $\mathbf{X}_0 = \begin{pmatrix} 1 \\ 0 \\ 0 \end{pmatrix}$ we see that $\mathbf{K} = \begin{pmatrix} 1 \\ 0 \\ 0 \end{pmatrix}$ is a dominant eigenvector with corresponding eigenvalue $\lambda = 3$.

9. Taking $\mathbf{X}_0 = \begin{pmatrix} 1 \\ 1 \\ 1 \end{pmatrix}$ and using scaling we obtain

$$\mathbf{X}_1 = \begin{pmatrix} 1 \\ 0 \\ 1 \end{pmatrix}, \quad \mathbf{X}_2 = \begin{pmatrix} 1 \\ -0.6667 \\ 1 \end{pmatrix}, \quad \mathbf{X}_3 = \begin{pmatrix} 1 \\ -0.9091 \\ 1 \end{pmatrix}, \quad \mathbf{X}_4 = \begin{pmatrix} 1 \\ -0.9767 \\ 1 \end{pmatrix},$$

$$\mathbf{X}_5 = \begin{pmatrix} 1 \\ -0.9942 \\ 1 \end{pmatrix}.$$

Taking $\mathbf{K} = \begin{pmatrix} 1 \\ -1 \\ 1 \end{pmatrix}$ as the dominant eigenvector we find $\lambda_1 = 4$. Now the normalized eigenvector is $\mathbf{K}_1 = \begin{pmatrix} 0.5774 \\ -0.5774 \\ 0.5774 \end{pmatrix}$ and $\mathbf{B} = \begin{pmatrix} 1.6667 & 0.3333 & -1.3333 \\ 0.3333 & 0.6667 & 0.3333 \\ -1.3333 & 0.3333 & 1.6667 \end{pmatrix}$. If $\mathbf{X}_0 = \begin{pmatrix} 1 \\ 1 \\ 1 \end{pmatrix}$ is now chosen only one more eigenvalue is found. Thus try $\mathbf{X}_0 = \begin{pmatrix} 1 \\ 1 \\ 0 \end{pmatrix}$. Using scaling we obtain

$$\mathbf{X}_1 = \begin{pmatrix} 1 \\ 0.5 \\ -0.5 \end{pmatrix}, \quad \mathbf{X}_2 = \begin{pmatrix} 1 \\ 0.2 \\ -0.8 \end{pmatrix}, \quad \mathbf{X}_3 = \begin{pmatrix} 1 \\ 0.0714 \\ -0.9286 \end{pmatrix}, \quad \mathbf{X}_4 = \begin{pmatrix} 1 \\ 0.0244 \\ -0.9756 \end{pmatrix},$$

$$\mathbf{X}_5 = \begin{pmatrix} 1 \\ 0.0082 \\ -0.9918 \end{pmatrix}.$$

Taking $\mathbf{K} = \begin{pmatrix} 1 \\ 0 \\ -1 \end{pmatrix}$ as the eigenvector we find $\lambda_2 = 3$. The normalized eigenvector in this case is $\mathbf{K}_2 = \begin{pmatrix} 0.7071 \\ 0 \\ -0.7071 \end{pmatrix}$ and $\mathbf{C} = \begin{pmatrix} 0.1667 & 0.3333 & 0.1667 \\ 0.3333 & 0.6667 & 0.3333 \\ 0.1667 & 0.3333 & 0.1667 \end{pmatrix}$. If $\mathbf{X}_0 = \begin{pmatrix} 1 \\ 1 \\ 1 \end{pmatrix}$ is chosen, and

scaling is used we obtain $\mathbf{X}_1 = \begin{pmatrix} 0.5 \\ 1 \\ 0.5 \end{pmatrix}$, $\mathbf{X}_2 = \begin{pmatrix} 0.5 \\ 1 \\ 0.5 \end{pmatrix}$. Taking $\mathbf{K} = \begin{pmatrix} 0.5 \\ 1 \\ 0.5 \end{pmatrix}$ we find $\lambda_3 = 1$.

The eigenvalues are 4, 3, and 1. The difficulty in choosing $\mathbf{X}_0 = \begin{pmatrix} 1 \\ 1 \\ 1 \end{pmatrix}$ to find the second

eigenvector results from the fact that this vector is a linear combination of the eigenvectors corresponding to the other two eigenvalues, with 0 contribution from the second eigenvector. When this occurs the development of the power method, shown in the text, breaks down.

12. The inverse matrix is $\begin{pmatrix} 1 & 3 \\ 4 & 2 \end{pmatrix}$. Taking $\mathbf{X}_0 = \begin{pmatrix} 1 \\ 1 \end{pmatrix}$ and using scaling we obtain

$$\mathbf{X}_1 = \begin{pmatrix} 0.6667 \\ 1 \end{pmatrix}, \quad \mathbf{X}_2 = \begin{pmatrix} 0.7857 \\ 1 \end{pmatrix}, \ldots, \mathbf{X}_{10} = \begin{pmatrix} 0.75 \\ 1 \end{pmatrix}.$$

Using $\mathbf{K} = \begin{pmatrix} 0.75 \\ 1 \end{pmatrix}$ we find $\lambda = 5$. The minimum eigenvalue of $\begin{pmatrix} -0.2 & 0.3 \\ 0.4 & -0.1 \end{pmatrix}$ is $1/5 = 0.2$.

8.12   Diagonalization

3. For $\lambda_1 = \lambda_2 = 1$ we obtain the single eigenvector $\mathbf{K}_1 = \begin{pmatrix} 1 \\ 1 \end{pmatrix}$. Hence $\mathbf{A}$ is not diagonalizable.

6. Distinct eigenvalues $\lambda_1 = -4$, $\lambda_2 = 10$ imply $\mathbf{A}$ is diagonalizable.

$$\mathbf{P} = \begin{pmatrix} -3 & 1 \\ 1 & -5 \end{pmatrix}, \quad \mathbf{D} = \begin{pmatrix} -4 & 0 \\ 0 & 10 \end{pmatrix}$$

9. Distinct eigenvalues $\lambda_1 = -i$, $\lambda_2 = i$ imply $\mathbf{A}$ is diagonalizable.

$$\mathbf{P} = \begin{pmatrix} 1 & 1 \\ -i & i \end{pmatrix}, \quad \mathbf{D} = \begin{pmatrix} -i & 0 \\ 0 & i \end{pmatrix}$$

12. Distinct eigenvalues $\lambda_1 = 3$, $\lambda_2 = 4$, $\lambda_3 = 5$ imply $\mathbf{A}$ is diagonalizable.

$$\mathbf{P} = \begin{pmatrix} 1 & 2 & 0 \\ 0 & 2 & 1 \\ 1 & 1 & -1 \end{pmatrix}, \quad \mathbf{D} = \begin{pmatrix} 3 & 0 & 0 \\ 0 & 4 & 0 \\ 0 & 0 & 5 \end{pmatrix}$$

15. The eigenvalues are $\lambda_1 = \lambda_2 = 1$, $\lambda_3 = 2$. For $\lambda_1 = \lambda_2 = 1$ we obtain the single eigenvector $\mathbf{K}_1 = \begin{pmatrix} 1 \\ 0 \\ 0 \end{pmatrix}$. Hence $\mathbf{A}$ is not diagonalizable.

18. For $\lambda_1 = \lambda_2 = \lambda_3 = 1$ we obtain the single eigenvector $\mathbf{K}_1 = \begin{pmatrix} 1 \\ -2 \\ 1 \end{pmatrix}$. Hence $\mathbf{A}$ is not diagonalizable.

21. $\lambda_1 = 0$, $\lambda_2 = 2$, $\mathbf{K}_1 = \begin{pmatrix} 1 \\ -1 \end{pmatrix}$, $\mathbf{K}_2 = \begin{pmatrix} 1 \\ 1 \end{pmatrix}$, $\mathbf{P} = \begin{pmatrix} \frac{1}{\sqrt{2}} & \frac{1}{\sqrt{2}} \\ -\frac{1}{\sqrt{2}} & \frac{1}{\sqrt{2}} \end{pmatrix}$, $\mathbf{D} = \begin{pmatrix} 0 & 0 \\ 0 & 2 \end{pmatrix}$

24. $\lambda_1 = -1$, $\lambda_2 = 3$, $\mathbf{K}_1 = \begin{pmatrix} 1 \\ 1 \end{pmatrix}$, $\mathbf{K}_2 = \begin{pmatrix} 1 \\ -1 \end{pmatrix}$, $\mathbf{P} = \begin{pmatrix} \frac{1}{\sqrt{2}} & \frac{1}{\sqrt{2}} \\ \frac{1}{\sqrt{2}} & -\frac{1}{\sqrt{2}} \end{pmatrix}$, $\mathbf{D} = \begin{pmatrix} -1 & 0 \\ 0 & 3 \end{pmatrix}$

27. $\lambda_1 = 3$, $\lambda_2 = 6$, $\lambda_3 = 9$, $\mathbf{K}_1 = \begin{pmatrix} 2 \\ 2 \\ 1 \end{pmatrix}$, $\mathbf{K}_2 = \begin{pmatrix} 2 \\ -1 \\ -2 \end{pmatrix}$, $\mathbf{K}_3 = \begin{pmatrix} 1 \\ -2 \\ 2 \end{pmatrix}$, $\mathbf{P} = \begin{pmatrix} \frac{2}{3} & \frac{2}{3} & \frac{1}{3} \\ \frac{2}{3} & -\frac{1}{3} & -\frac{2}{3} \\ \frac{1}{3} & -\frac{2}{3} & \frac{2}{3} \end{pmatrix}$,

$\mathbf{D} = \begin{pmatrix} 3 & 0 & 0 \\ 0 & 6 & 0 \\ 0 & 0 & 9 \end{pmatrix}$

30. $\lambda_1 = \lambda_2 = 0$, $\lambda_3 = -2$, $\lambda_4 = 2$, $\mathbf{K}_1 = \begin{pmatrix} -1 \\ 0 \\ 1 \\ 0 \end{pmatrix}$, $\mathbf{K}_2 = \begin{pmatrix} 0 \\ -1 \\ 0 \\ 1 \end{pmatrix}$, $\mathbf{K}_3 = \begin{pmatrix} 1 \\ -1 \\ 1 \\ -1 \end{pmatrix}$, $\mathbf{K}_4 = \begin{pmatrix} 1 \\ 1 \\ 1 \\ 1 \end{pmatrix}$,

$\mathbf{P} = \begin{pmatrix} -\frac{1}{\sqrt{2}} & 0 & \frac{1}{2} & \frac{1}{2} \\ 0 & -\frac{1}{\sqrt{2}} & -\frac{1}{2} & \frac{1}{2} \\ \frac{1}{\sqrt{2}} & 0 & \frac{1}{2} & \frac{1}{2} \\ 0 & \frac{1}{\sqrt{2}} & -\frac{1}{2} & \frac{1}{2} \end{pmatrix}$, $\mathbf{D} = \begin{pmatrix} 0 & 0 & 0 & 0 \\ 0 & 0 & 0 & 0 \\ 0 & 0 & -2 & 0 \\ 0 & 0 & 0 & 2 \end{pmatrix}$

33. The given equation can be written as $\mathbf{X}^T \mathbf{A} \mathbf{X} = 20$:

$(x \quad y) \begin{pmatrix} -3 & 4 \\ 4 & 3 \end{pmatrix} \begin{pmatrix} x \\ y \end{pmatrix} = 20$. Using $\lambda_1 = 5$, $\lambda_2 = -5$, $\mathbf{K}_1 = \begin{pmatrix} 1 \\ 2 \end{pmatrix}$,

$\mathbf{K}_2 = \begin{pmatrix} -2 \\ 1 \end{pmatrix}$, $\mathbf{P} = \begin{pmatrix} \frac{1}{\sqrt{5}} & -\frac{2}{\sqrt{5}} \\ \frac{2}{\sqrt{5}} & \frac{1}{\sqrt{5}} \end{pmatrix}$ and $\mathbf{X} = \mathbf{P}\mathbf{X}'$ we find

$$(X \quad Y) \begin{pmatrix} 5 & 0 \\ 0 & -5 \end{pmatrix} \begin{pmatrix} X \\ Y \end{pmatrix} = 20 \quad \text{or} \quad 5X^2 - 5Y^2 = 20.$$

The conic section is a hyperbola. Now from $\mathbf{X}' = \mathbf{P}^T\mathbf{X}$ we see that the XY-coordinates of $(1, 2)$ and $(-2, 1)$ are $(\sqrt{5}, 0)$ and $(0, \sqrt{5})$, respectively. From this we conclude that the X-axis and Y-axis are as shown in the accompanying figure.

36. Since eigenvectors are mutually orthogonal we use an orthogonal matrix $\mathbf{P}$ and $\mathbf{A} = \mathbf{PDP}^T$.

$$\mathbf{A} = \begin{pmatrix} \frac{1}{\sqrt{3}} & \frac{1}{\sqrt{2}} & \frac{1}{\sqrt{6}} \\ -\frac{1}{\sqrt{3}} & 0 & \frac{2}{\sqrt{6}} \\ \frac{1}{\sqrt{3}} & -\frac{1}{\sqrt{2}} & \frac{1}{\sqrt{6}} \end{pmatrix} \begin{pmatrix} 1 & 0 & 0 \\ 0 & 3 & 0 \\ 0 & 0 & 5 \end{pmatrix} \begin{pmatrix} \frac{1}{\sqrt{3}} & -\frac{1}{\sqrt{3}} & \frac{1}{\sqrt{3}} \\ \frac{1}{\sqrt{2}} & 0 & -\frac{1}{\sqrt{2}} \\ \frac{1}{\sqrt{6}} & \frac{2}{\sqrt{6}} & \frac{1}{\sqrt{6}} \end{pmatrix} = \begin{pmatrix} \frac{8}{3} & \frac{4}{3} & -\frac{1}{3} \\ \frac{4}{3} & \frac{11}{3} & \frac{4}{3} \\ -\frac{1}{3} & \frac{4}{3} & \frac{8}{3} \end{pmatrix}$$

39. $\lambda_1 = 2$, $\lambda_2 = -1$, $\mathbf{K}_1 = \begin{pmatrix} 1 \\ 1 \end{pmatrix}$, $\mathbf{K}_2 = \begin{pmatrix} -1 \\ 2 \end{pmatrix}$, $\mathbf{P} = \begin{pmatrix} 1 & -1 \\ 1 & 2 \end{pmatrix}$, $\mathbf{P}^{-1} = \begin{pmatrix} \frac{2}{3} & \frac{1}{3} \\ -\frac{1}{3} & \frac{1}{3} \end{pmatrix}$

$$\mathbf{A}^5 = \begin{pmatrix} 1 & -1 \\ 1 & 2 \end{pmatrix} \begin{pmatrix} 32 & 0 \\ 0 & -1 \end{pmatrix} \begin{pmatrix} \frac{2}{3} & \frac{1}{3} \\ -\frac{1}{3} & \frac{1}{3} \end{pmatrix} = \begin{pmatrix} 21 & 11 \\ 22 & 10 \end{pmatrix}$$

42. If $\mathbf{A}$ is a diagnalizable matrix, then $\mathbf{D} = \mathbf{P}^{-1}\mathbf{AP}$. The matrix $\mathbf{P}$ is not unique because interchanging columns of $\mathbf{P}$ or multiplying the columns by a nonzero constant still diagonalizes $\mathbf{A}$. For example, in Problem 1 we used $\mathbf{P} = \begin{pmatrix} -3 & 1 \\ 1 & 1 \end{pmatrix}$. But if we take $\begin{pmatrix} 1 & -3 \\ 1 & 1 \end{pmatrix}$,

$$\mathbf{P}_1^{-1} = \frac{1}{4} \begin{pmatrix} 1 & 3 \\ -1 & 1 \end{pmatrix}$$

$$\mathbf{P}_1^{-1}\mathbf{AP} = \frac{1}{4} \begin{pmatrix} 1 & 3 \\ -1 & 1 \end{pmatrix} \begin{pmatrix} 2 & 3 \\ 1 & 4 \end{pmatrix} \begin{pmatrix} 1 & -3 \\ 1 & 1 \end{pmatrix} = \begin{pmatrix} 5 & 0 \\ 0 & 1 \end{pmatrix}.$$

8.13   LU-Factorization

3. Proceeding as in Example 2, $\begin{pmatrix} -1 & 4 \\ 2 & 2 \end{pmatrix} = \begin{pmatrix} 1 & 0 \\ l_{21} & 1 \end{pmatrix} \begin{pmatrix} u_{11} & u_{12} \\ 0 & u_{22} \end{pmatrix} = \begin{pmatrix} u_{11} & u_{12} \\ u_{11}l_{21} & u_{12}l_{21} + u_{22} \end{pmatrix}.$

Thus

$$\begin{cases} u_{11} = -1 & u_{12} = 4 \\ u_{11}l_{21} = 2 & u_{12}l_{21} + u_{22} = 2. \end{cases}$$

Now $u_{11}l_{21} = (-1)l_{21} = 2$ and so $l_{21} = -2$. Also, $u_{12}l_{21} + u_{22} = (4)(-2) + u_{22} = 2$ so $u_{22} = 10$.

$$\mathbf{LU} = \begin{pmatrix} 1 & 0 \\ -2 & 1 \end{pmatrix} \begin{pmatrix} -1 & 4 \\ 0 & 10 \end{pmatrix} = \begin{pmatrix} -1 & 4 \\ 2 & 2 \end{pmatrix} = \mathbf{A}$$

6. Proceeding as in Example 2,

$$\begin{pmatrix} -3 & 2 & 1 \\ 9 & 3 & 2 \\ 3 & 1 & -1 \end{pmatrix} = \begin{pmatrix} 1 & 0 & 0 \\ l_{21} & 1 & 0 \\ l_{31} & l_{32} & 1 \end{pmatrix} \begin{pmatrix} u_{11} & u_{12} & u_{13} \\ 0 & u_{22} & u_{23} \\ 0 & 0 & u_{33} \end{pmatrix}$$

$$= \begin{pmatrix} u_{11} & u_{12} & u_{13} \\ l_{21}u_{11} & l_{21}u_{12} + u_{22} & l_{21}u_{13} + u_{23} \\ l_{31}u_{11} & l_{31}u_{12} + l_{32}u_{22} & l_{31}u_{13} + l_{32}u_{23} + u_{33} \end{pmatrix}.$$

Equating corresponding entries, we get

$$u_{11} = -3 \qquad u_{12} = 2 \qquad u_{13} = 1$$
$$l_{21}u_{11} = 9 \qquad l_{21}u_{12} + u_{22} = 3 \qquad l_{21}u_{13} + u_{23} = 2$$
$$l_{31}u_{11} = 3 \qquad l_{31}u_{12} + l_{32}u_{22} = 1 \qquad l_{31}u_{13} + l_{32}u_{23} + u_{33} = -1.$$

Thus

$$\mathbf{LU} = \begin{pmatrix} 1 & 0 & 0 \\ -3 & 1 & 0 \\ -1 & \frac{1}{3} & 1 \end{pmatrix} \begin{pmatrix} -3 & 2 & 1 \\ 0 & 9 & 5 \\ 0 & 0 & -\frac{5}{3} \end{pmatrix} = \mathbf{A}$$

9. Proceeding as in Example 2,

$$\begin{pmatrix} 1 & -2 & 1 \\ 0 & 1 & 2 \\ 2 & 6 & 1 \end{pmatrix} = \begin{pmatrix} 1 & 0 & 0 \\ l_{21} & 1 & 0 \\ l_{31} & l_{32} & 1 \end{pmatrix} \begin{pmatrix} u_{11} & u_{12} & u_{13} \\ 0 & u_{22} & u_{23} \\ 0 & 0 & u_{33} \end{pmatrix}$$

$$= \begin{pmatrix} u_{11} & u_{12} & u_{13} \\ l_{21}u_{11} & l_{21}u_{12} + u_{22} & l_{21}u_{13} + u_{23} \\ l_{31}u_{11} & l_{31}u_{12} + l_{32}u_{22} & l_{31}u_{13} + l_{32}u_{23} + u_{33} \end{pmatrix}.$$

Equating corresponding entries, we get

$$u_{11} = 1 \qquad u_{12} = -2 \qquad u_{13} = 1$$
$$l_{21}u_{11} = 0 \qquad l_{21}u_{12} + u_{22} = 1 \qquad l_{21}u_{13} + u_{23} = 2$$
$$l_{31}u_{11} = 2 \qquad l_{31}u_{12} + l_{32}u_{22} = 6 \qquad l_{31}u_{13} + l_{32}u_{23} + u_{33} = 1.$$

Thus

$$\mathbf{LU} = \begin{pmatrix} 1 & 0 & 0 \\ 0 & 1 & 0 \\ 2 & 10 & 1 \end{pmatrix} \begin{pmatrix} 1 & -2 & 1 \\ 0 & 1 & 2 \\ 0 & 0 & -21 \end{pmatrix} = \mathbf{A}.$$

12. Proceeding as in Examples 3 and 4,

$$\mathbf{A} = \begin{pmatrix} -2 & 10 \\ 1 & -4 \end{pmatrix} = \begin{pmatrix} 1 & -4 \\ -2 & 10 \end{pmatrix} \xrightarrow{2R_1 + R_2} \begin{pmatrix} 1 & -4 \\ 0 & 2 \end{pmatrix}$$

$$\mathbf{I} = \begin{pmatrix} 1 & 0 \\ 0 & 1 \end{pmatrix} = \begin{pmatrix} 1 & 0 \\ 0 & 1 \end{pmatrix} \xrightarrow{\text{Record } -2} \begin{pmatrix} 1 & 0 \\ -2 & 1 \end{pmatrix}.$$

Thus

$$\mathbf{LU} = \begin{pmatrix} 1 & 0 \\ -2 & 1 \end{pmatrix} \begin{pmatrix} 1 & -4 \\ 0 & 2 \end{pmatrix} = \mathbf{A}.$$

15. Proceeding as in Examples 3 and 4,

$$\mathbf{A} = \begin{pmatrix} 1 & 1 & 1 \\ 3 & 1 & 2 \\ 1 & -1 & 1 \end{pmatrix} \xrightarrow[\substack{-3R_1+R_2 \\ -1R_1+R_3}]{} \begin{pmatrix} 1 & 1 & 1 \\ 0 & -2 & -1 \\ 0 & -2 & 0 \end{pmatrix} \xrightarrow[]{-1R_2+R_3} \begin{pmatrix} 1 & 1 & 1 \\ 0 & -2 & -1 \\ 0 & 0 & 1 \end{pmatrix}$$

$$\mathbf{I} = \begin{pmatrix} 1 & 0 & 0 \\ 0 & 1 & 0 \\ 0 & 0 & 1 \end{pmatrix} \xrightarrow[\substack{\text{Record } 3 \\ \text{Record } 1}]{} \begin{pmatrix} 1 & 0 & 0 \\ 3 & 1 & 0 \\ 1 & 0 & 1 \end{pmatrix} \xrightarrow[]{\text{Record } 1} \begin{pmatrix} 1 & 0 & 0 \\ 3 & 1 & 0 \\ 1 & 1 & 1 \end{pmatrix}.$$

Thus

$$\mathbf{LU} = \begin{pmatrix} 1 & 0 & 0 \\ 3 & 1 & 0 \\ 1 & 1 & 1 \end{pmatrix} \begin{pmatrix} 1 & 1 & 1 \\ 0 & -2 & -1 \\ 0 & 0 & 1 \end{pmatrix} = \mathbf{A}.$$

18. Proceeding as in Examples 3 and 4,

$$\mathbf{A} = \begin{pmatrix} 16 & 4 & 20 \\ 4 & 5 & 3 \\ 20 & 3 & 29 \end{pmatrix} \xrightarrow[\substack{-\frac{1}{4}R_1+R_2 \\ -\frac{5}{4}R_1+R_3}]{} \begin{pmatrix} 16 & 4 & 20 \\ 0 & 4 & -2 \\ 0 & -2 & 4 \end{pmatrix} \xrightarrow[]{\frac{1}{2}R_2+R_3} \begin{pmatrix} 16 & 4 & 20 \\ 0 & 4 & -2 \\ 0 & 0 & 3 \end{pmatrix}$$

$$\mathbf{I} = \begin{pmatrix} 1 & 0 & 0 \\ 0 & 1 & 0 \\ 0 & 0 & 1 \end{pmatrix} \xrightarrow[\substack{\text{Record } \frac{1}{4} \\ \text{Record } \frac{5}{4}}]{} \begin{pmatrix} 1 & 0 & 0 \\ \frac{1}{4} & 1 & 0 \\ \frac{5}{4} & 0 & 1 \end{pmatrix} \xrightarrow[]{\text{Record } -\frac{1}{2}} \begin{pmatrix} 1 & 0 & 0 \\ \frac{1}{4} & 1 & 0 \\ \frac{5}{4} & -\frac{1}{2} & 1 \end{pmatrix}.$$

Thus

$$\mathbf{LU} = \begin{pmatrix} 1 & 0 & 0 \\ \frac{1}{4} & 1 & 0 \\ \frac{5}{4} & \frac{1}{2} & 1 \end{pmatrix} \begin{pmatrix} 16 & 4 & 20 \\ 0 & 4 & -2 \\ 0 & 0 & 3 \end{pmatrix} = \mathbf{A}.$$

21. Using the results from Problem 1, $\begin{pmatrix} 2 & -2 \\ 1 & 2 \end{pmatrix} \begin{pmatrix} x_1 \\ x_2 \end{pmatrix} = \begin{pmatrix} 1 \\ -2 \end{pmatrix}.$

Thus $\begin{pmatrix} 1 & 0 \\ \frac{1}{2} & 1 \end{pmatrix} \begin{pmatrix} 2 & -2 \\ 0 & 3 \end{pmatrix} \begin{pmatrix} x_1 \\ x_2 \end{pmatrix} = \begin{pmatrix} 1 \\ -2 \end{pmatrix}.$

Now let $\mathbf{UX} = \mathbf{Y}$ where $\mathbf{Y} = \begin{pmatrix} y_1 \\ y_2 \end{pmatrix}$ so we get

$$\begin{pmatrix} 1 & 0 \\ \frac{1}{2} & 1 \end{pmatrix} \begin{pmatrix} y_1 \\ y_2 \end{pmatrix} = \begin{pmatrix} 1 \\ -2 \end{pmatrix} \quad \text{or} \quad \mathbf{Y} = \begin{pmatrix} 1 \\ -\frac{5}{2} \end{pmatrix}.$$

Now solve $\mathbf{UX} = \mathbf{Y}$ to get

$$\begin{pmatrix} 2 & -2 \\ 0 & 3 \end{pmatrix} \begin{pmatrix} x_1 \\ x_2 \end{pmatrix} = \begin{pmatrix} 1 \\ -\frac{5}{2} \end{pmatrix} \quad \text{or} \quad \mathbf{X} = \begin{pmatrix} x_1 \\ x_2 \end{pmatrix} = \begin{pmatrix} -\frac{1}{3} \\ -\frac{5}{6} \end{pmatrix}.$$

24. Using the results from Problem 4, $\begin{pmatrix} 5 & -4 \\ 15 & 2 \end{pmatrix} \begin{pmatrix} x_1 \\ x_2 \end{pmatrix} = \begin{pmatrix} 1 \\ 7 \end{pmatrix}.$

Thus

$$\begin{pmatrix} 1 & 0 \\ 3 & 1 \end{pmatrix} \begin{pmatrix} 5 & -4 \\ 0 & 14 \end{pmatrix} \begin{pmatrix} x_1 \\ x_2 \end{pmatrix} = \begin{pmatrix} 1 \\ 7 \end{pmatrix}.$$

Now let $\mathbf{UX} = \mathbf{Y}$ where $\mathbf{Y} = \begin{pmatrix} y_1 \\ y_2 \end{pmatrix}$ so we get

$$\begin{pmatrix} 1 & 0 \\ 3 & 1 \end{pmatrix} \begin{pmatrix} y_1 \\ y_2 \end{pmatrix} = \begin{pmatrix} 1 \\ 7 \end{pmatrix} \quad \text{or} \quad \mathbf{Y} = \begin{pmatrix} 1 \\ 4 \end{pmatrix}.$$

Now solve $\mathbf{UX} = \mathbf{Y}$ to get

$$\begin{pmatrix} 5 & -4 \\ 0 & 14 \end{pmatrix} \begin{pmatrix} x_1 \\ x_2 \end{pmatrix} = \begin{pmatrix} 1 \\ 4 \end{pmatrix} \quad \text{or} \quad \mathbf{X} = \begin{pmatrix} x_1 \\ x_2 \end{pmatrix} = \begin{pmatrix} \frac{3}{7} \\ \frac{2}{7} \end{pmatrix}.$$

27. Using the results from Problem 7, $\begin{pmatrix} 1 & 2 & 7 \\ 2 & 5 & 6 \\ 7 & 6 & 4 \end{pmatrix} \begin{pmatrix} x_1 \\ x_2 \\ x_3 \end{pmatrix} = \begin{pmatrix} 109 \\ 109 \\ 218 \end{pmatrix}.$

Thus

$$\begin{pmatrix} 1 & 0 & 0 \\ 2 & 1 & 0 \\ 7 & -8 & 1 \end{pmatrix} \begin{pmatrix} 1 & 2 & 7 \\ 0 & 1 & -8 \\ 0 & 0 & -109 \end{pmatrix} \begin{pmatrix} x_1 \\ x_2 \\ x_3 \end{pmatrix} = \begin{pmatrix} 109 \\ 109 \\ 218 \end{pmatrix}.$$

Now let $\mathbf{UX} = \mathbf{Y}$ where $\mathbf{Y} = \begin{pmatrix} y_1 \\ y_2 \\ y_3 \end{pmatrix}$ so we get

$$\begin{pmatrix} 1 & 0 & 0 \\ 2 & 1 & 0 \\ 7 & -8 & 1 \end{pmatrix} \begin{pmatrix} y_1 \\ y_2 \\ y_3 \end{pmatrix} = \begin{pmatrix} 109 \\ 109 \\ 218 \end{pmatrix} \quad \text{or} \quad \mathbf{Y} = \begin{pmatrix} 109 \\ -109 \\ -1417 \end{pmatrix}.$$

Now solve $\mathbf{UX} = \mathbf{Y}$ to get

$$\begin{pmatrix} 1 & 2 & 7 \\ 0 & 1 & -8 \\ 0 & 0 & -109 \end{pmatrix} \begin{pmatrix} x_1 \\ x_2 \\ x_3 \end{pmatrix} = \begin{pmatrix} 109 \\ -109 \\ -1417 \end{pmatrix} \quad \text{or} \quad \mathbf{X} = \begin{pmatrix} x_1 \\ x_2 \\ x_3 \end{pmatrix} = \begin{pmatrix} 28 \\ -5 \\ 13 \end{pmatrix}.$$

30. Using the results from Problem 10, $\begin{pmatrix} 1 & 0 & 1 \\ 1 & 9 & 1 \\ 1 & 0 & -1 \end{pmatrix} \begin{pmatrix} x_1 \\ x_2 \\ x_3 \end{pmatrix} = \begin{pmatrix} 18 \\ 27 \\ -12 \end{pmatrix}.$

Thus

$$\begin{pmatrix} 1 & 0 & 0 \\ 1 & 1 & 0 \\ 1 & 0 & 1 \end{pmatrix} \begin{pmatrix} 1 & 0 & 1 \\ 0 & 9 & 0 \\ 0 & 0 & -2 \end{pmatrix} \begin{pmatrix} x_1 \\ x_2 \\ x_3 \end{pmatrix} = \begin{pmatrix} 18 \\ 27 \\ -12 \end{pmatrix}.$$

Now let $\mathbf{UX} = \mathbf{Y}$ where $\mathbf{Y} = \begin{pmatrix} y_1 \\ y_2 \\ y_3 \end{pmatrix}$ so we get

$$\begin{pmatrix} 1 & 0 & 0 \\ 1 & 1 & 0 \\ 1 & 0 & 1 \end{pmatrix} \begin{pmatrix} y_1 \\ y_2 \\ y_3 \end{pmatrix} = \begin{pmatrix} 18 \\ 27 \\ -12 \end{pmatrix} \quad \text{or} \quad \mathbf{Y} = \begin{pmatrix} 18 \\ 9 \\ -30 \end{pmatrix}.$$

Now solve $\mathbf{UX} = \mathbf{Y}$ to get

$$\begin{pmatrix} 1 & 0 & 1 \\ 0 & 9 & 0 \\ 0 & 0 & -2 \end{pmatrix} \begin{pmatrix} x_1 \\ x_2 \\ x_3 \end{pmatrix} = \begin{pmatrix} 18 \\ 9 \\ -30 \end{pmatrix} \quad \text{or} \quad \mathbf{X} = \begin{pmatrix} x_1 \\ x_2 \\ x_3 \end{pmatrix} = \begin{pmatrix} 3 \\ 1 \\ 15 \end{pmatrix}.$$

33. Proceeding as in Problems 30 and 31, $\begin{pmatrix} 1 & 1 & 1 \\ 1 & 2 & 2 \\ 1 & 2 & 3 \end{pmatrix} \begin{pmatrix} x_1 \\ x_2 \\ x_3 \end{pmatrix} = \begin{pmatrix} \frac{1}{2} \\ \frac{3}{4} \\ -\frac{1}{2} \end{pmatrix}$

Thus

$$\begin{pmatrix} 1 & 0 & 0 \\ 1 & 1 & 0 \\ 1 & 1 & 1 \end{pmatrix} \begin{pmatrix} 1 & 1 & 1 \\ 0 & 1 & 1 \\ 0 & 0 & 1 \end{pmatrix} \begin{pmatrix} x_1 \\ x_2 \\ x_3 \end{pmatrix} = \begin{pmatrix} \frac{1}{2} \\ \frac{3}{4} \\ -\frac{1}{2} \end{pmatrix}.$$

Now let $\mathbf{UX} = \mathbf{Y}$ where $\mathbf{Y} = \begin{pmatrix} y_1 \\ y_2 \\ y_3 \end{pmatrix}$ so we get

$$\begin{pmatrix} 1 & 0 & 0 \\ 1 & 1 & 0 \\ 1 & 1 & 1 \end{pmatrix} \begin{pmatrix} y_1 \\ y_2 \\ y_3 \end{pmatrix} = \begin{pmatrix} \frac{1}{2} \\ \frac{3}{4} \\ -\frac{1}{2} \end{pmatrix} \quad \text{or} \quad \mathbf{Y} = \begin{pmatrix} \frac{1}{2} \\ \frac{1}{4} \\ -\frac{5}{4} \end{pmatrix}.$$

Now solve $\mathbf{UX} = \mathbf{Y}$ to get

$$\begin{pmatrix} 1 & 1 & 1 \\ 0 & 1 & 1 \\ 0 & 0 & 1 \end{pmatrix} \begin{pmatrix} x_1 \\ x_2 \\ x_3 \end{pmatrix} = \begin{pmatrix} \frac{1}{2} \\ \frac{1}{4} \\ -\frac{5}{4} \end{pmatrix} \quad \text{or} \quad \mathbf{X} = \begin{pmatrix} x_1 \\ x_2 \\ x_3 \end{pmatrix} = \begin{pmatrix} \frac{1}{4} \\ \frac{3}{2} \\ -\frac{5}{4} \end{pmatrix}.$$

36. Using the results of Problem 16 and Theorem 8.5.8,

$$\det \mathbf{A} = \det \begin{pmatrix} 1 & 0 & 0 \\ -2 & 1 & 0 \\ -3 & -7 & 1 \end{pmatrix} \cdot \det \begin{pmatrix} -1 & 2 & -4 \\ 0 & -1 & 2 \\ 0 & 0 & 8 \end{pmatrix} = (1)(8) = 8.$$

39. Using the results of Problem 19 and Theorem 8.5.8,

$$\det \mathbf{A} = \det \begin{pmatrix} 1 & 0 & 0 & 0 \\ -2 & 1 & 0 & 0 \\ 1 & -4 & 1 & 0 \\ 5 & -11 & -2 & 1 \end{pmatrix} \cdot \det \begin{pmatrix} 1 & -2 & 1 & 0 \\ 0 & -1 & 0 & 1 \\ 0 & 0 & 3 & 7 \\ 0 & 0 & 0 & 26 \end{pmatrix} = (1)(-78) = -78.$$

42.
$$\begin{pmatrix} -1 & 2 & -4 \\ 2 & -5 & 10 \\ 3 & 1 & 6 \end{pmatrix} = \begin{pmatrix} l_{11} & 0 & 0 \\ l_{21} & l_{22} & 0 \\ l_{31} & l_{32} & l_{33} \end{pmatrix} \begin{pmatrix} 1 & u_{12} & u_{13} \\ 0 & 1 & u_{23} \\ 0 & 0 & 1 \end{pmatrix}$$

$$= \begin{pmatrix} l_{11} & l_{11}u_{12} & l_{11}u_{13} \\ l_{21} & l_{21}u_{12} + l_{22} & l_{21}u_{13} + l_{22}u_{23} \\ l_{31} & l_{31}u_{12} + l_{32} & l_{31}u_{13} + l_{32}u_{23} + l_{33} \end{pmatrix}$$

Equating corresponding entries, we get

$$\begin{pmatrix} l_{11} = -1 & l_{11}u_{12} = 2 & l_{11}u_{13} = -4 \\ l_{21} = 2 & l_{21}u_{12} + l_{22} = -5 & l_{21}u_{13} + l_{22}u_{23} = 10 \\ l_{31} = 3 & l_{31}u_{12} + l_{32} = 1 & l_{31}u_{13} + l_{32}u_{23} + l_{33} = 6 \end{pmatrix}.$$

Thus

$$\mathbf{LU} = \begin{pmatrix} -1 & 0 & 0 \\ 2 & -1 & 0 \\ 3 & 7 & 8 \end{pmatrix} \begin{pmatrix} 1 & -2 & 4 \\ 0 & 1 & -2 \\ 0 & 0 & 1 \end{pmatrix} = \mathbf{A}.$$

8.14 Cryptography

3. (a) The message is $\mathbf{M} = \begin{pmatrix} 16 & 8 & 15 & 14 & 5 \\ 0 & 8 & 15 & 13 & 5 \end{pmatrix}$. The encoded message is

$$\mathbf{B} = \mathbf{AM} = \begin{pmatrix} 3 & 5 \\ 2 & 3 \end{pmatrix} \begin{pmatrix} 16 & 8 & 15 & 14 & 5 \\ 0 & 8 & 15 & 13 & 5 \end{pmatrix} = \begin{pmatrix} 48 & 64 & 120 & 107 & 40 \\ 32 & 40 & 75 & 67 & 25 \end{pmatrix}.$$

(b) The decoded message is

$$\mathbf{M} = \mathbf{A}^{-1}\mathbf{B} = \begin{pmatrix} -3 & 5 \\ 2 & -3 \end{pmatrix} = \begin{pmatrix} 48 & 64 & 120 & 107 & 40 \\ 32 & 40 & 75 & 67 & 25 \end{pmatrix} = \begin{pmatrix} 16 & 8 & 15 & 14 & 5 \\ 0 & 8 & 15 & 13 & 5 \end{pmatrix}.$$

6. (a) The message is $\mathbf{M} = \begin{pmatrix} 4 & 18 & 0 & 10 & 15 & 8 \\ 14 & 0 & 9 & 19 & 0 & 20 \\ 8 & 5 & 0 & 19 & 16 & 25 \end{pmatrix}$. The encoded message is

$$\mathbf{B} = \mathbf{AM} = \begin{pmatrix} 5 & 3 & 0 \\ 4 & 3 & -1 \\ 5 & 2 & 2 \end{pmatrix} \begin{pmatrix} 4 & 18 & 0 & 10 & 15 & 8 \\ 14 & 0 & 9 & 19 & 0 & 20 \\ 8 & 5 & 0 & 19 & 16 & 25 \end{pmatrix}$$

$$= \begin{pmatrix} 62 & 90 & 27 & 107 & 75 & 100 \\ 50 & 67 & 27 & 78 & 44 & 67 \\ 64 & 100 & 18 & 126 & 107 & 130 \end{pmatrix}.$$

(b) The decoded message is

$$\mathbf{M} = \mathbf{A}^{-1}\mathbf{B} = \begin{pmatrix} 8 & -6 & -3 \\ -13 & 10 & 5 \\ -7 & 5 & 3 \end{pmatrix} \begin{pmatrix} 62 & 90 & 27 & 107 & 75 & 100 \\ 50 & 67 & 27 & 78 & 44 & 67 \\ 64 & 100 & 18 & 126 & 107 & 130 \end{pmatrix}$$

$$= \begin{pmatrix} 4 & 18 & 0 & 10 & 15 & 8 \\ 14 & 0 & 9 & 19 & 0 & 20 \\ 8 & 5 & 0 & 19 & 16 & 25 \end{pmatrix}.$$

9. The decoded message is

$$\mathbf{M} = \mathbf{A}^{-1}\mathbf{B} = \begin{pmatrix} 0 & 0 & 1 \\ 0 & 1 & 0 \\ 1 & 0 & -1 \end{pmatrix} \begin{pmatrix} 31 & 21 & 21 & 22 & 20 & 9 \\ 19 & 0 & 9 & 13 & 16 & 15 \\ 13 & 1 & 20 & 8 & 0 & 9 \end{pmatrix} = \begin{pmatrix} 13 & 1 & 20 & 8 & 0 & 9 \\ 19 & 0 & 9 & 13 & 16 & 15 \\ 18 & 20 & 1 & 14 & 20 & 0 \end{pmatrix}.$$

From correspondence (1) we obtain: MATH_IS_IMPORTANT.

12. (a) $\mathbf{M}^T = \begin{pmatrix} 22 & 8 & 19 & 27 & 21 & 3 & 3 & 27 & 21 & 18 & 21 \\ 13 & 3 & 21 & 22 & 3 & 25 & 27 & 6 & 7 & 14 & 23 \\ 2 & 27 & 21 & 7 & 27 & 5 & 21 & 17 & 2 & 25 & 7 \end{pmatrix}$

(b) $\mathbf{B}^T = \mathbf{M} = \begin{pmatrix} 1 & 1 & 0 \\ 1 & 0 & 1 \\ 1 & 1 & -1 \end{pmatrix} = \begin{pmatrix} 37 & 38 & 61 & 56 & 51 & 33 & 51 & 50 & 30 & 57 & 51 \\ 24 & 35 & 40 & 34 & 48 & 8 & 24 & 44 & 23 & 43 & 28 \\ 11 & -24 & 0 & 15 & -24 & 20 & 6 & -11 & 5 & -11 & 16 \end{pmatrix}$

(c) $\mathbf{BA}^{-1} = \mathbf{B} \begin{pmatrix} -1 & 1 & 1 \\ 2 & -1 & -1 \\ 1 & 0 & -1 \end{pmatrix} = \mathbf{M}$

8.15   Error-Correcting Code

3. $\begin{pmatrix} 0 & 0 & 0 & 1 & 1 \end{pmatrix}$ **6.** $\begin{pmatrix} 0 & 1 & 1 & 0 & 1 & 0 & 1 & 0 \end{pmatrix}$

9. Parity error   **12.** Parity error

In Problems 13–18, $\mathbf{D} = \begin{pmatrix} c_1 & c_2 & c_3 \end{pmatrix}$ *and* $\mathbf{P} = \begin{pmatrix} 1 & 1 & 0 & 1 \\ 1 & 0 & 1 & 1 \\ 0 & 1 & 1 & 1 \end{pmatrix}$.

15. $\mathbf{D}^T = \mathbf{P} \begin{pmatrix} 0 & 1 & 0 & 1 \end{pmatrix}^T = \begin{pmatrix} 0 & 1 & 0 \end{pmatrix}^T$; $\mathbf{C} = \begin{pmatrix} 0 & 1 & 0 & 0 & 1 & 0 & 1 \end{pmatrix}$

18. $\mathbf{D}^T = \mathbf{P} \begin{pmatrix} 1 & 1 & 0 & 0 \end{pmatrix}^T = \begin{pmatrix} 0 & 1 & 1 \end{pmatrix}^T$; $\mathbf{C} = \begin{pmatrix} 0 & 1 & 1 & 1 & 1 & 0 & 0 \end{pmatrix}$

In Problems 19–28, $\mathbf{W}$ *represents the correctly decoded message.*

21. $\mathbf{S} = \mathbf{H}\mathbf{R}^T = \mathbf{H} \begin{pmatrix} 1 & 1 & 0 & 1 & 1 & 0 & 1 \end{pmatrix} = \begin{pmatrix} 1 & 0 & 1 \end{pmatrix}^T$; not a code word. The error is in the fifth bit. $\mathbf{W} = \begin{pmatrix} 0 & 0 & 0 & 1 \end{pmatrix}$

24. $\mathbf{S} = \mathbf{H}\mathbf{R}^T = \mathbf{H} \begin{pmatrix} 1 & 1 & 0 & 0 & 1 & 1 & 0 \end{pmatrix} = \begin{pmatrix} 0 & 0 & 0 \end{pmatrix}^T$; a code word. $\mathbf{W} = \begin{pmatrix} 0 & 1 & 1 & 0 \end{pmatrix}$

27. $\mathbf{S} = \mathbf{H}\mathbf{R}^T = \mathbf{H} \begin{pmatrix} 1 & 0 & 1 & 1 & 0 & 1 & 1 \end{pmatrix} = \begin{pmatrix} 1 & 1 & 1 \end{pmatrix}^T$; not a code word. The error is in the seventh bit. $\mathbf{W} = \begin{pmatrix} 1 & 0 & 1 & 0 \end{pmatrix}$

30. (a) $c_4 = 0, c_3 = 1, c_2 = 1, c_1 = 0$; $\begin{pmatrix} 0 & 1 & 1 & 0 & 0 & 1 & 1 & 0 \end{pmatrix}$

(b) $\mathbf{H} = \begin{pmatrix} 0 & 0 & 0 & 0 & 1 & 1 & 1 & 1 \\ 0 & 0 & 1 & 1 & 0 & 0 & 1 & 1 \\ 0 & 1 & 0 & 1 & 0 & 1 & 0 & 1 \\ 1 & 1 & 1 & 1 & 1 & 1 & 1 & 1 \end{pmatrix}$

(c) $\mathbf{S} = \mathbf{H}\mathbf{R}^T = \mathbf{H} \begin{pmatrix} 0 & 0 & 1 & 1 & 1 & 1 & 0 & 0 \end{pmatrix}^T = \begin{pmatrix} 0 & 0 & 0 & 0 \end{pmatrix}^T$

8.16 | Method of Least Squares

3. We have $\mathbf{Y}^T = \begin{pmatrix} 1 & 1.5 & 3 & 4.5 & 5 \end{pmatrix}$ and $\mathbf{A}^T = \begin{pmatrix} 1 & 2 & 3 & 4 & 5 \\ 1 & 1 & 1 & 1 & 1 \end{pmatrix}$.

Now $\mathbf{A}^T\mathbf{A} = \begin{pmatrix} 55 & 15 \\ 15 & 5 \end{pmatrix}$ and $(\mathbf{A}^T\mathbf{A})^{-1} = \dfrac{1}{50}\begin{pmatrix} 5 & -15 \\ -15 & 55 \end{pmatrix}$

so $\mathbf{X} = (\mathbf{A}^T\mathbf{A})^{-1}\mathbf{A}^T\mathbf{Y} = \begin{pmatrix} 1.1 \\ -0.3 \end{pmatrix}$ and the least squares line is $y = 1.1x - 0.3$.

6. We have $\mathbf{Y}^T = \begin{pmatrix} 2 & 2.5 & 1 & 1.5 & 2 & 3.2 & 5 \end{pmatrix}$ and $\mathbf{A}^T = \begin{pmatrix} 1 & 2 & 3 & 4 & 5 & 6 & 7 \\ 1 & 1 & 1 & 1 & 1 & 1 & 1 \end{pmatrix}$.

Now $\mathbf{A}^T\mathbf{A} = \begin{pmatrix} 140 & 28 \\ 28 & 7 \end{pmatrix}$ and $(\mathbf{A}^T\mathbf{A})^{-1} = \dfrac{1}{196}\begin{pmatrix} 7 & -28 \\ -28 & 140 \end{pmatrix}$

so $\mathbf{X} = (\mathbf{A}^T\mathbf{A})^{-1}\mathbf{A}^T\mathbf{Y} = \begin{pmatrix} 0.407143 \\ 0.828571 \end{pmatrix}$ and the least squares line is $y = 0.407143x + 0.828571$.

9. The data yields

$$\mathbf{Y} = \begin{pmatrix} 1 \\ 1 \\ 2 \\ 5 \end{pmatrix}, \qquad \mathbf{A} = \begin{pmatrix} 1 & 1 & 1 \\ 4 & 2 & 1 \\ 9 & 3 & 1 \\ 16 & 4 & 1 \end{pmatrix}, \qquad \mathbf{A}^T = \begin{pmatrix} 1 & 4 & 9 & 16 \\ 1 & 2 & 3 & 4 \\ 1 & 1 & 1 & 1 \end{pmatrix}$$

and

$$\mathbf{A}^T\mathbf{A} = \begin{pmatrix} 1 & 4 & 9 & 16 \\ 1 & 2 & 3 & 4 \\ 1 & 1 & 1 & 1 \end{pmatrix} \begin{pmatrix} 1 & 1 & 1 \\ 4 & 2 & 1 \\ 9 & 3 & 1 \\ 16 & 4 & 1 \end{pmatrix} = \begin{pmatrix} 354 & 100 & 30 \\ 100 & 30 & 10 \\ 30 & 10 & 4 \end{pmatrix},$$

$$\left(\mathbf{A}^T\mathbf{A}\right)^{-1} = \frac{1}{20} \begin{pmatrix} 5 & -25 & 25 \\ -25 & 129 & -135 \\ 25 & -135 & 155 \end{pmatrix}.$$

Hence,

$$\mathbf{X} = \left(\mathbf{A}^T\mathbf{A}\right)^{-1}\mathbf{A}^T\mathbf{Y} = \frac{1}{20} \begin{pmatrix} 5 & -25 & 25 \\ -25 & 129 & -135 \\ 25 & -135 & 155 \end{pmatrix} \begin{pmatrix} 1 & 4 & 9 & 16 \\ 1 & 2 & 3 & 4 \\ 1 & 1 & 1 & 1 \end{pmatrix} \begin{pmatrix} 1 \\ 1 \\ 2 \\ 5 \end{pmatrix} = \begin{pmatrix} 0.75 \\ -2.45 \\ 2.75 \end{pmatrix}.$$

Therefore, $f(x) = 0.75x^2 - 2.45x + 2.75$.

8.17 Discrete Compartmental Models

In Problems 1-5 we use the fact that the element τ_{ij} in the transfer matrix $\mathbf{T}$ is the rate of transfer from compartment j to compartment i, and the fact that the sum of each column in $\mathbf{T}$ is 1.

3. (a) The initial state and the transfer matrix are

$$\mathbf{X}_0 = \begin{pmatrix} 100 \\ 0 \\ 0 \end{pmatrix} \quad \text{and} \quad \mathbf{T} = \begin{pmatrix} 0.2 & 0.5 & 0 \\ 0.3 & 0.1 & 0 \\ 0.5 & 0.4 & 1 \end{pmatrix}.$$

(b) We have

$$\mathbf{X}_1 = \mathbf{T}\mathbf{X}_0 = \begin{pmatrix} 20 \\ 30 \\ 50 \end{pmatrix} \quad \text{and} \quad \mathbf{X}_2 = \mathbf{T}\mathbf{X}_1 = \begin{pmatrix} 19 \\ 9 \\ 72 \end{pmatrix}.$$

(c) From $\mathbf{T}\hat{\mathbf{X}} - \hat{\mathbf{X}} = (\mathbf{T} - \mathbf{I})\hat{\mathbf{X}} = \mathbf{0}$ and the fact that the system is closed we obtain

$$\begin{aligned} -0.8x_1 + 0.5x_2 \quad &= 0 \\ 0.3x_1 - 0.9x_2 \quad &= 0 \\ x_1 + \quad x_2 + x_3 &= 100. \end{aligned}$$

The solution is $x_1 = x_2 = 0$, $x_3 = 100$, so the equilibrium state is $\hat{\mathbf{X}} = \begin{pmatrix} 0 \\ 0 \\ 100 \end{pmatrix}$.

Chapter 8 in Review

3. $\mathbf{AB} = \begin{pmatrix} 3 & 4 \\ 6 & 8 \end{pmatrix}$; $\mathbf{BA} = \begin{pmatrix} 11 \end{pmatrix}$ **6.** True

9. 0   **12.** True

15. False; if the characteristic equation of an $n \times n$ matrix has repeated roots, there may not be n linearly independent eigenvectors.

18. True

21. $\mathbf{A} = \frac{1}{2}(\mathbf{A}+\mathbf{A}^T)+\frac{1}{2}(\mathbf{A}-\mathbf{A}^T)$ where $\frac{1}{2}(\mathbf{A}+\mathbf{A}^T)$ is symmetric and $\frac{1}{2}(\mathbf{A}-\mathbf{A}^T)$ is skew-symmetric.

24. (a) $\sigma_x\sigma_y = \begin{pmatrix} i & 0 \\ 0 & -i \end{pmatrix} = -\sigma_y\sigma_x$; $\sigma_x\sigma_z = \begin{pmatrix} 0 & -1 \\ 1 & 0 \end{pmatrix} = -\sigma_z\sigma_x$; $\sigma_y\sigma_z = \begin{pmatrix} 0 & i \\ i & 0 \end{pmatrix} = -\sigma_z\sigma_y$

(b) We first note that for anticommuting matrices $\mathbf{AB} = -\mathbf{BA}$, so $\mathbf{C} = 2\mathbf{AB}$. Then

$$\mathbf{C}_{xy} = \begin{pmatrix} 2i & 0 \\ 0 & -2i \end{pmatrix}, \quad \mathbf{C}_{yz} = \begin{pmatrix} 0 & 2i \\ 2i & 0 \end{pmatrix}, \quad \text{and } \mathbf{C}_{zx} = \begin{pmatrix} 0 & 2 \\ -2 & 0 \end{pmatrix}.$$

27. Multiplying the second row by abc we obtain the third row. Thus the determinant is 0.

30. $(-3)(6)(9)(1) = -162$

33. From $x_1 I_2 + x_2 HNO_3 \rightarrow x_3 HIO_3 + x_4 NO_2 + x_5 H_2O$ we obtain the system
$2x_1 = x_3$, $x_2 = x_3+2x_5$, $x_2 = x_4$, $3x_2 = 3x_3+2x_4+x_5$. Letting $x_4 = x_2$ in the fourth equation we obtain $x_2 = 3x_3 + x_5$. Taking $x_1 = t$ we see that $x_3 = 2t$, $x_2 = 2t + 2x_5$, and $x_2 = 6t + x_5$. From the latter two equations we get $x_5 = 4t$. Taking $t = 1$ we have $x_1 = 1$, $x_2 = 10$, $x_3 = 2$, $x_4 = 10$, and $x_5 = 4$. The balanced equation is $I_2 + 10HNO_3 \rightarrow 2HIO_3 + 10NO_2 + 4H_2O$.

36. $\det \mathbf{A} = 4$, $\det \mathbf{A}_1 = 16$, $\det \mathbf{A}_2 = -4$, $\det \mathbf{A}_3 = 0$; $x_1 = \dfrac{16}{4} = 4$, $x_2 = \dfrac{-4}{4} = -1$, $x_3 = \dfrac{0}{4} = 0$

39. $\mathbf{AX} = \mathbf{B}$ is $\begin{pmatrix} 2 & 3 & -1 \\ 1 & -2 & 0 \\ -2 & 0 & 1 \end{pmatrix} \begin{pmatrix} x_1 \\ x_2 \\ x_3 \end{pmatrix} = \begin{pmatrix} 6 \\ -3 \\ 9 \end{pmatrix}$. Since $\mathbf{A}^{-1} = -\dfrac{1}{3}\begin{pmatrix} -2 & -3 & -2 \\ -1 & 0 & -1 \\ -4 & -6 & -7 \end{pmatrix}$, we have

$$\mathbf{X} = \mathbf{A}^{-1}\mathbf{B} = \begin{pmatrix} 7 \\ 5 \\ 23 \end{pmatrix}.$$

42. From the characteristic equation $\lambda^2 = 0$ we see that the eigenvalues are $\lambda_1 = \lambda_2 = 0$. For $\lambda_1 = \lambda_2 = 0$ we have $4k_1 = 0$ and $\mathbf{K}_1 = \begin{pmatrix} 0 \\ 1 \end{pmatrix}$ is a single eigenvector.

45. From the characteristic equation $-\lambda^3 - \lambda^2 + 21\lambda + 45 = -(\lambda+3)^2(\lambda-5) = 0$ we see that the eigenvalues are $\lambda_1 = \lambda_2 = -3$ and $\lambda_3 = 5$. For $\lambda_1 = \lambda_2 = -3$ we have

$$\begin{pmatrix} 1 & 2 & -3 & | & 0 \\ 2 & 4 & -6 & | & 0 \\ -1 & -2 & 3 & | & 0 \end{pmatrix} \xrightarrow[\text{operations}]{\text{row}} \begin{pmatrix} 1 & 2 & -3 & | & 0 \\ 0 & 0 & 0 & | & 0 \\ 0 & 0 & 0 & | & 0 \end{pmatrix}.$$

Thus $\mathbf{K}_1 = \begin{pmatrix} -2 & 1 & 0 \end{pmatrix}^T$ and $\mathbf{K}_2 = \begin{pmatrix} 3 & 0 & 1 \end{pmatrix}^T$. For $\lambda_3 = 5$ we have

$$\begin{pmatrix} -7 & 2 & -3 & | & 0 \\ 2 & -4 & -6 & | & 0 \\ -1 & -2 & -5 & | & 0 \end{pmatrix} \xrightarrow[\text{operations}]{\text{row}} \begin{pmatrix} 1 & -\frac{2}{7} & \frac{3}{7} & | & 0 \\ 0 & 1 & 2 & | & 0 \\ 0 & 0 & 0 & | & 0 \end{pmatrix}.$$

Thus $\mathbf{K}_3 = \begin{pmatrix} -1 & -2 & 1 \end{pmatrix}^T$.

48. (a) Eigenvalues are $\lambda_1 = \lambda_2 = 0$ and $\lambda_3 = 5$ with corresponding eigenvectors $\mathbf{K}_1 = \begin{pmatrix} 0 & 1 & 0 \end{pmatrix}^T$, $\mathbf{K}_2 = \begin{pmatrix} 2 & 0 & 1 \end{pmatrix}^T$, and $\mathbf{K}_3 = \begin{pmatrix} -1 & 0 & 2 \end{pmatrix}^T$. Since $\|\mathbf{K}_1\| = 1$, $\|\mathbf{K}_2\| = \sqrt{5}$, and $\|\mathbf{K}_3\| = \sqrt{5}$, we have

$$\mathbf{P} = \begin{pmatrix} 0 & \frac{2}{\sqrt{5}} & -\frac{1}{\sqrt{5}} \\ 1 & 0 & 0 \\ 0 & \frac{1}{\sqrt{5}} & \frac{2}{\sqrt{5}} \end{pmatrix} \quad \text{and} \quad \mathbf{P}^{-1} = \mathbf{P}^T = \begin{pmatrix} 0 & 1 & 0 \\ \frac{2}{\sqrt{5}} & 0 & \frac{1}{\sqrt{5}} \\ -\frac{1}{\sqrt{5}} & 0 & \frac{2}{\sqrt{5}} \end{pmatrix}.$$

(b) $\mathbf{P}^{-1}\mathbf{A}\mathbf{P} = \begin{pmatrix} 0 & 0 & 0 \\ 0 & 0 & 0 \\ 0 & 0 & 5 \end{pmatrix}$

51. The encoded message is

$$\mathbf{B} = \mathbf{A}\mathbf{M} = \begin{pmatrix} 10 & 1 \\ 9 & 1 \end{pmatrix} \begin{pmatrix} 19 & 1 & 20 & 5 & 12 & 12 & 9 & 20 & 5 & 0 & 12 & 1 & 21 \\ 14 & 3 & 8 & 5 & 4 & 0 & 15 & 14 & 0 & 6 & 18 & 9 & 0 \end{pmatrix}$$

$$= \begin{pmatrix} 204 & 13 & 208 & 55 & 124 & 120 & 105 & 214 & 50 & 6 & 138 & 19 & 210 \\ 185 & 12 & 188 & 50 & 112 & 108 & 96 & 194 & 45 & 6 & 126 & 18 & 189 \end{pmatrix}.$$

54. The decoded message is

$$\mathbf{M} = \mathbf{A}^{-1}\mathbf{B} = \begin{pmatrix} -3 & 2 & -1 \\ 1 & 0 & 0 \\ 2 & -1 & 1 \end{pmatrix} \begin{pmatrix} 5 & 2 & 21 \\ 27 & 17 & 40 \\ 21 & 13 & -2 \end{pmatrix} = \begin{pmatrix} 18 & 15 & 19 \\ 5 & 2 & 21 \\ 4 & 0 & 0 \end{pmatrix}.$$

From correspondence (1) we obtain: ROSEBUD__.

57. $\begin{pmatrix} 1 & 1 & 1 \\ 1 & -2 & 3 \\ 2 & 0 & -3 \end{pmatrix} \xrightarrow[-2R_1+R_3]{-R_1+R_2} \begin{pmatrix} 1 & 1 & 1 \\ 0 & -3 & 2 \\ 0 & -2 & -5 \end{pmatrix} \xrightarrow{-\frac{2}{3}R_2+R_3} \begin{pmatrix} 1 & 1 & 1 \\ 0 & -3 & 2 \\ 0 & 0 & -6.333 \end{pmatrix} = \mathbf{U}$

$$\begin{pmatrix} 1 & 0 & 0 \\ 0 & 1 & 0 \\ 0 & 0 & 1 \end{pmatrix} \xrightarrow[\substack{2R_1+R_3 \\ -\frac{2}{3}R_2+R_3}]{R_1+R_2} \begin{pmatrix} 1 & 0 & 0 \\ 1 & 1 & 0 \\ 2 & 0.667 & 1 \end{pmatrix} = \mathbf{L}$$

$$\mathbf{A} = \mathbf{LU} = \begin{pmatrix} 1 & 0 & 0 \\ 1 & 1 & 0 \\ 2 & 0.667 & 1 \end{pmatrix} \begin{pmatrix} 1 & 1 & 1 \\ 0 & -3 & 2 \\ 0 & 0 & -6.333 \end{pmatrix}$$

$$\mathbf{LY} = \mathbf{B} = \begin{pmatrix} 1 & 0 & 0 \\ 1 & 1 & 0 \\ 2 & 0.667 & 1 \end{pmatrix} \begin{pmatrix} y_1 \\ y_2 \\ y_3 \end{pmatrix} = \begin{pmatrix} 6 \\ 2 \\ 3 \end{pmatrix}, \quad \text{hence} \quad y_1 = 6, y_2 = -4, y_3 = -6.333$$

$$\mathbf{UX} = \mathbf{Y} = \begin{pmatrix} 1 & 1 & 1 \\ 0 & -3 & 2 \\ 0 & 0 & -6.333 \end{pmatrix} \begin{pmatrix} x_1 \\ x_2 \\ x_3 \end{pmatrix} = \begin{pmatrix} 6 \\ -4 \\ -6.333 \end{pmatrix}, \quad \text{hence} \quad x_1 = 3, x_2 = 2, x_3 = 1$$

60. The data yields

$$\mathbf{Y} = \begin{pmatrix} -2 \\ 0 \\ 5 \\ -1 \end{pmatrix}, \qquad \mathbf{A} = \begin{pmatrix} 1 & 1 & 1 \\ 4 & 2 & 1 \\ 9 & 3 & 1 \\ 16 & 4 & 1 \end{pmatrix}, \qquad \mathbf{A}^T = \begin{pmatrix} 1 & 4 & 9 & 16 \\ 1 & 2 & 3 & 4 \\ 1 & 1 & 1 & 1 \end{pmatrix}$$

and

$$\mathbf{A}^T\mathbf{A} = \begin{pmatrix} 1 & 4 & 9 & 16 \\ 1 & 2 & 3 & 4 \\ 1 & 1 & 1 & 1 \end{pmatrix} \begin{pmatrix} 1 & 1 & 1 \\ 4 & 2 & 1 \\ 9 & 3 & 1 \\ 16 & 4 & 1 \end{pmatrix} = \begin{pmatrix} 354 & 100 & 30 \\ 100 & 30 & 10 \\ 30 & 10 & 4 \end{pmatrix},$$

$$(\mathbf{A}^T\mathbf{A})^{-1} = \frac{1}{20} \begin{pmatrix} 5 & -25 & 25 \\ -25 & 129 & -135 \\ 25 & -135 & 155 \end{pmatrix}.$$

Hence,

$$\mathbf{X} = (\mathbf{A}^T\mathbf{A})^{-1}\mathbf{A}^T\mathbf{Y} = \frac{1}{20} \begin{pmatrix} 5 & -25 & 25 \\ -25 & 129 & -135 \\ 25 & -135 & 155 \end{pmatrix} \begin{pmatrix} 1 & 4 & 9 & 16 \\ 1 & 2 & 3 & 4 \\ 1 & 1 & 1 & 1 \end{pmatrix} \begin{pmatrix} -2 \\ 0 \\ 5 \\ -1 \end{pmatrix} = \begin{pmatrix} -2 \\ 10.8 \\ -11.5 \end{pmatrix}.$$

Therefore $f(x) = -2x^2 + 10.8x - 11.5$.

Chapter 9

Vector Calculus

9.1 │ Vector Functions

3.

6.

9.

Note: the scale is distorted in this graph. For $t = 0$, the graph starts at $(1, 0, 1)$. The upper loop shown intersects the xz-plane at about $(286751, 0, 286751)$.

12. $x = t$, $y = 2t$, $z = \pm\sqrt{t^2 + 4t^2 + 1} = \pm\sqrt{5t^2 - 1}$; $\mathbf{r}(t) = t\mathbf{i} + 2t\mathbf{j} \pm \sqrt{5t^2 - 1}\,\mathbf{k}$

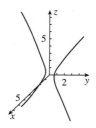

15. $\mathbf{r}(t) = \dfrac{\sin 2t}{t}\mathbf{i} + (t - 2)^5\mathbf{j} + \dfrac{\ln t}{1/t}\mathbf{k}$. Using L'Hôpital's Rule,

$$\lim_{t \to 0^+} \mathbf{r}(t) = \left[\frac{2\cos 2t}{1}\mathbf{i} + (t - 2)^5\mathbf{j} + \frac{1/t}{-1/t^2}\mathbf{k}\right] = 2\mathbf{i} - 32\mathbf{j}.$$

18. $\mathbf{r}'(t) = \langle -t \sin t, 1 - \sin t \rangle$; $\quad \mathbf{r}''(t) = \langle -t \cos t - \sin t, -\cos t \rangle$

21. $\mathbf{r}'(t) = -2 \sin t\mathbf{i} + 6 \cos t\mathbf{j}$

$\mathbf{r}'(\pi/6) = -\mathbf{i} + 3\sqrt{3}\,\mathbf{j}$

24. $\mathbf{r}'(t) = -3 \sin t\mathbf{i} + 3 \cos t\mathbf{j} + 2\mathbf{k}$

$\mathbf{r}'(\pi/4) = \dfrac{-3\sqrt{2}}{2}\,\mathbf{i} + \dfrac{3\sqrt{2}}{2}\,\mathbf{j} + 2\mathbf{k}$

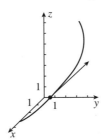

27. $\dfrac{d}{dt}[\mathbf{r}(t) \times \mathbf{r}'(t)] = \mathbf{r}(t) \times \mathbf{r}''(t) + \mathbf{r}'(t) \times \mathbf{r}'(t) = \mathbf{r}(t) \times \mathbf{r}''(t)$

30. $\dfrac{d}{dt}[\mathbf{r}_1(t) \times (\mathbf{r}_2(t) \times \mathbf{r}_3(t))] = \mathbf{r}_1(t) \times \dfrac{d}{dt}(\mathbf{r}_2(t) \times \mathbf{r}_3(t)) + \mathbf{r}_1'(t) \times (\mathbf{r}_2(t) \times \mathbf{r}_3(t))$

$$= \mathbf{r}_1(t) \times (\mathbf{r}_2(t) \times \mathbf{r}_3'(t) + \mathbf{r}_2'(t) \times \mathbf{r}_3(t)) + \mathbf{r}_1'(t) \times (\mathbf{r}_2(t) \times \mathbf{r}_3(t))$$

$$= \mathbf{r}_1(t) \times (\mathbf{r}_2(t) \times \mathbf{r}_3'(t)) + \mathbf{r}_1(t) \times (\mathbf{r}_2'(t) \times \mathbf{r}_3(t)) + \mathbf{r}_1(t) \times (\mathbf{r}_2(t) \times \mathbf{r}_3(t))$$

33. $\displaystyle\int_{-1}^{2} \mathbf{r}(t)\, dt = \left[\int_{-1}^{2} t\, dt \right]\mathbf{i} + \left[\int_{-1}^{2} 3t^2\, dt \right]\mathbf{j} + \left[\int_{-1}^{2} 4t^3\, dt \right]\mathbf{k} = \frac{1}{2}t^2 \Big|_{-1}^{2}\mathbf{i} + t^3 \Big|_{-1}^{2}\mathbf{j} + t^4 \Big|_{-1}^{2}\mathbf{k}$

$$= \frac{3}{2}\mathbf{i} + 9\mathbf{j} + 15\mathbf{k}$$

36. $\displaystyle\int \mathbf{r}(t)\, dt = \left[\int \frac{1}{1+t^2}\, dt \right]\mathbf{i} + \left[\int \frac{t}{1+t^2}\, dt \right]\mathbf{j} + \left[\int \frac{t^2}{1+t^2}\, dt \right]\mathbf{k}$

$$= [\tan^{-1} t + c_1]\mathbf{i} + \left[\frac{1}{2}\ln(1+t^2) + c_2 \right]\mathbf{j} + \left[\int \left(1 - \frac{1}{1+t^2} \right) dt \right]\mathbf{k}$$

$$= [\tan^{-1} t + c_1]\mathbf{i} + \left[\frac{1}{2}\ln(1+t^2) + c_2 \right]\mathbf{j} + [t - \tan^{-1} t + c_3]\mathbf{k}$$

$$= \tan^{-1} t\mathbf{i} + \frac{1}{2}\ln(1+t^2)\mathbf{j} + (t - \tan^{-1} t)\mathbf{k} + \mathbf{c},$$

where $\mathbf{c} = c_1\mathbf{i} + c_2\mathbf{j} + c_3\mathbf{k}$.

39. $\mathbf{r}'(t) = \displaystyle\int \mathbf{r}''(t)\, dt = \left[\int 12t\, dt \right]\mathbf{i} + \left[\int -3t^{-1/2}\, dt \right]\mathbf{j} + \left[\int 2\, dt \right]\mathbf{k}$

$$= \left[6t^2 + c_1 \right]\mathbf{i} + \left[-6t^{1/2} + c_2 \right]\mathbf{j} + [2t + c_3]\mathbf{k}$$

Since $\mathbf{r}'(1) = \mathbf{j} = (6 + c_1)\mathbf{i} + (-6 + c_2)\mathbf{j} + (2 + c_3)\mathbf{k}$, $c_1 = -6$, $c_2 = 7$, and $c_3 = -2$. Thus,

$$\mathbf{r}'(t) = (6t^2 - 6)\mathbf{i} + (-6t^{1/2} + 7)\mathbf{j} + (2t - 2)\mathbf{k}.$$

$$\mathbf{r}(t) = \int \mathbf{r}'(t)\, dt = \left[\int (6t^2 - 6)\, dt\right]\mathbf{i} + \left[\int (-6t^{1/2} + 7)\, dt\right]\mathbf{j} + \left[\int (2t - 2)\, dt\right]\mathbf{k}$$

$$= \left[2t^3 - 6t + c_4\right]\mathbf{i} + \left[-4t^{3/2} + 7t + c_5\right]\mathbf{j} + \left[t^2 - 2t + c_6\right]\mathbf{k}.$$

Since

$$\mathbf{r}(1) = 2\mathbf{i} - \mathbf{k} = (-4 + c_4)\mathbf{i} + (3 + c_5)\mathbf{j} + (-1 + c_6)\mathbf{k},$$

$c_4 = 6$, $c_5 = -3$, and $c_6 = 0$. Thus,

$$\mathbf{r}(t) = (2t^3 - 6t + 6)\mathbf{i} + (-4t^{3/2} + 7t - 3)\mathbf{j} + (t^2 - 2t)\mathbf{k}.$$

42. $\mathbf{r}'(t) = \mathbf{i} + (\cos t - t \sin t)\mathbf{j} + (\sin t + t \cos t)\mathbf{k}$

$\|\mathbf{r}'(t)\| = \sqrt{1^2 + (\cos t - t \sin t)^2 + (\sin t + t \cos t)^2} = \sqrt{2 + t^2}$

$$s = \int_0^\pi \sqrt{2 + t^2}\, dt = \left. \left(\frac{t}{2}\sqrt{2 + t^2} + \ln\left|t + \sqrt{2 + t^2}\right|\right)\right|_0^\pi$$

$$= \frac{\pi}{2}\sqrt{2 + \pi^2} + \ln\left(\pi + \sqrt{2 + \pi^2}\right) - \ln\sqrt{2}$$

45. $\mathbf{r}'(t) = -a \sin t\,\mathbf{i} + a \cos t\,\mathbf{j}$; $\|\mathbf{r}'(t)\| = \sqrt{a^2 \sin^2 t + a^2 \cos^2 t} = a$, $a > 0$; $s = \int_0^t a\, du = at$

$\mathbf{r}(s) = a \cos(s/a)\mathbf{i} + a \sin(s/a)\mathbf{j}$; $\mathbf{r}'(s) = -\sin(s/a)\mathbf{i} + \cos(s/a)\mathbf{j}$

$\|\mathbf{r}'(s)\| = \sqrt{\sin^2(s/a) + \cos^2(s/a)} = 1$

48. Since $\|\mathbf{r}(t)\|$ is the length of $\mathbf{r}(t)$, $\|\mathbf{r}(t)\| = c$ represents a curve lying on a sphere of radius c centered at the origin.

51. $\dfrac{d}{dt}[\mathbf{r}_1(t) \times \mathbf{r}_2(t)] = \lim\limits_{h \to 0} \dfrac{\mathbf{r}_1(t+h) \times \mathbf{r}_2(t+h) - \mathbf{r}_1(t) \times \mathbf{r}_2(t)}{h}$

$$= \lim_{h \to 0} \frac{\mathbf{r}_1(t+h) \times \mathbf{r}_2(t+h) - \mathbf{r}_1(t+h) \times \mathbf{r}_2(t) + \mathbf{r}_1(t+h) \times \mathbf{r}_2(t) - \mathbf{r}_1(t) \times \mathbf{r}_2(t)}{h}$$

$$= \lim_{h \to 0} \frac{\mathbf{r}_1(t+h) \times [\mathbf{r}_2(t+h) - \mathbf{r}_2(t)]}{h} + \lim_{h \to 0} \frac{[\mathbf{r}_1(t+h) - \mathbf{r}_1(t)] \times \mathbf{r}_2(t)}{h}$$

$$= \mathbf{r}_1(t) \times \left(\lim_{h \to 0} \frac{\mathbf{r}_2(t+h) - \mathbf{r}_2(t)}{h}\right) + \left(\lim_{h \to 0} \frac{\mathbf{r}_1(t+h) - \mathbf{r}_1(t)}{h}\right) \times \mathbf{r}_2(t)$$

$$= \mathbf{r}_1(t) \times \mathbf{r}_2'(t) + \mathbf{r}_1'(t) \times \mathbf{r}_2(t)$$

9.2   Motion on a Curve

3. $\mathbf{v}(t) = -2\sinh 2t\,\mathbf{i} + 2\cosh 2t\,\mathbf{j}$; $\mathbf{v}(0) = 2\mathbf{j}$; $\|\mathbf{v}(0)\| = 2$;

$\mathbf{a}(t) = -4\cosh 2t\,\mathbf{i} + 4\sinh 2t\,\mathbf{j}$; $\mathbf{a}(0) = -4\mathbf{i}$

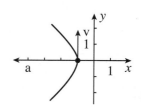

6. $\mathbf{v}(t) = \mathbf{i} + \mathbf{j} + 3t^2\mathbf{k}$; $\mathbf{v}(2) = \mathbf{i} + \mathbf{j} + 12\mathbf{k}$; $\|\mathbf{v}(2)\| = \sqrt{1+1+144} = \sqrt{146}$;

$\mathbf{a}(t) = 6t\mathbf{k}$; $\mathbf{a}(2) = 12\mathbf{k}$

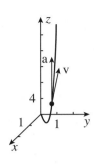

9. The particle passes through the xy-plane when $z(t) = t^2 - 5t = 0$ or $t = 0, 5$ which gives us the points $(0,0,0)$ and $(25, 115, 0)$. $\mathbf{v}(t) = 2t\mathbf{i} + (3t^2 - 2)\mathbf{j} + (2t - 5)\mathbf{k}$; $\mathbf{v}(0) = -2\mathbf{j} - 5\mathbf{k}$, $\mathbf{v}(5) = 10\mathbf{i} + 73\mathbf{j} + 5\mathbf{k}$; $\mathbf{a}(t) = 2\mathbf{i} + 6t\mathbf{j} + 2\mathbf{k}$; $\mathbf{a}(0) = 2\mathbf{i} + 2\mathbf{k}$, $\mathbf{a}(5) = 2\mathbf{i} + 30\mathbf{j} + 2\mathbf{k}$

12. Initially we are given $\mathbf{s}_0 = 1600\mathbf{j}$ and $\mathbf{v}_0 = (480\cos 30°)\mathbf{i} + (480\sin 30°)\mathbf{j} = 240\sqrt{3}\,\mathbf{i} + 240\mathbf{j}$. Using $\mathbf{a}(t) = -32\mathbf{j}$ we find

$$\mathbf{v}(t) = \int \mathbf{a}(t)\, dt = -32t\mathbf{j} + \mathbf{c}$$

$$240\sqrt{3}\,\mathbf{i} + 240\mathbf{j} = \mathbf{v}(0) = \mathbf{c}$$

$$\mathbf{v}(t) = -32t\mathbf{j} + 240\sqrt{3}\,\mathbf{i} + 240\mathbf{j} = 240\sqrt{3}\,\mathbf{i} + (240 - 32t)\mathbf{j}$$

$$\mathbf{r}(t) = \int \mathbf{v}(t)\, dt = 240\sqrt{3}\,t\mathbf{i} + (240t - 16t^2)\mathbf{j} + \mathbf{b}$$

$$1600\mathbf{j} = \mathbf{r}(0) = \mathbf{b}.$$

(a) The shell's trajectory is given by $\mathbf{r}(t) = 240\sqrt{3}\,t\mathbf{i} + (240t - 16t^2 + 1600)\mathbf{j}$ or $x = 240\sqrt{3}\,t$, $y = 240t - 16t^2 + 1600$.

(b) Solving $dy/dt = 240 - 32t = 0$, we see that y is maximum when $t = 15/2$. The maximum altitude is $y(15/2) = 2500$ ft.

(c) Solving $y(t) = -16t^2 + 240t + 1600 = -16(t - 20)(t + 5) = 0$, we see that the shell hits the ground when $t = 20$. The range of the shell is $x(20) = 4800\sqrt{3} \approx 8314$ ft.

(d) From **(c)**, impact is when $t = 20$. The speed at impact is

$$\|\mathbf{v}(20)\| = |240\sqrt{3}\,\mathbf{i} + (240 - 32 \cdot 20)\mathbf{j}| = \sqrt{240^2 \cdot 3 + (-400)^2} = 160\sqrt{13} \approx 577\,\text{ft/s}.$$

15. Let s be the initial speed. Then $\mathbf{v}(0) = s\cos 45°\mathbf{i} + s\sin 45°\mathbf{j} = \dfrac{s\sqrt{2}}{2}\mathbf{i} + \dfrac{s\sqrt{2}}{2}\mathbf{j}$. Using $\mathbf{a}(t) = -32\mathbf{j}$, we have

$$\mathbf{v}(t) = \int \mathbf{a}(t)\,dt = -32t\mathbf{j} + \mathbf{c}$$

$$\frac{s\sqrt{2}}{2}\mathbf{i} + \frac{s\sqrt{2}}{2}\mathbf{j} = \mathbf{v}(0) = \mathbf{c}$$

$$\mathbf{v}(t) = \frac{s\sqrt{2}}{2}\mathbf{i} + \left(\frac{s\sqrt{2}}{2} - 32t\right)\mathbf{j}$$

$$\mathbf{r}(t) = \frac{s\sqrt{2}}{2}t\mathbf{i} + \left(\frac{s\sqrt{2}}{2}t - 16t^2\right)\mathbf{j} + \mathbf{b}.$$

Since $\mathbf{r}(0) = \mathbf{0}$, $\mathbf{b} = \mathbf{0}$ and

$$\mathbf{r}(t) = \frac{s\sqrt{2}}{2}t\mathbf{i} + \left(\frac{s\sqrt{2}}{2}t - 16t^2\right)\mathbf{j}.$$

Setting $y(t) = s\sqrt{2}\,t/2 - 16t^2 = t(s\sqrt{2}/2 - 16t) = 0$ we see that the ball hits the ground when $t = \sqrt{2}\,s/32$. Thus, using $x(t) = s\sqrt{2}\,t/2$ and the fact that 100 yd = 300 ft,

$$300 = x(t) = \frac{s\sqrt{2}}{2}(\sqrt{2}\,s/32) = \frac{s^2}{32} \text{ and } s = \sqrt{9600} \approx 97.98 \text{ ft/s}.$$

18. (a) Using the vector function $\mathbf{r}(t) = (v_0 \cos\theta)t\mathbf{i} + [-\frac{1}{2}gt^2 + (v_0 \sin\theta)t + s_0]\mathbf{j}$, with $v_0 = 480$ ft/s, $\theta = 45°$ and $s_0 = 1600$ ft.

$$\mathbf{r}(t) = (480\cos 45°)t\,\mathbf{i} + \left[-\frac{1}{2}(32)t^2 + (480\sin 45°)t + 1600\right]\mathbf{j}$$

$$= (240\sqrt{2})t\,\mathbf{i} + [-16t^2 + 240\sqrt{2}\,t + 1600]\mathbf{j}$$

(b) For $\theta = 39.76°$, $\mathbf{r}(t) = (480\cos 39.76°)t\,\mathbf{i} + [-16t^2 + (480\sin 39.76°)t + 1600]\mathbf{j}$.

Impact is when $y(t) = -16t^2 + (480\sin 39.76°)t + 1600 = 0$, that is, when $t = 23.4513$ s; therefore the range is $x(23.4513) = 480(\cos 39.76°)(23.4513) = 8653.31$ ft.

(c)

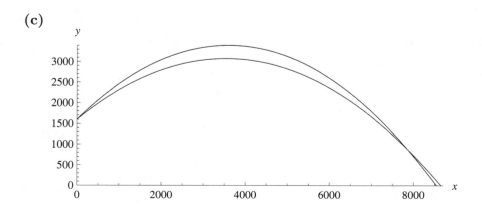

21. Let the initial speed of the projectile be s and let the target be at (x_0, y_0). Then $\mathbf{v}_p(0) = s\cos\theta\mathbf{i} + s\sin\theta\mathbf{j}$ and $\mathbf{v}_t(0) = \mathbf{0}$. Using $\mathbf{a}(t) = -32\mathbf{j}$, we have

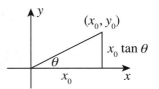

$$\mathbf{v}_p(t) = \int \mathbf{a}(t)\,dt = -32t\mathbf{j} + \mathbf{c}$$

$$s\cos\theta\mathbf{i} + s\sin\theta\mathbf{j} = \mathbf{v}_p(0) = \mathbf{c}$$

$$\mathbf{v}_p(t) = s\cos\theta\mathbf{i} + (s\sin\theta - 32t)\mathbf{j}$$

$$\mathbf{r}_p(t) = st\cos\theta\mathbf{i} + (st\sin\theta - 16t^2)\mathbf{j} + \mathbf{b}.$$

Since $\mathbf{r}_p(0) = \mathbf{0}$, $\mathbf{b} = \mathbf{0}$ and $\mathbf{r}_p(t) = st\cos\theta\mathbf{i} + (st\sin\theta - 16t^2)\mathbf{j}$. Also, $\mathbf{v}_t(t) = -32t\mathbf{j} + \mathbf{c}$ and since $\mathbf{v}_t(0) = \mathbf{0}$, $\mathbf{c} = \mathbf{0}$ and $\mathbf{v}_t(t) = -32t\mathbf{j}$. Then $\mathbf{r}_t(t) = -16t^2\mathbf{j} + \mathbf{b}$. Since $\mathbf{r}_t(0) = x_0\mathbf{i} + y_0\mathbf{j}$, $\mathbf{b} = x_0\mathbf{i} + y_0\mathbf{j}$ and $\mathbf{r}_t(t) = x_0\mathbf{i} + (y_0 - 16t^2)\mathbf{j}$. Now, the horizontal component of $\mathbf{r}_p(t)$ will be x_0 when $t = x_0/s\cos\theta$ at which time the vertical component of $\mathbf{r}_p(t)$ will be

$$(sx_0/s\cos\theta)\sin\theta - 16(x_0/s\cos\theta)^2 = x_0\tan\theta - 16(x_0/s\cos\theta)^2 = y_0 - 16(x_0/s\cos\theta)^2.$$

Thus, $\mathbf{r}_p(x_0/s\cos\theta) = \mathbf{r}_t(x_0/s\cos\theta)$ and the projectile will strike the target as it falls.

24. **(a)** $\mathbf{v}(t) = -b\sin t\mathbf{i} + b\cos t\mathbf{j} + c\mathbf{k}$; $\|\mathbf{v}(t)\| = \sqrt{b^2\sin^2 t + b^2\cos^2 t + c^2} = \sqrt{b^2 + c^2}$

(b) $s = \int_0^t \|\mathbf{v}(u)\|\,du \int_0^t \sqrt{b^2 + c^2}\,du = t\sqrt{b^2 + c^2}$; $\dfrac{ds}{dt} = \sqrt{b^2 + c^2}$

(c) $\dfrac{d^2 s}{dt^2} = 0$; $\mathbf{a}(t) = -b\cos t\mathbf{i} - b\sin t\mathbf{j}$; $\|\mathbf{a}(t)\| = \sqrt{b^2\cos^2 t + b^2\sin^2 t} = |b|$.

Thus, $d^2 s/dt^2 \neq \|\mathbf{a}(t)\|$.

27. Letting $\mathbf{r}(t) = x(t)\mathbf{i} + y(t)\mathbf{j} + z(t)\mathbf{k}$, the equation $d\mathbf{r}/dt = \mathbf{v}$ is equivalent to $dx/dt = 6t^2 x$, $dy/dt = -4ty^2$, $dz/dt = 2t(z + 1)$. Separating variables and integrating, we obtain $dx/x = 6t^2\,dt$, $dy/y^2 = -4t\,dt$, $dz/(z + 1) = 2t\,dt$, and $\ln x = 2t^3 + c_1$, $-1/y = -2t^2 + c_2$, $\ln(z + 1) = t^2 + c_3$. Thus,

$$\mathbf{r}(t) = k_1 e^{2t^3}\mathbf{i} + \frac{1}{2t^2 + k_2}\mathbf{j} + (k_3 e^{t^2} - 1)\mathbf{k}.$$

9.3 | Curvature

3. We assume $a > 0$. $\mathbf{r}'(t) = -a\sin t\mathbf{i} + a\cos t\mathbf{j} + c\mathbf{k}$;

$|\mathbf{r}'(t)| = \sqrt{a^2\sin^2 t + a^2\cos^2 t + c^2} = \sqrt{a^2 + c^2}$;

$\mathbf{T}(t) = -\dfrac{a\sin t}{\sqrt{a^2 + c^2}}\mathbf{i} + \dfrac{a\cos t}{\sqrt{a^2 + c^2}}\mathbf{j} + \dfrac{c}{\sqrt{a^2 + c^2}}\mathbf{k}$; $\dfrac{d\mathbf{T}}{dt} = -\dfrac{a\cos t}{\sqrt{a^2 + c^2}}\mathbf{i} - \dfrac{a\sin t}{\sqrt{a^2 + c^2}}\mathbf{j}$,

$\left|\dfrac{d\mathbf{T}}{dt}\right| = \sqrt{\dfrac{a^2\cos^2 t}{a^2 + c^2} + \dfrac{a^2\sin^2 t}{a^2 + c^2}} = \dfrac{a}{\sqrt{a^2 + c^2}}$; $\mathbf{N} = -\cos t\mathbf{i} - \sin t\mathbf{j}$;

$$\mathbf{b} = \mathbf{T} \times \mathbf{N} = \begin{vmatrix} \mathbf{i} & \mathbf{j} & \mathbf{k} \\ -\dfrac{a\sin t}{\sqrt{a^2+c^2}} & \dfrac{a\cos t}{\sqrt{a^2+c^2}} & \dfrac{c}{\sqrt{a^2+c^2}} \\ -\cos t & -\sin t & 0 \end{vmatrix} = \dfrac{c\sin t}{\sqrt{a^2+c^2}}\,\mathbf{i} - \dfrac{c\cos t}{\sqrt{a^2+c^2}}\,\mathbf{j} + \dfrac{a}{\sqrt{a^2+c^2}}\,\mathbf{k};$$

$$\kappa = \dfrac{|d\mathbf{T}/dt|}{|\mathbf{r}'(t)|} = \dfrac{a/\sqrt{a^2+c^2}}{\sqrt{a^2+c^2}} = \dfrac{a}{a^2+c^2}$$

6. From Problem 4, a normal to the osculating plane is $\mathbf{B}(1) = \frac{1}{\sqrt{6}}(\mathbf{i} - 2\mathbf{j} + \mathbf{k})$. The point on the curve when $t = 1$ is $(1, 1/2, 1/3)$. An equation of the plane is $(x-1) - 2(y-1/2) + (z-1/3) = 0$ or $x - 2y + z = 1/3$.

9. $\mathbf{v}(t) = 2t\mathbf{i} + 2t\mathbf{j} + 4t\mathbf{k}$, $\ |\mathbf{v}(t)| = 2\sqrt{6}\,t$, $t > 0$; $\ \mathbf{a}(t) = 2\mathbf{i} + 2\mathbf{j} + 4\mathbf{k}$; $\ \mathbf{v}\cdot\mathbf{a} = 24t$, $\ \mathbf{v}\times\mathbf{a} = \mathbf{0}$;

$$a_T = \dfrac{24t}{2\sqrt{6}\,t} = 2\sqrt{6}\,, \ \ a_N = 0,\ t > 0$$

12. $\mathbf{v}(t) = \dfrac{1}{1+t^2}\mathbf{i} + \dfrac{t}{1+t^2}\mathbf{j}$, $\ |\mathbf{v}(t)| = \dfrac{\sqrt{1+t^2}}{1+t^2}$; $\ \mathbf{a}(t) = -\dfrac{2t}{(1+t^2)^2}\mathbf{i} + \dfrac{1-t^2}{(1+t^2)^2}\mathbf{j}$;

$$\mathbf{v}\cdot\mathbf{a} = -\dfrac{2t}{(1+t^2)^3} + \dfrac{t-t^3}{(1+t^2)^3} = -\dfrac{t}{(1+t^2)^2}; \ \ \mathbf{v}\times\mathbf{a} = \dfrac{1}{(1+t^2)^2}\mathbf{k},\ |\mathbf{v}\times\mathbf{a}| = \dfrac{1}{(1+t^2)^2};$$

$$a_T = -\dfrac{t/(1+t^2)^2}{\sqrt{1+t^2}/(1+t^2)} = -\dfrac{t}{(1+t^2)^{3/2}}\,, \ \ a_N = \dfrac{1/(1+t^2)^2}{\sqrt{1+t^2}/(1+t^2)} = \dfrac{1}{(1+t^2)^{3/2}}$$

15. $\mathbf{v}(t) = -e^{-t}(\mathbf{i}+\mathbf{j}+\mathbf{k})$, $\ |\mathbf{v}(t)| = \sqrt{3}\,e^{-t}$; $\ \mathbf{a}(t) = e^{-t}(\mathbf{i}+\mathbf{j}+\mathbf{k})$; $\ \mathbf{v}\cdot\mathbf{a} = -3e^{-2t}$; $\ \mathbf{v}\times\mathbf{a} = \mathbf{0}$, $|\mathbf{v}\times\mathbf{a}| = 0$; $\ a_T = -\sqrt{3}\,e^{-t}$, $\ a_N = 0$

18. **(a)** $\mathbf{v}(t) = -a\sin t\mathbf{i} + b\cos t\mathbf{j}$, $\ |\mathbf{v}(t)| = \sqrt{a^2\sin^2 t + b^2\cos^2 t}$; $\ \mathbf{a}(t) = -a\cos t\mathbf{i} - b\sin t\mathbf{j}$;

$$\mathbf{v}\times\mathbf{a} = ab\mathbf{k}; \ \ |\mathbf{v}\times\mathbf{a}| = ab; \ \ \kappa = \dfrac{ab}{(a^2\sin^2 t + b^2\cos^2 t)^{3/2}}$$

(b) When $a = b$, $|\mathbf{v}(t)| = a$, $|\mathbf{v}\times\mathbf{a}| = a^2$, and $\kappa = a^2/a^3 = 1/a$.

21. $\mathbf{v}(t) = f'(t)\mathbf{i} + g'(t)\mathbf{j}$, $\ |\mathbf{v}(t)| = \sqrt{[f'(t)]^2 + [g'(t)]^2}$; $\ \mathbf{a}(t) = f''(t)\mathbf{i} + g''(t)\mathbf{j}$;

$$\mathbf{v}\times\mathbf{a} = [f'(t)g''(t) - g'(t)f''(t)]\mathbf{k}, \ \ |\mathbf{v}\times\mathbf{a}| = |f'(t)g''(t) - g'(t)f''(t)|;$$

$$\kappa = \dfrac{|\mathbf{v}\times\mathbf{a}|}{|\mathbf{v}|^3} = \dfrac{|f'(t)g''(t) - g'(t)f''(t)|}{([f'(t)]^2 + [g'(t)]^2)^{3/2}}$$

24. $F(x) = x^3$, $\ F(-1) = -1$, $\ F(1/2) = 1/8$; $\ F'(x) = 3x^2$, $\ F'(-1) = 3$, $\ F'(1/2) = 3/4$;

$F''(x) = 6x$, $\ F''(-1) = -6$, $\ F''(1/2) = 3$; $\ \kappa(-1) = \dfrac{|-6|}{(1+3^2)^{3/2}} = \dfrac{6}{10\sqrt{10}} = \dfrac{3}{5\sqrt{10}} \approx 0.19$;

$$\rho(-1) = \dfrac{5\sqrt{10}}{3} \approx 5.27; \ \ \kappa\left(\dfrac{1}{2}\right) = \dfrac{3}{[1+(3/4)^2]^{3/2}} = \dfrac{3}{125/64} = \dfrac{192}{125} \approx 1.54;$$

$$\rho\left(\dfrac{1}{2}\right) = \dfrac{125}{192} \approx 0.65$$

Since $1.54 > 0.19$, the curve is "sharper" at $(1/2, 1/8)$.

9.4 | **Partial Derivatives**

3. $x^2 - y^2 = 1 + c^2$

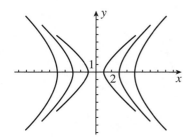

6. $y = x + \tan c, \; -\pi/x < c < \pi/2$

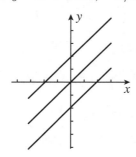

9. $x^2 + 3y^2 + 6z^2 = c$; ellipsoid

12. Setting $x = -4$, $y = 2$, and $z = -3$ in $x^2/16 + y^2/4 + z^2/9 = c$ we obtain $c = 3$. The equation of the surface is $x^2/16 + y^2/4 + z^2/9 = 3$. Setting $y = z = 0$ we find the x-intercepts are $\pm 4\sqrt{3}$. Similarly, the y-intercepts are $\pm 2\sqrt{3}$ and the z-intercepts are $\pm 3\sqrt{3}$.

15. $z_x = 20x^3y^3 - 2xy^6 + 30x^4$; $\;z_y = 15x^4y^2 - 6x^2y^5 - 4$

18. $z_x = 12x^2 - 10x + 8$; $\;z_y = 0$

21. $z_x = 2(\cos 5x)(-\sin 5x)(5) = -10\sin 5x\cos 5x$; $\;z_y = 2(\sin 5y)(\cos 5y)(5) = 10\sin 5y\cos 5y$

24. $f_\theta = \phi^2\left(\cos\dfrac{\theta}{\phi}\right)\left(\dfrac{1}{\phi}\right) = \phi\cos\dfrac{\theta}{\phi}$; $\;f_\phi = \phi^2\left(\cos\dfrac{\theta}{\phi}\right)\left(-\dfrac{\theta}{\phi^2}\right) + 2\phi\sin\dfrac{\theta}{\phi} = -\theta\cos\dfrac{\theta}{\phi} + 2\phi\sin\dfrac{\theta}{\phi}$

27. $g_u = \dfrac{8u}{4u^2 + 5v^3}$; $\;g_v = \dfrac{15v^2}{4u^2 + 5v^3}$

30. $w_x = xy\left(\dfrac{1}{x}\right) + (\ln xz)y = y + y\ln xz$; $\;w_y = x\ln xz$; $\;w_z = \dfrac{xy}{z}$

33. $\dfrac{\partial z}{\partial x} = \dfrac{2x}{x^2 + y^2}$, $\;\dfrac{\partial^2 z}{\partial x^2} = \dfrac{(x^2 + y^2)2 - 2x(2x)}{(x^2 + y^2)^2} = \dfrac{2y^2 - 2x^2}{(x^2 + y^2)^2}$; $\;\dfrac{\partial z}{\partial y} = \dfrac{2y}{x^2 + y^2}$,

$\dfrac{\partial^2 z}{\partial y^2} = \dfrac{(x^2 + y^2)2 - 2y(2y)}{(x^2 + y^2)^2} = \dfrac{2x^2 - 2y^2}{(x^2 + y^2)^2}$; $\;\dfrac{\partial^2 z}{\partial x^2} + \dfrac{\partial^2 z}{\partial y^2} = \dfrac{2y^2 - 2x^2 + 2x^2 - 2y^2}{(x^2 + y^2)^2} = 0$

36. $\dfrac{\partial u}{\partial x} = -\sin(x + at) + \cos(x - at)$, $\;\dfrac{\partial^2 u}{\partial x^2} = -\cos(x + at) - \sin(x - at)$;

$\dfrac{\partial u}{\partial t} = -a\sin(x + at) - a\cos(x - at)$, $\;\dfrac{\partial^2 u}{\partial t^2} = -a^2\cos(x + at) - a^2\sin(x - at)$;

$a^2\dfrac{\partial^2 u}{\partial x^2} = -a^2\cos(x + at) - a^2\sin(x - at) = \dfrac{\partial^2 u}{\partial t^2}$

39. $z_x = v^2 e^{uv^2}(3x^2) + 2uve^{uv^2}(1) = 3x^2v^2e^{uv^2} + 2uve^{uv^2}$;

$z_y = v^2e^{uv^2}(0) + 2uve^{uv^2}(-2y) = -4yuve^{uv^2}$

42. $z_u = \dfrac{2y}{(x+y)^2}\left(\dfrac{1}{v}\right) + \dfrac{-2x}{(x+y)^2}\left(-\dfrac{v^2}{u^2}\right) = \dfrac{2y}{v(x+y)^2} + \dfrac{2xv^2}{u^2(x+y)^2}$

$\quad\; z_v = \dfrac{2y}{(x+y)^2}\left(-\dfrac{u}{v^2}\right) + \dfrac{-2x}{(x+y)^2}\left(\dfrac{2v}{u}\right) = -\dfrac{2yu}{v^2(x+y)^2} - \dfrac{4xv}{u(x+y)^2}$

45. $R_u = s^2t^4\left(e^{v^2}\right) + 2rst^4\left(-2uve^{-u^2}\right) + 4rs^2t^3\left(2uv^2e^{u^2v^2}\right)$

$\qquad = s^2t^4e^{v^2} - 4uvrst^4e^{-u^2} + 8uv^2rs^2t^3e^{u^2v^2}$

$\quad\; R_v = s^2t^4\left(2uve^{v^2}\right) + 2rst^4\left(e^{-u^2}\right) + 4rs^2t^3\left(2u^2ve^{u^2v^2}\right)$

$\qquad = 2s^2t^4uve^{v^2} + 2rst^4e^{-u^2} + 8rs^2t^3u^2ve^{u^2v^2}$

48. $s_\phi = 2pe^{3\theta} + 2q[-\sin(\phi+\theta)] - 2r\theta^2 + 4(2) = 2pe^{3\theta} - 2q\sin(\phi+\theta) - 2r\theta^2 + 8$

$\quad\; s_\theta = 2p(3\phi e^{3\theta}) + 2q[-\sin(\phi+\theta)] - 2r(2\phi\theta) + 4(8) = 6p\phi e^{3\theta} - 2q\sin(\phi+\theta) - 4r\phi\theta + 32$

51. $\dfrac{dw}{dt} = -3\sin(3u+4v)(2) - 4\sin(3u+4v)(-1); \quad u(\pi) = 5\pi/2, \quad v(\pi) = -5\pi/4$

$\quad\; \left.\dfrac{dw}{dt}\right|_\pi = -6\sin\left(\dfrac{15\pi}{2} - 5\pi\right) + 4\sin\left(\dfrac{15\pi}{2} - 5\pi\right) = -2\sin\dfrac{5\pi}{2} = -2$

54. $\dfrac{dP}{dt} = \dfrac{(V - 0.0427)(0.08)dT/dt}{(V-0.0427)^2} - \dfrac{0.08T(dV/dt)}{(V-0.0427)^2} + \dfrac{3.6}{V^3}\dfrac{dV}{dt}$

$\qquad = \dfrac{0.08}{V-0.0427}\dfrac{dT}{dt} + \left(\dfrac{3.6}{V^3} - \dfrac{0.08T}{(V-0.0427)^2}\right)\dfrac{dV}{dt}$

57. Since the height of the triangle is $x\sin\theta$, the area is given by $A = \frac{1}{2}xy\sin\theta$. Then

$$\dfrac{dA}{dt}\dfrac{\partial A}{\partial x}\dfrac{dx}{dt} + \dfrac{\partial A}{\partial y}\dfrac{dy}{dt} + \dfrac{\partial A}{\partial \theta}\dfrac{d\theta}{dt} = \dfrac{1}{2}y\sin\theta\dfrac{dx}{dt} + \dfrac{1}{2}x\sin\theta\dfrac{dy}{dt} + \dfrac{1}{2}xy\cos\theta\dfrac{d\theta}{dt}.$$

When $x = 10$, $y = 8$, $\theta = \pi/6$, $dx/dt = 0.3$, $dy/dt = 0.5$, and $d\theta/dt = 0.1$,

$$\dfrac{dA}{dt} = \dfrac{1}{2}(8)\left(\dfrac{1}{2}\right)(0.3) + \dfrac{1}{2}(10)\left(\dfrac{1}{2}\right)(0.5) + \dfrac{1}{2}(10)(8)\left(\dfrac{\sqrt{3}}{2}\right)(0.1)$$

$$= 0.6 + 1.25 + 2\sqrt{3} = 1.85 + 2\sqrt{3} \approx 5.31 \text{ cm}^2/\text{s}.$$

9.5 | Directional Derivative

3. $\nabla F = \dfrac{y^2}{z^3}\mathbf{i} + \dfrac{2xy}{z^3}\mathbf{j} - \dfrac{3xy^2}{z^4}\mathbf{k}$

6. $\nabla f = \dfrac{3x^2}{2\sqrt{x^3y - y^4}}\mathbf{i} + \dfrac{x^3 - 4y^3}{2\sqrt{x^3y - y^4}}\mathbf{j}; \quad \nabla f(3,2) = \dfrac{27}{\sqrt{38}}\mathbf{i} - \dfrac{5}{2\sqrt{38}}\mathbf{j}$

9. $D_{\mathbf{u}}f(x,y) = \lim\limits_{h\to0} \dfrac{f(x+h\sqrt{3}/2, y+h/2) - f(x,y)}{h}$

$\qquad\qquad = \lim\limits_{h\to0} \dfrac{(x+h\sqrt{3}/2)^2 + (y+h/2)^2 - x^2 - y^2}{h}$

$\qquad\qquad = \lim\limits_{h\to0} \dfrac{h\sqrt{3}\,x + 3h^2/4 + hy + h^2/4}{h} = \lim\limits_{h\to0}\,(\sqrt{3}\,x + 3h/4 + y + h/4) = \sqrt{3}\,x + y$

12. $\mathbf{u} = \dfrac{\sqrt{2}}{2}\mathbf{i} + \dfrac{\sqrt{2}}{2}\mathbf{j};\ \ \nabla f = (4+y^2)\mathbf{i} + (2xy-5)\mathbf{j};\ \ \nabla f(3,-1) = 5\mathbf{i} - 11\mathbf{j};$

$\qquad D_{\mathbf{u}}f(3,-1) = \dfrac{5\sqrt{2}}{2} - \dfrac{11\sqrt{2}}{2} = -3\sqrt{2}$

15. $\mathbf{u} = (2\mathbf{i}+\mathbf{j})/\sqrt{5};\ \ \nabla f = 2y(xy+1)\mathbf{i} + 2x(xy+1)\mathbf{j};\ \ \nabla f(3,2) = 28\mathbf{i} + 42\mathbf{j}$

$\qquad D_{\mathbf{u}}f(3,2) = \dfrac{2(28)}{\sqrt{5}} + \dfrac{42}{\sqrt{5}} = \dfrac{98}{\sqrt{5}}$

18. $\mathbf{u} = \dfrac{1}{\sqrt{6}}\mathbf{i} - \dfrac{2}{\sqrt{6}}\mathbf{j} + \dfrac{1}{\sqrt{6}}\mathbf{k};\ \ \nabla F = \dfrac{2x}{z^2}\mathbf{i} - \dfrac{2y}{z^2}\mathbf{j} + \dfrac{2y^2 - 2x^2}{z^3}\mathbf{k};\ \ \nabla F(2,4,-1) = 4\mathbf{i} - 8\mathbf{j} - 24\mathbf{k}$

$\qquad D_{\mathbf{u}}F(2,4,-1) = \dfrac{4}{\sqrt{6}} - \dfrac{16}{\sqrt{6}} - \dfrac{24}{\sqrt{6}} = -6\sqrt{6}$

21. $\mathbf{u} = (-4\mathbf{i}-\mathbf{j})/\sqrt{17};\ \ \nabla f = 2(x-y)\mathbf{i} - 2(x-y)\mathbf{j};\ \ \nabla f(4,2) = 4\mathbf{i} - 4\mathbf{j};$

$\qquad D_{\mathbf{u}}F(4,2) = -\dfrac{16}{\sqrt{17}}\dfrac{4}{\sqrt{17}} = -\dfrac{12}{\sqrt{17}}$

24. $\nabla f = (xye^{x-y} + ye^{x-y})\mathbf{i} + (-xye^{x-y} + xe^{x-y})\mathbf{j};\ \ \nabla f(5,5) = 30\mathbf{i} - 20\mathbf{j}$

The maximum $D_{\mathbf{u}}$ is $[30^2 + (-20)^2]^{1/2} = 10\sqrt{13}$ in the direction $30\mathbf{i} - 20\mathbf{j}$.

27. $\nabla f = 2x\sec^2(x^2+y^2)\mathbf{i} + 2y\sec^2(x^2+y^2)\mathbf{j};$

$\qquad \nabla f(\sqrt{\pi/6}, \sqrt{\pi/6}) = 2\sqrt{\pi/6}\sec^2(\pi/3)(\mathbf{i}+\mathbf{j}) = 8\sqrt{\pi/6}\,(\mathbf{i}+\mathbf{j})$

The minimum $D_{\mathbf{u}}$ is $-8\sqrt{\pi/6}\,(1^2 + 1^2)^{1/2} = -8\sqrt{\pi/3}$ in the direction $-(\mathbf{i}+\mathbf{j})$.

30. $\nabla F = \dfrac{1}{x}\mathbf{i} + \dfrac{1}{y}\mathbf{j} - \dfrac{1}{z}\mathbf{k};\ \ \nabla F(1/2, 1/6, 1/3) = 2\mathbf{i} + 6\mathbf{j} - 3\mathbf{k}$

The minimum $D_{\mathbf{u}}$ is $-[2^2 + 6^2 + (-3)^2]^{1/2} = -7$ in the direction $-2\mathbf{i} - 6\mathbf{j} + 3\mathbf{k}$.

33. (a) Vectors perpendicular to $4\mathbf{i} + 3\mathbf{j}$ are $\pm(3\mathbf{i} - 4\mathbf{j})$. Take $\mathbf{u} = \pm\left(\dfrac{3}{5}\mathbf{i} - \dfrac{4}{5}\mathbf{j}\right)$.

(b) $\mathbf{u} = (4\mathbf{i}+3\mathbf{j})/\sqrt{16+9} = \dfrac{4}{5}\mathbf{i} + \dfrac{3}{5}\mathbf{j}$

(c) $\mathbf{u} = -\dfrac{4}{5}\mathbf{i} - \dfrac{3}{5}\mathbf{j}$

36. $\nabla U = \dfrac{Gmx}{(x^2+y^2)^{3/2}}\mathbf{i} + \dfrac{Gmy}{(x^2+y^2)^{3/2}}\mathbf{j} = \dfrac{Gm}{(x^2+y^2)^{3/2}}(x\mathbf{i}+y\mathbf{j})$

The maximum and minimum values of $D_\mathbf{u}U(x,y)$ are obtained when $\mathbf{u}$ is in the directions ∇U and $-\nabla U$, respectively. Thus, at a point (x,y), not $(0,0)$, the directions of maximum and minimum increase in U are $x\mathbf{i}+y\mathbf{j}$ and $-x\mathbf{i}-y\mathbf{j}$, respectively. A vector at (x,y) in the direction $\pm(x\mathbf{i}+y\mathbf{j})$ lies on a line through the origin.

39. $\nabla T = 4x\mathbf{i}+2y\mathbf{j}$; $\nabla T(4,2) = 16\mathbf{i}+4\mathbf{j}$. The minimum change in temperature (that is, the maximum decrease in temperature) is in the direction $-\nabla T(4,3) = -16\mathbf{i}-4\mathbf{j}$.

42. Substituting $x=0$, $y=0$, $z=1$, and $T=500$ into $T = k/(x^2+y^2+z^2)$ we see that $k=500$ and $T(x,y,z) = 500/(x^2+y^2+z^2)$. To find the rate of change of T at $\langle 2,3,3\rangle$ in the direction of $\langle 3,1,1\rangle$ we first compute $\langle 3,1,1\rangle - \langle 2,3,3\rangle = \langle 1,-2,-2\rangle$. Then $\mathbf{u} = \frac{1}{3}\langle 1,-2,-2\rangle = \frac{1}{3}\mathbf{i} - \frac{2}{3}\mathbf{j} - \frac{2}{3}\mathbf{k}$. Now

$$\nabla T = -\frac{1000x}{(x^2+y^2+z^2)^2}\mathbf{i} - \frac{1000y}{(x^2+y^2+z^2)^2}\mathbf{j} - \frac{1000z}{(x^2+y^2+z^2)^2}\mathbf{k}$$

and

$$\nabla T(2,3,3) = -\frac{500}{121}\mathbf{i} - \frac{750}{121}\mathbf{j} - \frac{750}{121}\mathbf{k},$$

so

$$D_\mathbf{u}T(2,3,3) = \frac{1}{3}\left(-\frac{500}{121}\right) - \frac{2}{3}\left(-\frac{750}{121}\right) - \frac{2}{3}\left(-\frac{750}{121}\right) = \frac{2500}{363}.$$

The direction of maximum increase is $\nabla T(2,3,3) = -\frac{500}{121}\mathbf{i} - \frac{750}{121}\mathbf{j} - \frac{750}{121}\mathbf{k} = \frac{250}{121}(-2\mathbf{i}-3\mathbf{j}-3\mathbf{k})$, and the maximum rate of change of T is $|\nabla T(2,3,3)| = \frac{250}{121}\sqrt{4+9+9} = \frac{250}{121}\sqrt{22}$.

45. $\nabla(cf) = \dfrac{\partial}{\partial x}(cf)\mathbf{i} + \dfrac{\partial}{\partial y}(cf)\mathbf{j} = cf_x\mathbf{i}+cf_y\mathbf{j} = c(f_x\mathbf{i}+f_y\mathbf{j}) = c\nabla f$

48. $\nabla(f/g) = [(gf_x-fg_x)/g^2]\mathbf{i} + [(gf_y-fg_y)/g^2]\mathbf{j} = g(f_x\mathbf{i}+f_y\mathbf{j})/g^2 - f(g_x\mathbf{i}+g_y\mathbf{j})/g^2$
$\qquad = g\nabla f/g^2 - f\nabla g/g^2 = (g\nabla f - f\nabla g)/g^2$

9.6 Tangent Planes and Normal Lines

3. Since $f(2,5)=1$, the level curve is $y = x^2+1$.
$\nabla f = -2x\mathbf{i}+\mathbf{j}$; $\nabla f(2,5) = -10\mathbf{i}+\mathbf{j}$

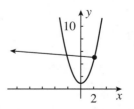

6. Since $f(2,2)=2$, the level curve is $y^2 = 2x$, $x \neq 0$. $\nabla f = -\dfrac{y^2}{x^2}\mathbf{i} + \dfrac{2y}{x}\mathbf{j}$;
$\nabla f(2,2) = -\mathbf{i}+2\mathbf{j}$

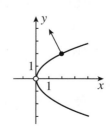

9. Since $F(3, 1, 1) = 2$, the level surface is $y + z = 2$. $\nabla F = \mathbf{j} + \mathbf{k}$;

$\nabla F(3, 1, 1) = \mathbf{j} + \mathbf{k}$

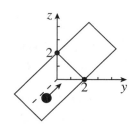

12. Since $F(0, -1, 1) = 0$, the level surface is $x^2 - y^2 + z = 0$ or $z = y^2 - x^2$.

$\nabla F = 2x\mathbf{i} - 2y\mathbf{j} + \mathbf{k}$; $\nabla F(0, -1, 1) = 2\mathbf{j} + \mathbf{k}$

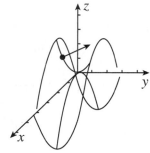

15. $F(x, y, z) = x^2 + y^2 + z^2$; $\nabla F = 2x\mathbf{i} + 2y\mathbf{j} + 2z\mathbf{k}$. $\nabla F(-2, 2, 1) = -4\mathbf{i} + 4\mathbf{j} + 2\mathbf{k}$. The equation of the tangent plane is $-4(x + 2) + 4(y - 2) + 2(z - 1) = 0$ or $-2x + 2y + z = 9$.

18. $F(x, y, z) = xy + yz + zx$; $\nabla F = (y + z)\mathbf{i} + (x + z)\mathbf{j} + (y + x)\mathbf{k}$; $\nabla F(1, -3, -5) = -8\mathbf{i} - 4\mathbf{j} - 2\mathbf{k}$. The equation of the tangent plane is $-8(x - 1) - 4(y + 3) - 2(z + 5) = 0$ or $4x + 2y + z = -7$.

21. $F(x, y, z) = \cos(2x + y) - z$ and

$\nabla F = -2\sin(2x + y)\mathbf{i} - \sin(2x + y)\mathbf{j} - \mathbf{k}$; $\nabla F(\pi/2, \pi/4, -1/\sqrt{2}) = \sqrt{2}\,\mathbf{i} + \dfrac{\sqrt{2}}{2}\mathbf{j} - \mathbf{k}$.

The equation of the tangent plane is $\sqrt{2}\left(x - \dfrac{\pi}{2}\right) + \dfrac{\sqrt{2}}{2}\left(y - \dfrac{\pi}{4}\right) - \left(z + \dfrac{1}{\sqrt{2}}\right) = 0$,

$2\left(x - \dfrac{\pi}{2}\right) + \left(y - \dfrac{\pi}{4}\right) - \sqrt{2}\left(z + \dfrac{1}{\sqrt{2}}\right) = 0$, or $2x + y - \sqrt{2}\,z = \dfrac{5\pi}{4} + 1$.

24. $F(x, y, z) = 8e^{-2y}\sin 4x - z$; $\nabla F = 32e^{-2y}\cos 4x\,\mathbf{i} - 16e^{-2y}\sin 4x\,\mathbf{j} - \mathbf{k}$;

$\nabla F(\pi/24, 0, 4) = 16\sqrt{3}\,\mathbf{i} - 8\mathbf{j} - \mathbf{k}$.

The equation of the tangent plane is

$$16\sqrt{3}(x - \pi/24) - 8(y - 0) - (z - 4) = 0 \quad \text{or} \quad 16\sqrt{3}\,x - 8y - z = \dfrac{2\sqrt{3}\,\pi}{3} - 4.$$

27. The gradient of $F(x, y, z) = x^2 + 4x + y^2 + z^2 - 2z$ is $\nabla F = (2x + 4)\mathbf{i} + 2y\mathbf{j} + (2z - 2)\mathbf{k}$, so a normal to the surface at (x_0, y_0, z_0) is $(2x_0 + 4)\mathbf{i} + 2y_0\mathbf{j} + (2z_0 - 2)\mathbf{k}$. A horizontal plane has normal $c\mathbf{k}$ for $c \neq 0$. Thus, we want $2x_0 + 4 = 0$, $2y_0 = 0$, $2z_0 - 2 = c$ or $x_0 = -2$, $y_0 = 0$, $z_0 = c + 1$. since (x_0, y_0, z_0) is on the surface, $(-2)^2 + 4(-2) + (c + 1)^2 - 2(c + 1) = c^2 - 5 = 11$ and $c = \pm 4$. The points on the surface are $(-2, 0, 5)$ and $(-2, 0, -3)$.

30. If (x_0, y_0, z_0) is on $x^2/a^2 - y^2/b^2 + z^2/c^2 = 1$, then $x_0^2/a^2 - y_0^2/b^2 + z_0^2/c^2 = 1$ and (x_0, y_0, z_0) is on the plane $xx_0/a^2 - yy_0/b^2 + zz_0/c^2 = 1$. A normal to the surface at (x_0, y_0, z_0) is

$$\nabla F(x_0, y_0, z_0) = (2x_0/a^2)\mathbf{i} - (2y_0/b^2)\mathbf{j} + (2z_0/c^2)\mathbf{k}.$$

A normal to the plane is $(x_0/a^2)\mathbf{i} - (y_0/b^2)\mathbf{j} + (z_0/c^2)\mathbf{k}$. Since the normal to the surface is a multiple of the normal to the plane, the normal vectors are parallel, and the plane is tangent to the surface.

33. $F(x,y,z) = x^2 + 2y^2 + z^2$; $\nabla F = 2x\mathbf{i} + 4y\mathbf{j} + 2z\mathbf{k}$; $\nabla F(1,-1,1) = 2\mathbf{i} - 4\mathbf{j} + 2\mathbf{k}$. Parametric equations of the line are $x = 1 + 2t$, $y = -1 - 4t$, $z = 1 + 2t$.

36. $F(x,y,z) = x^2 + y^2 - z^2$; $\nabla F = 2x\mathbf{i} + 2y\mathbf{j} - 2z\mathbf{k}$; $\nabla F(3,4,5) = 6\mathbf{i} + 8\mathbf{j} - 10\mathbf{k}$. Symmetric equations of the line are $\dfrac{x-3}{6} = \dfrac{y-4}{8} = \dfrac{z-5}{-10}$.

39. Let $F(x,y,z) = x^2 + y^2 + z^2 - 25$ and $G(x,y,z) = -x^2 + y^2 + z^2$. Then

$$F_x G_x + F_y G_y + F_z g_z = (2x)(-2x) + (2y)(2y) + (2z)(2z) = 4(-x^2 + y^2 + z^2).$$

For (x,y,z) on both surfaces, $F(x,y,z) = G(x,y,z) = 0$. Thus,
$F_x G_x + F_y G_y + F_z G_z = 4(0) = 0$ and the surfaces are orthogonal at points of intersection.

9.7   Curl and Divergence

3.

6.

9. curl $\mathbf{F} = \mathbf{0}$; div $\mathbf{F} = 4y + 8z$

12. curl $\mathbf{F} = -x^3 z\mathbf{i} + (3x^2 yz - z)\mathbf{j} + \left(\frac{3}{2}x^2 y^2 - y - 15y^2\right)\mathbf{k}$; div $\mathbf{F} = (x^3 y - x) - (x^3 y - x) = 0$

15. curl $\mathbf{F} = (xy^2 e^y + 2xy e^y + x^3 y e^z + x^3 yz e^z)\mathbf{i} - y^2 e^y\mathbf{j} + (-3x^2 yz e^z - xe^x)\mathbf{k}$;
   div $\mathbf{F} = xy e^x + y e^x - x^3 z e^z$

18. curl $\mathbf{r} = \begin{vmatrix} \mathbf{i} & \mathbf{j} & \mathbf{k} \\ \partial/\partial x & \partial/\partial y & \partial/\partial z \\ x & y & z \end{vmatrix} = 0\mathbf{i} - 0\mathbf{j} + 0\mathbf{k} = \mathbf{0}$

21. $\nabla \cdot (\mathbf{a} \times \mathbf{r}) = \begin{vmatrix} \partial/\partial x & \partial/\partial y & \partial/\partial z \\ a_1 & a_2 & a_3 \\ x & y & z \end{vmatrix} = \dfrac{\partial}{\partial x}(a_2 z - a_3 y) - \dfrac{\partial}{\partial y}(a_1 z - a_3 x) + \dfrac{\partial}{\partial z}(a_1 y - a_2 x) = 0$

24. $\mathbf{r} \cdot \mathbf{a} = a_1 x + a_2 y + a_3 z; \; \mathbf{r} \cdot \mathbf{r} = x^2 + y^2 + z^2; \; \nabla \cdot [(\mathbf{r} \cdot \mathbf{r})\mathbf{a}] = 2xa_1 + 2ya_2 + 2za_3 = 2(\mathbf{r} \cdot \mathbf{a})$

27. $\nabla \cdot (f\mathbf{F}) = \nabla \cdot (fP\mathbf{i} + fQ\mathbf{j} + fR\mathbf{k}) = fP_x + Pf_x + fQ_y + Qf_y + fR_z + Rf_z$

$$= f(P_x + Q_y + R_z) + (Pf_x + Qf_y + Rf_z) = f(\nabla \cdot \mathbf{F}) + \mathbf{F} \cdot (\nabla f)$$

30. Assuming continuous second partial derivatives,

$$\text{div (curl } \mathbf{F}) = \nabla \cdot [(R_y - Q_z)\mathbf{i} - (R_x - P_z)\mathbf{j} + (Q_x - P_y)\mathbf{k}]$$
$$= (R_{yx} - Q_{zx} - (R_{xy} - P_{zy}) + (Q_{xz} - P_{yz}) = 0.$$

33. $\nabla \cdot \nabla f = \nabla \cdot (f_x \mathbf{i} + f_y \mathbf{j} + f_z \mathbf{k}) = f_{xx} + f_{yy} + f_{zz}$

36. (a) For $\mathbf{F} = P\mathbf{i} + Q\mathbf{j} + R\mathbf{k}$,

$$\text{curl (curl } \mathbf{F}) = (Q_{xy} - P_{yy} - P_{zz} + R_{xz})\mathbf{i} + (R_{yz} - Q_{zz} - Q_{xx} + P_{yx})\mathbf{j}$$
$$+ (P_{zx} - R_{xx} - R_{yy} + Q_{zy})\mathbf{k}$$

and

$$-\nabla^2 \mathbf{F} + \text{grad (div } \mathbf{F}) = -(P_{xx} + P_{yy} + P_{zz})\mathbf{i} - (Q_{xx} + Q_{yy} + Q_{zz})\mathbf{j}$$
$$- (R_{xx} + R_{yy} + R_{zz})\mathbf{k} + \text{grad } (P_x + Q_y + R_z)$$
$$= -P_{xx}\mathbf{i} - Q_{yy}\mathbf{j} - R_{zz}\mathbf{k} + (-P_{yy} - P_{zz})\mathbf{i} + (-Q_{xx} - Q_{zz})\mathbf{j}$$
$$+ (-R_{xx} - R_{yy})\mathbf{k} + (P_{xx} + Q_{yx} + R_{zx})\mathbf{i}$$
$$+ (P_{xy} + Q_{yy} + R_{zy})\mathbf{j} + (P_{xz} + Q_{yz} + R_{zz})\mathbf{k}$$
$$= (-P_{yy} - P_{zz} + Q_{yx} + R_{zx})\mathbf{i} + (-Q_{xx} - Q_{zz} + P_{xy} + R_{zy})\mathbf{j}$$
$$+ (-R_{xx} - R_{yy} + P_{xz} + Q_{yz})\mathbf{k}.$$

Thus, curl (curl $\mathbf{F}$) $= -\nabla^2 \mathbf{F} + \text{grad (div } \mathbf{F})$.

(b) For $\mathbf{F} = xy\mathbf{i} + 4yz^2\mathbf{j} + 2xz\mathbf{k}$, $\nabla^2 \mathbf{F} = 0\mathbf{i} + 8y\mathbf{j} + 0\mathbf{k}$, div $\mathbf{F} = y + 4z^2 + 2x$, and grad (div $\mathbf{F}$) $= 2\mathbf{i} + \mathbf{j} + 8z\mathbf{k}$. Then curl (curl $\mathbf{F}$) $= -8y\mathbf{j} + 2\mathbf{i} + \mathbf{j} + 8z\mathbf{k} = 2\mathbf{i} + (1 - 8y)\mathbf{j} + 8z\mathbf{k}$.

39. curl $\mathbf{F} = -Gm_1m_2 \begin{vmatrix} \mathbf{i} & \mathbf{j} & \mathbf{k} \\ \partial/\partial x & \partial/\partial y & \partial/\partial z \\ x/|\mathbf{r}|^3 & y/|\mathbf{r}|^3 & z/|\mathbf{r}|^3 \end{vmatrix}$

$$= -Gm_1m_2[(-3yz/|\mathbf{r}|^5 + 3yz/|\mathbf{r}|^5)\mathbf{i} - (-3xz/|\mathbf{r}|^5 + 3xz/|\mathbf{r}|^5)\mathbf{j}$$
$$+ (-3xy/|\mathbf{r}|^5 + 3xy/|\mathbf{r}|^5)\mathbf{k}]$$
$$= \mathbf{0}$$

$$\text{div } \mathbf{F} = -Gm_1m_2 \left[\frac{-2x^2 + y^2 + z^2}{|\mathbf{r}|^{5/2}} + \frac{x^2 - 2y^2 + z^2}{|\mathbf{r}|^{5/2}} + \frac{x^2 + y^2 - 2z^2}{|\mathbf{r}|^{5/2}} \right] = 0$$

42. Recall that $\mathbf{a} \cdot (\mathbf{a} \times \mathbf{b}) = 0$. Then, using Problems 31, 29, and 28,

$$\nabla \cdot \mathbf{F} = \text{div }(\nabla f \times f\nabla g) = f\nabla g \cdot (\text{curl }\nabla f) - \nabla f \cdot (\text{curl }f\nabla g) = f\nabla g \cdot \mathbf{0} - \nabla f \cdot (\nabla \times f\nabla g)$$
$$= -\nabla f \cdot [f(\nabla \times \nabla g) + (\nabla f \times \nabla g)] = -\nabla f \cdot [f\text{curl }\nabla g + (\nabla f \times \nabla g)]$$
$$= -\nabla f \cdot [f\mathbf{0} + (\nabla f \times \nabla g)] = -\nabla f \cdot (\nabla f \times \nabla g) = 0.$$

45. We note that div $\mathbf{F} = 2xyz - 2xyz + 1 = 1 \neq 0$. If $\mathbf{F} = \text{curl }\mathbf{G}$, then div $(\text{curl }\mathbf{G}) = \text{div }\mathbf{F} = 1$. But, by Problem 30, for any vector field $\mathbf{G}$, div(curl $\mathbf{G}) = 0$. Thus, $\mathbf{F}$ cannot be the curl of $\mathbf{G}$.

9.8 | Line Integrals

3.
$$\int_C (3x^2 + 6y^2)\,dx = \int_{-1}^{0} [3x^2 + 6(2x+1)^2]\,dx$$
$$= \int_{-1}^{0} (27x^2 + 24x + 6)\,dx = (9x^3 + 12x^2 + 6x)\Big|_{-1}^{0} = -(-9 + 12 - 6) = 3$$

$$\int_C (3x^2 + 6y^2)\,dy = \int_{-1}^{0} [3x^2 + 6(2x+1)^2]2\,dx = 6$$

$$\int_C (3x^2 + 6y^2)\,ds = \int_{-1}^{0} [3x^2 + 6(2x+1)^2]\sqrt{1+4}\,dx = 3\sqrt{5}$$

6.
$$\int_C 4xyz\,dx = \int_0^1 4\left(\frac{1}{3}t^3\right)(t^2)(2t)t^2\,dt = \frac{8}{3}\int_0^1 t^8\,dt = \frac{8}{27}t^9\Big|_0^1 = \frac{8}{27}$$

$$\int_C 4xyz\,dy = \int_0^1 4\left(\frac{1}{3}t^3\right)(t^2)(2t)2t\,dt = \frac{16}{3}\int_0^1 t^7\,dt = \frac{2}{3}t^8\Big|_0^1 = \frac{2}{3}$$

$$\int_C 4xyz\,dz = \int_0^1 4\left(\frac{1}{3}t^3\right)(t^2)(2t)2\,dt = \frac{16}{3}\int_0^1 t^6\,dt = \frac{16}{21}t^7\Big|_0^1 = \frac{16}{21}$$

$$\int_C 4xyz\,ds = \int_0^1 4\left(\frac{1}{3}t^3\right)(t^2)(2t)\sqrt{t^4 + 4t^2 + 4}\,dt = \frac{8}{3}\int_0^1 t^6(t^2 + 2)\,dt = \frac{8}{3}\left(\frac{1}{9}t^9 + \frac{2}{7}t^7\right)\Big|_0^1$$
$$= \frac{200}{189}$$

9. From $(-1, 2)$ to $(2, 2)$ we use x as a parameter with $y = 2$ and $dy = 0$. From $(2, 2)$ to $(2, 5)$ we use y as a parameter with $x = 2$ and $dx = 0$.

$$\int_C (2x + y)\,dx + xy\,dy = \int_{-1}^{2} (2x + 2)\,dx + \int_2^5 2y\,dy = (x^2 + 2x)\Big|_{-1}^{2} + y^2\Big|_2^5 = 9 + 21 = 30$$

12. Using x as the parameter, $dy = dx$.

$$\int_C y\,dx + x\,dy = \int_0^1 x\,dx + \int_0^1 x\,dx = \int_0^1 2x\,dx = x^2 \Big|_0^1 = 1$$

15. $\displaystyle\int_C (6x^2 + 2y^2)\,dx + 4xy\,dy = \int_4^9 (6t + 2t^2)\frac{1}{2}\,t^{-1/2}\,dt + \int_4^9 4\sqrt{t}\,t\,dt = \int_4^9 (3t^{1/2} + 5t^{3/2})\,dt$

$$= (2t^{3/2} + 2t^{5/2})\Big|_4^9 = 460$$

18. $\displaystyle\int_C 4x\,dx + 2y\,dy = \int_{-1}^2 4(y^3 + 1)3y^2\,dy + \int_{-1}^2 2y\,dy = \int_{-1}^2 (12y^5 + 12y^2 + 2y)\,dy$

$$= (2y^6 + 4y^3 + y^2)\Big|_{-1}^2 = 165$$

21. From $(1,1)$ to $(-1,1)$ and $(-1,-1)$ to $(1,-1)$ we use x as a parameter with $y = 1$ and $y = -1$, respectively, and $dy = 0$. From $(-1,1)$ to $(-1,-1)$ and $(1,-1)$ to $(1,1)$ we use y as a parameter with $x = -1$ and $x = 1$, respectively, and $dx = 0$.

$$\oint_C x^2y^3\,dx - xy^2\,dy = \int_1^{-1} x^2(1)\,dx + \int_1^{-1} -(-1)y^2\,dy + \int_{-1}^1 x^2(-1)^3\,dx + \int_{-1}^1 -(1)y^2\,dy$$

$$= \frac{1}{3}x^3\Big|_1^{-1} + \frac{1}{3}y^3\Big|_1^{-1} - \frac{1}{3}x^3\Big|_{-1}^1 - \frac{1}{3}y^3\Big|_{-1}^1 = -\frac{8}{3}$$

24. $\displaystyle\oint_C y\,dx - x\,dy = \int_0^\pi 3\sin t(-2\sin t)\,dt - \int_0^\pi 2\cos t(3\cos t)\,dt = -6\int_0^\pi (\sin^2 t + \cos^2 t)\,dt$

$$= -6\int_0^\pi dt = -6\pi$$

Thus, $\displaystyle\int_{-C} y\,dx - x\,dy = 6\pi$.

27. From $(0,0,0)$ to $(6,0,0)$ we use x as a parameter with $y = dy = 0$ and $z = dz = 0$. From $(6,0,0)$ to $(6,0,5)$ we use z as a parameter with $x = 6$ and $dx = 0$ and $y = dy = 0$. From $(6,0,5)$ to $(6,8,5)$ we use y as a parameter with $x = 6$ and $dx = 0$ and $z = 5$ and $dz = 0$.

$$\int_C y\,dx + z\,dy + x\,dz = \int_0^6 0\,dx + \int_0^5 6\,dz + \int_0^8 5\,dy = 70$$

30. $\mathbf{F} = e^t\mathbf{i} + te^{t^3}\mathbf{j} + t^3e^{t^6}\mathbf{k}$; $d\mathbf{r} = (\mathbf{i} + 2t\mathbf{j} + 3t^2\mathbf{k})\,dt$;

$$\int_C \mathbf{F}\cdot d\mathbf{r} = \int_0^1 (e^t + 2t^2e^{t^3} + 3t^5e^{t^6})\,dt = \left(e^t + \frac{2}{3}e^{t^3} + \frac{1}{2}e^{t^6}\right)\Big|_0^1 = \frac{13}{6}(e - 1)$$

33. Let $\mathbf{r}_1 = (1 + 2t)\mathbf{i} + \mathbf{j}$, $\mathbf{r}_2 = 3\mathbf{i} + (1 + t)\mathbf{j}$, and $\mathbf{r}_3 = (3 - 2t)\mathbf{i} + (2 - t)\mathbf{j}$ for $0 \le t \le 1$. Then

$$d\mathbf{r}_1 = 2\mathbf{i}, \qquad d\mathbf{r}_2 = \mathbf{j}, \qquad d\mathbf{r}_3 = -2\mathbf{i} - \mathbf{j},$$
$$\mathbf{F}_1 = (1 + 2t + 2)\mathbf{i} + (6 - 2 - 4t)\mathbf{j} = (3 + 2t)\mathbf{i} + (4 - 4t)\mathbf{j},$$
$$\mathbf{F}_2 = (3 + 2 + 2t)\mathbf{i} + (6 + 6t - 6)\mathbf{j} = (5 + 2t)\mathbf{i} + 6t\mathbf{j},$$
$$\mathbf{F}_3 = (3 - 2t + 4 - 2t)\mathbf{i} + (12 - 6t - 6 + 4t)\mathbf{j} = (7 - 4t)\mathbf{i} + (6 - 2t)\mathbf{j},$$

and

$$W = \int_{C_1} \mathbf{F}_1 \cdot d\mathbf{r}_1 + \int_{C_2} \mathbf{F}_2 \cdot d\mathbf{r}_2 + \int_{C_3} \mathbf{F}_3 \cdot d\mathbf{r}_3$$

$$= \int_0^1 (6 + 4t)\, dt + \int_0^1 6t\, dt + \int_0^1 (-14 + 8t - 6 + 2t)\, dt$$

$$= \int_0^1 (-14 + 20t)\, dt = (-14t + 10t^2)\Big|_0^1 = -4.$$

36. Let $\mathbf{r} = t\mathbf{i} + t\mathbf{j} + t\mathbf{k}$ for $1 \le t \le 3$. Then $d\mathbf{r} = \mathbf{i} + \mathbf{j} + \mathbf{k}$, and

$$\mathbf{F} = \frac{c}{|\mathbf{r}|^3}(t\mathbf{i} + t\mathbf{j} + t\mathbf{k}) = \frac{ct}{(\sqrt{3t^2})^3}(\mathbf{i} + \mathbf{j} + \mathbf{k}) = \frac{c}{3\sqrt{3}\, t^2}(\mathbf{i} + \mathbf{j} + \mathbf{k}),$$

$$W = \int_C \mathbf{F} \cdot d\mathbf{r} = \int_1^3 \frac{c}{3\sqrt{3}\, t^2}(1 + 1 + 1)\, dt = \frac{c}{\sqrt{3}} \int_1^3 \frac{1}{t^2}\, dt = \frac{c}{\sqrt{3}} \left(-\frac{1}{t}\right)\Big|_1^3 = \frac{c}{\sqrt{3}} \left(-\frac{1}{3} + 1\right)$$

$$= \frac{2c}{3\sqrt{3}}.$$

39. Since $\mathbf{v} \cdot \mathbf{v} = v^2$, $\dfrac{d}{dt} v^2 = \dfrac{d}{dt}(\mathbf{v} \cdot \mathbf{v}) = \mathbf{v} \cdot \dfrac{d\mathbf{v}}{dt} + \dfrac{d\mathbf{v}}{dt} \cdot \mathbf{v} = 2\dfrac{d\mathbf{v}}{dt} \cdot \mathbf{v}$. Then

$$W = \int_C \mathbf{F} \cdot d\mathbf{r} = \int_a^b m\mathbf{a} \cdot \left(\frac{d\mathbf{r}}{dt}\, dt\right) = m \int_a^b \frac{d\mathbf{v}}{dt} \cdot \mathbf{v}\, dt = m \int_a^b \frac{1}{2}\left(\frac{d}{dt} v^2\right) dt$$

$$= \frac{1}{2} m(v^2)\Big|_a^b = \frac{1}{2} m[v(b)]^2 - \frac{1}{2} m[v(a)]^2.$$

42. On C_1, $\mathbf{T} = \mathbf{i}$ and $\mathbf{F} \cdot \mathbf{T} = \text{comp}_{\mathbf{T}}\mathbf{F} \approx 1$. On C_2, $\mathbf{T} = -\mathbf{j}$ and $\mathbf{F} \cdot \mathbf{T} = \text{comp}_{\mathbf{T}}\mathbf{F} \approx 2$. On C_3, $\mathbf{T} = -\mathbf{i}$ and $\mathbf{F} \cdot \mathbf{T} = \text{comp}_{\mathbf{T}}\mathbf{F} \approx 1.5$. Using the fact that the lengths of C_1, C_2, and C_3 are 4, 5, and 5, respectively, we have

$$W = \int_C \mathbf{F} \cdot \mathbf{T}\, ds = \int_{C_1} \mathbf{F} \cdot \mathbf{T}\, ds + \int_{C_2} \mathbf{F} \cdot \mathbf{T}\, ds + \int_{C_3} \mathbf{F} \cdot \mathbf{T}\, ds \approx 1(4) + 2(5) + 1.5(5) = 21.5 \text{ ft-lb.}$$

9.9 | Independence of Path

3. (a) $P_y = 2 = Q_x$ and the integral is independent of path. $\phi_x = x + 2y$,

$$\phi = \frac{1}{2}x^2 + 2xy + g(y), \ \phi_y = 2x + g'(y) = 2x - y, \ g(y) = -\frac{1}{2}y^2, \ \phi = \frac{1}{2}x^2 + 2xy - \frac{1}{2}y^2,$$

$$\int_{(1,0)}^{(3,2)} (x + 2y)\, dx + (2x - y)\, dy = \left(\frac{1}{2}x^2 + 2xy - \frac{1}{2}y^2\right)\Bigg|_{(1,0)}^{(3,2)} = 14$$

(b) Use $y = x - 1$ for $1 \le x \le 3$.

$$\int_{(1,0)}^{(3,2)} (x + 2y)\, dx + (2x - y)\, dy = \int_1^3 [x + 2(x - 1) + 2x - (x - 1)\, dx$$

$$= \int_1^3 (4x - 1)\, dx = (2x^2 - x)\Bigg|_1^3 = 14$$

6. (a) $P_y = -xy(x^2 + y^2)^{-3/2} = Q_x$ and the integral is independent of path. $\phi_x = \dfrac{x}{\sqrt{x^2 + y^2}}$,

$$\phi = \sqrt{x^2 + y^2} + g(y), \ \phi_y = \frac{y}{\sqrt{x^2 + y^2}} + g'(y) = \frac{y}{\sqrt{x^2 + y^2}}, \ g(y) = 0, \ \phi = \sqrt{x^2 + y^2},$$

$$\int_{(1,0)}^{(3,4)} \frac{x\, dx + y\, dy}{\sqrt{x^2 + y^2}} = \sqrt{x^2 + y^2}\,\Bigg|_{(1,0)}^{(3,4)} = 4$$

(b) Use $y = 2x - 2$ for $1 \le x \le 3$.

$$\int_{(1,0)}^{(3,4)} \frac{x\, dx + y\, dy}{\sqrt{x^2 + y^2}} = \int_1^3 \frac{x + (2x - 2)2}{\sqrt{x^2 + (2x - 2)^2}}\, dx = \int_1^3 \frac{5x - 4}{\sqrt{5x^2 - 8x + 4}}$$

$$= \sqrt{5x^2 - 8x + 4}\,\Bigg|_1^3 = 4$$

9. (a) $P_y = 3y^2 + 3x^2 = Q_x$ and the integral is independent of path. $\phi_x = y^3 + 3x^2 y$,

$\phi = xy^3 + x^3 y + g(y), \ \phi_y = 3xy^2 + x^3 + g'(y) = x^3 + 3y^2 x + 1, \ g(y) = y, \ \phi = xy^3 + x^3 y + y,$

$$\int_{(0,0)}^{(2,8)} (y^3 + 3x^2 y)\, dx + (x^3 + 3y^2 x + 1)\, dy = (xy^3 + x^3 y + y)\Bigg|_{(0,0)}^{(2,8)} = 1096$$

(b) Use $y = 4x$ for $0 \le x \le 2$.

$$\int_{(0,0)}^{(2,8)} (y^3 + 3x^2 y)\, dx + (x^3 + 3y^2 x + 1)\, dy = \int_0^2 [(64x^3 + 12x^3) + (x^3 + 48x^3 + 1)(4)]\, dx$$

$$= \int_0^2 (272x^3 + 4)\, dx = (68x^4 + 4x)\Bigg|_0^2 = 1096$$

12. $P_y = 6xy^2 = Q_x$ and the vector field is a gradient field. $\phi_x = 2xy^3$, $\phi = x^2y^3 + g(y)$,
$\phi_y = 3x^2y^2 + g'(y) = 3x^2y^2 + 3y^2$, $g(y) = y^3$, $\phi = x^2y^3 + y^3$

15. $P_y = 1 = Q_x$ and the vector field is a gradient field. $\phi_x = x^3 + y$, $\phi = \dfrac{1}{4}x^4 + xy + g(y)$,
$\phi_y = x + g'(y) = x + y^3$, $g(y) - \dfrac{1}{4}y^4$, $\phi = \dfrac{1}{4}x^4 + xy + \dfrac{1}{4}y^4$

18. Since $P_y = -e^{-y} = Q_x$, $\mathbf{F}$ is conservative and $\displaystyle\int_C \mathbf{F} \cdot d\mathbf{r}$ is independent of the path. Thus, instead of the given curve we may use the simpler curve C_1: $y = 0$, $-2 \le -x \le 2$. Then $dy = 0$ and

$$W = \int_{C_1} (2x + e^{-y})\, dx + (4y - xe^{-y})\, dy = \int_2^{-2} (2x + 1)\, dx$$

$$= (x^2 + x)\Big|_2^{-2} = (4 - 2) - (4 + 2) = -4.$$

21. $P_y = 2x\cos y = Q_x$, $Q_z = 0 = R_y$, $R_x = 3e^{3z} = P_z$, and the integral is independent of path. Integrating $\phi_x = 2x\sin y + e^{3z}$ we find $\phi = x^2\sin y + xe^{3z} + g(y,z)$. Then $\phi_y = x^2\cos y + g_y = Q = x^2\cos y$, so $g_y = 0$, $g(y,z) = h(z)$, and $\phi = x^2\sin y + xe^{3z} + h(z)$. Now $\phi_z = 3xe^{3z} + h'(z) = R = 3xe^{3z} + 5$, so $h'(z) = 5$ and $h(z) = 5z$. Thus $\phi = x^2\sin y + xe^{3z} + 5z$ and

$$\int_{(1,0,0)}^{(2,\pi/2,1)} (2x\sin y + e^{3z})\, dx + x^2\cos y\, dy + (3xe^{3z} + 5)\, dz$$

$$= (x^2\sin y + xe^{3z} + 5z)\Big|_{(1,0,0)}^{(2,\pi/2,1)} = [4(1) + 2e^3 + 5] - [0 + 1 + 0] = 8 + 2e^3.$$

24. $P_y = 0 = Q_x$, $Q_z = 2y = R_y$, $R_x = 2x = P_z$ and the integral is independent of path. Parameterize the line segment between the points by $x = -2(1 - t)$, $y = 3(1 - t)$, $z = 1 - t$, for $0 \le t \le 1$. Then $dx = 2\, dt$, $dy = -3\, dt$, $dz = -dt$, and

$$\int_{(-2,3,1)}^{(0,0,0)} 2xz\, dx + 2yz\, dy + (x^2 + y^2)\, dz$$

$$= \int_0^1 [-4(1 - t)^2(2) + 6(1 - t)^2(-3) + 4(1 - t)^2(-1) + 9(1 - t)^2(-1)]\, dt$$

$$= \int_0^1 -39(1 - t)^2\, dt = 13(1 - t)^3\Big|_0^1 = -13.$$

27. Since $P_y = Gm_1m_2(2xy/|\mathbf{r}|^5) = Q_x$, $Q_z = Gm_1m_2(2yz/|\mathbf{r}|^5) = R_y$, and $R_x = Gm_1m_2(2xz/|\mathbf{r}|^5) = P_z$, the force field is conservative.

$$\phi_x = -Gm_1m_2\frac{x}{(x^2 + y^2 + z^2)^{3/2}}, \quad \phi = Gm_1m_2(x^2 + y^2 + z^2)^{-1/2} + g(y,z),$$

$$\phi_y = -Gm_1m_2\frac{y}{(x^2+y^2+z^2)^{3/2}} + g_y(y,z) = -Gm_1m_2\frac{y}{(x^2+y^2+z^2)^{3/2}}, \quad g(y,z) = h(z),$$

$$\phi = Gm_1m_2(x^2+y^2+z^2)^{-1/2} + h(z),$$

$$\phi_z = -Gm_1m_2\frac{z}{(x^2+y^2+z^2)^{3/2}} + h'(z) = -Gm_1m_2\frac{z}{(x^2+y^2+z^2)^{3/2}},$$

$$h(z) = 0, \quad \phi = \frac{Gm_1m_2}{\sqrt{x^2+y^2+z^2}} = \frac{Gm_1m_2}{|\mathbf{r}|}$$

30. From $\mathbf{F} = (x^2+y^2)^{n/2}(x\mathbf{i}+y\mathbf{j})$ we obtain $P_y = nxy(x^2+y^2)^{n/2-1} = Q_x$, so that $\mathbf{F}$ is conservative. From $\phi_x = x(x^2+y^2)^{n/2}$ we obtain the potential function $\phi = (x^2+y^2)^{(n+2)/2}/(n+2)$. Then

$$W = \int_{(x_1,y_1)}^{(x_2,y_2)} \mathbf{F}\cdot d\mathbf{r} = \left(\frac{(x^2+y^2)^{(n+2)/2}}{n+2}\right)\Bigg|_{(x_1,y_1)}^{(x_2,y_2)}$$

$$= \frac{1}{n+2}\left[(x_2^2+y_2^2)^{(n+2)/2} - (x_1^2+y_1^2)^{(n+2)/2}\right].$$

9.10 Double Integrals

3. $\displaystyle\int_1^{3x} x^3 e^{xy}\,dy = x^2 e^{xy}\Big|_1^{3x} = x^2(e^{3x^2} - e^x)$

6. $\displaystyle\int_{x^3}^{x} e^{2y/x}\,dy = \frac{x}{2}e^{2y/x}\Big|_{x^3}^{x} = \frac{x}{2}(e^{2x/x} - e^{2x^3/x}) = \frac{x}{2}(e^2 - e^{2x^2})$

9.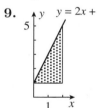

12.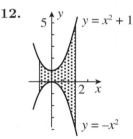

15. $\displaystyle\iint_R (2x+4y+1)\,dA = \int_0^1\int_{x^3}^{x^2}(2x+4y+1)\,dy\,dx$

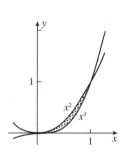

$$= \int_0^1 (2xy+2y^2+y)\Big|_{x^3}^{x^2}\,dx$$

$$= \int_0^1 [(2x^3+2x^4+x^2) - (2x^4+2x^6+x^3)]\,dx$$

$$= \int_0^1 (x^3+x^2-2x^6)\,dx = \left(\frac{1}{4}x^4+\frac{1}{3}x^3-\frac{2}{7}x^7\right)\Big|_0^1$$

$$= \frac{1}{4}+\frac{1}{3}-\frac{2}{7} = \frac{25}{84}$$

18. $\displaystyle\iint_R \frac{x}{\sqrt{y}}\, dA = \int_{-1}^{1}\int_{x^2+1}^{3-x^2} xy^{-1/2}\, dy\, dx = \int_{-1}^{1} 2x\sqrt{y}\,\Big|_{x^2+1}^{3-x^2}\, dx$

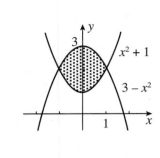

$$= 2\int_{-1}^{1}\left(x\sqrt{3-x^2} - x\sqrt{x^2+1}\,\right) dx$$

$$= 2\left[-\frac{1}{3}(3-x^2)^{3/2} - \frac{1}{3}(x^2+1)^{3/2}\right]\Bigg|_{-1}^{1}$$

$$= -\frac{2}{3}[(2^{3/2} + 2^{3/2}) - (2^{3/2} + 2^{3/2})] = 0$$

21. $\displaystyle\iint_R \sqrt{x^2+1}\, dA = \int_{0}^{\sqrt{3}}\int_{-x}^{x} \sqrt{x^2+1}\, dy\, dx = \int_{0}^{\sqrt{3}} y\sqrt{x^2+1}\,\Big|_{-x}^{x}\, dx$

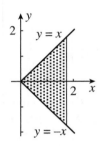

$$= \int_{0}^{\sqrt{3}}\left(x\sqrt{x^2+1} + x\sqrt{x^2+1}\,\right) dx = \int_{0}^{\sqrt{3}} 2x\sqrt{x^2+1}\, dx$$

$$= \frac{2}{3}(x^2+1)^{3/2}\,\Bigg|_{0}^{\sqrt{3}} = \frac{2}{3}(4^{3/2} - 1^{3/2}) = \frac{14}{3}$$

24. The correct integral is **(b)**.

$$V = 8\int_{0}^{2}\int_{0}^{\sqrt{4-y^2}} (4-y^2)^{1/2}\, dx\, dy = 8\int_{0}^{2}(4-y^2)^{1/2}x\,\Big|_{0}^{\sqrt{4-y^2}}\, dy = 8\int_{0}^{2}(4-y^2)\, dy$$

$$= 8\left(4y - \frac{1}{3}y^3\right)\Bigg|_{0}^{2} = \frac{128}{3}$$

27. Solving for z, we have $x = 2 - \frac{1}{2}x + \frac{1}{2}y$. Setting $z = 0$, we see that this surface (plane) intersects the xy-plane in the line $y = x - 4$. Since $z(0,0) = 2 > 0$, the surface lies above the xy-plane over the quarter-circular region.

$$V = \int_{0}^{2}\int_{0}^{\sqrt{4-x^2}}\left(2 - \frac{1}{2}x + \frac{1}{2}y\right) dy\, dx = \int_{0}^{2}\left(2y - \frac{1}{2}xy + \frac{1}{4}y^2\right)\Bigg|_{0}^{\sqrt{4-x^2}}\, dx$$

$$= \int_{0}^{2}\left(2\sqrt{4-x^2} - \frac{1}{2}x\sqrt{4-x^2} + 1 - \frac{1}{4}x^2\right) dx$$

$$= \left[x\sqrt{4-x^2} + 4\sin^{-1}\frac{x}{2} + \frac{1}{6}(4-x^2)^{3/2} + x - \frac{1}{12}x^3\right]\Bigg|_{0}^{2} = \left(2\pi + 2 - \frac{2}{3}\right) - \frac{4}{3} = 2\pi$$

30. In the first octant, $z = x + y$ is nonnegative. Then

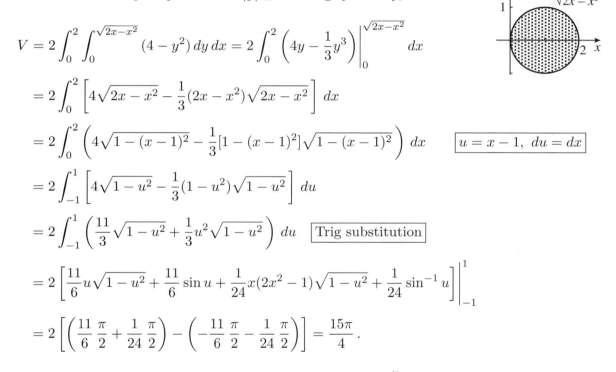

$$V = \int_0^3 \int_0^{\sqrt{9-x^2}} (x+y)\,dy\,dx = \int_0^3 \left(xy + \frac{1}{2}y^2 \right) \Big|_0^{\sqrt{9-x^2}} dx$$

$$= \int_0^3 \left(x\sqrt{9-x^2} + \frac{9}{2} - \frac{1}{2}x^2 \right) dx = \left[-\frac{1}{3}(9-x^2)^{3/2} + \frac{9}{2}x - \frac{1}{6}x^3 \right]\Big|_0^3 = \left(\frac{27}{2} - \frac{9}{2} \right) - (-9)$$

$$= 18.$$

33. Note that $z = 4 - y^2$ is positive for $|y| \le 1$. Using symmetry,

$$V = 2\int_0^2 \int_0^{\sqrt{2x-x^2}} (4-y^2)\,dy\,dx = 2\int_0^2 \left(4y - \frac{1}{3}y^3 \right)\Big|_0^{\sqrt{2x-x^2}} dx$$

$$= 2\int_0^2 \left[4\sqrt{2x-x^2} - \frac{1}{3}(2x-x^2)\sqrt{2x-x^2} \right] dx$$

$$= 2\int_0^2 \left(4\sqrt{1-(x-1)^2} - \frac{1}{3}[1-(x-1)^2]\sqrt{1-(x-1)^2} \right) dx \qquad \boxed{u = x-1,\ du = dx}$$

$$= 2\int_{-1}^1 \left[4\sqrt{1-u^2} - \frac{1}{3}(1-u^2)\sqrt{1-u^2} \right] du$$

$$= 2\int_{-1}^1 \left(\frac{11}{3}\sqrt{1-u^2} + \frac{1}{3}u^2\sqrt{1-u^2} \right) du \qquad \boxed{\text{Trig substitution}}$$

$$= 2\left[\frac{11}{6}u\sqrt{1-u^2} + \frac{11}{6}\sin u + \frac{1}{24}x(2x^2-1)\sqrt{1-u^2} + \frac{1}{24}\sin^{-1}u \right]\Big|_{-1}^1$$

$$= 2\left[\left(\frac{11}{6}\frac{\pi}{2} + \frac{1}{24}\frac{\pi}{2} \right) - \left(-\frac{11}{6}\frac{\pi}{2} - \frac{1}{24}\frac{\pi}{2} \right) \right] = \frac{15\pi}{4}.$$

36. $\displaystyle \int_0^1 \int_{2y}^2 e^{-y/x}\,dx\,dy = \int_0^2 \int_0^{x/2} e^{-y/x}\,dy\,dx = \int_0^2 -xe^{-y/x}\Big|_0^{x/2} dx$

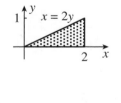

$$= \int_0^2 (-xe^{-1/2} + x)\,dx = \int_0^2 (1 - e^{-1/2})x\,dx$$

$$= \frac{1}{2}(1-e^{-1/2})x^2\Big|_0^2 = 2(1-e^{-1/2})$$

39. $\displaystyle \int_0^1 \int_x^1 \frac{1}{1+y^4}\,dy\,dx = \int_0^1 \int_0^y \frac{1}{1+y^4}\,dx\,dy = \int_0^1 \frac{x}{1+y^4}\Big|_0^y dy$

$$= \int_0^1 \frac{y}{1+y^4}\,dy = \frac{1}{2}\tan^{-1}y^2\Big|_0^1 = \frac{\pi}{8}$$

42. $m = \int_0^2 \int_0^{4-2x} x^2 \, dy \, dx = \int_0^2 x^2 y \Big|_0^{4-2x} dx = \int_0^2 x^2(4 - 2x) \, dx$

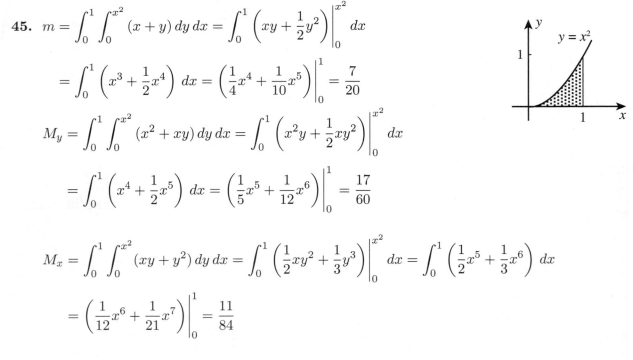

$= \int_0^2 (4x^2 - 2x^3) \, dx = \left(\frac{4}{3}x^3 - \frac{1}{2}x^4\right)\Big|_0^2 = \frac{32}{3} - 8 = \frac{8}{3}$

$M_y = \int_0^2 \int_0^{4-2x} x^3 \, dy \, dx = \int_0^2 x^3 y \Big|_0^{4-2x} dx = \int_0^2 x^3(4 - 2x) \, dx$

$= \int_0^2 (4x^3 - 2x^4) \, dx = \left(x^4 - \frac{2}{5}x^5\right)\Big|_0^2 = 16 - \frac{64}{5} = \frac{16}{5}$

$M_x = \int_0^2 \int_0^{4-2x} x^2 y \, dy \, dx = \int_0^2 \frac{1}{2}x^2 y^2 \Big|_0^{4-2x} dx = \frac{1}{2}\int_0^2 x^2(4 - 2x)^2 \, dx$

$= \frac{1}{2}\int_0^2 (16x^2 - 16x^3 + 4x^4) \, dx = 2\int_0^2 (4x^2 - 4x^3 + x^4) \, dx = 2\left(\frac{4}{3}x^3 - x^4 + \frac{1}{5}x^5\right)\Big|_0^2$

$= 2\left(\frac{32}{3} - 16 + \frac{32}{5}\right) = \frac{32}{15}$

$\bar{x} = M_y/m = \dfrac{16/5}{8/3} = 6/5; \quad \bar{y} = M_x/m = \dfrac{32/15}{8/3} = 4/5.$ The center of mass is $(6/5, 4/5)$.

45. $m = \int_0^1 \int_0^{x^2} (x + y) \, dy \, dx = \int_0^1 \left(xy + \frac{1}{2}y^2\right)\Big|_0^{x^2} dx$

$= \int_0^1 \left(x^3 + \frac{1}{2}x^4\right) dx = \left(\frac{1}{4}x^4 + \frac{1}{10}x^5\right)\Big|_0^1 = \frac{7}{20}$

$M_y = \int_0^1 \int_0^{x^2} (x^2 + xy) \, dy \, dx = \int_0^1 \left(x^2 y + \frac{1}{2}xy^2\right)\Big|_0^{x^2} dx$

$= \int_0^1 \left(x^4 + \frac{1}{2}x^5\right) dx = \left(\frac{1}{5}x^5 + \frac{1}{12}x^6\right)\Big|_0^1 = \frac{17}{60}$

$M_x = \int_0^1 \int_0^{x^2} (xy + y^2) \, dy \, dx = \int_0^1 \left(\frac{1}{2}xy^2 + \frac{1}{3}y^3\right)\Big|_0^{x^2} dx = \int_0^1 \left(\frac{1}{2}x^5 + \frac{1}{3}x^6\right) dx$

$= \left(\frac{1}{12}x^6 + \frac{1}{21}x^7\right)\Big|_0^1 = \frac{11}{84}$

$\bar{x} = M_y/m = \dfrac{17/60}{7/20} = 17/21; \quad \bar{y} = M_x/m = \dfrac{11/84}{7/20} = 55/147.$ The center of mass is $(17/21, 55/147)$.

48. The density is $\rho = kx$.

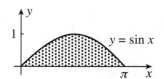

$$m = 2 \int_0^1 \int_0^{1-x^2} ky \, dy \, dx = 2k \int_0^1 \frac{1}{2} y^2 \Big|_0^{1-x^2} dx$$

$$M_y = \int_0^\pi \int_0^{\sin x} kx^2 \, dy \, dx = \int_0^\pi kx^2 y \Big|_0^{\sin x} dx = \int_0^\pi kx^2 \sin x \, dx \qquad \boxed{\text{Integration by parts}}$$

$$= k(-x^2 \cos x + 2 \cos x + 2x \sin x) \Big|_0^\pi = k[(\pi^2 - 2) - 2] = k(\pi^2 - 4)$$

$$M_x = \int_0^\pi \int_0^{\sin x} kxy \, dy \, dx = \int_0^\pi \frac{1}{2} kxy^2 \Big|_0^{\sin x} dx = \int_0^\pi \frac{1}{2} kx \sin^2 x \, dx = \int_0^\pi \frac{1}{4} kx(1 - \cos 2x) \, dx$$

$$= \frac{1}{4} k \left[\int_0^\pi x \, dx - \int_0^\pi x \cos 2x \, dx \right] \qquad \boxed{\text{Integration by parts}}$$

$$= \frac{1}{4} k \left[\frac{1}{2} x^2 \Big|_0^\pi - \frac{1}{4}(\cos 2x + 2x \sin 2x) \Big|_0^\pi \right] = \frac{1}{4} k \left(\frac{1}{2} \pi^2 \right) = \frac{1}{8} k\pi^2$$

$\bar{x} = M_y/m = \dfrac{k(\pi^2 - 4)}{k\pi} = \pi - 4/\pi; \quad \bar{y} = M_x/m = \dfrac{k\pi^2/8}{k\pi} = \pi/8.$ The center of mass is $(\pi - 4/\pi, \pi/8)$.

51. Since both the region and ρ are symmetric with respect to the y-axis, $\bar{x} = 0$. Using symmetry,

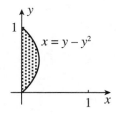

$$I_x = \int_0^1 \int_0^{y-y^2} 2xy^2 \, dx \, dy = \int_0^1 x^2 y^2 \Big|_0^{y-y^2} dy = \int_0^1 (y - y^2)^2 y^2 \, dy$$

$$= \int_0^1 (y^4 - 2y^5 + y^6) \, dy = \left(\frac{1}{5} y^5 - \frac{1}{3} y^6 + \frac{1}{7} y^7 \right) \Big|_0^1 = \frac{1}{105}$$

54.

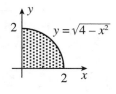

$$I_x = \int_0^2 \int_0^{\sqrt{4-x^2}} y^3 \, dy \, dx = \int_0^2 \frac{1}{4} y^4 \Big|_0^{\sqrt{4-x^2}} dx = \frac{1}{4} \int_0^2 (4 - x^2)^2 \, dx$$

$$= \frac{1}{4} \int_0^2 (16 - 8x^2 + x^4) \, dx = \frac{1}{4} \left(16x - \frac{8}{3} x^3 + \frac{1}{5} x^5 \right) \Big|_0^2$$

$$= \frac{1}{4} \left(32 - \frac{64}{3} + \frac{32}{5} \right) = 8 \left(1 - \frac{2}{3} + \frac{1}{5} \right) = \frac{64}{15}$$

57.

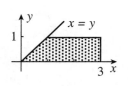

$$I_y = \int_0^1 \int_y^3 (4x^3 + 3x^2 y) \, dx \, dy = \int_0^1 (x^4 + x^3 y) \Big|_y^3 dy$$

$$= \int_0^1 (81 + 27y - 2y^4) \, dy = \left(81y + \frac{27}{2} y^2 - \frac{2}{5} y^5 \right) \Big|_0^1 = \frac{941}{10}$$

60. $m = \int_0^a \int_0^{a-x} k\, dy\, dx = \int_0^a ky \Big|_0^{a-x} dx = k \int_0^a (a-x)\, dx = k\left(ax - \frac{1}{2}x^2\right)\Big|_0^a$

$= \frac{1}{2}ka^2$

$I_x = \int_0^a \int_0^{a-x} ky^2\, dy\, dx = \int_0^a \frac{1}{3}ky^3 \Big|_0^{a-x} dx = \frac{1}{3}k \int_0^a (a-x)^3\, dx$

$= \frac{1}{3}k \int_0^a (a^3 - 3a^2x + 3ax^2 - x^3)\, dx = \frac{1}{3}k\left(a^3x - \frac{3}{2}a^2x^2 + ax^3 - \frac{1}{4}x^4\right)\Big|_0^a = \frac{1}{12}ka^4$

$R_g = \sqrt{\frac{I_x}{m}} = \sqrt{\frac{ka^4/12}{ka^2/2}} = \sqrt{\frac{1}{6}}\, a$

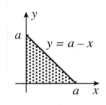

63. From the solution to Problem 60, $m = \frac{1}{2}ka^2$ and $I_x = \frac{1}{12}ka^4$.

$I_y = \int_0^a \int_0^{a-x} kx^2\, dy\, dx = \int_0^a kx^2 y \Big|_0^{a-x} dx = k \int_0^a x^2(a-x)\, dx$

$= k\left(\frac{1}{3}ax^3 - \frac{1}{4}x^4\right)\Big|_0^a = \frac{1}{12}$

$I_0 = I_x + I_y = \frac{1}{12}ka^4 + \frac{1}{12}ka^4 = \frac{1}{6}ka^4$

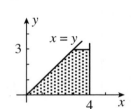

66. $I_0 = \int_0^3 \int_y^4 k(x^2 + y^2)\, dx\, dy = k \int_0^3 \left(\frac{1}{3}x^3 + xy^2\right)\Big|_y^4 dy$

$= k \int_0^3 \left(\frac{64}{3} + 4y^2 - \frac{1}{3}y^3 - y^3\right) dy = k\left(\frac{64}{3}y + \frac{4}{3}y^3 - \frac{1}{3}y^4\right)\Big|_0^3 = 73k$

9.11 Double Integrals in Polar Coordinates

3. Solving $r = 2\sin\theta$ and $r = 1$, we obtain $\sin\theta = 1/2$ or $\theta = \pi/6$. Using symmetry,

$A = 2\int_0^{\pi/6} \int_0^{2\sin\theta} r\, dr\, d\theta + 2\int_{\pi/6}^{\pi/2} \int_0^1 r\, dr\, d\theta$

$= 2\int_0^{\pi/6} \frac{1}{2}r^2 \Big|_0^{2\sin\theta} d\theta + 2\int_{\pi/6}^{\pi/2} \frac{1}{2}r^2 \Big|_0^1 d\theta$

$= \int_0^{\pi/6} 4\sin^2\theta\, d\theta + \int_{\pi/6}^{\pi/2} d\theta = (2\theta - \sin 2\theta)\Big|_0^{\pi/6} + \left(\frac{\pi}{2} - \frac{\pi}{6}\right) = \frac{\pi}{3} - \frac{\sqrt{3}}{2} + \frac{\pi}{3} = \frac{4\pi - 3\sqrt{3}}{6}$

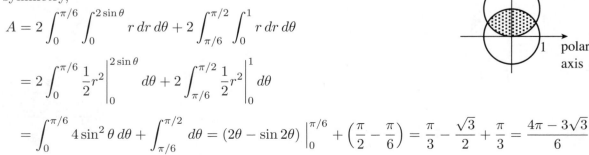

6. $V = \displaystyle\int_0^{2\pi}\int_0^2 \sqrt{9-r^2}\,r\,dr\,d\theta = \int_0^{2\pi} -\frac{1}{3}(9-r^2)^{3/2}\Big|_0^2 \, d\theta$

$= -\dfrac{1}{3}\displaystyle\int_0^{2\pi}(5^{3/2}-27)\,d\theta = \dfrac{1}{3}(27-5^{3/2})2\pi = \dfrac{2\pi(27-5\sqrt{5})}{3}$

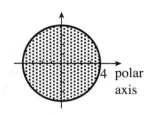

9. $V = \displaystyle\int_0^{\pi/2}\int_0^{1+\cos\theta}(r\sin\theta)r\,dr\,d\theta = \int_0^{\pi/2}\frac{1}{3}r^3\sin\theta\Big|_0^{1+\cos\theta} d\theta$

$= \dfrac{1}{3}\displaystyle\int_0^{\pi/2}(1+\cos\theta)^3\sin\theta\,d\theta = \dfrac{1}{3}\left[-\frac{1}{4}(1+\cos\theta)^4\right]\Big|_0^{\pi/2}$

$= -\dfrac{1}{12}(1-2^4) = \dfrac{5}{4}$

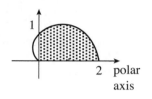

12. The interior of the upper-half circle is traced from $\theta = 0$ to $\pi/2$. The density is kr. Since both the region and the density are symmetric about the polar axis, $\bar{y} = 0$.

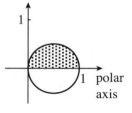

$m = \displaystyle\int_0^{\pi/2}\int_0^{\cos\theta} kr^2\,dr\,d\theta = k\int_0^{\pi/2}\frac{1}{3}r^3\Big|_0^{\cos\theta} d\theta = \frac{k}{3}\int_0^{\pi/2}\cos^3\theta\,d\theta$

$= \dfrac{k}{3}\left(\dfrac{2}{3}+\dfrac{1}{3}\cos^2\theta\right)\sin\theta\Big|_0^{\pi/2} = \dfrac{2k}{9}$

$M_y = k\displaystyle\int_0^{\pi/2}\int_0^{\cos\theta}(r\cos\theta)(r)(r\,dr\,d\theta) = k\int_0^{\pi/2}\int_0^{\cos\theta} r^3\cos\theta\,dr\,d\theta$

$= k\displaystyle\int_0^{\pi/2}\frac{1}{4}r^4\cos\theta\Big|_0^{\cos\theta} d\theta = \frac{k}{4}\int_0^{\pi/2}\cos^5\theta\,d\theta = \frac{k}{4}\left(\sin\theta - \frac{2}{3}\sin^3\theta + \frac{1}{5}\sin^5\theta\right)\Big|_0^{\pi/2} = \frac{2k}{15}$

Thus, $\bar{x} = \dfrac{2k/15}{2k/9} = 3/5$ and the center of mass is $(3/5, 0)$.

15. The density is $\rho = k/r$.

$m = \displaystyle\int_0^{\pi/2}\int_2^{2+2\cos\theta}\frac{k}{r}\,r\,dr\,d\theta = k\int_0^{\pi/2}\int_2^{2+2\cos\theta} dr\,d\theta$

$= k\displaystyle\int_0^{\pi/2} 2\cos\theta\,d\theta = 2k(\sin\theta)\Big|_0^{\pi/2} = 2k$

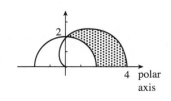

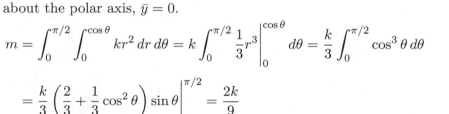

$$M_y = \int_0^{\pi/2} \int_2^{2+2\cos\theta} x\frac{k}{r} r\, dr\, d\theta = k \int_0^{\pi/2} \int_2^{2+2\cos\theta} r\cos\theta\, dr\, d\theta$$

$$= k \int_0^{\pi/2} \frac{1}{2} r^2 \Big|_2^{2+2\cos\theta} \cos\theta\, d\theta = \frac{1}{2} k \int_0^{\pi/2} (8\cos\theta + 4\cos^2\theta)\cos\theta\, d\theta$$

$$= 2k \int_0^{\pi/2} (2\cos^2\theta + \cos\theta - \sin^2\theta\cos\theta)\, d\theta = 2k \left(\theta + \frac{1}{2}\sin 2\theta + \sin\theta - \frac{1}{3}\sin^3\theta\right)\Big|_0^{\pi/2}$$

$$= 2k \left(\frac{\pi}{2} + \frac{2}{3}\right) = \frac{3\pi + 4}{3} k$$

$$M_x = \int_0^{\pi/2} \int_2^{2+2\cos\theta} y\frac{k}{r} r\, dr\, d\theta = k \int_0^{\pi/2} \int_2^{2+2\cos\theta} r\sin\theta\, dr\, d\theta$$

$$= k \int_0^{\pi/2} \frac{1}{2} r^2 \Big|_2^{2+2\cos\theta} \sin\theta\, d\theta = \frac{1}{2} k \int_0^{\pi/2} (8\cos\theta + 4\cos^2\theta)\sin\theta\, d\theta$$

$$= \frac{1}{2} k \left(-4\cos^2\theta - \frac{4}{3}\cos^3\theta\right)\Big|_0^{\pi/2} = \frac{1}{2} k \left[-\left(-4 - \frac{4}{3}\right)\right] = \frac{8}{3} k$$

$$\bar{x} = M_y/m = \frac{(3\pi+4)k/3}{2k} = \frac{3\pi+4}{6}; \quad \bar{y} = M_x/m = \frac{8k/3}{2k} = \frac{4}{3}. \text{ The center of mass is}$$
$$((3\pi+4)/6, 4/3).$$

18. $I_x = \int_0^{2\pi} \int_0^a y^2 \frac{1}{1+r^4} r\, dr\, d\theta = \int_0^{2\pi} \int_0^a \frac{r^3}{1+r^4} \sin^2\theta\, dr\, d\theta$

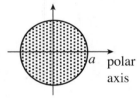

$$= \int_0^{2\pi} \frac{1}{4}\ln(1+r^4)\Big|_0^a \sin^2\theta\, d\theta = \frac{1}{4}\ln(1+a^4)\left(\frac{1}{2}\theta - \frac{1}{4}\sin 2\theta\right)\Big|_0^{2\pi}$$

$$= \frac{\pi}{4}\ln(1+a^4)$$

21. From the solution to Problem 17, $I_x = k\pi a^4/4$. By symmetry, $I_y = I_x$.
Thus $I_0 = k\pi a^4/2$.

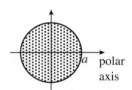

24. $I_0 = \int_0^{\pi} \int_0^{2a\cos\theta} r^2 kr\, dr\, d\theta = k \int_0^{\pi} \frac{1}{4} r^4 \Big|_0^{2a\cos\theta} d\theta = 4ka^4 \int_0^{\pi} \cos^4\theta\, d\theta$

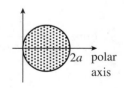

$$= 4ka^4 \left(\frac{3}{8}\theta + \frac{1}{4}\sin 2\theta + \frac{1}{32}\sin 4\theta\right)\Big|_0^{\pi} = 4ka^4 \left(\frac{3\pi}{8}\right) = \frac{3k\pi a^4}{2}$$

27. $\int_0^1 \int_0^{\sqrt{1-y^2}} e^{x^2+y^2}\, dx\, dy = \int_0^{\pi/2} \int_0^1 e^{r^2} r\, dr\, d\theta = \int_0^{\pi/2} \frac{1}{2} e^{r^2}\Big|_0^1 d\theta$

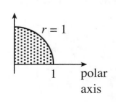

$$= \frac{1}{2}\int_0^{\pi/2} (e-1)\, d\theta = \frac{\pi(e-1)}{4}$$

30. $\displaystyle\int_0^1 \int_0^{\sqrt{2y-y^2}} (1 - x^2 - y^2)\, dx\, dy$

$= \displaystyle\int_0^{\pi/4} \int_0^{2\sin\theta} (1 - r^2) r\, dr\, d\theta + \int_{\pi/4}^{\pi/2} \int_0^{\csc\theta} (1 - r^2) r\, dr\, d\theta$

$= \displaystyle\int_0^{\pi/4} \left(\frac{1}{2}r^2 - \frac{1}{4}r^4 \right) \Big|_0^{2\sin\theta} d\theta + \int_{\pi/4}^{\pi/2} \left(\frac{1}{2}r^2 - \frac{1}{4}r^4 \right) \Big|_0^{\csc\theta} d\theta$

$= \displaystyle\int_0^{\pi/4} (2\sin^2\theta - 4\sin^4\theta)\, d\theta + \int_{\pi/4}^{\pi/2} \left(\frac{1}{2}\csc^2\theta - \frac{1}{4}\csc^4\theta \right) d\theta$

$= \left[\theta - \frac{1}{2}\sin 2\theta - \left(\frac{3}{2}\theta - \sin 2\theta + \frac{1}{8}\sin 4\theta \right) \right] + \left[-\frac{1}{2}\cot\theta - \frac{1}{4}\left(-\cot\theta - \frac{1}{3}\cot^3\theta \right) \right] \Big|_{\pi/4}^{\pi/2}$

$= \left(-\frac{\pi}{8} + \frac{1}{2} \right) + \left[0 - \left(-\frac{1}{4} + \frac{1}{12} \right) \right] = \dfrac{16 - 3\pi}{24}$

33. The volume of the cylindrical portion of the tank is $V_c = \pi(4.2)^2 19.3 \approx 1069.56$ m^3. We take the equation of the ellipsoid to be

$$\frac{x^2}{(4.2)^2} + \frac{z^2}{(5.15)^2} = 1 \quad \text{or} \quad z = \pm\frac{5.15}{4.2}\sqrt{(4.2)^2 - x^2 - y^2}\,.$$

The volume of the ellipsoid is

$$V_e = 2\left(\frac{5.15}{4.2} \right) \iint_R \sqrt{(4.2)^2 - x^2 - y^2}\, dx\, dy = \frac{10.3}{4.2} \int_0^{2\pi} \int_0^{4.2} [(4.2)^2 - r^2]^{1/2} r\, dr\, d\theta$$

$$= \frac{10.3}{4.2} \int_0^{2\pi} \left[\left(-\frac{1}{2} \right) \frac{2}{3} [(4.2)^2 - r^2]^{3/2} \Big|_0^{4.2} \right] d\theta = \frac{10.3}{4.2} \frac{1}{3} \int_0^{2\pi} (4.2)^3\, d\theta$$

$$= \frac{2\pi}{3} \frac{10.3}{4.2} (4.2)^3 \approx 380.53.$$

The volume of the tank is approximately $1069.56 + 380.53 = 1450.09$ m^3.

9.12   Green's Theorem

3. $\displaystyle\oint_C -y^2\, dx + x^2\, dy = \int_0^{2\pi} (-9\sin^2 t)(-3\sin t)\, dt + \int_0^{2\pi} 9\cos^2 t(3\cos t)\, dt$

$= 27 \displaystyle\int_0^{2\pi} [(1 - \cos^2 t)\sin t + (1 - \sin^2 t)\cos t]\, dt$

$= 27 \left(-\cos t + \frac{1}{3}\cos^3 t + \sin t - \frac{1}{3}\sin^3 t \right) \Big|_0^{2\pi} = 27(0) = 0$

$$\iint_R (2x + 2y)\, dA = 2 \int_0^{2\pi} \int_0^3 (r\cos\theta + r\sin\theta)r\, dr\, d\theta = 2 \int_0^{2\pi} \int_0^3 r^2(\cos\theta + \sin\theta)\, dr\, d\theta$$

$$= 2 \int_0^{2\pi} \left[\frac{1}{3}r^3(\cos\theta + \sin\theta) \right]\Big|_0^3 d\theta = 18 \int_0^{2\pi} (\cos\theta + \sin\theta)\, d\theta$$

$$= 18(\sin\theta - \cos\theta)\Big|_0^{2\pi} = 18(0) = 0$$

6. $P = x + y^2$, $P_y = 2y$, $Q = 2x^2 - y$, $Q_x = 4x$

$$\oint_C (x + y^2)\, dx + (2x^2 - y)\, dy = \iint_R (4x - 2y)\, dA = \int_{-2}^2 \int_{x^2}^4 (4x - 2y)\, dy\, dx$$

$$= \int_{-2}^2 (4xy - y^2)\Big|_{x^2}^4 dx = \int_{-2}^2 (16x - 16 - 4x^3 + x^4)\, dx$$

$$= \left(8x^2 - 16x - x^4 + \frac{1}{5}x^5 \right)\Big|_{-2}^2 = -\frac{96}{5}$$

9. $P = 2xy$, $P_y = 2x$, $Q = 3xy^2$, $Q_x = 3y^2$

$$\oint_C 2xy\, dx + 3xy^2\, dy = \iint_R (3y^2 - 2x)\, dA = \int_1^2 \int_2^{2x} (3y^2 - 2x)\, dy\, dx$$

$$= \int_1^2 (y^3 - 2xy)\Big|_2^{2x} dx = \int_1^2 (8x^3 - 4x^2 - 8 + 4x)\, dx$$

$$= \left(2x^4 - \frac{4}{3}x^3 - 8x + 2x^2 \right)\Big|_1^2 = \frac{40}{3} - \left(-\frac{16}{3} \right) = \frac{56}{3}$$

12. $P = e^{x^2}$, $P_y = 0$, $Q = 2\tan^{-1} x$, $Q_x = \dfrac{2}{1 + x^2}$

$$\oint_C e^{x^2}\, dx + 2\tan^{-1} x\, dy = \iint_R \frac{2}{1 + x^2}\, dA = \int_{-1}^0 \int_{-x}^1 \frac{2}{1 + x^2}\, dy\, dx$$

$$= \int_{-1}^0 \left(\frac{2y}{1 + x^2} \right)\Big|_{-x}^1 dx = \int_{-1}^0 \left(\frac{2}{1 + x^2} + \frac{2x}{1 + x^2} \right) dx$$

$$= [2\tan^{-1} x + \ln(1 + x^2)]\Big|_{-1}^0 = 0 - \left(-\frac{\pi}{2} + \ln 2 \right) = \frac{\pi}{2} - \ln 2$$

15. $P = ay$, $P_y = a$, $Q = bx$, $Q_x = b$.

$$\oint_C ay\, dx + bx\, dy = \iint_R (b - a)\, dA = (b - a) \times (\text{area bounded by } C)$$

18. $P = -y$, $P_y = -1$, $Q = x$, $Q_x = 1$. $\dfrac{1}{2}\oint_C -y\, dx + x\, dy = \dfrac{1}{2}\iint_R 2\, dA = \iint_R dA = \text{area of } R$

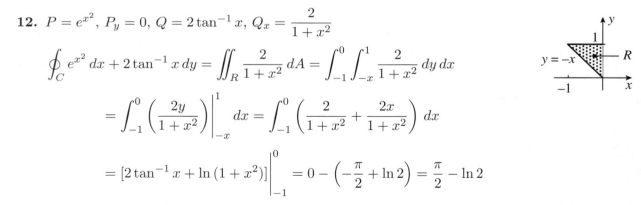

21. (a) Parameterize C by $x = x_1 + (x_2 - x_1)t$ and $y = y_1 + (y_2 - y_1)t$ for $0 \le t \le 1$. Then

$$\int_C -y\,dx + x\,dy = \int_0^1 -[y_1 + (y_2 - y_1)t](x_2 - x_1)\,dt$$

$$+ \int_0^1 [x_1 + (x_2 - x_1)t](y_2 - y_1)\,dt$$

$$= -(x_2 - x_1)\left[y_1 t + \frac{1}{2}(y_2 - y_1)t^2\right]\Big|_0^1 + (y_2 - y_1)\left[x_1 t + \frac{1}{2}(x_2 - x_1)t^2\right]\Big|_0^1$$

$$= -(x_2 - x_1)\left[y_1 + \frac{1}{2}(y_2 - y_1)\right] + (y_2 - y_1)\left[x_1 + \frac{1}{2}(x_2 - x_1)\right] = x_1 y_2 - x_2 y_1.$$

(b) Let C_i be the line segment from (x_i, y_i) to (x_{i+1}, y_{i+1}) for $i = 1, 2, \ldots, n-1$, and C_2 the line segment from (x_n, y_n) to (x_1, y_1). Then

$$A = \frac{1}{2}\oint_C -y\,dx + x\,dy \qquad \boxed{\text{Using Problem 18}}$$

$$= \frac{1}{2}\left[\int_{C_1} -y\,dx + x\,dy + \int_{C_2} -y\,dx + x\,dy + \cdots\right.$$

$$\left. + \int_{C_{n-1}} -y\,dx + x\,dy + \int_{C_n} -y\,dx + x\,dy\right]$$

$$= \frac{1}{2}(x_1 y_2 - x_2 y_1) + \frac{1}{2}(x_2 y_3 - x_3 y_2) + \frac{1}{2}(x_{n-1} y_n - x_n y_{n-1}) + \frac{1}{2}(x_n y_1 - x_1 y_n).$$

24. $P = \cos x^2 - y$, $P_y = -1$; $Q = \sqrt{y^3 + 1}$, $Q_x = 0$

$$\oint_C (\cos x^2 - y)\,dx + \sqrt{y^3 + 1}\,dy = \iint_R (0 + 1)\,dA = \iint_R dA = (6\sqrt{2})^2 - \pi(2)(4) = 72 - 8\pi$$

27. Writing $\iint_R x^2\,dA = \iint_R (Q_x - P_y)\,dA$ we identify $Q = 0$ and $P = -x^2 y$. Then, with C: $x = 3\cos t$, $y = 2\sin t$, $0 \le t \le 2\pi$, we have

$$\iint_R x^2\,dA = \oint_C P\,dx + Q\,dy = \oint_C -x^2 y\,dx = -\int_0^{2\pi} 9\cos^2 t(2\sin t)(-3\sin t)\,dt$$

$$= \frac{54}{4}\int_0^{2\pi} 4\sin^2 t\cos^2 t\,dt = \frac{27}{2}\int_0^{2\pi} \sin^2 2t\,dt = \frac{27}{4}\int_0^{2\pi} (1 - \cos 4t)\,dt$$

$$= \frac{27}{4}\left(t - \frac{1}{4}\sin 4t\right)\Big|_0^{2\pi} = \frac{27\pi}{2}.$$

30. $P = -xy^2$, $P_y = -2xy$, $Q = x^2 y$, $Q_x = 2xy$. Using polar coordinates,

$$W = \oint_C \mathbf{F} \cdot d\mathbf{r} = \iint_R 4xy\, dA = \int_0^{\pi/2} \int_1^2 4(r\cos\theta)(r\sin\theta)r\, dr\, d\theta = \int_0^{\pi/2} (r^4 \cos\theta \sin\theta)\Big|_1^2 d\theta$$

$$= 15 \int_0^{\pi/2} \sin\theta \cos\theta\, d\theta = \frac{15}{2} \sin^2\theta\Big|_0^{\pi/2} = \frac{15}{2}.$$

33. Using Green's Theorem,

$$W = \oint_C \mathbf{F} \cdot d\mathbf{r} = \oint_C -y\, dx + x\, dy = \iint_R 2\, dA = 2\int_0^{2\pi} \int_0^{1+\cos\theta} r\, dr\, d\theta$$

$$= 2\int_0^{2\pi} \left(\frac{1}{2}r^2\right)\Big|_0^{1+\cos\theta} d\theta = \int_0^{2\pi} (1 + 2\cos\theta + \cos^2\theta)\, d\theta$$

$$= \left(\theta + 2\sin\theta + \frac{1}{2}\theta + \frac{1}{4}\sin 2\theta\right)\Big|_0^{2\pi} = 3\pi.$$

9.13 | Surface Integrals

3. Using $f(x,y) = z = \sqrt{16 - x^2}$ we see that for $0 \le x \le 2$ and $0 \le y \le 5$, $z > 0$.
Thus, the surface is entirely above the region. Now $f_x = -\dfrac{x}{\sqrt{16 - x^2}}$, $f_y = 0$,

$$1 + f_x^2 + f_y^2 = 1 + \frac{x^2}{16 - x^2} = \frac{16}{16 - x^2} \text{ and}$$

$$A = \int_0^5 \int_0^2 \frac{4}{\sqrt{16 - x^2}}\, dx\, dy = 4\int_0^5 \sin^{-1}\frac{x}{4}\Big|_0^2 dy = 4\int_0^5 \frac{\pi}{6}\, dy = \frac{10\pi}{3}.$$

6. The surfaces $x^2 + y^2 + z^2 = 2$ and $z^2 = x^2 + y^2$ intersect on the cylinder $2x^2 + 2y^2 = 2$ or $x^2 + y^2 = 1$. There are portions of the sphere within the cone both above and below the xy-plane. Using $f(x,y) = \sqrt{2 - x^2 - y^2}$ we have $f_x = -\dfrac{x}{\sqrt{2 - x^2 - y^2}}$, $f_y = -\dfrac{y}{\sqrt{2 - x^2 - y^2}}$,

$$1 + f_x^2 + f_y^2 = \frac{2}{2 - x^2 - y^2}. \text{ Then}$$

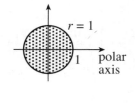

$$A = 2\left[\int_0^{2\pi} \int_0^1 \frac{\sqrt{2}}{\sqrt{2 - r^2}} r\, dr\, d\theta\right] = 2\sqrt{2}\int_0^{2\pi} -\sqrt{2 - r^2}\Big|_0^1 d\theta = 2\sqrt{2}\int_0^{2\pi} (\sqrt{2} - 1)\, d\theta$$

$$= 4\pi\sqrt{2}(\sqrt{2} - 1).$$

9. There are portions of the sphere within the cylinder both above and below the xy-plane. Using $f(x,y) = z = \sqrt{a^2 - x^2 - y^2}$ we have

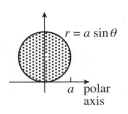

$$f_x = -\frac{x}{\sqrt{1^2 - x^2 - y^2}}, \; f_y = -\frac{y}{\sqrt{a^2 - x^2 - y^2}},$$

$$1 + f_x^2 + f_y^2 = \frac{a^2}{a^2 - x^2 - y^2}. \text{ Then, using symmetry,}$$

$$A = 2\left[2\int_0^{\pi/2}\int_0^{a\sin\theta} \frac{a}{\sqrt{a^2 - r^2}} r \, dr \, d\theta\right] = 4a\int_0^{\pi/2} -\sqrt{a^2 - r^2}\,\Big|_0^{a\sin\theta} \, d\theta$$

$$= 4a\int_0^{\pi/2}(a - a\sqrt{1 - \sin^2\theta})\, d\theta = 4a^2\int_0^{\pi/2}(1 - \cos 2\theta)\, d\theta$$

$$= 4a^2(\theta - \sin\theta)\,\Big|_0^{\pi/2} = 4a^2\left(\frac{\pi}{2} - 1\right) = 2a^2(\pi - 2).$$

12. From Example 1, the area of the portion of the hemisphere within $x^2 + y^2 = b^2$ is $2\pi a(a - \sqrt{a^2 - b^2})$. Thus, the area of the sphere is

$$A = 2\lim_{b\to a} 2\pi a(a - \sqrt{a^2 - b^2}) = 2(2\pi a^2) = 4\pi a^2.$$

15. $z_x = -2x, \; z_y = 0; \; dS = \sqrt{1 + 4x^2}\, dA$

$$\iint_S x \, dS = \int_0^4\int_0^{\sqrt{2}} x\sqrt{1 + 4x^2}\, dx\, dy$$

$$= \int_0^4 \frac{1}{12}(1 + 4x^2)^{3/2}\,\Big|_0^{\sqrt{2}} \, dy = \int_0^4 \frac{13}{6}\, dy = \frac{26}{3}$$

18. $z_x = \frac{x}{\sqrt{x^2 + y^2}}, \; z_y = \frac{y}{\sqrt{x^2 + y^2}}; \; dS = \sqrt{2}\, dA.$

Using polar coordinates,

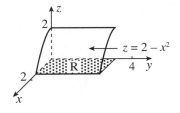

$$\iint_S (x + y + z)\, dS = \iint_R (x + y + \sqrt{x^2 + y^2})\sqrt{2}\, dA$$

$$= \sqrt{2}\int_0^{2\pi}\int_1^4 (r\cos 2\theta + r\sin\theta + r)r\, dr\, d\theta$$

$$= \sqrt{2}\int_0^{2\pi}\int_1^4 r^2(1 + \cos 2\theta + \sin\theta)\, dr\, d\theta = \sqrt{2}\int_0^{2\pi} \frac{1}{3}r^3(1 + \cos 2\theta + \sin\theta)\,\Big|_1^4 \, d\theta$$

$$= \frac{63\sqrt{2}}{3}\int_0^{2\pi}(1 + \cos 2\theta + \sin\theta)\, d\theta = 21\sqrt{2}(\theta + \sin\theta - \cos 2\theta)\,\Big|_0^{2\pi} = 42\sqrt{2}\,\pi.$$

21. $z_x = -x$, $z_y = -y$; $dS = \sqrt{1 + x^2 + y^2}\, dA$

$$\iint_S xy\, dS = \int_0^1 \int_0^1 xy\sqrt{1 + x^2 + y^2}\, dx\, dy$$

$$= \int_0^1 \frac{1}{3}y(1 + x^2 + y^2)^{3/2}\Big|_0^1 dy$$

$$= \int_0^1 \left[\frac{1}{3}y(2 + y^2)^{3/2} - \frac{1}{3}y(1 + y^2)^{3/2}\right] dy = \left[\frac{1}{15}(2 + y^2)^{5/2} - \frac{1}{15}(1 + y^2)^{5/2}\right]\Big|_0^1$$

$$= \frac{1}{15}(3^{5/2} - 2^{7/2} + 1)$$

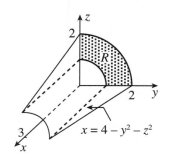

24. $x_y = -2y$, $x_z = -2z$; $dS = \sqrt{1 + 4y^2 + 4z^2}\, dA$

Using polar coordinates,

$$\iint_S (1 + 4y^2 + 4z^2)^{1/2}\, dS = \int_0^{\pi/2} \int_1^2 (1 + 4r^2)r\, dr\, d\theta$$

$$= \int_0^{\pi/2} \frac{1}{16}(1 + 4r^2)^2\Big|_1^2 d\theta$$

$$= \frac{1}{16}\int_0^{\pi/2} 12\, d\theta = \frac{3\pi}{8}.$$

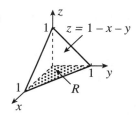

27. The density is $\rho = kx^2$. The surface is $z = 1 - x - y$. Then $z_x = -1$, $z_y = -1$; $dS = \sqrt{3}\, dA$.

$$m = \iint_S kx^2\, dS = k\int_0^1 \int_0^{1-x} x^2\sqrt{3}\, dy\, dx = \sqrt{3}\, k\int_0^1 \frac{1}{3}x^3\Big|_0^{1-x} dx$$

$$= \frac{\sqrt{3}}{3}k\int_0^1 (1 - x)^3\, dx = \frac{\sqrt{3}}{3}k\left[-\frac{1}{4}(1 - x)^4\right]\Big|_0^1 = \frac{\sqrt{3}}{12}k$$

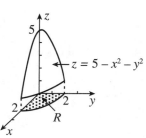

30. The surface is $g(x, y, z) = x^2 + y^2 + z - 5 = 0$. $\nabla g = 2x\mathbf{i} + 2y\mathbf{j} + \mathbf{k}$,

$|\nabla g| = \sqrt{1 + 4x^2 + 4y^2}$; $\mathbf{n} = \dfrac{2x\mathbf{i} + 2y\mathbf{j} + \mathbf{k}}{\sqrt{1 + 4x^2 + 4y^2}}$;

$\mathbf{F} \cdot \mathbf{n} = \dfrac{z}{\sqrt{1 + 4x^2 + 4y^2}}$; $z_x = -2x$, $z_y = -2y$,

$dS = \sqrt{1 + 4x^2 + 4y^2}\, dA$. Using polar coordinates,

$$\text{Flux} = \iint_S \mathbf{F} \cdot \mathbf{n}\, dS = \iint_R \frac{z}{\sqrt{1 + 4x^2 + 4y^2}}\sqrt{1 + 4x^2 + 4y^2}\, dA = \iint_R (5 - x^2 - y^2)\, dA$$

$$= \int_0^{2\pi} \int_0^2 (5 - r^2)r\, dr\, d\theta = \int_0^{2\pi} \left(\frac{5}{2}r^2 - \frac{1}{4}r^4\right)\Big|_0^2 d\theta = \int_0^{2\pi} 6\, d\theta = 12\pi.$$

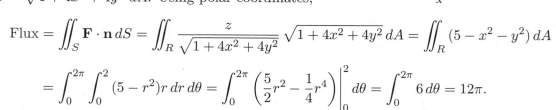

33. The surface is $g(x, y, z) = x^2 + y^2 + z - 4$. $\nabla g = 2x\mathbf{i} + 2y\mathbf{j} + \mathbf{k}$,

$|\nabla g| = \sqrt{4x^2 + 4y^2 + 1}$; $\mathbf{n} = \dfrac{2x\mathbf{i} + 2y\mathbf{j} + \mathbf{k}}{\sqrt{4x^2 + 4y^2 + 1}}$;

$\mathbf{F} \cdot \mathbf{n} = \dfrac{x^3 + y^3 + z}{\sqrt{4x^2 + 4y^2 + 1}}$; $z_x = -2x$, $z_y = -2y$,

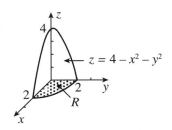

$dS = \sqrt{1 + 4x^2 + 4y^2}\, dA$. Using polar coordinates,

$$\text{Flux} = \iint_S \mathbf{F} \cdot \mathbf{n}\, dS = \iint_R (x^3 + y^3 + z)\, dA = \iint_R (4 - x^2 - y^2 + x^3 + y^3)\, dA$$

$$= \int_0^{2\pi} \int_0^2 (4 - r^2 + r^3 \cos^3 \theta + r^3 \sin^3 \theta)\, r\, dr\, d\theta$$

$$= \int_0^{2\pi} \left(2r^2 - \frac{1}{4}r^4 + \frac{1}{5}r^5 \cos^3 \theta + \frac{1}{5}r^5 \sin^3 \theta \right)\Bigg|_0^2 d\theta$$

$$= \int_0^{2\pi} \left(4 + \frac{32}{5}\cos^3 \theta + \frac{32}{5}\sin^3 \theta \right) d\theta = 4\theta \Big|_0^{2\pi} + 0 + 0 = 8\pi.$$

36. For S_1: $g(x, y, z) = x^2 + y^2 + z - 4$, $\nabla g = 2x\mathbf{i} + 2y\mathbf{j} + \mathbf{k}$, $|\nabla g| = \sqrt{4x^2 + 4y^2 + 1}$;

$\mathbf{n}_1 = \dfrac{2x\mathbf{i} + 2y\mathbf{j} + \mathbf{k}}{\sqrt{4x^2 + 4y^2 + 1}}$; $\mathbf{F} \cdot \mathbf{n}_1 = 6z^2/\sqrt{4x^2 + 4y^2 + 1}$; $z_x = -2x$, $z_y = -2y$,

$dS_1 = \sqrt{1 + 4x^2 + 4y^2}\, dA$. For S_2: $g(x, y, z) = x^2 + y^2 - z$,

$\nabla g = 2x\mathbf{i} + 2y\mathbf{j} - \mathbf{k}$, $|\nabla g| = \sqrt{4x^2 + 4y^2 + 1}$; $\mathbf{n}_2 = \dfrac{2x\mathbf{i} + 2y\mathbf{j} - \mathbf{k}}{\sqrt{4x^2 + y^2 + 1}}$;

$\mathbf{F} \cdot \mathbf{n}_2 = -6z^2/\sqrt{4x^2 + 4y^2 + 1}$; $z_x = 2x$, $z_y = 2y$, $dS_2 = \sqrt{1 + 4x^2 + 4y^2}\, dA$.

Using polar coordinates and R: $x^2 + y^2 \le 2$ we have

$$\text{Flux} = \iint_{S_1} \mathbf{F} \cdot \mathbf{n}_1\, dS_1 + \iint_{S_1} \mathbf{F} \cdot \mathbf{n}_2\, dS_2 = \iint_R 6z^2\, dA + \iint_R -6z^2\, dA$$

$$= \iint_R [6(4 - x^2 - y^2)^2 - 6(x^2 + y^2)^2]\, dA = 6 \int_0^{2\pi} \int_0^{\sqrt{2}} [(4 - r^2)^2 - r^4]\, r\, dr\, d\theta$$

$$= 6 \int_0^{2\pi} \left[-\frac{1}{6}(4 - r^2)^3 - \frac{1}{6}r^6 \right]\Bigg|_0^{\sqrt{2}} d\theta = -\int_0^{2\pi} [(2^3 - 4^3) + (\sqrt{2})^6]\, d\theta = \int_0^{2\pi} 48\, d\theta = 96\pi.$$

39. Referring to the solution to Problem 37, we find $\mathbf{n} = \dfrac{x\mathbf{i} + y\mathbf{j} + z\mathbf{k}}{\sqrt{x^2 + y^2 + z^2}}$ and

$$dS = \frac{a}{\sqrt{a^2 - x^2 - y^2}}\, dA.$$

Now

$$\mathbf{F} \cdot \mathbf{n} = kq\, \frac{\mathbf{r}}{|\mathbf{r}|^3} \cdot \frac{\mathbf{r}}{|\mathbf{r}|} = \frac{kq}{|\mathbf{r}|^4}\, |\mathbf{r}|^2 = \frac{kq}{|\mathbf{r}|^2} = \frac{kq}{x^2 + y^2 + z^2} = \frac{kq}{a^2}$$

and

$$\text{Flux} = \iint_S \mathbf{F} \cdot \mathbf{n}\, dS = \iint_S \frac{kq}{a^2}\, dS = \frac{kq}{a^2} \times \text{area} = \frac{kq}{a^2}(4\pi a^2) = 4\pi kq.$$

42. The area of the hemisphere is $A(s) = 2\pi a^2$. By symmetry, $\bar{x} = \bar{y} = 0$.

$$z_x = -\frac{x}{\sqrt{a^2 - x^2 - y^2}}, \ z_y = -\frac{y}{\sqrt{a^2 - x^2 - y^2}};$$

$$dS = \sqrt{1 + \frac{x^2}{a^2 - x^2 - y^2} + \frac{y^2}{a^2 - x^2 - y^2}}\, dA = \frac{a}{\sqrt{a^2 - x^2 - y^2}}\, dA$$

Using polar coordinates,

$$z = \iint_S \frac{z\, dS}{2\pi a^2} = \frac{1}{2\pi a^2} \iint_R \sqrt{a^2 - x^2 - y^2}\, \frac{a}{\sqrt{a^2 - x^2 - y^2}}\, dA = \frac{1}{2\pi a} \int_0^{2\pi} \int_0^a r\, dr\, d\theta$$

$$= \frac{1}{2\pi a} \int_0^{2\pi} \frac{1}{2} r^2 \Big|_0^a\, d\theta = \frac{1}{2\pi a} \int_0^{2\pi} \frac{1}{2} s^2\, d\theta = \frac{a}{2}.$$

The centroid is $(0, 0, a/2)$.

9.14   Stokes' Theorem

3. Surface Integral: $\text{curl}\,\mathbf{F} = \mathbf{i} + \mathbf{j} + \mathbf{k}$. Letting $g(x, y, z) = 2x + y + 2z - 6$, we have $\nabla g = 2\mathbf{i} + \mathbf{j} + 2\mathbf{k}$ and $\mathbf{n} = (2\mathbf{i} + \mathbf{j} + 2\mathbf{k})/3$. Then $\iint_S (\text{curl}\,\mathbf{F}) \cdot \mathbf{n}\, dS = \iint_S \frac{5}{3}\, dS$. Letting the surface be $z = 3 - \frac{1}{2}y - x$ we have $z_x = -1$, $z_y = -\frac{1}{2}$, and $dS = \sqrt{1 + (-1)^2 + (-\frac{1}{2})^2}\, dA = \frac{3}{2}\, dA$. Then

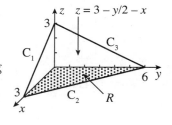

$$\iint_S (\text{curl}\,\mathbf{F}) \cdot \mathbf{n}\, dS = \iint_R \frac{5}{3}\left(\frac{3}{2}\right) dA = \frac{5}{2} \times (\text{area of } R) = \frac{5}{2}(9) = \frac{45}{2}.$$

Line Integral: C_1: $z = 3 - x$, $0 \le x \le 3$, $y = 0$; C_2: $y = 6 - 2x$, $3 \ge x \ge 0$, $z = 0$; C_3: $z = 3 - y/2$, $6 \ge y \ge 0$, $x = 0$.

$$\oint_C z\, dx + x\, dy + y\, dz = \int_{C_1} z\, dx + \int_{C_2} x\, dy + \int_{C_3} y\, dz$$

$$= \int_0^3 (3 - x)\, dx + \int_3^0 x(-2\, dx) + \int_6^0 y(-dy/2)$$

$$= \left(3x - \frac{1}{2}x^2\right)\Big|_0^3 - x^2\Big|_3^0 - \frac{1}{4}y^2\Big|_6^0 = \frac{9}{2} - (0 - 9) - \frac{1}{4}(0 - 36) = \frac{45}{2}$$

6. curl $\mathbf{F} = -2xz\mathbf{i} + z^2\mathbf{k}$. A unit vector normal to the plane is $\mathbf{n} = (\mathbf{j}+\mathbf{k})/\sqrt{2}$. From $z = 1 - y$, we have $z_x = 0$ and $z_y = -1$. Thus, $dS = \sqrt{1+1}\,dA = \sqrt{2}\,dA$ and

$$\oint_C \mathbf{F}\cdot d\mathbf{r} = \iint_S (\text{curl }\mathbf{F})\cdot\mathbf{n}\,dS = \iint_R \frac{1}{\sqrt{2}}z^2\sqrt{2}\,dA = \iint_R (1-y)^2\,dA$$

$$= \int_0^2 \int_0^1 (1-y)^2\,dy\,dx = \int_0^2 -\frac{1}{3}(1-y)^3\Big|_0^1\,dx = \int_0^2 \frac{1}{3}\,dx = \frac{2}{3}.$$

9. curl $\mathbf{F} = (-3x^2 - 3y^2)\mathbf{k}$. A unit vector normal to the plane is $\mathbf{n} = (\mathbf{i}+\mathbf{j}+\mathbf{k})/\sqrt{3}$. From $z = 1 - x - y$, we have $z_x = z_y = -1$ and $dS = \sqrt{3}\,dA$. Then, using polar coordinates,

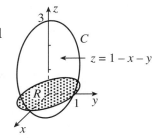

$$\oint_C \mathbf{F}\cdot d\mathbf{r} = \iint_S (\text{curl }\mathbf{F})\cdot\mathbf{n}\,dS = \iint_R (-\sqrt{3}\,x^2 - \sqrt{3}\,y^2)\sqrt{3}\,dA$$

$$= 3\iint_R (-x^2 - y^2)\,dA = 3\int_0^{2\pi}\int_0^1 (-r^2)r\,dr\,d\theta$$

$$= 3\int_0^{2\pi} -\frac{1}{4}r^4\Big|_0^1\,d\theta = 3\int_0^{2\pi} -\frac{1}{4}\,d\theta = -\frac{3\pi}{2}.$$

12. curl $\mathbf{F} = \mathbf{i}+\mathbf{j}+\mathbf{k}$. Taking the surface S bounded by C to be the portion of the plane $x+y+z = 0$ inside C, we have $\mathbf{n} = (\mathbf{i}+\mathbf{j}+\mathbf{k})/\sqrt{3}$ and $dS = \sqrt{3}\,dA$.

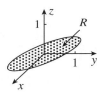

$$\oint_C \mathbf{F}\cdot d\mathbf{r} = \iint_S (\text{curl }\mathbf{F})\cdot\mathbf{n}\,dS = \iint_S \sqrt{3}\,dS = \sqrt{3}\iint_R \sqrt{3}\,dA = 3\times(\text{area of }R)$$

The region R is obtained by eliminating z from the equations of the plane and the sphere. This gives $x^2 + xy + y^2 = \frac{1}{2}$. Rotating axes, we see that R is enclosed by the ellipse $X^2/(1/3) + Y^2/1 = 1$ in a rotated coordinate system. Thus,

$$\oint_C \mathbf{F}\cdot d\mathbf{r} = 3\times(\text{area of }R) = 3\left(\pi\frac{1}{\sqrt{3}}1\right) = \sqrt{3}\,\pi.$$

15. Parameterize C by C_1: $x = 0$, $z = 0$, $2 \ge y \ge 0$; C_2: $z = x$, $y = 0$, $0 \le x \le 2$; C_3: $x = 2$, $z = 2$, $0 \le y \le 2$; C_4: $z = x$, $y = 2$, $2 \ge x \ge 0$. Then

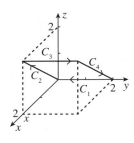

$$\iint_S (\text{curl }\mathbf{F})\cdot\mathbf{n}\,dS = \oint_C \mathbf{F}\cdot\mathbf{r} = \oint_C 3x^2\,dx + 8x^3 y\,dy + 3x^2 y\,dz$$

$$= \int_{C_1} 0\,dx + 0\,dy + 0\,dz + \int_{C_2} 3x^2\,dx + \int_{C_3} 64\,dy$$

$$\qquad\qquad + \int_{C_4} 3x^2\,dx + 6x^2\,dx$$

$$= \int_0^2 3x^2\,dx + \int_0^2 64\,dy + \int_2^0 9x^2\,dx$$

$$= x^3\Big|_0^2 + 64y\Big|_0^2 + 3x^3\Big|_2^0 = 112.$$

18. (a) curl $\mathbf{F} = xz\mathbf{i} - yz\mathbf{j}$. A unit vector normal to the surface is $\mathbf{n} = \dfrac{2x\mathbf{i} + 2y\mathbf{j} + \mathbf{k}}{\sqrt{4x^2 + 4y^2 + 1}}$ and $dS = \sqrt{1 + 4x^2 + 4y^2}\, dA$. Then, using $x = \cos t$, $y = \sin t$, $0 \le t \le 2\pi$, we have

$$\iint_S (\text{curl }\mathbf{F}) \cdot \mathbf{n}\, dS = \iint_R (2x^2 z - 2y^2 z)\, dA = \iint_R (2x^2 - 2y^2)(1 - x^2 - y^2)\, dA$$

$$= \iint_R (2x^2 - 2y^2 - 2x^4 + 2y^4)\, dA$$

$$= \int_0^{2\pi} \int_0^1 (2r^2 \cos^2\theta - 2r^2 \sin^2\theta - 2r^4 \cos^4\theta + 2r^4 \cos^4\theta)\, r\, dr\, d\theta$$

$$= 2 \int_0^{2\pi} \int_0^1 [r^3 \cos 2\theta - r^5(\cos^2\theta - \sin^2\theta)(\cos^2\theta + \sin^2\theta)]\, dr\, d\theta$$

$$= 2 \int_0^{2\pi} \int_0^1 (r^3 \cos 2\theta - r^5 \cos 2\theta)\, dr\, d\theta = 2 \int_0^{2\pi} \cos 2\theta \left(\frac{1}{4}r^4 - \frac{1}{6}r^6 \right) \Big|_0^1 d\theta$$

$$= \frac{1}{6} \int_0^{2\pi} \cos 2\theta\, d\theta = 0.$$

(b) We take the surface to be $z = 0$. Then $\mathbf{n} = \mathbf{k}$, curl $\mathbf{F} \cdot \mathbf{n} = $ curl $\mathbf{F} \cdot \mathbf{k} = 0$ and
$$\iint_S (\text{curl }\mathbf{F}) \cdot \mathbf{n}\, dS = 0.$$

(c) By Stokes' Theorem, using $z = 0$, we have

$$\iint_S (\text{curl }\mathbf{F}) \cdot \mathbf{n}\, dS = \oint_C \mathbf{F} \cdot d\mathbf{r} = \oint_C xyz\, dz = \oint_C xy(0)\, dz = 0.$$

9.15   Triple Integrals

3. $\displaystyle \int_0^6 \int_0^{6-x} \int_0^{6-x-z} dy\, dz\, dx = \int_0^6 \int_0^{6-x} (6 - x - z)\, dz\, dx = \int_0^6 \left(6z - xz - \frac{1}{2}z^2 \right) \Big|_0^{6-x} dx$

$$= \int_0^6 \left[6(6 - x) - x(6 - x) - \frac{1}{2}(6 - x)^2 \right] dx = \int_0^6 \left(18 - 6x + \frac{1}{2}x^2 \right) dx$$

$$= \left(18x - 3x^2 + \frac{1}{6}x^3 \right) \Big|_0^6 = 36$$

6. $\displaystyle \int_0^{\sqrt{2}} \int_{\sqrt{y}}^2 \int_0^{e^{x^2}} x\, dz\, dx\, dy = \int_0^{\sqrt{2}} \int_{\sqrt{y}}^2 xe^{x^2}\, dx\, dy = \int_0^{\sqrt{2}} \frac{1}{2}e^{x^2} \Big|_{\sqrt{y}}^2 dy = \frac{1}{2} \int_0^{\sqrt{2}} (e^4 - e^y)\, dy$

$$= \frac{1}{2}(ye^4 - e^y) \Big|_0^{\sqrt{2}} = \frac{1}{2}[(e^4\sqrt{2} - e^{\sqrt{2}}) - (-1)] = \frac{1}{2}(1 + e^4\sqrt{2} - e^{\sqrt{2}})$$

9. $\displaystyle\iiint_D z\,dV = \int_0^5\int_1^3\int_y^{y+2} z\,dx\,dy\,dz = \int_0^5\int_1^3 xz\Big|_y^{y+2}\,dy\,dz$

$\displaystyle = \int_0^5\int_1^3 2z\,dy\,dz = \int_0^5 2yz\Big|_1^3\,dz = \int_0^5 4z\,dz = 2z^2\Big|_0^5 = 50$

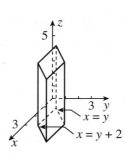

12. The other five integrals are

$\displaystyle\int_0^3\int_0^{\sqrt{36-4y^2}/3}\int_1^3 F(x,y,z)\,dz\,dx\,dy,$

$\displaystyle\int_1^3\int_0^2\int_0^{\sqrt{36-9x^2}/2} F(x,y,z)\,dy\,dx\,dz,$

$\displaystyle\int_1^3\int_0^3\int_0^{\sqrt{36-4y^2}/3} F(x,y,z)\,dx\,dy\,dz,$

$\displaystyle\int_0^3\int_1^3\int_0^{\sqrt{36-4y^2}/3} F(x,y,z)\,dx\,dz\,dy,\quad \int_0^2\int_1^3\int_0^{\sqrt{36-9x^2}/2} F(x,y,z)\,dy\,dz\,dx.$

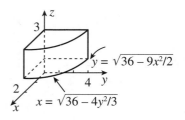

15.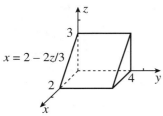

18.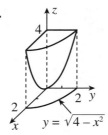

21. Solving $x=y^2$ and $4-x=y^2$, we obtain $x=2$, $y=\pm\sqrt{2}$. Using symmetry,

$\displaystyle V = 2\int_0^3\int_0^{\sqrt{2}}\int_{y^2}^{4-y^2} dx\,dy\,dz = 2\int_0^3\int_0^{\sqrt{2}}(4-2y^2)\,dy\,dz$

$\displaystyle = 2\int_0^3\left(4y-\frac{1}{3}y^3\right)\Big|_0^{\sqrt{2}}\,dz = 2\int_0^3 \frac{8\sqrt{2}}{3}\,dz = 16\sqrt{2}.$

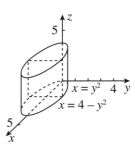

24. Solving $x=2$, $y=x$, and $z=x^2+y^2$, we obtain the point $(2,2,8)$.

$\displaystyle V = \int_0^2\int_0^x\int_0^{x^2+y^2} dz\,dy\,dx = \int_0^2\int_0^x (x^2+y^2)\,dy\,dx$

$\displaystyle = \int_0^2\left(x^2 y+\frac{1}{3}y^3\right)\Big|_0^x\,dx = \int_0^2 \frac{4}{3}x^3\,dx = \frac{1}{3}x^4\Big|_0^2 = \frac{16}{3}.$

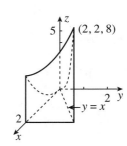

27. The density is $\rho(x, y, z) = ky$. Since both the region and the density function are symmetric with respect to the xy-and yz-planes, $\bar{x} = \bar{z} = 0$. Using symmetry,

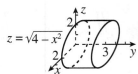

$$m = 4\int_0^3 \int_0^2 \int_0^{\sqrt{4-x^2}} ky\, dz\, dx\, dy = 4k\int_0^3 \int_0^2 yz\Big|_0^{\sqrt{4-x^2}} dx\, dy = 4k\int_0^3 \int_0^2 y\sqrt{4-x^2}\, dx\, dy$$

$$= 4k\int_0^3 y\left(\frac{x}{2}\sqrt{4-x^2} + 2\sin^{-1}\frac{x}{2}\right)\Big|_0^2 dy = 4k\int_0^3 \pi y\, dy = 4\pi k\left(\frac{1}{2}y^2\right)\Big|_0^3 = 18\pi k$$

$$M_{xz} = 4\int_0^3 \int_0^2 \int_0^{\sqrt{4-x^2}} ky^2\, dz\, dx\, dy = 4k\int_0^3 \int_0^2 y^2 z\Big|_0^{\sqrt{4-x^2}} dx\, dy = 4k\int_0^3 \int_0^2 y^2\sqrt{4-x^2}\, dx\, dy$$

$$= 4k\int_0^3 y^2\left(\frac{x}{2}\sqrt{4-x^2} + 2\sin^{-1}\frac{x}{2}\right)\Big|_0^2 dy = 4k\int_0^3 \pi y^2\, dy = 4\pi k\left(\frac{1}{3}y^3\right)\Big|_0^3 = 36\pi k.$$

$\bar{y} = M_{xz}/m = \dfrac{36\pi k}{18\pi k} = 2$. The center of mass is $(0, 2, 0)$.

30. Both the region and the density function are symmetric with respect to the xz- and yz-planes. Thus,

$$m = 4\int_{-1}^2 \int_0^{\sqrt{1+z^2}} \int_0^{\sqrt{1+z^2-y^2}} z^2\, dx\, dy\, dz.$$

33. $I_z = k\displaystyle\int_0^1 \int_0^{1-x} \int_0^{1-x-y} (x^2 + y^2)\, dz\, dy\, dx$

$= k\displaystyle\int_0^1 \int_0^{1-x} (x^2 + y^2)(1 - x - y)\, dy\, dx$

$= k\displaystyle\int_0^1 \int_0^{1-x} (x^2 - x^3 - x^2 y + y^2 - xy^2 - y^3)\, dy\, dx$

$= k\displaystyle\int_0^1 \left[(x^2 - x^3)y - \frac{1}{2}x^2 y^2 + \frac{1}{3}(1-x)y^3 - \frac{1}{4}y^4\right]\Big|_0^{1-x} dx$

$= k\displaystyle\int_0^1 \left[\frac{1}{2}x^2 - x^3 + \frac{1}{2}x^4 + \frac{1}{12}(1-x)^4\right] dx = k\left[\frac{1}{6}x^6 - \frac{1}{4}x^4 + \frac{1}{10}x^5 - \frac{1}{60}(1-x)^5\right]\Big|_0^1 = \frac{k}{30}$

36. $x = 2\cos 5\pi/6 = -\sqrt{3}$; $y = 2\sin 5\pi/6 = 1$; $(-\sqrt{3}, 1, -3)$

39. With $x = 1$ and $y = -1$ we have $r^2 = 2$ and $\tan\theta = -1$. The point is $(\sqrt{2}, -\pi/4, -9)$.

42. With $x = 1$ and $y = 2$ we have $r^2 = 5$ and $\tan\theta = 2$. The point is $(\sqrt{5}, \tan^{-1} 2, 7)$.

45. $r^2 - z^2 = 1$ **48.** $z = 2y$

51. The equations are $r^2 = 4$, $r^2 + z^2 = 16$, and $z = 0$.

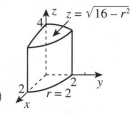

$$V = \int_0^{2\pi} \int_0^2 \int_0^{\sqrt{16-r^2}} r\, dz\, dr\, d\theta = \int_0^{2\pi} \int_0^2 r\sqrt{16-r^2}\, dr\, d\theta$$

$$= \int_0^{2\pi} -\frac{1}{3}(16-r^2)^{3/2}\Big|_0^2 \, d\theta = \int_0^{2\pi} \frac{1}{3}(64 - 24\sqrt{3})\, d\theta = \frac{2\pi}{3}(64 - 24\sqrt{3})$$

54. Substituting the first equation into the second, we see that the surfaces intersect in the plane $y = 4$. Using polar coordinates in the xz-plane, the equations of the surfaces become $y = r^2$ and $y = \frac{1}{2}r^2 + 2$.

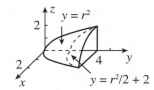

$$V = \int_0^{2\pi} \int_0^2 \int_{r^2}^{r^2/2+2} r\, dy\, dr\, d\theta = \int_0^{2\pi} \int_0^2 r\left(\frac{r^2}{2} + 2 - r^2\right) dr\, d\theta$$

$$= \int_0^{2\pi} \int_0^2 \left(2r - \frac{1}{2}r^3\right) dr\, d\theta = \int_0^{2\pi} \left(r^2 - \frac{1}{8}r^4\right)\Big|_0^2 \, d\theta = \int_0^{2\pi} 2\, d\theta = 4\pi$$

57. The equation is $z = \sqrt{9-r^2}$ and the density is $\rho = k/r^2$. When $z = 2$, $r = \sqrt{5}$.

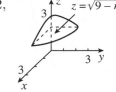

$$I_z = \int_0^{2\pi} \int_0^{\sqrt{5}} \int_2^{\sqrt{9-r^2}} r^2(k/r^2) r\, dz\, dr\, d\theta = k\int_0^{2\pi} \int_0^{\sqrt{5}} rz\Big|_2^{\sqrt{9-r^2}} \, dr\, d\theta$$

$$= k\int_0^{2\pi} \int_0^{\sqrt{5}} (r\sqrt{9-r^2} - 2r)\, dr\, d\theta = k\int_0^{2\pi} \left[-\frac{1}{3}(9-r^2)^{3/2} - r^2\right]\Big|_0^{\sqrt{5}} \, d\theta$$

$$= k\int_0^{2\pi} \frac{4}{3}\, d\theta = \frac{8}{3}\pi k$$

60. **(a)** $x = 5\sin(5\pi/4)\cos(2\pi/3) = 5\sqrt{2}/4$; $\quad y = 5\sin(5\pi/4)\sin(2\pi/3) = -5\sqrt{6}/4$;
$z = 5\cos(5\pi/4) = -5\sqrt{2}/2$; $\quad (5\sqrt{2}/4, -5\sqrt{6}/4, -5\sqrt{2}/2)$

(b) With $x = 5\sqrt{2}/4$ and $y = -5\sqrt{6}/4$ we have $r^2 = 25/2$ and $\tan\theta = -\sqrt{3}$. The point is $(5/\sqrt{2}, 2\pi/3, -5\sqrt{2}/2)$.

63. With $x = -5$, $y = -5$, and $z = 0$, we have $\rho^2 = 50$, $\tan\theta = 1$, and $\cos\phi = 0$. The point is $(5\sqrt{2}, \pi/2, 5\pi/4)$.

66. With $x = -\sqrt{3}/2$, $y = 0$, and $z = -1/2$, we have $\rho^2 = 1$, $\tan\theta = 0$, and $\cos\phi = -1/2$. The point is $(1, 2\pi/3, 0)$.

69. $4z^2 = 3x^2 + 3y^2 + 3z^2$; $\quad 4\rho^2\cos^2\phi = 3\rho^2$; $\quad \cos\phi = \pm\sqrt{3}/2$; $\quad \phi = \pi/6$ or equivalently, $\phi = 5\pi/6$

72. $\cos\phi = 1/2$; $\quad \rho^2\cos^2\phi = \rho^2/4$; $\quad 4z^2 = x^2 + y^2 + z^2$; $\quad x^2 + y^2 = 3z^2$

75. The equations are $\phi = \pi/4$ and $\rho = 3$.

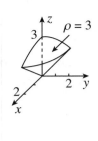

$$V = \int_0^{2\pi} \int_0^{\pi/4} \int_0^3 \rho^2 \sin\phi \, d\rho \, d\phi \, d\theta = \int_0^{2\pi} \int_0^{\pi/4} \frac{1}{3}\rho^3 \sin\phi \Big|_0^3 \, d\phi \, d\theta$$

$$= \int_0^{2\pi} \int_0^{\pi/4} 9 \sin\phi \, d\phi \, d\theta = \int_0^{2\pi} -9\cos\phi \Big|_0^{\pi/4} \, d\theta = -9 \int_0^{2\pi} \left(\frac{\sqrt{2}}{2} - 1 \right) d\theta$$

$$= 9\pi(2 - \sqrt{2})$$

78. The equations are $\rho = 1$ and $\phi = \pi/4$. We find the volume above the xy-plane and double.

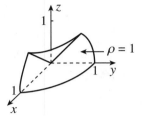

$$V = 2 \int_0^{2\pi} \int_{\pi/4}^{\pi/2} \int_0^1 \rho^2 \sin\phi \, d\rho \, d\phi \, d\theta = 2 \int_0^{2\pi} \int_{\pi/4}^{\pi/2} \frac{1}{3}\rho^3 \sin\phi \Big|_0^1 \, d\phi \, d\theta$$

$$= \frac{2}{3} \int_0^{2\pi} \int_{\pi/4}^{\pi/2} \sin\phi \, d\phi \, d\theta = \frac{2}{3} \int_0^{2\pi} -\cos\phi \Big|_{\pi/4}^{\pi/2} \, d\theta = \frac{2}{3} \int_0^{2\pi} \frac{\sqrt{2}}{2} \, d\theta = \frac{2\pi\sqrt{2}}{3}$$

81. We are given density $= k/\rho$.

$$m = \int_0^{2\pi} \int_0^{\cos^{-1} 4/5} \int_{4\sec\phi}^5 \frac{k}{\rho}\rho^2 \sin\phi \, d\rho \, d\phi \, d\theta$$

$$= k \int_0^{2\pi} \int_0^{\cos^{-1} 4/5} \frac{1}{2}\rho^2 \sin\phi \Big|_{4\sec\phi}^5 \, d\phi \, d\theta$$

$$= \frac{1}{2}k \int_0^{2\pi} \int_0^{\cos^{-1} 4/5} (25\sin\phi - 16\tan\phi\sec\phi) \, d\phi \, d\theta$$

$$= \frac{1}{2}k \int_0^{2\pi} (-25\cos\phi - 16\sec\phi) \Big|_0^{\cos^{-1} 4/5} \, d\theta = \frac{1}{2}k \int_0^{2\pi} [-25(4/5) - 16(5/4) - (-25 - 16)] \, d\theta$$

$$= \frac{1}{2}k \int_0^{2\pi} d\theta = k\pi$$

9.16   Divergence Theorem

3. div $\mathbf{F} = 3x^2 + 3y^2 + 3z^2$. Using spherical coordinates,

$$\iint_S \mathbf{F} \cdot \mathbf{n} \, dS = \iiint_D 3(x^2 + y^2 + z^2) \, dV = \int_0^{2\pi} \int_0^{\pi} \int_0^a 3\rho^2 \rho^2 \sin\phi \, d\rho \, d\phi \, d\theta$$

$$= \int_0^{2\pi} \int_0^{\pi} \frac{3}{5}\rho^5 \sin\phi \Big|_0^a \, d\phi \, d\theta = \frac{3a^5}{5} \int_0^{2\pi} \int_0^{\pi} \sin\phi \, d\phi \, d\theta$$

$$= \frac{3a^5}{5} \int_0^{2\pi} -\cos\phi \Big|_0^{\pi} \, d\theta = \frac{6a^5}{5} \int_0^{2\pi} d\theta = \frac{12\pi a^5}{5}.$$

6. div $\mathbf{F} = 2x + 2z + 12z^2$.

$$\iint_S \mathbf{F} \cdot \mathbf{n}\, dS = \iiint_D \text{div } \mathbf{F}\, dV = \int_0^3 \int_0^2 \int_0^1 (2x + 2z + 12z^2)\, dx\, dy\, dz$$

$$= \int_0^3 \int_0^2 (x^2 + 2xz + 12xz^2) \Big|_0^1 dy\, dz$$

$$= \int_0^3 \int_0^2 (1 + 2z + 12z^2)\, dy\, dz = \int_0^3 2(1 + 2z + 12z^2)\, dz = (2z + 2z^2 + 8z^3) \Big|_0^3 = 240$$

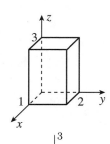

9. div $\mathbf{F} = \dfrac{1}{x^2 + y^2 + z^2}$. Using spherical coordinates,

$$\iint_S \mathbf{F} \cdot \mathbf{n}\, dS = \iiint_D \text{div } \mathbf{F}\, dV = \int_0^{2\pi} \int_0^{\pi} \int_a^b \frac{1}{\rho^2} \rho^2 \sin\phi\, d\rho\, d\phi\, d\theta$$

$$= \int_0^{2\pi} \int_0^{\pi} (b - a) \sin\phi\, d\phi\, d\theta = (b - a) \int_0^{2\pi} -\cos\phi \Big|_0^{\pi} d\theta$$

$$= (b - a) \int_0^{2\pi} 2\, d\theta = 4\pi(b - a).$$

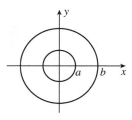

12. div $\mathbf{F} = 30xy$.

$$\iint_S \mathbf{F} \cdot \mathbf{n}\, dS = \iiint_D 30xy\, dV = \int_0^2 \int_0^{2-x} \int_{x+y}^3 30xy\, dz\, dy\, dx$$

$$= \int_0^2 \int_0^{2-x} 30xyz \Big|_{x+y}^3 dy\, dx$$

$$= \int_0^2 \int_0^{2-x} (90xy - 30x^2y - 30xy^2)\, dy\, dx$$

$$= \int_0^2 (45xy^2 - 15x^2y^2 - 10xy^3) \Big|_0^{2-x} dx$$

$$= \int_0^2 (-5x^4 + 45x^3 - 120x^2 + 100x)\, dx = \left(-x^5 + \frac{45}{4}x^4 - 40x^3 + 50x^2\right) \Big|_0^2 = 28$$

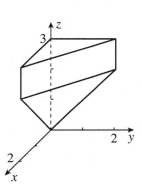

15. (a) div $\mathbf{E} = q \left[\dfrac{-2x^2 + y^2 + z^2}{(x^2 + y^2 + z^2)^{5/2}} + \dfrac{x^2 - 2y^2 + z^2}{(x^2 + y^2 + z^2)^{5/2}} + \dfrac{x^2 + y^2 - 2z^2}{(x^2 + y^2 + z^2)^{5/2}} \right] = 0$

$$\iint_{S \cup S_a} (\mathbf{E} \cdot \mathbf{n})\, dS = \iiint_D \text{div } \mathbf{E}\, dV = \iiint_D 0\, dV = 0$$

(b) From **(a)**, $\displaystyle\iint_S (\mathbf{E} \cdot \mathbf{n})\, dS + \iint_{S_a} (\mathbf{E} \cdot \mathbf{n})\, dS = 0$ and $\displaystyle\iint_S (\mathbf{E} \cdot \mathbf{n})\, dS = -\iint_{S_a} (\mathbf{E} \cdot \mathbf{n})\, dS$. On

S_a, $|\mathbf{r}| = a$, $\mathbf{n} = -(x\mathbf{i} + y\mathbf{j} + z\mathbf{k})/a = -\mathbf{r}/a$ and

$\mathbf{E} \cdot \mathbf{n} = (q\mathbf{r}/a^3) \cdot (-\mathbf{r}/a) = -qa^2/a^4 = -q/a^2.$ Thus

$$\iint_S (\mathbf{E} \cdot \mathbf{n})\, dS = -\iint_{S_a} \left(-\frac{q}{a^2}\right) dS = \frac{q}{a^2} \iint_{S_a} dS = \frac{q}{a^2} \times (\text{area of } S_a) = \frac{q}{a^2}(4\pi a^2) = 4\pi q.$$

18. By the Divergence Theorem and Problem 30 in Section 9.7,

$$\iint_S (\text{curl } \mathbf{F} \cdot \mathbf{n})\, dS = \iiint_D \text{div (curl } \mathbf{F})\, dV = \iiint_D 0\, dV = 0.$$

21. If $G(x, y, z)$ is a vector valued function then we define surface integrals and triple integrals of **G** component-wise. In this case, if **a** is a constant vector it is easily shown that

$$\iint_S \mathbf{a} \cdot \mathbf{G}\, dS = \mathbf{a} \cdot \iint_S \mathbf{G}\, dS \quad \text{and} \quad \iiint_D \mathbf{a} \cdot \mathbf{G}\, dV = \mathbf{a} \cdot \iiint_D \mathbf{G}\, dV.$$

Now let $\mathbf{F} = f\mathbf{a}$. Then

$$\iint_S \mathbf{F} \cdot \mathbf{n}\, dS = \iint_S (f\mathbf{a}) \cdot \mathbf{n}\, dS = \iint_S \mathbf{a} \cdot (f\mathbf{n})\, dS$$

and, using Problem 27 in Section 9.7 and the fact that $\nabla \cdot \mathbf{a} = 0$, we have

$$\iiint_D \text{div } \mathbf{F}\, dV = \iiint_D \nabla \cdot (f\mathbf{a})\, dV = \iiint_D [f(\nabla \cdot \mathbf{a}) + \mathbf{a} \cdot \nabla f]\, dV = \iiint_D \mathbf{a} \cdot \nabla f\, dV.$$

By the Divergence Theorem,

$$\iint_S \mathbf{a} \cdot (f\mathbf{n})\, dS = \iint_S \mathbf{F} \cdot \mathbf{n}\, dS = \iiint_D \text{div } \mathbf{F}\, dV = \iiint_D \mathbf{a} \cdot \nabla f\, dV$$

and

$$\mathbf{a} \cdot \left(\iint_S f\mathbf{n}\, dS\right) = \mathbf{a} \cdot \left(\iiint_D \nabla f\, dV\right) \quad \text{or} \quad \mathbf{a} \cdot \left(\iint_S f\mathbf{n}\, dS - \iiint_D \nabla f\, dV\right) = 0.$$

Since **a** is arbitrary,

$$\iint_S f\mathbf{n}\, dS - \iiint_D \nabla f\, dV = 0 \quad \text{and} \quad \iint_S f\mathbf{n}\, dS = \iiint_D \nabla f\, dV.$$

9.17   Change of Variables in Multiple Integrals

3. The uv-corner points $(0,0)$, $(2,0)$, $(2,2)$ correspond to xy-points $(0,0)$, $(4,2)$, $(6,-4)$.

$v = 0$:
$x = 2u$, $y = u$ implies $y = x/2$

$u = 2$:
$x = 4 + v$, $y = 2 - 3v$ implies
$y = 2 - 3(x - 4) = -3x + 14$

$v = u$:
$x = 3u$, $y = -2u$ implies $y = -2x/3$

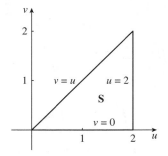

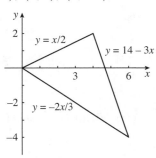

6. The uv-corner points $(1,1)$, $(2,1)$, $(2,2)$, $(1,2)$ correspond to the xy-points $(1,1)$, $(2,1)$, $(4,4)$, $(2,4)$.

$v = 1$: $x = u$, $y = 1$ implies $y = 1$, $1 \le x \le 2$

$u = 2$: $x = 2v$, $y = v^2$ implies $y = x^2/4$

$v = 2$: $x = 2u$, $y = 4$ implies $y = 4$, $2 \le x \le 4$

$u = 1$: $x = v$, $y = v^2$ implies $y = x^2$

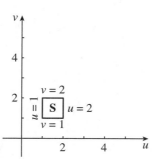

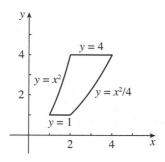

9. $\dfrac{\partial(u,v)}{\partial(x,y)} = \begin{vmatrix} -2y/x^3 & 1/x^2 \\ -y^2/x^2 & 2y/x \end{vmatrix} = -\dfrac{3y^2}{x^4} = -3\left(\dfrac{y}{x^2}\right)^2 = -3u^2; \quad \dfrac{\partial(x,y)}{\partial(u,v)} = \dfrac{1}{-3u^2} = -\dfrac{1}{3u^2}$

12. $\dfrac{\partial(x,y)}{\partial(u,v)} = \begin{vmatrix} 1-v & v \\ -u & u \end{vmatrix} = u$. The transformation is 0 when u is 0, for $0 \le v \le 1$.

15. $R1$: $y = x^2$ implies $u = 1$

$R2$: $x = y^2$ implies $v = 1$

$R3$: $y = \frac{1}{2}x^2$ implies $u = 2$

$R4$: $x = \frac{1}{2}y^2$ implies $v = 2$

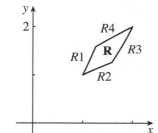

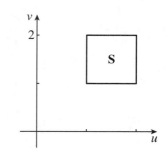

$\dfrac{\partial(u,v)}{\partial(x,y)} = \begin{vmatrix} 2x/y & -x^2/y^2 \\ -y^2/x^2 & 2y/x \end{vmatrix} = 3 \quad \text{or} \quad \dfrac{\partial(x,y)}{\partial(u,v)} = \dfrac{1}{3}$

$$\iint_R \frac{y^2}{x}\, dA = \iint_S v\left(\frac{1}{3}\right) dA' = \frac{1}{3}\int_1^2 \int_1^2 v\, du\, dv = \frac{1}{3}\int_1^2 v\, dv = \frac{1}{6}v^2 \Big|_1^2 = \frac{1}{2}$$

18. $R1$: $xy = -2$ implies $v = -2$

$R2$: $x^2 - y^2 = 9$ implies $u = 9$

$R3$: $xy = 2$ implies $v = 2$

$R4$: $x^2 - y^2 = 1$ implies $u = 1$

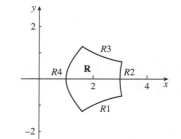

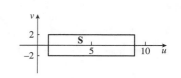

$$\frac{\partial(u, v)}{\partial(x, y)} = \begin{vmatrix} 2x & -2y \\ y & x \end{vmatrix} = 2(x^2 + y^2) \quad \text{or} \quad \frac{\partial(x, y)}{\partial(u, v)} = \frac{1}{2(x^2 + y^2)}$$

$$\iint_R (x^2 + y^2) \sin xy \, dA = \iint_S (x^2 + y^2) \sin v \left(\frac{1}{2(x^2 + y^2)} \right) dA' = \frac{1}{2} \int_{-2}^{2} \int_{1}^{9} \sin v \, du \, dv$$

$$= \frac{1}{2} \int_{-2}^{2} 8 \sin v \, dv = 0$$

21. R1: $y = 1/x$ implies $u = 1$

R2: $y = x$ implies $v = 1$

R3: $y = 4/x$ implies $u = 4$

R4: $y = 4x$ implies $v = 4$

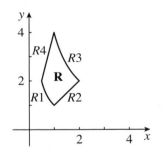
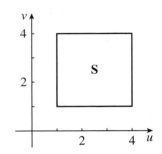

$$\frac{\partial(u, v)}{\partial(x, y)} = \begin{vmatrix} y & x \\ -y/x^2 & 1/x \end{vmatrix} = \frac{2y}{x} \quad \text{or} \quad \frac{\partial(x, y)}{\partial(u, v)} = \frac{x}{2y}$$

$$\iint_R y^4 \, dA = \iint_S u^2 v^2 \left(\frac{1}{2v} \right) du \, dv = \frac{1}{2} \int_{1}^{4} \int_{1}^{4} u^2 v \, du \, dv = \frac{1}{2} \int_{1}^{4} \frac{1}{3} u^3 v \Big|_{1}^{4} dv = \frac{1}{6} \int_{1}^{4} 63 v \, dv$$

$$= \frac{21}{4} v^2 \Big|_{1}^{4} = \frac{315}{4}$$

24. We let $u = y - x$ and $v = y$.

R1: $y = 0$ implies $v = 0$, $u = -x$ implies $v = 0$, $0 \le u \le 2$

R2: $x = 0$ implies $v = u$

R3: $y = x + 2$ implies $u = 2$

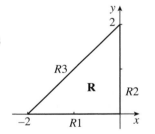
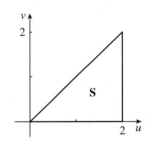

$$\frac{\partial(u, v)}{\partial(x, y)} = \begin{vmatrix} -1 & 1 \\ 0 & 1 \end{vmatrix} = -1 \quad \text{or} \quad \frac{\partial(x, y)}{\partial(u, v)} = -1$$

$$\iint_R e^{y^2 - 2xy + x^2} \, dA = \iint_S e^{u^2} |-1| \, dA' = \int_{0}^{2} \int_{0}^{u} e^{u^2} \, dv \, du = \int_{0}^{2} u e^{u^2} \, du = \frac{1}{2} e^{u^2} \Big|_{0}^{2} = \frac{1}{2}(e^4 - 1)$$

27. Let $u = xy$ and $v = xy^{1.4}$. Then $xy^{1.4} = c$ implies $v = c$; $xy = b$ implies $u = b$; $xy^{1.4} = d$ implies $v = d$; $xy = a$ implies $u = a$.

$$\frac{\partial(u, v)}{\partial(x, y)} = \begin{vmatrix} y & x \\ y^{1.4} & 1.4xy^{0.4} \end{vmatrix} = 0.4xy^{1.4} = 0.4v \quad \text{or} \quad \frac{\partial(x, y)}{\partial(u, v)} = \frac{5}{2v}$$

$$\iint_R dA = \iint_S \frac{5}{2v}\, dA' = \int_c^d \int_a^b \frac{5}{2v}\, du\, dv = \frac{5}{2}(b-a)\int_c^d \frac{dv}{v} = \frac{5}{2}(b-a)(\ln d - \ln c)$$

30. $$\frac{\partial(x,y,z)}{\partial(\rho,\phi,\theta)} = \begin{vmatrix} \sin\phi\cos\theta & \rho\cos\phi\cos\theta & -\rho\sin\phi\sin\theta \\ \sin\phi\sin\theta & \rho\cos\phi\sin\theta & \rho\sin\phi\cos\theta \\ \cos\phi & -\rho\sin\phi & 0 \end{vmatrix}$$

$$= \cos\phi(\rho^2\sin\phi\cos\phi\cos^2\theta + \rho^2\sin\phi\cos\phi\sin^2\theta)$$

$$+ \rho\sin\phi(\rho\sin^2\phi\cos^2\theta + \rho\sin^2\phi\sin^2\theta)$$

$$= \rho^2\sin\phi\cos^2\phi(\cos^2\theta + \sin^2\theta) + \rho^2\sin^3\phi(\cos^2\theta + \sin^2\theta)$$

$$= \rho^2\sin\phi(\cos^2\phi + \sin^2\phi) = \rho^2\sin\phi$$

Chapter 9 in Review

3. True

6. False; consider $f(x,y) = xy$ at $(0,0)$.

9. False; $\displaystyle\int_C x\,dx + x^2\,dy = 0$ from $(-1,0)$ to $(1,0)$ along the x-axis and along the semicircle $y = \sqrt{1-x^2}$, but since $x\,dx + x^2\,dy$ is not exact, the integral is not independent of path.

12. True   **15.** True   **18.** True

21. $\mathbf{v}(t) = 6\mathbf{i} + \mathbf{j} + 2t\mathbf{k}$; $\mathbf{a}(t) = 2\mathbf{k}$. To find when the particle passes through the plane, we solve $-6t + t + t^2 = -4$ or $t^2 - 5t + 4 = 0$. This gives $t = 1$ and $t = 4$. $\mathbf{v}(1) = 6\mathbf{i} + \mathbf{j} + 2\mathbf{k}$, $\mathbf{a}(1) = 2\mathbf{k}$; $\mathbf{v}(4) = 6\mathbf{i} + \mathbf{j} + 8\mathbf{k}$, $\mathbf{a}(4) = 2\mathbf{k}$

24. $\mathbf{v}(t) = t\mathbf{i} + t^2\mathbf{j} - t\mathbf{k}$; $|\mathbf{v}| = t\sqrt{t^2+2}, t > 0$; $\mathbf{a}(t) = \mathbf{i} + 2t\mathbf{j} - \mathbf{k}$; $\mathbf{v}\cdot\mathbf{a} = t + 2t^3 + t = 2t + 2t^3$;

$\mathbf{v}\times\mathbf{a} = t^2\mathbf{i} + t^2\mathbf{k}$, $|\mathbf{v}\times\mathbf{a}| = t^2\sqrt{2}$; $a_T = \dfrac{2t+2t^3}{t\sqrt{t^2+2}} = \dfrac{2+2t^2}{\sqrt{t^2+2}}$, $a_N = \dfrac{t^2\sqrt{2}}{t\sqrt{t^2+2}} = \dfrac{\sqrt{2}\,t}{\sqrt{t^2+2}}$;

$\kappa = \dfrac{t^2\sqrt{2}}{t^3(t^2+2)^{3/2}} = \dfrac{\sqrt{2}}{t(t^2+2)^{3/2}}$

27. $\nabla f = (2xy - y^2)\mathbf{i} + (x^2 - 2xy)\mathbf{j}$; $\mathbf{u} = \dfrac{2}{\sqrt{40}}\mathbf{i} + \dfrac{6}{\sqrt{40}}\mathbf{j} = \dfrac{1}{\sqrt{10}}(\mathbf{i} + 3\mathbf{j})$;

$D_{\mathbf{u}}f = \dfrac{1}{\sqrt{10}}(2xy - y^2 + 3x^2 - 6xy) = \dfrac{1}{\sqrt{10}}(3x^2 - 4xy - y^2)$

30. (a) $\dfrac{dw}{dt} = \dfrac{\partial w}{\partial x}\dfrac{dx}{dt} + \dfrac{\partial w}{\partial y}\dfrac{dy}{dt} + \dfrac{\partial w}{\partial z}\dfrac{dz}{dt}$

$\qquad = \dfrac{x}{\sqrt{x^2+y^2+z^2}}\,6\cos 2t + \dfrac{y}{\sqrt{x^2+y^2+z^2}}\,(-8\sin 2t) + \dfrac{z}{\sqrt{x^2+y^2+z^2}}\,15t^2$

$\qquad = \dfrac{(6x\cos 2t - 8y\sin 2t + 15zt^2)}{\sqrt{x^2+y^2+z^2}}$

(b) $\dfrac{\partial w}{\partial t} = \dfrac{\partial w}{\partial x}\dfrac{\partial x}{\partial t} + \dfrac{\partial w}{\partial y}\dfrac{\partial y}{\partial t} + \dfrac{\partial w}{\partial z}\dfrac{\partial z}{\partial t}$

$\qquad = \dfrac{x}{\sqrt{x^2+y^2+z^2}}\,\dfrac{6}{r}\cos\dfrac{2t}{r} + \dfrac{y}{\sqrt{x^2+y^2+z^2}}\left(\dfrac{8r}{t^2}\sin\dfrac{2r}{t}\right) + \dfrac{z}{\sqrt{x^2+y^2+z^2}}\,15t^2r^3$

$\qquad = \dfrac{\left(\dfrac{6x}{r}\cos\dfrac{2t}{r} + \dfrac{8yr}{t^2}\sin\dfrac{2r}{t} + 15zt^2r^3\right)}{\sqrt{x^2+y^2+z^2}}$

33. (a) $V = \displaystyle\int_0^1 \int_x^{2x} \sqrt{1-x^2}\,dy\,dx = \int_0^1 y\sqrt{1-x^2}\,\Big|_x^{2x}\,dx = \int_0^1 x\sqrt{1-x^2}\,dx$

$\qquad = -\dfrac{1}{3}(1-x^2)^{3/2}\,\Big|_0^1 = \dfrac{1}{3}$

(b) $V = \displaystyle\int_0^1 \int_{y/2}^{y} \sqrt{1-x^2}\,dx\,dy + \int_1^2 \int_{y/2}^{1} \sqrt{1-x^2}\,dx\,dy$

36. (a) Using symmetry,

$\qquad V = 8\displaystyle\int_0^a \int_0^{\sqrt{a^2-x^2}} \int_0^{\sqrt{a^2-x^2-y^2}} dz\,dy\,dx$

$\qquad = 8\displaystyle\int_0^a \int_0^{\sqrt{a^2-x^2}} \sqrt{a^2-x^2-y^2}\,dy\,dx \qquad \boxed{\text{Trig substitution}}$

$\qquad = 8\displaystyle\int_0^a \left(\dfrac{y}{2}\sqrt{a^2-x^2-y^2} + \dfrac{a^2-x^2}{2}\sin^{-1}\dfrac{y}{\sqrt{a^2-x^2}}\right]\Bigg|_0^{\sqrt{a^2-x^2}}\,dx$

$\qquad = 8\displaystyle\int_0^a \dfrac{\pi}{2}\dfrac{a^2-x^2}{2}\,dx = 2\pi\left(a^2x - \dfrac{1}{3}x^3\right)\Big|_0^a = \dfrac{4}{3}\pi a^3.$

(b) Using symmetry,

$\qquad V = 2\displaystyle\int_0^{2\pi} \int_0^a \int_0^{\sqrt{a^2-r^2}} r\,dz\,dr\,d\theta = 2\int_0^{2\pi}\int_0^a r\sqrt{a^2-r^2}\,dr\,d\theta$

$\qquad = 2\displaystyle\int_0^{2\pi} -\dfrac{1}{3}(a^2-r^2)^{3/2}\,\Big|_0^a\,d\theta = \dfrac{2}{3}\int_0^{2\pi} a^3\,d\theta = \dfrac{4}{3}\pi a^3.$

(c) $V = \int_0^{2\pi} \int_0^{\pi} \int_0^a \rho^2 \sin\phi \, d\rho \, d\phi \, d\theta = \int_0^{2\pi} \int_0^{\pi} \frac{1}{3}\rho^3 \sin\phi \Big|_0^a \, d\phi \, d\theta$

$= \frac{1}{3} \int_0^{2\pi} \int_0^{\pi} a^3 \sin\phi \, d\phi \, d\theta = \frac{1}{3} \int_0^{2\pi} -a^3 \cos\phi \Big|_0^{\pi} \, d\theta = \frac{1}{3} \int_0^{2\pi} 2a^3 \, d\theta = \frac{4}{3}\pi a^3$

39. $2xy + 2xy + 2xy = 6xy$

42. $\nabla(6xy) = 6y\mathbf{i} + 6x\mathbf{j}$

45. Since $P_y = 6x^2 y = Q_x$, the integral is independent of path.

$\phi_x = 3x^2 y^2, \quad \phi = x^3 y^2 + g(y), \quad \phi_y = 2x^3 y + g'(y) = 2x^3 y - 3y^2; \quad g(y) = -y^3; \quad \phi = x^3 y^2 - y^3;$

$\int_{(0,0)}^{(1,-2)} 3x^2 y^2 \, dx + (2x^3 y - 3y^2) \, dy = (x^3 y^2 - y^3) \Big|_{(0,0)}^{(1,-2)} = 12$

48. Parameterize C by $x = \cos t$, $y = \sin t$; $0 \le t \le 2\pi$. Then

$\oint_C \mathbf{F} \cdot d\mathbf{r} = \int_0^{2\pi} [4\sin t(-\sin t \, dt) + 6\cos t(\cos t) \, dt] = \int_0^{2\pi} (6\cos^2 t - 4\sin^2 t) \, dt$

$= \int_0^{2\pi} (10\cos^2 t - 4) \, dt = \left(5t + \frac{5}{2}\sin 2t - 4t \right) \Big|_0^{2\pi} = 2\pi.$

Using Green's Theorem, $Q_x - P_y = 6 - 4 = 2$ and $\oint_C \mathbf{F} \cdot d\mathbf{r} = \iint_R 2 \, dA = 2(\pi \cdot 1^2) = 2\pi.$

51. $z_x = 2x$, $z_y = 0$; $dS = \sqrt{1 + 4x^2} \, dA$

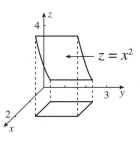

$\iint_S \frac{z}{xy} \, dS = \int_1^3 \int_1^2 \frac{x^2}{xy} \sqrt{1 + 4x^2} \, dx \, dy = \int_1^3 \frac{1}{y} \left[\frac{1}{12}(1 + 4x^2)^{3/2} \right] \Big|_1^2 \, dy$

$= \frac{1}{12} \int_1^3 \frac{17^{3/2} - 5^{3/2}}{y} \, dy = \frac{17\sqrt{17} - 5\sqrt{5}}{12} \ln y \Big|_1^3$

$= \frac{17\sqrt{17} - 5\sqrt{5}}{12} \ln 3$

54. In Problem 53, $\mathbf{F}$ is not continuous at $(0,0,0)$ which is in any acceptable region containing the sphere.

57. Identify $\mathbf{F} = -2y\mathbf{i} + 3x\mathbf{j} + 10z\mathbf{k}$. Then curl $\mathbf{F} = 5\mathbf{k}$. The curve C lies in the plane $z = 3$, so $\mathbf{n} = \mathbf{k}$ and $dS = dA$. Thus,

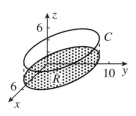

$\oint_C \mathbf{F} \cdot d\mathbf{r} = \iint_S (\text{curl } \mathbf{F}) \cdot \mathbf{n} \, dS = \iint_R 5 \, dA$

$= 5 \times (\text{area of })R = 5(25\pi) = 125\pi.$

60. div $\mathbf{F} = x^2 + y^2 + z^2$. Using cylindrical coordinates,

$$\iint_S \mathbf{F} \cdot \mathbf{n}\, dS = \iiint_D \operatorname{div} \mathbf{F}\, dV = \iiint_D (x^2 + y^2 + z^2)\, dV = \int_0^{2\pi} \int_0^1 \int_0^1 (r^2 + z^2) r\, dz\, dr\, d\theta$$

$$= \int_0^{2\pi} \int_0^1 \left. \left(r^3 z + \frac{1}{3} r z^3 \right) \right|_0^1 dr\, d\theta = \int_0^{2\pi} \int_0^1 \left(r^3 + \frac{1}{3} r \right) dr\, d\theta$$

$$= \int_0^{2\pi} \left. \left(\frac{1}{4} r^4 + \frac{1}{6} r^2 \right) \right|_0^1 d\theta = \int_0^{2\pi} \frac{5}{12}\, d\theta = \frac{5\pi}{6}.$$

63. $x = 0$ implies $u = 0$, $v = -y^2$ implies $u = 0$, $-1 \le v \le 0$

$x = 1$ implies $u = 2y$, $v = 1 - y^2 = 1 - u^2/4$

$y = 0$ implies $u = 0$, $v = x^2$ implies $u = 0$, $0 \le v \le 1$

$y = 1$ implies $u = 2x$, $v = x^2 - 1 = u^2/4 - 1$

$$\frac{\partial(u,v)}{\partial(x,y)} = \begin{vmatrix} 2y & 2x \\ 2x & -2y \end{vmatrix} = -4(x^2 + y^2) \qquad \text{or} \qquad \frac{\partial(x,y)}{\partial(u,v)} = -\frac{1}{4(x^2 + y^2)}$$

$$\iint_R (x^2 + y^2) \sqrt[3]{x^2 - y^2}\, dA = \iint_S (x^2 + y^2) \sqrt[3]{v} \left| -\frac{1}{4(x^2 + y^2)} \right| dA'$$

$$= \frac{1}{4} \int_0^2 \int_{u^2/4 - 1}^{1 - u^2/4} v^{1/3}\, dv\, du = \frac{1}{4} \int_0^2 \left. \frac{3}{4} v^{4/3} \right|_{u^2/4-1}^{1-u^2/4} du$$

$$= \frac{3}{16} \int_0^2 \left[(1 - u^2/4)^{4/3} - (u^2/4 - 1)^{4/3} \right] du$$

$$= \frac{3}{16} \int_0^2 \left[(1 - u^2/4)^{4/3} - (1 - u^2/4)^{4/3} \right] du = 0$$

66. (a) Both states span 7 degrees of longitude and 4 degrees of latitude, but Colorado is larger because it lies to the south of Wyoming. Lines of longitude converge as they go north, so the east-west dimensions of Wyoming are shorter than those of Colorado.

(b) We use the function $f(x, y) = \sqrt{R^2 - x^2 - y^2}$ to describe the northern hemisphere, where $R \approx 3960$ miles is the radius of the Earth. We need to compute the surface area over a polar rectangle P of the form $\theta_1 \le \theta \le \theta_2$, $R \cos \phi_2 \le r \le R \cos \phi_1$. We have

$$f_x = \frac{-x}{\sqrt{R^2 - x^2 - y^2}} \quad \text{and} \quad f_y = \frac{-y}{\sqrt{R^2 - x^2 - y^2}}$$

so that

$$\sqrt{1 + f_x^2 + f_y^2} = \sqrt{1 + \frac{x^2 + y^2}{R^2 - x^2 - y^2}} = \frac{R}{\sqrt{R^2 - r^2}}.$$

Thus

$$A = \iint_P \sqrt{1 + f_x^2 + f_y^2}\, dA = \int_{\theta_1}^{\theta_2} \int_{R\cos\phi_2}^{R\cos\phi_1} \frac{R}{\sqrt{R^2 - r^2}}\, r\, dr\, d\theta$$

$$= (\theta_2 - \theta_1)R\sqrt{R^2 - r^2}\,\Big|_{R\cos\phi_1}^{R\cos\phi_2} = (\theta_2 - \theta_1)R^2(\sin\phi_2 - \sin\phi_1).$$

The ratio of Wyoming to Colorado is then $\dfrac{\sin 45° - \sin 41°}{\sin 41° - \sin 37°} \approx 0.941$. Thus Wyoming is about 6% smaller than Colorado.

(c) $97{,}914/104{,}247 \approx 0.939$, which is close to the theoretical value of 0.941. (Our formula for the area says that the area of Colorado is approximately 103,924 square miles, while the area of Wyoming is approximately 97,801 square miles.)

Chapter 10

Systems of Linear Differential Equations

10.1 | Theory of Linear Systems

3. Let $\mathbf{X} = \begin{pmatrix} x \\ y \end{pmatrix}$. Then $\mathbf{X}' = \begin{pmatrix} \dfrac{5}{2} & -7 \\ -\dfrac{1}{4} & \dfrac{3}{2} \end{pmatrix} \mathbf{X} + \begin{pmatrix} \dfrac{9}{2}t \\ -\dfrac{1}{2} \end{pmatrix}$.

6. Let $\mathbf{X} = \begin{pmatrix} x \\ y \\ z \end{pmatrix}$. Then $\mathbf{X}' = \begin{pmatrix} 1 & -1 & 0 \\ 1 & 0 & 2 \\ -1 & 0 & 1 \end{pmatrix} \mathbf{X}$.

9. $\dfrac{dx}{dt} = 4x + 2y + e^t$; $\quad \dfrac{dy}{dt} = -x + 3y - e^t$

12. $\dfrac{dx}{dt} = 3x - 7y + 4\sin t + (t-4)e^{4t}$; $\quad \dfrac{dy}{dt} = x + y + 8\sin t + (2t+1)e^{4t}$

15. Let $y = x_1$, $y' = x_2$, and $y'' = x_3$, then $y' = x_1' = x_2$, $y'' = x_2' = x_3$, and $y''' = x_3' = y - 10y' - 6y'' + t^2 = x_1 - 10x_2 - 6x_3 + t^2$. Thus,

$$x_1' = x_2$$
$$x_2' = x_3$$
$$x_3' = x_1 - 10x_2 - 6x_3 + t^2$$

yields

$$\begin{pmatrix} x_1' \\ x_2' \\ x_3' \end{pmatrix} = \begin{pmatrix} 0 & 1 & 0 \\ 0 & 0 & 1 \\ 1 & -10 & -6 \end{pmatrix} \begin{pmatrix} x_1 \\ x_2 \\ x_3 \end{pmatrix} + \begin{pmatrix} 0 \\ 0 \\ t^2 \end{pmatrix}.$$

18. Since

$$\mathbf{X}' = \begin{pmatrix} 5\cos t - 5\sin t \\ 2\cos t - 4\sin t \end{pmatrix} e^t \quad \text{and} \quad \begin{pmatrix} -2 & 5 \\ -2 & 4 \end{pmatrix} \mathbf{X} = \begin{pmatrix} 5\cos t - 5\sin t \\ 2\cos t - 4\sin t \end{pmatrix} e^t$$

we see that

$$\mathbf{X}' = \begin{pmatrix} -2 & 5 \\ -2 & 4 \end{pmatrix} \mathbf{X}.$$

21. Since

$$\mathbf{X}' = \begin{pmatrix} 0 \\ 0 \\ 0 \end{pmatrix} \quad \text{and} \quad \begin{pmatrix} 1 & 2 & 1 \\ 6 & -1 & 0 \\ -1 & -2 & -1 \end{pmatrix} \mathbf{X} = \begin{pmatrix} 0 \\ 0 \\ 0 \end{pmatrix}$$

we see that

$$\mathbf{X}' = \begin{pmatrix} 1 & 2 & 1 \\ 6 & -1 & 0 \\ -1 & -2 & -1 \end{pmatrix} \mathbf{X}.$$

24. Yes, since $W(\mathbf{X}_1, \mathbf{X}_2) = 8e^{2t} \neq 0$ the set $\mathbf{X}_1$, $\mathbf{X}_2$ is linearly independent on the interval $(-\infty, \infty)$.

27. Since

$$\mathbf{X}'_p = \begin{pmatrix} 2 \\ -1 \end{pmatrix} \quad \text{and} \quad \begin{pmatrix} 1 & 4 \\ 3 & 2 \end{pmatrix} \mathbf{X}_p + \begin{pmatrix} 2 \\ -4 \end{pmatrix} t + \begin{pmatrix} -7 \\ -18 \end{pmatrix} = \begin{pmatrix} 2 \\ -1 \end{pmatrix}$$

we see that

$$\mathbf{X}'_p = \begin{pmatrix} 1 & 4 \\ 3 & 2 \end{pmatrix} \mathbf{X}_p + \begin{pmatrix} 2 \\ -4 \end{pmatrix} t + \begin{pmatrix} -7 \\ -18 \end{pmatrix}.$$

30. Since

$$\mathbf{X}'_p = \begin{pmatrix} 3\cos 3t \\ 0 \\ -3\sin 3t \end{pmatrix} \quad \text{and} \quad \begin{pmatrix} 1 & 2 & 3 \\ -4 & 2 & 0 \\ -6 & 1 & 0 \end{pmatrix} \mathbf{X}_p + \begin{pmatrix} -1 \\ 4 \\ 3 \end{pmatrix} \sin 3t = \begin{pmatrix} 3\cos 3t \\ 0 \\ -3\sin 3t \end{pmatrix}$$

we see that

$$\mathbf{X}'_p = \begin{pmatrix} 1 & 2 & 3 \\ -4 & 2 & 0 \\ -6 & 1 & 0 \end{pmatrix} \mathbf{X}_p + \begin{pmatrix} -1 \\ 4 \\ 3 \end{pmatrix} \sin 3t.$$

10.2   Homogeneous Linear Systems

3. The system is

$$\mathbf{X}' = \begin{pmatrix} -4 & 2 \\ -\frac{5}{2} & 2 \end{pmatrix} \mathbf{X}$$

and $\det(\mathbf{A} - \lambda\mathbf{I}) = (\lambda - 1)(\lambda + 3) = 0$. For $\lambda_1 = 1$ we obtain

$$\begin{pmatrix} -5 & 2 & | & 0 \\ -\frac{5}{2} & 1 & | & 0 \end{pmatrix} \longrightarrow \begin{pmatrix} -5 & 2 & | & 0 \\ 0 & 0 & | & 0 \end{pmatrix} \quad \text{so that} \quad \mathbf{K}_1 = \begin{pmatrix} 2 \\ 5 \end{pmatrix}.$$

For $\lambda_2 = -3$ we obtain

$$\begin{pmatrix} -1 & 2 & | & 0 \\ -\frac{5}{2} & 5 & | & 0 \end{pmatrix} \longrightarrow \begin{pmatrix} -1 & 2 & | & 0 \\ 0 & 0 & | & 0 \end{pmatrix} \quad \text{so that} \quad \mathbf{K}_2 = \begin{pmatrix} 2 \\ 1 \end{pmatrix}.$$

Then

$$\mathbf{X} = c_1 \begin{pmatrix} 2 \\ 5 \end{pmatrix} e^t + c_2 \begin{pmatrix} 2 \\ 1 \end{pmatrix} e^{-3t}.$$

6. The system is

$$\mathbf{X}' = \begin{pmatrix} -6 & 2 \\ -3 & 1 \end{pmatrix} \mathbf{X}$$

and $\det(\mathbf{A} - \lambda\mathbf{I}) = \lambda(\lambda + 5) = 0$. For $\lambda_1 = 0$ we obtain

$$\begin{pmatrix} -6 & 2 & | & 0 \\ -3 & 1 & | & 0 \end{pmatrix} \longrightarrow \begin{pmatrix} 1 & -\frac{1}{3} & | & 0 \\ 0 & 0 & | & 0 \end{pmatrix} \quad \text{so that} \quad \mathbf{K}_1 = \begin{pmatrix} 1 \\ 3 \end{pmatrix}.$$

For $\lambda_2 = -5$ we obtain

$$\begin{pmatrix} -1 & 2 & | & 0 \\ -3 & 6 & | & 0 \end{pmatrix} \longrightarrow \begin{pmatrix} 1 & -2 & | & 0 \\ 0 & 0 & | & 0 \end{pmatrix} \quad \text{so that} \quad \mathbf{K}_2 = \begin{pmatrix} 2 \\ 1 \end{pmatrix}.$$

Then

$$\mathbf{X} = c_1 \begin{pmatrix} 1 \\ 3 \end{pmatrix} + c_2 \begin{pmatrix} 2 \\ 1 \end{pmatrix} e^{-5t}.$$

9. We have $\det(\mathbf{A} - \lambda\mathbf{I}) = -(\lambda + 1)(\lambda - 3)(\lambda + 2) = 0$. For $\lambda_1 = -1$, $\lambda_2 = 3$, and $\lambda_3 = -2$ we obtain

$$\mathbf{K}_1 = \begin{pmatrix} -1 \\ 0 \\ 1 \end{pmatrix}, \quad \mathbf{K}_2 = \begin{pmatrix} 1 \\ 4 \\ 3 \end{pmatrix}, \quad \text{and} \quad \mathbf{K}_3 = \begin{pmatrix} 1 \\ -1 \\ 3 \end{pmatrix},$$

so that

$$\mathbf{X} = c_1 \begin{pmatrix} -1 \\ 0 \\ 1 \end{pmatrix} e^{-t} + c_2 \begin{pmatrix} 1 \\ 4 \\ 3 \end{pmatrix} e^{3t} + c_3 \begin{pmatrix} 1 \\ -1 \\ 3 \end{pmatrix} e^{-2t}.$$

12. We have $\det(\mathbf{A} - \lambda\mathbf{I}) = (\lambda - 3)(\lambda + 5)(6 - \lambda) = 0$. For $\lambda_1 = 3$, $\lambda_2 = -5$, and $\lambda_3 = 6$ we obtain

$$\mathbf{K}_1 = \begin{pmatrix} 1 \\ 1 \\ 0 \end{pmatrix}, \quad \mathbf{K}_2 = \begin{pmatrix} 1 \\ -1 \\ 0 \end{pmatrix}, \quad \text{and} \quad \mathbf{K}_3 = \begin{pmatrix} 2 \\ -2 \\ 11 \end{pmatrix},$$

so that

$$\mathbf{X} = c_1 \begin{pmatrix} 1 \\ 1 \\ 0 \end{pmatrix} e^{3t} + c_2 \begin{pmatrix} 1 \\ -1 \\ 0 \end{pmatrix} e^{-5t} + c_3 \begin{pmatrix} 2 \\ -2 \\ 11 \end{pmatrix} e^{6t}.$$

15. (a) From the discussion in Section 2.9 we get the system

$$\begin{cases} \dfrac{dx_1}{dt} = -\dfrac{3}{100}x_1 + \dfrac{1}{100}x_2 \\[2mm] \dfrac{dx_2}{dt} = \dfrac{1}{50}x_1 - \dfrac{1}{50}x_2 \end{cases}. \qquad \text{Thus} \qquad \mathbf{X}' = \begin{pmatrix} -\frac{3}{100} & \frac{1}{100} \\ \frac{1}{50} & -\frac{1}{50} \end{pmatrix} \mathbf{X}.$$

(b) The eigenvalues and eigenvectors of the coefficient matrix are found by solving $\det(\mathbf{A} - \lambda\mathbf{I}) = 0$ to get

$$\lambda_1 = -\frac{1}{25} \quad \text{so} \quad \mathbf{K}_1 = \begin{pmatrix} -1 \\ 1 \end{pmatrix} \quad \text{and} \quad \lambda_2 = -\frac{1}{100} \quad \text{so} \quad \mathbf{K}_2 = \begin{pmatrix} 1 \\ 2 \end{pmatrix}.$$

The general solution is then

$$\mathbf{X}(t) = c_1\mathbf{K}_1 e^{\lambda_1 t} + c_2\mathbf{K}_2 e^{\lambda_2 t} = c_1 \begin{pmatrix} -1 \\ 1 \end{pmatrix} e^{-t/25} + c_2 \begin{pmatrix} 1 \\ 2 \end{pmatrix} e^{-t/100}.$$

Using the initial conditions, we get

$$\mathbf{X}(0) = c_1 \begin{pmatrix} -1 \\ 1 \end{pmatrix} + c_2 \begin{pmatrix} 1 \\ 2 \end{pmatrix} = \begin{pmatrix} -c_1 + c_2 \\ c_1 + c_2 \end{pmatrix} = \begin{pmatrix} 20 \\ 5 \end{pmatrix} \quad \text{so} \quad c_1 = -\frac{35}{3} \quad \text{and} \quad c_2 = \frac{25}{3}.$$

Thus

$$\mathbf{X}(t) = -\frac{35}{3} \begin{pmatrix} -1 \\ 1 \end{pmatrix} e^{-t/25} + \frac{25}{3} \begin{pmatrix} 1 \\ 2 \end{pmatrix} e^{-t/100}$$

Hence

$$x_1(t) = \frac{35}{3} e^{-t/25} + \frac{25}{3} e^{-t/100} \quad \text{and} \quad x_2(t) = -\frac{35}{3} e^{-t/25} + \frac{50}{3} e^{-t/100}.$$

(c)

(d) Set $x_1(t) = x_2(t)$ and, using a scientific calculator, we find that $t \approx 34.3277$ min.

21. We have $\det(\mathbf{A} - \lambda\mathbf{I}) = \lambda^2 = 0$. For $\lambda_1 = 0$ we obtain

$$\mathbf{K} = \begin{pmatrix} 1 \\ 3 \end{pmatrix}.$$

A solution of $(\mathbf{A} - \lambda_1\mathbf{I})\mathbf{P} = \mathbf{K}$ is

$$\mathbf{P} = \begin{pmatrix} 1 \\ 2 \end{pmatrix}$$

so that

$$\mathbf{X} = c_1 \begin{pmatrix} 1 \\ 3 \end{pmatrix} + c_2 \left[\begin{pmatrix} 1 \\ 3 \end{pmatrix} t + \begin{pmatrix} 1 \\ 2 \end{pmatrix} \right].$$

24. We have $\det(\mathbf{A} - \lambda\mathbf{I}) = (\lambda - 6)^2 = 0$. For $\lambda_1 = 6$ we obtain

$$\mathbf{K} = \begin{pmatrix} 3 \\ 2 \end{pmatrix}.$$

A solution of $(\mathbf{A} - \lambda_1\mathbf{I})\mathbf{P} = \mathbf{K}$ is

$$\mathbf{P} = \begin{pmatrix} \frac{1}{2} \\ 0 \end{pmatrix}$$

so that

$$\mathbf{X} = c_1 \begin{pmatrix} 3 \\ 2 \end{pmatrix} e^{6t} + c_2 \left[\begin{pmatrix} 3 \\ 2 \end{pmatrix} te^{6t} + \begin{pmatrix} \frac{1}{2} \\ 0 \end{pmatrix} e^{6t} \right].$$

27. We have $\det(\mathbf{A} - \lambda\mathbf{I}) = -\lambda(5 - \lambda)^2 = 0$. For $\lambda_1 = 0$ we obtain

$$\mathbf{K}_1 = \begin{pmatrix} -4 \\ -5 \\ 2 \end{pmatrix}.$$

For $\lambda_2 = 5$ we obtain

$$\mathbf{K} = \begin{pmatrix} -2 \\ 0 \\ 1 \end{pmatrix}.$$

A solution of $(\mathbf{A} - \lambda_1\mathbf{I})\mathbf{P} = \mathbf{K}$ is

$$\mathbf{P} = \begin{pmatrix} \frac{5}{2} \\ \frac{1}{2} \\ 0 \end{pmatrix}$$

so that

$$\mathbf{X} = c_1 \begin{pmatrix} -4 \\ -5 \\ 2 \end{pmatrix} + c_2 \begin{pmatrix} -2 \\ 0 \\ 1 \end{pmatrix} e^{5t} + c_3 \left[\begin{pmatrix} -2 \\ 0 \\ 1 \end{pmatrix} te^{5t} + \begin{pmatrix} \frac{5}{2} \\ \frac{1}{2} \\ 0 \end{pmatrix} e^{5t} \right].$$

30. We have $\det(\mathbf{A} - \lambda\mathbf{I}) = (\lambda - 4)^3 = 0$. For $\lambda_1 = 4$ we obtain

$$\mathbf{K} = \begin{pmatrix} 1 \\ 0 \\ 0 \end{pmatrix}.$$

Solutions of $(\mathbf{A} - \lambda_1\mathbf{I})\mathbf{P} = \mathbf{K}$ and $(\mathbf{A} - \lambda_1\mathbf{I})\mathbf{Q} = \mathbf{P}$ are

$$\mathbf{P} = \begin{pmatrix} 0 \\ 1 \\ 0 \end{pmatrix} \quad \text{and} \quad \mathbf{Q} = \begin{pmatrix} 0 \\ 0 \\ 1 \end{pmatrix}$$

so that

$$\mathbf{X} = c_1 \begin{pmatrix} 1 \\ 0 \\ 0 \end{pmatrix} e^{4t} + c_2 \left[\begin{pmatrix} 1 \\ 0 \\ 0 \end{pmatrix} te^{4t} + \begin{pmatrix} 0 \\ 1 \\ 0 \end{pmatrix} e^{4t} \right] + c_3 \left[\begin{pmatrix} 1 \\ 0 \\ 0 \end{pmatrix} \frac{t^2}{2} e^{4t} + \begin{pmatrix} 0 \\ 1 \\ 0 \end{pmatrix} te^{4t} + \begin{pmatrix} 0 \\ 0 \\ 1 \end{pmatrix} e^{4t} \right].$$

33. In this case $\det(\mathbf{A} - \lambda\mathbf{I}) = (2 - \lambda)^5$, and $\lambda_1 = 2$ is an eigenvalue of multiplicity 5. Linearly independent eigenvectors are

$$\mathbf{K}_1 = \begin{pmatrix} 1 \\ 0 \\ 0 \\ 0 \\ 0 \end{pmatrix}, \qquad \mathbf{K}_2 = \begin{pmatrix} 0 \\ 0 \\ 1 \\ 0 \\ 0 \end{pmatrix}, \qquad \text{and} \qquad \mathbf{K}_3 = \begin{pmatrix} 0 \\ 0 \\ 0 \\ 1 \\ 0 \end{pmatrix}.$$

In Problems 35–46 the form of the answer will vary according to the choice of eigenvector. For example, in Problem 35, if $\mathbf{K}_1$ is chosen to be $\begin{pmatrix} 1 \\ 2-i \end{pmatrix}$ the solution has the form

$$\mathbf{X} = c_1 \begin{pmatrix} \cos t \\ 2\cos t + \sin t \end{pmatrix} e^{4t} + c_2 \begin{pmatrix} \sin t \\ 2\sin t - \cos t \end{pmatrix} e^{4t}.$$

36. We have $\det(\mathbf{A} - \lambda\mathbf{I}) = \lambda^2 + 1 = 0$. For $\lambda_1 = i$ we obtain

$$\mathbf{K}_1 = \begin{pmatrix} -1 - i \\ 2 \end{pmatrix}$$

so that

$$\mathbf{X}_1 = \begin{pmatrix} -1 - i \\ 2 \end{pmatrix} e^{it} = \begin{pmatrix} \sin t - \cos t \\ 2\cos t \end{pmatrix} + i \begin{pmatrix} -\cos t - \sin t \\ 2\sin t \end{pmatrix}.$$

Then

$$\mathbf{X} = c_1 \begin{pmatrix} \sin t - \cos t \\ 2\cos t \end{pmatrix} + c_2 \begin{pmatrix} -\cos t - \sin t \\ 2\sin t \end{pmatrix}.$$

39. We have $\det(\mathbf{A} - \lambda\mathbf{I}) = \lambda^2 + 9 = 0$. For $\lambda_1 = 3i$ we obtain

$$\mathbf{K}_1 = \begin{pmatrix} 4 + 3i \\ 5 \end{pmatrix}$$

so that

$$\mathbf{X}_1 = \begin{pmatrix} 4 + 3i \\ 5 \end{pmatrix} e^{3it} = \begin{pmatrix} 4\cos 3t - 3\sin 3t \\ 5\cos 3t \end{pmatrix} + i \begin{pmatrix} 4\sin 3t + 3\cos 3t \\ 5\sin 3t \end{pmatrix}.$$

Then

$$\mathbf{X} = c_1 \begin{pmatrix} 4\cos 3t - 3\sin 3t \\ 5\cos 3t \end{pmatrix} + c_2 \begin{pmatrix} 4\sin 3t + 3\cos 3t \\ 5\sin 3t \end{pmatrix}.$$

42. We have $\det(\mathbf{A} - \lambda\mathbf{I}) = -(\lambda + 3)(\lambda^2 - 2\lambda + 5) = 0$. For $\lambda_1 = -3$ we obtain

$$\mathbf{K}_1 = \begin{pmatrix} 0 \\ -2 \\ 1 \end{pmatrix}.$$

For $\lambda_2 = 1 + 2i$ we obtain

$$\mathbf{K}_2 = \begin{pmatrix} -2 - i \\ -3i \\ 2 \end{pmatrix}$$

so that

$$\mathbf{X}_2 = \begin{pmatrix} -2\cos 2t + \sin 2t \\ 3\sin 2t \\ 2\cos 2t \end{pmatrix} e^t + i \begin{pmatrix} -\cos 2t - 2\sin 2t \\ -3\cos 2t \\ 2\sin 2t \end{pmatrix} e^t.$$

Then

$$\mathbf{X} = c_1 \begin{pmatrix} 0 \\ -2 \\ 1 \end{pmatrix} e^{-3t} + c_2 \begin{pmatrix} -2\cos 2t + \sin 2t \\ 3\sin 2t \\ 2\cos 2t \end{pmatrix} e^t + c_3 \begin{pmatrix} -\cos 2t - 2\sin 2t \\ -3\cos 2t \\ 2\sin 2t \end{pmatrix} e^t.$$

45. We have $\det(\mathbf{A} - \lambda\mathbf{I}) = (2 - \lambda)(\lambda^2 + 4\lambda + 13) = 0$. For $\lambda_1 = 2$ we obtain

$$\mathbf{K}_1 = \begin{pmatrix} 28 \\ -5 \\ 25 \end{pmatrix}.$$

For $\lambda_2 = -2 + 3i$ we obtain

$$\mathbf{K}_2 = \begin{pmatrix} 4 + 3i \\ -5 \\ 0 \end{pmatrix}$$

so that

$$\mathbf{X}_2 = \begin{pmatrix} 4 + 3i \\ -5 \\ 0 \end{pmatrix} e^{(-2+3i)t} = \begin{pmatrix} 4\cos 3t - 3\sin 3t \\ -5\cos 3t \\ 0 \end{pmatrix} e^{-2t} + i \begin{pmatrix} 4\sin 3t + 3\cos 3t \\ -5\sin 3t \\ 0 \end{pmatrix} e^{-2t}.$$

Then

$$\mathbf{X} = c_1 \begin{pmatrix} 28 \\ -5 \\ 25 \end{pmatrix} e^{2t} + c_2 \begin{pmatrix} 4\cos 3t - 3\sin 3t \\ -5\cos 3t \\ 0 \end{pmatrix} e^{-2t} + c_3 \begin{pmatrix} 4\sin 3t + 3\cos 3t \\ -5\sin 3t \\ 0 \end{pmatrix} e^{-2t}.$$

48. We have $\det(\mathbf{A} - \lambda\mathbf{I}) = \lambda^2 - 10\lambda + 29 = 0$. For $\lambda_1 = 5 + 2i$ we obtain

$$\mathbf{K}_1 = \begin{pmatrix} 1 \\ 1 - 2i \end{pmatrix}$$

so that

$$\mathbf{X}_1 = \begin{pmatrix} 1 \\ 1 - 2i \end{pmatrix} e^{(5+2i)t} = \begin{pmatrix} \cos 2t \\ \cos 2t + 2\sin 2t \end{pmatrix} e^{5t} + i \begin{pmatrix} \sin 2t \\ \sin 2t - 2\cos 2t \end{pmatrix} e^{5t}.$$

and

$$\mathbf{X} = c_1 \begin{pmatrix} \cos 2t \\ \cos 2t + 2\sin 2t \end{pmatrix} e^{5t} + c_2 \begin{pmatrix} \sin 2t \\ \sin 2t - 2\cos 2t \end{pmatrix} e^{5t}.$$

If $\mathbf{X}(0) = \begin{pmatrix} -2 \\ 8 \end{pmatrix}$, then $c_1 = -2$ and $c_2 = -5$. The solution is

$$\mathbf{X} = -2 \begin{pmatrix} \cos 2t \\ \cos 2t + 2\sin 2t \end{pmatrix} e^{5t} - 5 \begin{pmatrix} \sin 2t \\ \sin 2t - 2\cos 2t \end{pmatrix} e^{5t}.$$

10.3 Solution by Diagonalization

3. $\lambda_1 = \frac{1}{2}$, $\lambda_2 = \frac{3}{2}$, $\mathbf{K}_1 = \begin{pmatrix} 1 \\ -2 \end{pmatrix}$, $\mathbf{K}_2 = \begin{pmatrix} 1 \\ 2 \end{pmatrix}$, $\mathbf{P} = \begin{pmatrix} 1 & 1 \\ -2 & 2 \end{pmatrix}$;

$$\mathbf{X} = \mathbf{PY} = \begin{pmatrix} 1 & 1 \\ -2 & 2 \end{pmatrix} \begin{pmatrix} c_1 e^{t/2} \\ c_2 e^{3t/2} \end{pmatrix} = \begin{pmatrix} c_1 e^{t/2} + c_2 e^{3t/2} \\ -2c_1 e^{t/2} + 2c_2 e^{3t/2} \end{pmatrix} = c_1 \begin{pmatrix} 1 \\ -2 \end{pmatrix} e^{t/2} + c_2 \begin{pmatrix} 1 \\ 2 \end{pmatrix} e^{3t/2}$$

6. $\lambda_1 = -1$, $\lambda_2 = 1$, $\lambda_3 = 4$, $\mathbf{K}_1 = \begin{pmatrix} -1 \\ 0 \\ 1 \end{pmatrix}$, $\mathbf{K}_2 = \begin{pmatrix} 1 \\ -2 \\ 1 \end{pmatrix}$, $\mathbf{K}_3 = \begin{pmatrix} 1 \\ 1 \\ 1 \end{pmatrix}$, $\mathbf{P} = \begin{pmatrix} -1 & 1 & 1 \\ 0 & -2 & 1 \\ 1 & 1 & 1 \end{pmatrix}$;

$$\mathbf{X} = \mathbf{PY} = \begin{pmatrix} -1 & 1 & 1 \\ 0 & -2 & 1 \\ 1 & 1 & 1 \end{pmatrix} \begin{pmatrix} c_1 e^{-t} \\ c_2 e^{t} \\ c_3 e^{4t} \end{pmatrix} = \begin{pmatrix} -c_1 + c_2 e^{t} + c_3 e^{4t} \\ -2c_2 e^{t} + c_3 e^{4t} \\ c_1 e^{-t} + c_2 e^{t} + c_3 e^{4t} \end{pmatrix}$$

$$= c_1 \begin{pmatrix} -1 \\ 0 \\ 1 \end{pmatrix} e^{-t} + c_2 \begin{pmatrix} 1 \\ -2 \\ 1 \end{pmatrix} e^{t} + c_3 \begin{pmatrix} 1 \\ 1 \\ 1 \end{pmatrix} e^{4t}$$

9. $\lambda_1 = 1$, $\lambda_2 = 2$, $\lambda_3 = 3$, $\mathbf{K}_1 = \begin{pmatrix} 1 \\ 1 \\ 1 \end{pmatrix}$, $\mathbf{K}_2 = \begin{pmatrix} 2 \\ 2 \\ 3 \end{pmatrix}$, $\mathbf{K}_3 = \begin{pmatrix} 3 \\ 4 \\ 5 \end{pmatrix}$, $\mathbf{P} = \begin{pmatrix} 1 & 2 & 3 \\ 1 & 2 & 4 \\ 1 & 3 & 5 \end{pmatrix}$;

$$\mathbf{X} = \mathbf{PY} = \begin{pmatrix} 1 & 2 & 3 \\ 1 & 2 & 4 \\ 1 & 3 & 5 \end{pmatrix} \begin{pmatrix} c_1 e^{t} \\ c_2 e^{2t} \\ c_3 e^{3t} \end{pmatrix} = \begin{pmatrix} c_1 e^{t} + 2c_2 e^{2t} + 3c_3 e^{3t} \\ c_1 e^{t} + 2c_2 e^{2t} + 4c_3 e^{3t} \\ c_1 e^{t} + 3c_2 e^{2t} + 5c_3 e^{3t} \end{pmatrix}$$

$$= c_1 \begin{pmatrix} 1 \\ 1 \\ 1 \end{pmatrix} e^{t} + c_2 \begin{pmatrix} 2 \\ 2 \\ 3 \end{pmatrix} e^{2t} + c_3 \begin{pmatrix} 3 \\ 4 \\ 5 \end{pmatrix} e^{3t}$$

10.4 | Nonhomogeneous Linear Systems

3. Solving

$$\det(\mathbf{A} - \lambda\mathbf{I}) = \begin{vmatrix} 1 - \lambda & 3 \\ 3 & 1 - \lambda \end{vmatrix} = \lambda^2 - 2\lambda - 8 = (\lambda - 4)(\lambda + 2) = 0$$

we obtain eigenvalues $\lambda_1 = -2$ and $\lambda_2 = 4$. Corresponding eigenvectors are

$$\mathbf{K}_1 = \begin{pmatrix} 1 \\ -1 \end{pmatrix} \quad \text{and} \quad \mathbf{K}_2 = \begin{pmatrix} 1 \\ 1 \end{pmatrix}.$$

Thus

$$\mathbf{X}_c = c_1 \begin{pmatrix} 1 \\ -1 \end{pmatrix} e^{-2t} + c_2 \begin{pmatrix} 1 \\ 1 \end{pmatrix} e^{4t}.$$

Substituting

$$\mathbf{X}_p = \begin{pmatrix} a_3 \\ b_3 \end{pmatrix} t^2 + \begin{pmatrix} a_2 \\ b_2 \end{pmatrix} t + \begin{pmatrix} a_1 \\ b_1 \end{pmatrix}$$

into the system yields

$$a_3 + 3b_3 = 2 \qquad\qquad a_2 + 3b_2 = 2a_3 \qquad\qquad a_1 + 3b_1 = a_2$$

$$3a_3 + b_3 = 0 \qquad\qquad 3a_2 + b_2 + 1 = 2b_3 \qquad\qquad 3a_1 + b_1 + 5 = b_2$$

from which we obtain $a_3 = -1/4$, $b_3 = 3/4$, $a_2 = 1/4$, $b_2 = -1/4$, $a_1 = -2$, and $b_1 = 3/4$.
Then

$$\mathbf{X}(t) = c_1 \begin{pmatrix} 1 \\ -1 \end{pmatrix} e^{-2t} + c_2 \begin{pmatrix} 1 \\ 1 \end{pmatrix} e^{4t} + \begin{pmatrix} -\frac{1}{4} \\ \frac{3}{4} \end{pmatrix} t^2 + \begin{pmatrix} \frac{1}{4} \\ -\frac{1}{4} \end{pmatrix} t + \begin{pmatrix} -2 \\ \frac{3}{4} \end{pmatrix}.$$

6. Solving

$$\det(\mathbf{A} - \lambda\mathbf{I}) = \begin{vmatrix} -1 - \lambda & 5 \\ -1 & 1 - \lambda \end{vmatrix} = \lambda^2 + 4 = 0$$

we obtain the eigenvalues $\lambda_1 = 2i$ and $\lambda_2 = -2i$. Corresponding eigenvectors are

$$\mathbf{K}_1 = \begin{pmatrix} 5 \\ 1 + 2i \end{pmatrix} \quad \text{and} \quad \mathbf{K}_2 = \begin{pmatrix} 5 \\ 1 - 2i \end{pmatrix}.$$

Thus

$$\mathbf{X}_c = c_1 \begin{pmatrix} 5\cos 2t \\ \cos 2t - 2\sin 2t \end{pmatrix} + c_2 \begin{pmatrix} 5\sin 2t \\ 2\cos 2t + \sin 2t \end{pmatrix}.$$

Substituting

$$\mathbf{X}_p = \begin{pmatrix} a_2 \\ b_2 \end{pmatrix} \cos t + \begin{pmatrix} a_1 \\ b_1 \end{pmatrix} \sin t$$

into the system yields

$$-a_2 + 5b_2 - a_1 = 0$$
$$-a_2 + b_2 - b_1 - 2 = 0$$
$$-a_1 + 5b_1 + a_2 + 1 = 0$$
$$-a_1 + b_1 + b_2 = 0$$

from which we obtain $a_2 = -3$, $b_2 = -2/3$, $a_1 = -1/3$, and $b_1 = 1/3$. Then

$$\mathbf{X}(t) = c_1 \begin{pmatrix} 5\cos 2t \\ \cos 2t - 2\sin 2t \end{pmatrix} + c_2 \begin{pmatrix} 5\sin 2t \\ 2\cos 2t + \sin 2t \end{pmatrix} + \begin{pmatrix} -3 \\ -\frac{2}{3} \end{pmatrix} \cos t + \begin{pmatrix} -\frac{1}{3} \\ \frac{1}{3} \end{pmatrix} \sin t.$$

9. First solve the associated homogeneous system

$$\mathbf{X}' = \begin{pmatrix} -1 & -2 \\ 3 & 4 \end{pmatrix} \mathbf{X}$$

The eigenvalues and eigenvectors of the coefficient matrix are found by solving $\det(\mathbf{A} - \lambda\mathbf{I}) = 0$ to get

$$\lambda_1 = 1 \quad \text{so} \quad \mathbf{K}_1 = \begin{pmatrix} 1 \\ -1 \end{pmatrix} \quad \text{and} \quad \lambda_2 = 2 \quad \text{so} \quad \mathbf{K}_2 = \begin{pmatrix} 2 \\ -3 \end{pmatrix}.$$

The complementary solution is then

$$\mathbf{X}_c(t) = c_1 \mathbf{K}_1 e^{\lambda_1 t} + c_2 \mathbf{K}_2 e^{\lambda_2 t} = c_1 \begin{pmatrix} 1 \\ -1 \end{pmatrix} e^t + c_2 \begin{pmatrix} 2 \\ -3 \end{pmatrix} e^{2t}.$$

Based on the form of $\mathbf{F}(t)$, guess $\mathbf{X}_p = \begin{pmatrix} a_1 \\ b_1 \end{pmatrix}$ and force it into the original system to get $\mathbf{X}_p = \begin{pmatrix} -9 \\ 6 \end{pmatrix}$. The general solution is then

$$\mathbf{X} = \mathbf{X}_c + \mathbf{X}_p = c_1 \begin{pmatrix} 1 \\ -1 \end{pmatrix} e^t + c_2 \begin{pmatrix} 2 \\ -3 \end{pmatrix} e^{2t} + \begin{pmatrix} -9 \\ 6 \end{pmatrix}.$$

Next use the initial condition to solve for c_1 and c_2:

$$\mathbf{X}(0) = c_1 \begin{pmatrix} 1 \\ -1 \end{pmatrix} + c_2 \begin{pmatrix} 2 \\ -3 \end{pmatrix} + \begin{pmatrix} -9 \\ 6 \end{pmatrix} = \begin{pmatrix} -4 \\ 5 \end{pmatrix}$$

$$c_1 = 13 \quad \text{and} \quad c_2 = -4$$

$$\mathbf{X} = 13 \begin{pmatrix} 1 \\ -1 \end{pmatrix} e^t - 4 \begin{pmatrix} 2 \\ -3 \end{pmatrix} e^{2t} + \begin{pmatrix} -9 \\ 6 \end{pmatrix}.$$

12. (a) Let $\mathbf{X} = \begin{pmatrix} i_2 \\ i_3 \end{pmatrix}$ so that

$$\mathbf{X}' = \begin{pmatrix} -2 & -2 \\ -2 & -5 \end{pmatrix} \mathbf{X} + \begin{pmatrix} 60 \\ 60 \end{pmatrix} \quad \text{and} \quad \mathbf{X}_c = c_1 \begin{pmatrix} 2 \\ -1 \end{pmatrix} e^{-t} + c_2 \begin{pmatrix} 1 \\ 2 \end{pmatrix} e^{-6t}.$$

If $\mathbf{X}_p = \begin{pmatrix} a_1 \\ b_1 \end{pmatrix}$ then $\mathbf{X}_p = \begin{pmatrix} 30 \\ 0 \end{pmatrix}$ so that

$$\mathbf{X} = c_1 \begin{pmatrix} 2 \\ -1 \end{pmatrix} e^{-t} + c_2 \begin{pmatrix} 1 \\ 2 \end{pmatrix} e^{-6t} + \begin{pmatrix} 30 \\ 0 \end{pmatrix}.$$

For $\mathbf{X}(0) = \begin{pmatrix} 0 \\ 0 \end{pmatrix}$ we find $c_1 = -12$ and $c_2 = -6$.

(b) $i_1(t) = i_2(t) + i_3(t) = -12e^{-t} - 18e^{-6t} + 30$

15. From

$$\mathbf{X}' = \begin{pmatrix} 3 & -5 \\ \frac{3}{4} & -1 \end{pmatrix} \mathbf{X} + \begin{pmatrix} 1 \\ -1 \end{pmatrix} e^{t/2}$$

we obtain

$$\mathbf{X}_c = c_1 \begin{pmatrix} 10 \\ 3 \end{pmatrix} e^{3t/2} + c_2 \begin{pmatrix} 2 \\ 1 \end{pmatrix} e^{t/2}.$$

Then

$$\mathbf{\Phi} = \begin{pmatrix} 10e^{3t/2} & 2e^{t/2} \\ 3e^{3t/2} & e^{t/2} \end{pmatrix} \quad \text{and} \quad \mathbf{\Phi}^{-1} = \begin{pmatrix} \frac{1}{4}e^{-3t/2} & -\frac{1}{2}e^{-3t/2} \\ -\frac{3}{4}e^{-t/2} & \frac{5}{2}e^{-t/2} \end{pmatrix}$$

so that

$$\mathbf{U} = \int \mathbf{\Phi}^{-1} \mathbf{F}\, dt = \int \begin{pmatrix} \frac{3}{4}e^{-t} \\ -\frac{13}{4} \end{pmatrix} dt = \begin{pmatrix} -\frac{3}{4}e^{-t} \\ -\frac{13}{4}t \end{pmatrix}$$

and

$$\mathbf{X}_p = \mathbf{\Phi}\mathbf{U} = \begin{pmatrix} -\frac{13}{2} \\ -\frac{13}{4} \end{pmatrix} te^{t/2} + \begin{pmatrix} -\frac{15}{2} \\ -\frac{9}{4} \end{pmatrix} e^{t/2}.$$

The solution is

$$\mathbf{X} = \mathbf{X}_c + \mathbf{X}_p = c_1 \begin{pmatrix} 10 \\ 3 \end{pmatrix} e^{3t/2} + c_2 \begin{pmatrix} 2 \\ 1 \end{pmatrix} e^{t/2} - 11 \begin{pmatrix} \frac{13}{2} \\ \frac{13}{4} \end{pmatrix} te^{t/2} - \begin{pmatrix} \frac{15}{2} \\ \frac{9}{4} \end{pmatrix} e^{t/2}.$$

18. From

$$\mathbf{X}' = \begin{pmatrix} 0 & 2 \\ -1 & 3 \end{pmatrix} \mathbf{X} + \begin{pmatrix} 2 \\ e^{-3t} \end{pmatrix}$$

we obtain

$$\mathbf{X}_c = c_1 \begin{pmatrix} 2 \\ 1 \end{pmatrix} e^t + c_2 \begin{pmatrix} 1 \\ 1 \end{pmatrix} e^{2t}.$$

Then

$$\mathbf{\Phi} = \begin{pmatrix} 2e^t & e^{2t} \\ e^t & e^{2t} \end{pmatrix} \quad \text{and} \quad \mathbf{\Phi}^{-1} = \begin{pmatrix} e^{-t} & -e^{-t} \\ -e^{-2t} & 2e^{-2t} \end{pmatrix}$$

so that

$$\mathbf{U} = \int \mathbf{\Phi}^{-1}\mathbf{F}\,dt = \int \begin{pmatrix} 2e^{-t} - e^{-4t} \\ -2e^{-2t} + 2e^{-5t} \end{pmatrix} dt = \begin{pmatrix} -2e^{-t} + \frac{1}{4}e^{-4t} \\ e^{-2t} - \frac{2}{5}e^{-5t} \end{pmatrix}$$

and

$$\mathbf{X}_p = \mathbf{\Phi}\mathbf{U} = \begin{pmatrix} \frac{1}{10}e^{-3t} - 3 \\ -\frac{3}{20}e^{-3t} - 1 \end{pmatrix}.$$

The solution is

$$\mathbf{X} = \mathbf{X}_c + \mathbf{X}_p = c_1 \begin{pmatrix} 2 \\ 1 \end{pmatrix} e^t + c_2 \begin{pmatrix} 1 \\ 1 \end{pmatrix} e^{2t} + \begin{pmatrix} \frac{1}{10}e^{-3t} - 3 \\ -\frac{3}{20}e^{-3t} - 1 \end{pmatrix}.$$

21. From

$$\mathbf{X}' = \begin{pmatrix} 3 & 2 \\ -2 & -1 \end{pmatrix} \mathbf{X} + \begin{pmatrix} 2 \\ 1 \end{pmatrix} e^{-t}$$

we obtain

$$\mathbf{X}_c = c_1 \begin{pmatrix} 1 \\ -1 \end{pmatrix} e^t + c_2 \left[\begin{pmatrix} 1 \\ -1 \end{pmatrix} te^t + \begin{pmatrix} 0 \\ \frac{1}{2} \end{pmatrix} e^t \right].$$

Then

$$\mathbf{\Phi} = \begin{pmatrix} e^t & te^t \\ -e^t & \frac{1}{2}e^t - te^t \end{pmatrix} \quad \text{and} \quad \mathbf{\Phi}^{-1} = \begin{pmatrix} e^{-t} - 2te^{-t} & -2te^{-t} \\ 2e^{-t} & 2e^{-t} \end{pmatrix}$$

so that

$$\mathbf{U} = \int \mathbf{\Phi}^{-1}\mathbf{F}\,dt = \int \begin{pmatrix} 2e^{-2t} - 6te^{-2t} \\ 6e^{-2t} \end{pmatrix} dt = \begin{pmatrix} \frac{1}{2}e^{-2t} + 3te^{-2t} \\ -3e^{-2t} \end{pmatrix}$$

and

$$\mathbf{X}_p = \mathbf{\Phi}\mathbf{U} = \begin{pmatrix} \frac{1}{2} \\ -2 \end{pmatrix} e^{-t}.$$

The solution is

$$\mathbf{X} = \mathbf{X}_c + \mathbf{X}_p = c_1 \begin{pmatrix} 1 \\ -1 \end{pmatrix} e^t + c_2 \begin{pmatrix} t \\ \frac{1}{2} - t \end{pmatrix} e^t + \begin{pmatrix} \frac{1}{2} \\ -2 \end{pmatrix} e^{-t}.$$

24. From

$$\mathbf{X}' = \begin{pmatrix} 1 & -1 \\ 1 & 1 \end{pmatrix} \mathbf{X} + \begin{pmatrix} 3 \\ 3 \end{pmatrix} e^t$$

we obtain

$$\mathbf{X}_c = c_1 \begin{pmatrix} -\sin t \\ \cos t \end{pmatrix} e^t + c_2 \begin{pmatrix} \cos t \\ \sin t \end{pmatrix} e^t.$$

Then

$$\boldsymbol{\Phi} = \begin{pmatrix} -\sin t & \cos t \\ \cos t & \sin t \end{pmatrix} e^t \quad \text{and} \quad \boldsymbol{\Phi}^{-1} = \begin{pmatrix} -\sin t & \cos t \\ \cos t & \sin t \end{pmatrix} e^{-t}$$

so that

$$\mathbf{U} = \int \boldsymbol{\Phi}^{-1} \mathbf{F} \, dt = \int \begin{pmatrix} -3\sin t + 3\cos t \\ 3\cos t + 3\sin t \end{pmatrix} dt = \begin{pmatrix} 3\cos t + 3\sin t \\ 3\sin t - 3\cos t \end{pmatrix}$$

and

$$\mathbf{X}_p = \boldsymbol{\Phi}\mathbf{U} = \begin{pmatrix} -3 \\ 3 \end{pmatrix} e^t.$$

The solution is

$$\mathbf{X} = \mathbf{X}_c + \mathbf{X}_p = c_1 \begin{pmatrix} -\sin t \\ \cos t \end{pmatrix} e^t + c_2 \begin{pmatrix} \cos t \\ \sin t \end{pmatrix} e^t + \begin{pmatrix} -3 \\ 3 \end{pmatrix} e^t.$$

27. From

$$\mathbf{X}' = \begin{pmatrix} 0 & 1 \\ -1 & 0 \end{pmatrix} \mathbf{X} + \begin{pmatrix} 0 \\ \sec t \tan t \end{pmatrix}$$

we obtain

$$\mathbf{X}_c = c_1 \begin{pmatrix} \cos t \\ -\sin t \end{pmatrix} + c_2 \begin{pmatrix} \sin t \\ \cos t \end{pmatrix}.$$

Then

$$\boldsymbol{\Phi} = \begin{pmatrix} \cos t & \sin t \\ -\sin t & \cos t \end{pmatrix} t \quad \text{and} \quad \boldsymbol{\Phi}^{-1} = \begin{pmatrix} \cos t & -\sin t \\ \sin t & \cos t \end{pmatrix}$$

so that

$$\mathbf{U} = \int \boldsymbol{\Phi}^{-1} \mathbf{F} \, dt = \int \begin{pmatrix} -\tan^2 t \\ \tan t \end{pmatrix} dt = \begin{pmatrix} t - \tan t \\ -\ln|\cos t| \end{pmatrix}$$

and

$$\mathbf{X}_p = \boldsymbol{\Phi}\mathbf{U} = \begin{pmatrix} \cos t \\ -\sin t \end{pmatrix} t + \begin{pmatrix} -\sin t \\ \sin t \tan t \end{pmatrix} - \begin{pmatrix} \sin t \\ \cos t \end{pmatrix} \ln|\cos t|.$$

The solution is

$$\mathbf{X} = \mathbf{X}_c + \mathbf{X}_p = c_1 \begin{pmatrix} \cos t \\ -\sin t \end{pmatrix} + c_2 \begin{pmatrix} \sin t \\ \cos t \end{pmatrix} + \begin{pmatrix} \cos t \\ -\sin t \end{pmatrix} t + \begin{pmatrix} -\sin t \\ \sin t \tan t \end{pmatrix} - \begin{pmatrix} \sin t \\ \cos t \end{pmatrix} \ln|\cos t|.$$

30. From

$$\mathbf{X}' = \begin{pmatrix} 1 & -2 \\ 1 & -1 \end{pmatrix} \mathbf{X} + \begin{pmatrix} \tan t \\ 1 \end{pmatrix}$$

we obtain

$$\mathbf{X}_c = c_1 \begin{pmatrix} \cos t - \sin t \\ \cos t \end{pmatrix} + c_2 \begin{pmatrix} \cos t + \sin t \\ \sin t \end{pmatrix}.$$

Then

$$\mathbf{\Phi} = \begin{pmatrix} \cos t - \sin t & \cos t + \sin t \\ \cos t & \sin t \end{pmatrix} \quad \text{and} \quad \mathbf{\Phi}^{-1} = \begin{pmatrix} -\sin t & \cos t + \sin t \\ \cos t & \sin t - \cos t \end{pmatrix}$$

so that

$$\mathbf{U} = \int \mathbf{\Phi}^{-1}\mathbf{F}\,dt = \int \begin{pmatrix} 2\cos t + \sin t - \sec t \\ 2\sin t - \cos t \end{pmatrix} dt = \begin{pmatrix} 2\sin t - \cos t - \ln|\sec t + \tan t| \\ -2\cos t - \sin t \end{pmatrix}$$

and

$$\mathbf{X}_p = \mathbf{\Phi}\mathbf{U} = \begin{pmatrix} 3\sin t \cos t - \cos^2 t - 2\sin^2 t + (\sin t - \cos t)\ln|\sec t + \tan t| \\ \sin^2 t - \cos^2 t - \cos t(\ln|\sec t + \tan t|) \end{pmatrix}.$$

The solution is

$$\mathbf{X} = \mathbf{X}_c + \mathbf{X}_p = c_1 \begin{pmatrix} \cos t - \sin t \\ \cos t \end{pmatrix} + c_2 \begin{pmatrix} \cos t + \sin t \\ \sin t \end{pmatrix}$$

$$+ \begin{pmatrix} 3\sin t \cos t - \cos^2 t - 2\sin^2 t + (\sin t - \cos t)\ln|\sec t + \tan t| \\ \sin^2 t - \cos^2 t - \cos t(\ln|\sec t + \tan t|) \end{pmatrix}.$$

33. From

$$\mathbf{X}' = \begin{pmatrix} 3 & -1 \\ -1 & 3 \end{pmatrix} \mathbf{X} + \begin{pmatrix} 4e^{2t} \\ 4e^{4t} \end{pmatrix}$$

we obtain

$$\mathbf{\Phi} = \begin{pmatrix} -e^{4t} & e^{2t} \\ e^{4t} & e^{2t} \end{pmatrix}, \quad \mathbf{\Phi}^{-1} = \begin{pmatrix} -\frac{1}{2}e^{-4t} & \frac{1}{2}e^{-4t} \\ \frac{1}{2}e^{-2t} & \frac{1}{2}e^{-2t} \end{pmatrix},$$

and

$$\mathbf{X} = \mathbf{\Phi}\mathbf{\Phi}^{-1}(0)\mathbf{X}(0) + \mathbf{\Phi}\int_0^t \mathbf{\Phi}^{-1}\mathbf{F}\,ds = \mathbf{\Phi} \cdot \begin{pmatrix} 0 \\ 1 \end{pmatrix} + \mathbf{\Phi} \cdot \begin{pmatrix} e^{-2t} + 2t - 1 \\ e^{2t} + 2t - 1 \end{pmatrix}$$

$$= \begin{pmatrix} 2 \\ 2 \end{pmatrix} te^{2t} + \begin{pmatrix} -1 \\ 1 \end{pmatrix} e^{2t} + \begin{pmatrix} -2 \\ 2 \end{pmatrix} te^{4t} + \begin{pmatrix} 2 \\ 0 \end{pmatrix} e^{4t}.$$

39. $\lambda_1 = 0$, $\lambda_2 = 10$, $\mathbf{K}_1 = \begin{pmatrix} 1 \\ -1 \end{pmatrix}$, $\mathbf{K}_2 = \begin{pmatrix} 1 \\ 1 \end{pmatrix}$, $\mathbf{P} = \begin{pmatrix} 1 & 1 \\ -1 & 1 \end{pmatrix}$, $\mathbf{P}^{-1} = \dfrac{1}{2}\begin{pmatrix} 1 & -1 \\ 1 & 1 \end{pmatrix}$,

$$\mathbf{P}^{-1}\mathbf{F} = \begin{pmatrix} t-4 \\ t+4 \end{pmatrix};$$

$$\mathbf{Y}' = \begin{pmatrix} 0 & 0 \\ 0 & 10 \end{pmatrix}\mathbf{Y} + \begin{pmatrix} t-4 \\ t+4 \end{pmatrix}$$

$$y_1 = \frac{1}{2}t^2 - 4t + c_1, \qquad y_2 = -\frac{1}{10}t - \frac{41}{100} + c_2 e^{10t}$$

$$\mathbf{X} = \mathbf{P}\mathbf{Y} = \begin{pmatrix} 1 & 1 \\ -1 & 1 \end{pmatrix}\begin{pmatrix} \frac{1}{2}t^2 - 4t + c_1 \\ -\frac{1}{10}t - \frac{41}{100} + c_2 e^{10t} \end{pmatrix} = \begin{pmatrix} \frac{1}{2}t^2 - \frac{41}{10}t - \frac{41}{100} + c_1 + c_2 e^{10t} \\ -\frac{1}{2}t^2 + \frac{39}{10}t - \frac{41}{100} - c_1 + c_2 e^{10t} \end{pmatrix}$$

$$= c_1 \begin{pmatrix} 1 \\ -1 \end{pmatrix} + c_2 \begin{pmatrix} 1 \\ 1 \end{pmatrix}e^{10t} + \frac{1}{2}\begin{pmatrix} 1 \\ -1 \end{pmatrix}t^2 + \frac{1}{10}\begin{pmatrix} -41 \\ 39 \end{pmatrix}t - \frac{41}{100}\begin{pmatrix} 1 \\ 1 \end{pmatrix}$$

10.5 | Matrix Exponential

3. For

$$\mathbf{A} = \begin{pmatrix} 1 & 1 & 1 \\ 1 & 1 & 1 \\ -2 & -2 & -2 \end{pmatrix}$$

we have

$$\mathbf{A}^2 = \begin{pmatrix} 1 & 1 & 1 \\ 1 & 1 & 1 \\ -2 & -2 & -2 \end{pmatrix}\begin{pmatrix} 1 & 1 & 1 \\ 1 & 1 & 1 \\ -2 & -2 & -2 \end{pmatrix} = \begin{pmatrix} 0 & 0 & 0 \\ 0 & 0 & 0 \\ 0 & 0 & 0 \end{pmatrix}.$$

Thus, $\mathbf{A}^3 = \mathbf{A}^4 = \mathbf{A}^5 = \cdots = \mathbf{0}$ and

$$e^{\mathbf{A}t} = \mathbf{I} + \mathbf{A}t = \begin{pmatrix} 1 & 0 & 0 \\ 0 & 1 & 0 \\ 0 & 0 & 1 \end{pmatrix} + \begin{pmatrix} t & t & t \\ t & t & t \\ -2t & -2t & -2t \end{pmatrix} = \begin{pmatrix} t+1 & t & t \\ t & t+1 & t \\ -2t & -2t & -2t+1 \end{pmatrix}.$$

6. Using the result of Problem 2,

$$\mathbf{X} = \begin{pmatrix} \cosh t & \sinh t \\ \sinh t & \cosh t \end{pmatrix}\begin{pmatrix} c_1 \\ c_2 \end{pmatrix} = c_1\begin{pmatrix} \cosh t \\ \sinh t \end{pmatrix} + c_2\begin{pmatrix} \sinh t \\ \cosh t \end{pmatrix}.$$

9. To solve

$$\mathbf{X}' = \begin{pmatrix} 1 & 0 \\ 0 & 2 \end{pmatrix}\mathbf{X} + \begin{pmatrix} 3 \\ -1 \end{pmatrix}$$

we identify $t_0 = 0$, $\mathbf{F}(t) = \begin{pmatrix} 3 \\ -1 \end{pmatrix}$, and use the results of Problem 1 and Equation (6) in the text.

$$\mathbf{X}(t) = e^{\mathbf{A}t}\mathbf{C} + e^{\mathbf{A}t}\int_{t_0}^{t} e^{-\mathbf{A}s}\mathbf{F}(s)\,ds$$

$$= \begin{pmatrix} e^t & 0 \\ 0 & e^{2t} \end{pmatrix}\begin{pmatrix} c_1 \\ c_2 \end{pmatrix} + \begin{pmatrix} e^t & 0 \\ 0 & e^{2t} \end{pmatrix}\int_0^t \begin{pmatrix} e^{-s} & 0 \\ 0 & e^{-2s} \end{pmatrix}\begin{pmatrix} 3 \\ -1 \end{pmatrix}\,ds$$

$$= \begin{pmatrix} c_1 e^t \\ c_2 e^{2t} \end{pmatrix} + \begin{pmatrix} e^t & 0 \\ 0 & e^{2t} \end{pmatrix}\int_0^t \begin{pmatrix} 3e^{-s} \\ -e^{-2s} \end{pmatrix}\,ds$$

$$= \begin{pmatrix} c_1 e^t \\ c_2 e^{2t} \end{pmatrix} + \begin{pmatrix} e^t & 0 \\ 0 & e^{2t} \end{pmatrix}\begin{pmatrix} -3e^{-s} \\ \frac{1}{2}e^{-2s} \end{pmatrix}\Bigg|_0^t$$

$$= \begin{pmatrix} c_1 e^t \\ c_2 e^{2t} \end{pmatrix} + \begin{pmatrix} e^t & 0 \\ 0 & e^{2t} \end{pmatrix}\begin{pmatrix} -3e^{-t} + 3 \\ \frac{1}{2}e^{-2t} - \frac{1}{2} \end{pmatrix}$$

$$= \begin{pmatrix} c_1 e^t \\ c_2 e^{2t} \end{pmatrix} + \begin{pmatrix} -3 + 3e^t \\ \frac{1}{2} - \frac{1}{2}e^{2t} \end{pmatrix} = c_3 \begin{pmatrix} 1 \\ 0 \end{pmatrix} e^t + c_4 \begin{pmatrix} 0 \\ 1 \end{pmatrix} e^{2t} + \begin{pmatrix} -3 \\ \frac{1}{2} \end{pmatrix}$$

12. To solve

$$\mathbf{X}' = \begin{pmatrix} 0 & 1 \\ 1 & 0 \end{pmatrix}\mathbf{X} + \begin{pmatrix} \cosh t \\ \sinh t \end{pmatrix}$$

we identify $t_0 = 0$, $\mathbf{F}(t) = \begin{pmatrix} \cosh t \\ \sinh t \end{pmatrix}$, and use the results of Problem 2 and Equation (6) in the text.

$$\mathbf{X}(t) = e^{\mathbf{A}t}\mathbf{C} + e^{\mathbf{A}t}\int_{t_0}^{t} e^{-\mathbf{A}s}\mathbf{F}(s)\,ds$$

$$= \begin{pmatrix} \cosh t & \sinh t \\ \sinh t & \cosh t \end{pmatrix}\begin{pmatrix} c_1 \\ c_2 \end{pmatrix} + \begin{pmatrix} \cosh t & \sinh t \\ \sinh t & \cosh t \end{pmatrix}\int_0^t \begin{pmatrix} \cosh s & -\sinh s \\ -\sinh s & \cosh s \end{pmatrix}\begin{pmatrix} \cosh s \\ \sinh s \end{pmatrix}\,ds$$

$$= \begin{pmatrix} c_1 \cosh t + c_2 \sinh t \\ c_1 \sinh t + c_2 \cosh t \end{pmatrix} + \begin{pmatrix} \cosh t & \sinh t \\ \sinh t & \cosh t \end{pmatrix}\int_0^t \begin{pmatrix} 1 \\ 0 \end{pmatrix}\,ds$$

$$= \begin{pmatrix} c_1 \cosh t + c_2 \sinh t \\ c_1 \sinh t + c_2 \cosh t \end{pmatrix} + \begin{pmatrix} \cosh t & \sinh t \\ \sinh t & \cosh t \end{pmatrix}\begin{pmatrix} s \\ 0 \end{pmatrix}\Bigg|_0^t$$

$$= \begin{pmatrix} c_1 \cosh t + c_2 \sinh t \\ c_1 \sinh t + c_2 \cosh t \end{pmatrix} + \begin{pmatrix} \cosh t & \sinh t \\ \sinh t & \cosh t \end{pmatrix}\begin{pmatrix} t \\ 0 \end{pmatrix}$$

$$= \begin{pmatrix} c_1 \cosh t + c_2 \sinh t \\ c_1 \sinh t + c_2 \cosh t \end{pmatrix} + \begin{pmatrix} t \cosh t \\ t \sinh t \end{pmatrix} = c_1 \begin{pmatrix} \cosh t \\ \sinh t \end{pmatrix} + c_2 \begin{pmatrix} \sinh t \\ \cosh t \end{pmatrix} + t \begin{pmatrix} \cosh t \\ \sinh t \end{pmatrix}$$

15. From $s\mathbf{I} - \mathbf{A} = \begin{pmatrix} s - 4 & -3 \\ 4 & s + 4 \end{pmatrix}$ we find

$$(s\mathbf{I} - \mathbf{A})^{-1} = \begin{pmatrix} \dfrac{3/2}{s - 2} - \dfrac{1/2}{s + 2} & \dfrac{3/4}{s - 2} - \dfrac{3/4}{s + 2} \\[2mm] \dfrac{-1}{s - 2} + \dfrac{1}{s + 2} & \dfrac{-1/2}{s - 2} + \dfrac{3/2}{s + 2} \end{pmatrix}$$

and

$$e^{\mathbf{A}t} = \begin{pmatrix} \frac{3}{2}e^{2t} - \frac{1}{2}e^{-2t} & \frac{3}{4}e^{2t} - \frac{3}{4}e^{-2t} \\[2mm] -e^{2t} + e^{-2t} & -\frac{1}{2}e^{2t} + \frac{3}{2}e^{-2t} \end{pmatrix}.$$

The general solution of the system is then

$$\mathbf{X} = e^{\mathbf{A}t}\mathbf{C} = \begin{pmatrix} \frac{3}{2}e^{2t} - \frac{1}{2}e^{-2t} & \frac{3}{4}e^{2t} - \frac{3}{4}e^{-2t} \\[2mm] -e^{2t} + e^{-2t} & -\frac{1}{2}e^{2t} + \frac{3}{2}e^{-2t} \end{pmatrix} \begin{pmatrix} c_1 \\ c_2 \end{pmatrix}$$

$$= c_1 \begin{pmatrix} \frac{3}{2} \\ -1 \end{pmatrix} e^{2t} + c_1 \begin{pmatrix} -\frac{1}{2} \\ 1 \end{pmatrix} e^{-2t} + c_2 \begin{pmatrix} \frac{3}{4} \\ -\frac{1}{2} \end{pmatrix} e^{2t} + c_2 \begin{pmatrix} -\frac{3}{4} \\ \frac{3}{2} \end{pmatrix} e^{-2t}$$

$$= \left(\frac{1}{2}c_1 + \frac{1}{4}c_2 \right) \begin{pmatrix} 3 \\ -2 \end{pmatrix} e^{2t} + \left(-\frac{1}{2}c_1 - \frac{3}{4}c_2 \right) \begin{pmatrix} 1 \\ -2 \end{pmatrix} e^{-2t}$$

$$= c_3 \begin{pmatrix} 3 \\ -2 \end{pmatrix} e^{2t} + c_4 \begin{pmatrix} 1 \\ -2 \end{pmatrix} e^{-2t}.$$

18. From $s\mathbf{I} - \mathbf{A} = \begin{pmatrix} s & -1 \\ 2 & s + 2 \end{pmatrix}$ we find

$$(s\mathbf{I} - \mathbf{A})^{-1} = \begin{pmatrix} \dfrac{s + 1 + 1}{(s + 1)^2 + 1} & \dfrac{1}{(s + 1)^2 + 1} \\[3mm] \dfrac{-2}{(s + 1)^2 + 1} & \dfrac{s + 1 - 1}{(s + 1)^2 + 1} \end{pmatrix}$$

and

$$e^{\mathbf{A}t} = \begin{pmatrix} e^{-t}\cos t + e^{-t}\sin t & e^{-t}\sin t \\[2mm] -2e^{-t}\sin t & e^{-t}\cos t - e^{-t}\sin t \end{pmatrix}.$$

The general solution of the system is then

$$\mathbf{X} = e^{\mathbf{A}t}\mathbf{C} = \begin{pmatrix} e^{-t}\cos t + e^{-t}\sin t & e^{-t}\sin t \\[2mm] -2e^{-t}\sin t & e^{-t}\cos t - e^{-t}\sin t \end{pmatrix} \begin{pmatrix} c_1 \\ c_2 \end{pmatrix}$$

$$= c_1 \begin{pmatrix} 1 \\ 0 \end{pmatrix} e^{-t}\cos t + c_1 \begin{pmatrix} 1 \\ -2 \end{pmatrix} e^{-t}\sin t + c_2 \begin{pmatrix} 0 \\ 1 \end{pmatrix} e^{-t}\cos t + c_2 \begin{pmatrix} 1 \\ -1 \end{pmatrix} e^{-t}\sin t$$

$$= c_1 \begin{pmatrix} \cos t + \sin t \\ -2\sin t \end{pmatrix} e^{-t} + c_2 \begin{pmatrix} \sin t \\ \cos t - \sin t \end{pmatrix} e^{-t}.$$

21. The eigenvalues are $\lambda_1 = -1$ and $\lambda_2 = 3$. This leads to the system

$$e^{-t} = b_0 - b_1$$
$$e^{3t} = b_0 + 3b_1,$$

which has the solution $b_0 = \frac{3}{4}e^{-t} + \frac{1}{4}e^{3t}$ and $b_1 = -\frac{1}{4}e^{-t} + \frac{1}{4}e^{3t}$. Then

$$e^{\mathbf{A}t} = b_0\mathbf{I} + b_1\mathbf{A} = \begin{pmatrix} e^{3t} & -2e^{-t} + 2e^{3t} \\ 0 & e^{-t} \end{pmatrix}.$$

The general solution of the system is then

$$\mathbf{X} = e^{\mathbf{A}t}\mathbf{C} = \begin{pmatrix} e^{3t} & -2e^{-t} + 2e^{3t} \\ 0 & e^{-t} \end{pmatrix}\begin{pmatrix} c_1 \\ c_2 \end{pmatrix}$$

$$= c_1 \begin{pmatrix} 1 \\ 0 \end{pmatrix} e^{3t} + c_2 \begin{pmatrix} -2 \\ 1 \end{pmatrix} e^{-t} + c_2 \begin{pmatrix} 2 \\ 0 \end{pmatrix} e^{3t}$$

$$= c_2 \begin{pmatrix} -2 \\ 1 \end{pmatrix} e^{-t} + (c_1 + 2c_2) \begin{pmatrix} 1 \\ 0 \end{pmatrix} e^{3t}$$

$$= c_3 \begin{pmatrix} -2 \\ 1 \end{pmatrix} e^{-t} + c_4 \begin{pmatrix} 1 \\ 0 \end{pmatrix} e^{3t}.$$

24. From Equation (3) in the text

$$e^{\mathbf{D}t} = \begin{pmatrix} 1 & 0 & \cdots & 0 \\ 0 & 1 & \cdots & 0 \\ \vdots & \vdots & \ddots & \vdots \\ 0 & 0 & \cdots & 1 \end{pmatrix} + \begin{pmatrix} \lambda_1 & 0 & \cdots & 0 \\ 0 & \lambda_2 & \cdots & 0 \\ \vdots & \vdots & \ddots & \vdots \\ 0 & 0 & \cdots & \lambda_n \end{pmatrix} + \frac{1}{2!}t^2\begin{pmatrix} \lambda_1^2 & 0 & \cdots & 0 \\ 0 & \lambda_2^2 & \cdots & 0 \\ \vdots & \vdots & \ddots & \vdots \\ 0 & 0 & \cdots & \lambda_n^2 \end{pmatrix}$$

$$+ \frac{1}{3!}t^3\begin{pmatrix} \lambda_1^3 & 0 & \cdots & 0 \\ 0 & \lambda_2^3 & \cdots & 0 \\ \vdots & \vdots & \ddots & \vdots \\ 0 & 0 & \cdots & \lambda_n^3 \end{pmatrix} + \cdots$$

$$= \begin{pmatrix} 1 + \lambda_1 t + \frac{1}{2!}(\lambda_1 t)^2 + \cdots & 0 & \cdots & 0 \\ 0 & 1 + \lambda_2 t + \frac{1}{2!}(\lambda_2 t)^2 + \cdots & \cdots & 0 \\ \vdots & \vdots & \ddots & \vdots \\ 0 & 0 & \cdots & 1 + \lambda_n t + \frac{1}{2!}(\lambda_n t)^2 + \cdots \end{pmatrix}$$

$$= \begin{pmatrix} e^{\lambda_1 t} & 0 & \cdots & 0 \\ 0 & e^{\lambda_2 t} & \cdots & 0 \\ \vdots & \vdots & \ddots & \vdots \\ 0 & 0 & \cdots & e^{\lambda_n t} \end{pmatrix}.$$

Chapter 10 in Review

3. Since

$$\begin{pmatrix} 4 & 6 & 6 \\ 1 & 3 & 2 \\ -1 & -4 & -3 \end{pmatrix} \begin{pmatrix} 3 \\ 1 \\ -1 \end{pmatrix} = \begin{pmatrix} 12 \\ 4 \\ -4 \end{pmatrix} = 4 \begin{pmatrix} 3 \\ 1 \\ -1 \end{pmatrix},$$

we see that $\lambda = 4$ is an eigenvalue with eigenvector $\mathbf{K}_3$. The corresponding solution is $\mathbf{X}_3 = \mathbf{K}_3 e^{4t}$.

6. We have $\det(\mathbf{A} - \lambda\mathbf{I}) = (\lambda + 6)(\lambda + 2) = 0$ so that

$$\mathbf{X} = c_1 \begin{pmatrix} 1 \\ -1 \end{pmatrix} e^{-6t} + c_2 \begin{pmatrix} 1 \\ 1 \end{pmatrix} e^{-2t}.$$

9. We have $\det(\mathbf{A} - \lambda\mathbf{I}) = -(\lambda - 2)(\lambda - 4)(\lambda + 3) = 0$ so that

$$\mathbf{X} = c_1 \begin{pmatrix} -2 \\ 3 \\ 1 \end{pmatrix} e^{2t} + c_2 \begin{pmatrix} 0 \\ 1 \\ 1 \end{pmatrix} e^{4t} + c_3 \begin{pmatrix} 7 \\ 12 \\ -16 \end{pmatrix} e^{-3t}.$$

12. We have

$$\mathbf{X}_c = c_1 \begin{pmatrix} 2\cos t \\ -\sin t \end{pmatrix} e^t + c_2 \begin{pmatrix} 2\sin t \\ \cos t \end{pmatrix} e^t.$$

Then

$$\boldsymbol{\Phi} = \begin{pmatrix} 2\cos t & 2\sin t \\ -\sin t & \cos t \end{pmatrix} e^t, \quad \boldsymbol{\Phi}^{-1} = \begin{pmatrix} \frac{1}{2}\cos t & -\sin t \\ \frac{1}{2}\sin t & \cos t \end{pmatrix} e^{-t},$$

and

$$\mathbf{U} = \int \boldsymbol{\Phi}^{-1}\mathbf{F}\,dt = \int \begin{pmatrix} \cos t - \sec t \\ \sin t \end{pmatrix} dt = \begin{pmatrix} \sin t - \ln|\sec t + \tan t| \\ -\cos t \end{pmatrix},$$

so that

$$\mathbf{X}_p = \boldsymbol{\Phi}\mathbf{U} = \begin{pmatrix} -2\cos t \ln|\sec t + \tan t| \\ -1 + \sin t \ln|\sec t + \tan t| \end{pmatrix} e^t.$$

The solution is

$$\mathbf{X} = \mathbf{X}_c + \mathbf{X}_p = c_1 \begin{pmatrix} 2\cos t \\ -\sin t \end{pmatrix} e^t + c_2 \begin{pmatrix} 2\sin t \\ \cos t \end{pmatrix} e^t + \begin{pmatrix} -2\cos t \ln|\sec t + \tan t| \\ -1 + \sin t \ln|\sec t + \tan t| \end{pmatrix} e^t.$$

15. (a) Letting

$$\mathbf{K} = \begin{pmatrix} k_1 \\ k_2 \\ k_3 \end{pmatrix}$$

we note that $(\mathbf{A} - 2\mathbf{I})\mathbf{K} = \mathbf{0}$ implies that $3k_1 + 3k_2 + 3k_3 = 0$, so $k_1 = -(k_2 + k_3)$. Choosing $k_2 = 0$, $k_3 = 1$ and then $k_2 = 1$, $k_3 = 0$ we get

$$\mathbf{K}_1 = \begin{pmatrix} -1 \\ 0 \\ 1 \end{pmatrix} \quad \text{and} \quad \mathbf{K}_2 = \begin{pmatrix} -1 \\ 1 \\ 0 \end{pmatrix},$$

respectively. Thus,

$$\mathbf{X}_1 = \begin{pmatrix} -1 \\ 0 \\ 1 \end{pmatrix} e^{2t} \quad \text{and} \quad \mathbf{X}_2 = \begin{pmatrix} -1 \\ 1 \\ 0 \end{pmatrix} e^{2t}$$

are two solutions.

(b) From $\det(\mathbf{A} - \lambda\mathbf{I}) = \lambda^2(3 - \lambda) = 0$ we see that $\lambda_1 = 3$, and $\lambda_2 = 0$ is an eigenvalue of multiplicity two. Letting

$$\mathbf{K} = \begin{pmatrix} k_1 \\ k_2 \\ k_3 \end{pmatrix},$$

as in part $(\mathbf{A})$, we note that $(\mathbf{A} - 0\mathbf{I})\mathbf{K} = \mathbf{A}\mathbf{K} = \mathbf{0}$ implies that $k_1 + k_2 + k_3 = 0$, so $k_1 = -(k_2 + k_3)$. Choosing $k_2 = 0$, $k_3 = 1$, and then $k_2 = 1$, $k_3 = 0$ we get

$$\mathbf{K}_2 = \begin{pmatrix} -1 \\ 0 \\ 1 \end{pmatrix} \quad \text{and} \quad \mathbf{K}_3 = \begin{pmatrix} -1 \\ 1 \\ 0 \end{pmatrix},$$

respectively. Since an eigenvector corresponding to $\lambda_1 = 3$ is

$$\mathbf{K}_1 = \begin{pmatrix} 1 \\ 1 \\ 1 \end{pmatrix},$$

the general solution of the system is

$$\mathbf{X} = c_1 \begin{pmatrix} 1 \\ 1 \\ 1 \end{pmatrix} e^{3t} + c_2 \begin{pmatrix} -1 \\ 0 \\ 1 \end{pmatrix} + c_3 \begin{pmatrix} -1 \\ 1 \\ 0 \end{pmatrix}.$$

Chapter 11

Systems of Nonlinear Differential Equations

11.1 Autonomous Systems

3. The corresponding plane autonomous system is

$$x' = y, \quad y' = x^2 - y(1 - x^3).$$

If (x, y) is a critical point, $y = 0$ and so $x^2 - y(1 - x^3) = x^2 = 0$. Therefore $(0, 0)$ is the sole critical point.

6. The corresponding plane autonomous system is

$$x' = y, \quad y' = -x + \epsilon x |x|.$$

If (x, y) is a critical point, $y = 0$ and $-x + \epsilon x |x| = x(-1 + \epsilon |x|) = 0$. Hence $x = 0$, $1/\epsilon$, $-1/\epsilon$. The critical points are $(0, 0)$, $(1/\epsilon, 0)$ and $(-1/\epsilon, 0)$.

9. From $x - y = 0$ we have $y = x$. Substituting into $3x^2 - 4y = 0$ we obtain $3x^2 - 4x = x(3x - 4) = 0$. It follows that $(0, 0)$ and $(4/3, 4/3)$ are the critical points of the system.

12. Adding the two equations we obtain $10 - 15y/(y + 5) = 0$. It follows that $y = 10$, and from $-2x + y + 10 = 0$ we can conclude that $x = 10$. Therefore $(10, 10)$ is the sole critical point of the system.

15. From $x(1 - x^2 - 3y^2) = 0$ we have $x = 0$ or $x^2 + 3y^2 = 1$. If $x = 0$, then substituting into $y(3 - x^2 - 3y^2)$ gives $y(3 - 3y^2) = 0$. Therefore $y = 0$, 1, -1. Likewise $x^2 = 1 - 3y^2$ yields $2y = 0$ so that $y = 0$ and $x^2 = 1 - 3(0)^2 = 1$. The critical points of the system are therefore $(0, 0)$, $(0, 1)$, $(0, -1)$, $(1, 0)$, and $(-1, 0)$.

18. (a) From Exercises 10.2, Problem 6, $x = c_1 + 2c_2 e^{-5t}$ and $y = 3c_1 + c_2 e^{-5t}$, which is not periodic.

(b) From $\mathbf{X}(0) = (3, 4)$ it follows that $c_1 = c_2 = 1$. Therefore $x = 1 + 2e^{-5t}$ and $y = 3 + e^{-5t}$ gives $y = \frac{1}{2}(x - 1) + 3$.

(c)

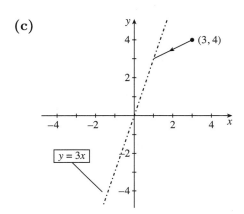

21. (a) From Exercises 10.2, Problem 37, $x = c_1(\sin t - \cos t)e^{4t} + c_2(-\sin t - \cos t)e^{4t}$ and $y = 2c_1(\cos t)e^{4t} + 2c_2(\sin t)e^{4t}$. Because of the presence of e^{4t}, there are no periodic solutions.

(b) From $\mathbf{X}(0) = (-1, 2)$ it follows that $c_1 = 1$ and $c_2 = 0$. Therefore $x = (\sin t - \cos t)e^{4t}$ and $y = 2(\cos t)e^{4t}$.

(c)

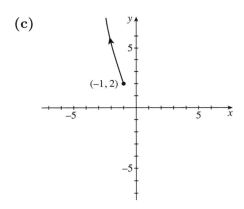

24. Switching to polar coordinates,

$$\frac{dr}{dt} = \frac{1}{r}\left(x\frac{dx}{dt} + y\frac{dy}{dt}\right) = \frac{1}{r}(xy - x^2r^2 - xy + y^2r^2) = r^3$$

$$\frac{d\theta}{dt} = \frac{1}{r^2}\left(-y\frac{dx}{dt} + x\frac{dy}{dt}\right) = \frac{1}{r^2}(-y^2 - xyr^2 - x^2 + xyr^2) = -1.$$

If we use separation of variables, it follows that

$$r = \frac{1}{\sqrt{-2t + c_1}} \quad \text{and} \quad \theta = -t + c_2.$$

Since $\mathbf{X}(0) = (4, 0)$, $r = 4$ and $\theta = 0$ when $t = 0$. It follows that $c_2 = 0$ and $c_1 = \frac{1}{16}$. The final solution can be written as

$$r = \frac{4}{\sqrt{1 - 32t}}, \qquad \theta = -t.$$

Note that $r \to \infty$ as $t \to \left(\frac{1}{32}\right)$. Because $0 \le t \le \frac{1}{32}$, the curve is not a spiral.

27. The system has no critical points, so there are no periodic solutions.

30. The system has no critical points, so there are no periodic solutions.

11.2   Stability of Linear Systems

3. (a) All solutions are unstable spirals which become unbounded as t increases.

(b)

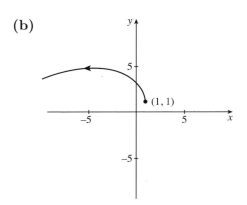

6. (a) All solutions become unbounded and $y = x/2$ serves as the asymptote.

(b)

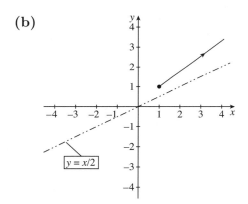

9. Since $\Delta = -41 < 0$, we can conclude from Figure 11.2.12 that $(0,0)$ is a saddle point.

12. Since $\Delta = 1$ and $\tau = -1$, $\tau^2 - 4\Delta = -3$ and so from Figure 11.2.12, $(0,0)$ is a stable spiral point.

15. Since $\Delta = 0.01$ and $\tau = -0.03$, $\tau^2 - 4\Delta < 0$ and so from Figure 11.2.12, $(0,0)$ is a stable spiral point.

18. Note that $\Delta = 1$ and $\tau = \mu$. Therefore we need both $\tau = \mu < 0$ and $\tau^2 - 4\Delta = \mu^2 - 4 < 0$ for $(0,0)$ to be a stable spiral point. These two conditions can be written as $-2 < \mu < 0$.

21. $\mathbf{AX_1} + \mathbf{F} = \mathbf{0}$ implies that $\mathbf{AX_1} = -\mathbf{F}$ or $\mathbf{X_1} = -\mathbf{A}^{-1}\mathbf{F}$. Since $\mathbf{X}_p(t) = -\mathbf{A}^{-1}\mathbf{F}$ is a particular solution, it follows from Theorem 10.1.6 that $\mathbf{X}(t) = \mathbf{X}_c(t) + \mathbf{X_1}$ is the general solution to $\mathbf{X}' = \mathbf{AX} + \mathbf{F}$. If $\tau < 0$ and $\Delta > 0$ then $\mathbf{X}_c(t)$ approaches $(0,0)$ by Theorem 11.2.1(a). It follows that $\mathbf{X}(t)$ approaches $\mathbf{X_1}$ as $t \to \infty$.

24. **(a)** The critical point is $\mathbf{X}_1 = (-1, -2)$.

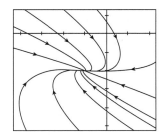

(b) From the graph, $\mathbf{X}_1$ appears to be a stable node or a degenerate stable node.

(c) Since $\tau = -16$, $\Delta = 64$, and $\tau^2 - 4\Delta = 0$, $(0, 0)$ is a degenerate stable node.

11.3 Linearization and Local Stability

3. The critical points are $x = 0$ and $x = n + 1$. Since $g'(x) = k(n+1) - 2kx$, $g'(0) = k(n+1) > 0$ and $g'(n+1) = -k(n+1) < 0$. Therefore $x = 0$ is unstable while $x = n+1$ is asymptotically stable. See Theorem 11.3.2.

6. The only critical point is $v = mg/k$. Now $g(v) = g - (k/m)v$ and so $g'(v) = -k/m < 0$. Therefore $v = mg/k$ is an asymptotically stable critical point by Theorem 11.3.2.

9. Critical points occur at $P = a/b$, c but not at $P = 0$. Since $g'(P) = (a - bP) + (P - c)(-b)$,

$$g'(a/b) = (a/b - c)(-b) = -a + bc \quad \text{and} \quad g'(c) = a - bc.$$

Since $a < bc$, $-a + bc > 0$ and $a - bc < 0$. Therefore $P = a/b$ is unstable while $P = c$ is asymptotically stable.

12. Critical points are $(1, 0)$ and $(-1, 0)$, and

$$\mathbf{g}'(\mathbf{X}) = \begin{pmatrix} 2x & -2y \\ 0 & 2 \end{pmatrix}.$$

At $\mathbf{X} = (1, 0)$, $\tau = 4$, $\Delta = 4$, and so $\tau^2 - 4\Delta = 0$. We can conclude that $(1, 0)$ is unstable but we are unable to classify this critical point any further. At $\mathbf{X} = (-1, 0)$, $\Delta = -4 < 0$ and so $(-1, 0)$ is a saddle point.

15. Since $x^2 - y^2 = 0$, $y^2 = x^2$ and so $x^2 - 3x + 2 = (x - 1)(x - 2) = 0$. It follows that the critical points are $(1, 1)$, $(1, -1)$, $(2, 2)$, and $(2, -2)$. We next use the Jacobian

$$\mathbf{g}'(\mathbf{X}) = \begin{pmatrix} -3 & 2y \\ 2x & -2y \end{pmatrix}$$

to classify these four critical points. For $\mathbf{X} = (1, 1)$, $\tau = -5$, $\Delta = 2$, and so $\tau^2 - 4\Delta = 17 > 0$. Therefore $(1, 1)$ is a stable node. For $\mathbf{X} = (1, -1)$, $\Delta = -2 < 0$ and so $(1, -1)$ is a saddle point. For $\mathbf{X} = (2, 2)$, $\Delta = -4 < 0$ and so we have another saddle point. Finally, if $\mathbf{X} = (2, -2)$, $\tau = 1$, $\Delta = 4$, and so $\tau^2 - 4\Delta = -15 < 0$. Therefore $(2, -2)$ is an unstable spiral point.

18. We found that $(0,0)$, $(0,1)$, $(0,-1)$, $(1,0)$ and $(-1,0)$ were the critical points in Problem 15, Section 11.1. The Jacobian is

$$\mathbf{g}'(\mathbf{X}) = \begin{pmatrix} 1 - 3x^2 - 3y^2 & -6xy \\ -2xy & 3 - x^2 - 9y^2 \end{pmatrix}.$$

For $\mathbf{X} = (0,0)$, $\tau = 4$, $\Delta = 3$ and so $\tau^2 - 4\Delta = 4 > 0$. Therefore $(0,0)$ is an unstable node. Both $(0,1)$ and $(0,-1)$ give $\tau = -8$, $\Delta = 12$, and $\tau^2 - 4\Delta = 16 > 0$. These two critical points are therefore stable nodes. For $\mathbf{X} = (1,0)$ or $(-1,0)$, $\Delta = -4 < 0$ and so saddle points occur.

21. The corresponding plane autonomous system is

$$\theta' = y, \quad y' = \left(\cos\theta - \frac{1}{2}\right)\sin\theta.$$

Since $|\theta| < \pi$, it follows that critical points are $(0,0)$, $(\pi/3, 0)$ and $(-\pi/3, 0)$. The Jacobian matrix is

$$\mathbf{g}'(\mathbf{X}) = \begin{pmatrix} 0 & 1 \\ \cos 2\theta - \frac{1}{2}\cos\theta & 0 \end{pmatrix}$$

and so at $(0,0)$, $\tau = 0$ and $\Delta = -1/2$. Therefore $(0,0)$ is a saddle point. For $\mathbf{X} = (\pm\pi/3, 0)$, $\tau = 0$ and $\Delta = 3/4$. It is not possible to classify either critical point in this borderline case.

24. The corresponding plane autonomous system is

$$x' = y, \quad y' = -\frac{4x}{1 + x^2} - 2y$$

and the only critical point is $(0,0)$. Since the Jacobian matrix is

$$\mathbf{g}'(\mathbf{X}) = \begin{pmatrix} 0 & 1 \\ -4\dfrac{1 - x^2}{(1 + x^2)^2} & -2 \end{pmatrix},$$

$\tau = -2$, $\Delta = 4$, $\tau^2 - 4\Delta = -12$, and so $(0,0)$ is a stable spiral point.

27. The corresponding plane autonomous system is

$$x' = y, \quad y' = -\frac{(\beta + \alpha^2 y^2)x}{1 + \alpha^2 x^2}$$

and the Jacobian matrix is

$$\mathbf{g}'(\mathbf{X}) = \begin{pmatrix} 0 & 1 \\ \dfrac{(\beta + \alpha y^2)(\alpha^2 x^2 - 1)}{(1 + \alpha^2 x^2)^2} & \dfrac{-2\alpha^2 yx}{1 + \alpha^2 x^2} \end{pmatrix}.$$

For $\mathbf{X} = (0,0)$, $\tau = 0$ and $\Delta = \beta$. Since $\beta < 0$, we can conclude that $(0,0)$ is a saddle point.

30. (a) The corresponding plane autonomous system is

$$x' = y, \quad y' = \epsilon\left(y - \frac{1}{3}y^3\right) - x$$

and so the only critical point is $(0,0)$. Since the Jacobian matrix is

$$\mathbf{g}'(\mathbf{X}) = \begin{pmatrix} 0 & 1 \\ -1 & \epsilon(1 - y^2) \end{pmatrix},$$

$\tau = \epsilon$, $\Delta = 1$, and so $\tau^2 - 4\Delta = \epsilon^2 - 4$ at the critical point $(0,0)$.

(b) When $\tau = \epsilon > 0$, $(0,0)$ is an unstable critical point.

(c) When $\epsilon < 0$ and $\tau^2 - 4\Delta = \epsilon^2 - 4 < 0$, $(0,0)$ is a stable spiral point. These two requirements can be written as $-2 < \epsilon < 0$.

(d) When $\epsilon = 0$, $x'' + x = 0$ and so $x = c_1 \cos t + c_2 \sin t$. Therefore all solutions are periodic (with period 2π) and so $(0,0)$ is a center.

33. (a) $x' = 2xy = 0$ implies that either $x = 0$ or $y = 0$. If $x = 0$, then from $1 - x^2 + y^2 = 0$, $y^2 = -1$ and there are no real solutions. If $y = 0$, $1 - x^2 = 0$ and so $(1,0)$ and $(-1,0)$ are critical points. The Jacobian matrix is

$$\mathbf{g}'(\mathbf{X}) = \begin{pmatrix} 2y & 2x \\ -2x & 2y \end{pmatrix}$$

and so $\tau = 0$ and $\Delta = 4$ at either $\mathbf{X} = (1,0)$ or $(-1,0)$. We obtain no information about these critical points in this borderline case.

(b) The differential equation is

$$\frac{dy}{dx} = \frac{y'}{x'} = \frac{1 - x^2 + y^2}{2xy}$$

or

$$2xy\frac{dy}{dx} = 1 - x^2 + y^2.$$

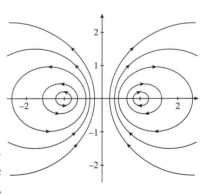

Letting $\mu = y^2/x$, it follows that $d\mu/dx = (1/x^2) - 1$ and so $\mu = -(1/x) - x + 2c$. Therefore $y^2/x = -(1/x) - x + 2c$ which can be put in the form $(x - c)^2 + y^2 = c^2 - 1$. The solution curves are shown and so both $(1,0)$ and $(-1,0)$ are centers.

36. The corresponding plane autonomous system is

$$x' = y, \quad y' = \epsilon x^2 - x + 1$$

and so the critical points must satisfy $y = 0$ and

$$x = \frac{1 \pm \sqrt{1 - 4\epsilon}}{2\epsilon}.$$

Therefore we must require that $\epsilon \leq \frac{1}{4}$ for real solutions to exist. We will use the Jacobian matrix

$$\mathbf{g}'(\mathbf{X}) = \begin{pmatrix} 0 & 1 \\ 2\epsilon x - 1 & 0 \end{pmatrix}$$

to attempt to classify $((1 \pm \sqrt{1 - 4\epsilon})/2\epsilon, 0)$ when $\epsilon \leq 1/4$. Note that $\tau = 0$ and $\Delta = \mp\sqrt{1 - 4\epsilon}$. For $\mathbf{X} = ((1 + \sqrt{1 - 4\epsilon})/2\epsilon, 0)$ and $\epsilon < 1/4$, $\Delta < 0$ and so a saddle point occurs. For $\mathbf{X} = ((1 - \sqrt{1 - 4\epsilon})/2\epsilon, 0)$, $\Delta \geq 0$ and we are not able to classify this critical point using linearization.

39. **(a)** Letting $x = \theta$ and $y = x'$ we obtain the system $x' = y$ and $y' = 1/2 - \sin x$. Since $\sin \pi/6 = \sin 5\pi/6 = 1/2$ we see that $(\pi/6, 0)$ and $(5\pi/6, 0)$ are critical points of the system.

(b) The Jacobian matrix is

$$\mathbf{g}'(\mathbf{X}) = \begin{pmatrix} 0 & 1 \\ -\cos x & 0 \end{pmatrix}$$

and so

$$\mathbf{A}_1 = \mathbf{g}' = ((\pi/6, 0)) = \begin{pmatrix} 0 & 1 \\ -\sqrt{3}/2 & 0 \end{pmatrix} \quad \text{and} \quad \mathbf{A}_2 = \mathbf{g}' = ((5\pi/6, 0)) = \begin{pmatrix} 0 & 1 \\ \sqrt{3}/2 & 0 \end{pmatrix}.$$

Since $\det \mathbf{A}_1 > 0$ and the trace of $\mathbf{A}_1$ is 0, no conclusion can be drawn regarding the critical point $(\pi/6, 0)$. Since $\det \mathbf{A}_2 < 0$, we see that $(5\pi/6, 0)$ is a saddle point.

(c) From the system in part **(A)** we obtain the first-order differential equation

$$\frac{dy}{dx} = \frac{1/2 - \sin x}{y}.$$

Separating variables and integrating we obtain

$$\int y \, dy = \int \left(\frac{1}{2} - \sin x \right) dx$$

and

$$\frac{1}{2}y^2 = \frac{1}{2}x + \cos x + c_1$$

or

$$y^2 = x + 2\cos x + c_2.$$

For x_0 near $\pi/6$, if $\mathbf{X}(0) = (x_0, 0)$ then $c_2 = -x_0 - 2\cos x_0$ and $y^2 = x + 2\cos x - x_0 - 2\cos x_0$. Thus, there are two values of y for each x in a sufficiently small interval around $\pi/6$. Therefore $(\pi/6, 0)$ is a center.

11.4 Autonomous Systems as Mathematical Models

3. The corresponding plane autonomous system is

$$x' = y, \quad y' = -g\frac{f'(x)}{1 + [f'(x)]^2} - \frac{\beta}{m}y$$

and

$$\frac{\partial}{\partial x}\left(-g\frac{f'(x)}{1 + [f'(x)]^2} - \frac{\beta}{m}y\right) = -g\frac{(1 + [f'(x)]^2)f''(x) - f'(x)2f'(x)f''(x)}{(1 + [f'(x)]^2)^2}.$$

If $\mathbf{X}_1 = (x_1, y_1)$ is a critical point, $y_1 = 0$ and $f'(x_1) = 0$. The Jacobian at this critical point is therefore

$$\mathbf{g}'(\mathbf{X}_1) = \begin{pmatrix} 0 & 1 \\ -gf''(x_1) & -\dfrac{\beta}{m} \end{pmatrix}.$$

6. (a) If $f(x) = \cosh x$, $f'(x) = \sinh x$ and $[f'(x)]^2 + 1 = \sinh^2 x + 1 = \cosh^2 x$. Therefore

$$\frac{dy}{dx} = \frac{y'}{x'} = -g\frac{\sinh x}{\cosh^2 x}\frac{1}{y}.$$

We can separate variables to show that $y^2 = 2g/\cosh x + c$. But $x(0) = x_0$ and $y(0) = x'(0) = v_0$. Therefore $c = v_0^2 - (2g/\cosh x_0)$ and so

$$y^2 = \frac{2g}{\cosh x} - \frac{2g}{\cosh x_0} + v_0^2.$$

Now

$$\frac{2g}{\cosh x} - \frac{2g}{\cosh x_0} + v_0^2 \geq 0 \quad \text{if and only if} \quad \cosh x \leq \frac{2g\cosh x_0}{2g - v_0^2\cosh x_0}$$

and the solution to this inequality is an interval $[-a, a]$. Therefore each x in $(-a, a)$ has two corresponding values of y and so the solution is periodic.

(b) Since $z = \cosh x$, the maximum height occurs at the largest value of x on the cycle. From (a), $x_{\max} = a$ where $\cosh a = 2g\cosh x_0/(2g - v_0^2\cosh x_0)$. Therefore

$$z_{\max} = \frac{2g\cosh x_0}{2g - v_0^2\cosh x_0}.$$

9. (a) In the Lotka–Volterra model the average number of predators is d/c and the average number of prey is a/b. But

$$x' = -ax + bxy - \epsilon_1 x = -(a + \epsilon_1)x + bxy$$
$$y' = -cxy + dy - \epsilon_2 y = -cxy + (d - \epsilon_2)y$$

and so the new critical point in the first quadrant is $(d/c - \epsilon_2/c, a/b + \epsilon_1/b)$.

(b) The average number of predators $d/c - \epsilon_2/c$ has decreased while the average number of prey $a/b + \epsilon_1/b$ has increased. The fishery science model is consistent with Volterra's principle.

12. $\Delta = r_1 r_2$, $\tau = r_1 + r_2$ and $\tau^2 - 4\Delta = (r_1 + r_2)^2 - 4r_1 r_2 = (r_1 - r_2)^2$. Therefore when $r_1 \neq r_2$, $(0,0)$ is an unstable node.

15. $K_1/\alpha_{12} < K_2 < K_1\alpha_{21}$ and so $\alpha_{12}\alpha_{21} > 1$. Therefore $\Delta = (1 - \alpha_{12}\alpha_{21})\hat{x}\hat{y}\, r_1 r_2/K_1 K_2 < 0$ and so $(\hat{x}, \hat{y})$ is a saddle point.

18. (a) The magnitude of the frictional force between the bead and the wire is $\mu(mg\cos\theta)$ for some $\mu > 0$. The component of this frictional force in the x-direction is

$$(\mu mg\cos\theta)\cos\theta = \mu mg\cos^2\theta.$$

But

$$\cos\theta = \frac{1}{\sqrt{1 + [f'(x)]^2}} \quad \text{and so} \quad \mu mg\cos^2\theta = \frac{\mu mg}{1 + [f'(x)]^2}.$$

It follows from Newton's Second Law that

$$mx'' = -mg\frac{f'(x)}{1 + [f'(x)]^2} - \beta x' + mg\frac{\mu}{1 + [f'(x)]^2}$$

and so

$$x'' = g\frac{\mu - f'(x)}{1 + [f'(x)]^2} - \frac{\beta}{m}x'.$$

(b) A critical point (x, y) must satisfy $y = 0$ and $f'(x) = \mu$. Therefore critical points occur at $(x_1, 0)$ where $f'(x_1) = \mu$. The Jacobian matrix of the plane autonomous system is

$$\mathbf{g}'(\mathbf{X}) = \begin{pmatrix} 0 & 1 \\ g\dfrac{(1 + [f'(x)]^2)(-f''(x)) - (\mu - f'(x))2f'(x)f''(x)}{(1 + [f'(x)]^2)^2} & -\dfrac{\beta}{m} \end{pmatrix}$$

and so at a critical point $\mathbf{X}_1$,

$$\mathbf{g}'(\mathbf{X}) = \begin{pmatrix} 0 & 1 \\ \dfrac{-gf''(x_1)}{1 + \mu^2} & -\dfrac{\beta}{m} \end{pmatrix}.$$

Therefore $\tau = -\beta/m < 0$ and $\Delta = gf''(x_1)/(1 + \mu^2)$. When $f''(x_1) < 0$, $\Delta < 0$ and so a saddle point occurs. When $f''(x_1) > 0$ and

$$\tau^2 - 4\Delta = \frac{\beta^2}{m^2} - 4g\frac{f''(x_1)}{1 + \mu^2} < 0,$$

$(x_1, 0)$ is a stable spiral point. This condition can also be written as

$$\beta^2 < 4gm^2\frac{f''(x_1)}{1 + \mu^2}.$$

21. The equation

$$x' = \alpha \frac{y}{1+y} x - x = x \left(\frac{\alpha y}{1+y} - 1 \right) = 0$$

implies that $x = 0$ or $y = 1/(\alpha - 1)$. When $\alpha > 0$, $\hat{y} = 1/(\alpha - 1) > 0$. If $x = 0$, then from the differential equation for y', $y = \beta$. On the other hand, if $\hat{y} = 1/(\alpha - 1)$, $\hat{y}/(1 + \hat{y}) = 1/\alpha$ and so $\hat{x}/\alpha - 1/(\alpha - 1) + \beta = 0$. It follows that

$$\hat{x} = \alpha \left(\beta - \frac{1}{\alpha - 1} \right) = \frac{\alpha}{\alpha - 1}[(\alpha - 1)\beta - 1]$$

and if $\beta(\alpha - 1) > 1$, $\hat{x} > 0$. Therefore $(\hat{x}, \hat{y})$ is the unique critical point in the first quadrant. The Jacobian matrix is

$$\mathbf{g}'(\mathbf{X}) = \begin{pmatrix} \alpha \dfrac{y}{y+1} - 1 & \dfrac{\alpha x}{(1+y)^2} \\ -\dfrac{y}{1+y} & \dfrac{-x}{(1+y)^2} - 1 \end{pmatrix}$$

and for $\mathbf{X} = (\hat{x}, \hat{y})$, the Jacobian can be written in the form

$$\mathbf{g}'((\hat{x}, \hat{y})) = \begin{pmatrix} 0 & \dfrac{(\alpha - 1)^2}{\alpha} \hat{x} \\ -\dfrac{1}{\alpha} & -\dfrac{(\alpha - 1)^2}{\alpha^2} - 1 \end{pmatrix}.$$

It follows that

$$\tau = -\left[\frac{(\alpha - 1)^2}{\alpha^2} \hat{x} + 1 \right] < 0, \quad \Delta = \frac{(\alpha - 1)^2}{\alpha^2} \hat{x}$$

and so $\tau = -(\Delta + 1)$. Therefore $\tau^2 - 4\Delta = (\Delta + 1)^2 - 4\Delta = (\Delta - 1)^2 > 0$. Therefore $(\hat{x}, \hat{y})$ is a stable node.

11.5 │ Periodic Solutions, Limit Cycles, and Global Stability

3. For $P = -x + y^2$ and $Q = x - y$, $\dfrac{\partial P}{\partial x} + \dfrac{\partial Q}{\partial y} = -2 < 0$. Therefore there are no periodic solutions by Theorem 11.5.2.

6. From $y' = xy - y = y(x - 1) = 0$ either $y = 0$ or $x = 1$. If $y = 0$, then from $2x + y^2 = 0$, $x = 0$. Likewise $x = 1$ implies that $2 + y^2 = 0$, which has no real solutions. Therefore $(0, 0)$ is the only critical point. But $\mathbf{g}'((0, 0))$ has determinant $\Delta = -2$. The single critical point is a saddle point and so, by the corollary to Theorem 11.5.1, there are no periodic solutions.

9. For $\delta(x, y) = e^{ax+by}$, $\dfrac{\partial}{\partial x}(\delta P) + \dfrac{\partial}{\partial y}(\delta Q)$ can be simplified to

$$e^{ax+by}[-bx^2 - 2ax + axy + (2b + 1)y].$$

Setting $a = 0$ and $b = -1/2$,

$$\frac{\partial}{\partial x}(\delta P) + \frac{\partial}{\partial y}(\delta Q) = \frac{1}{2}x^2 e^{-\frac{1}{2}y}$$

which does not change signs. Therefore by Theorem 11.5.3 there are no periodic solutions.

12. The corresponding plane autonomous system is $x' = y$, $y' = g(x, y)$ and so

$$\frac{\partial P}{\partial x} + \frac{\partial Q}{\partial y} = \frac{\partial g}{\partial y} = \frac{\partial g}{\partial x'} \neq 0$$

in the region R. Therefore $\dfrac{\partial P}{\partial x} + \dfrac{\partial Q}{\partial y}$ cannot change signs and so there are no periodic solutions by Theorem 11.5.2.

15. If $\mathbf{n} = (-2x, -2y)$,

$$\mathbf{V} \cdot \mathbf{n} = 2x^2 - 4xy + 2y^2 + 2y^4 = 2(x - y)^2 + 2y^4 \geq 0.$$

Therefore $x^2 + y^2 \leq r$ serves as an invariant region for any $r > 0$ by Theorem 11.5.4.

18. The corresponding plane autonomous system is

$$x' = y, \quad y' = y(1 - 3x^2 - 2y^2) - x$$

and it is easy to see that $(0, 0)$ is the only critical point. If $\mathbf{n} = (-2x, -2y)$ then

$$\mathbf{V} \cdot \mathbf{n} = -2xy - 2y^2(1 - 3x^2 - 2y^2) + 2xy = -2y^2(1 - 2r^2 - x^2).$$

If $r = \frac{1}{2}\sqrt{2}$, $2r^2 = 1$ and so $\mathbf{V} \cdot \mathbf{n} = 2x^2y^2 \geq 0$. Therefore $\frac{1}{4} \leq x^2 + y^2 \leq \frac{1}{2}$ serves as an invariant region. By Theorem 11.5.5(b) there is at least one periodic solution.

21. (a) $\dfrac{\partial P}{\partial x} + \dfrac{\partial Q}{\partial y} = 2xy - 1 - x^2 \leq 2x - 1 - x^2 = -(x - 1)^2 \leq 0$. Therefore there are no periodic solutions.

(b) If (x, y) is a critical point, $x^2y = \frac{1}{2}$ and so from $x' = x^2y - x + 1$, $\frac{1}{2} - x + 1 = 0$. Therefore $x = 3/2$ and so $y = 2/9$. For this critical point, $\tau = -31/12 < 0$, $\Delta = 9/4 > 0$, and $\tau^2 - 4\Delta < 0$. Therefore $(3/2, 2/9)$ is a stable spiral point and so, from Theorem 11.5.6(b), $\lim_{x \to \infty} \mathbf{X}(t) = (3/2, 2/9)$.

Chapter 11 in Review

3. A center or a saddle point

6. True

9. True

12. (a) If $\mathbf{X}(0) = \mathbf{X}_0$ lies on the line $y = -2x$, then $\mathbf{X}(t)$ approaches $(0,0)$ along this line. For all other initial conditions, $\mathbf{X}(t)$ approaches $(0,0)$ from the direction determined by the line $y = x$.

(b) If $\mathbf{X}(0) = \mathbf{X}_0$ lies on the line $y = -x$, then $\mathbf{X}(t)$ approaches $(0,0)$ along this line. For all other initial conditions, $\mathbf{X}(t)$ becomes unbounded and $y = 2x$ serves as an asymptote.

15. The corresponding plane autonomous system is $x' = y$, $y' = \mu(1-x^2)-x$ and so the Jacobian at the critical point $(0,0)$ is

$$\mathbf{g}'((0,0)) = \begin{pmatrix} 0 & 1 \\ -1 & \mu \end{pmatrix}.$$

Therefore $\tau = \mu$, $\Delta = 1$ and $\tau^2 - 4\Delta = \mu^2 - 4$. Now $\mu^2 - 4 < 0$ if and only if $-2 < \mu < 2$. We may therefore conclude that $(0,0)$ is a stable node for $\mu < -2$, a stable spiral point for $-2 < \mu < 0$, an unstable spiral point for $0 < \mu < 2$, and an unstable node for $\mu > 2$.

18. The corresponding plane autonomous system is

$$x' = y, \quad y' = -\frac{\beta}{m}y - \frac{k}{m}(s+x)^3 + g$$

and so the Jacobian is

$$\mathbf{g}'(\mathbf{X}) = \begin{pmatrix} 0 & 1 \\ -\dfrac{3k}{m}(s+x)^2 & -\dfrac{\beta}{m} \end{pmatrix}.$$

For $\mathbf{X} = (0,0)$, $\tau = -\dfrac{\beta}{m} < 0$, $\Delta = \dfrac{3k}{m}s^2 > 0$.
Therefore

$$\tau^2 - 4\Delta = \frac{\beta^2}{m^2} - \frac{12k}{m}s^2 = \frac{1}{m^2}(\beta^2 - 12kms^2).$$

Therefore $(0,0)$ is a stable node if $\beta^2 > 12kms^2$ and a stable spiral point provided $\beta^2 < 12kms^2$, where $ks^3 = mg$.

21. (a) If $x = \theta$ and $y = x' = \theta'$, the corresponding plane autonomous system is

$$x' = y, \quad y' = \omega^2 \sin x \cos x - \frac{g}{l}\sin x - \frac{\beta}{ml}y.$$

Therefore $\dfrac{\partial P}{\partial x} + \dfrac{\partial Q}{\partial y} = -\dfrac{\beta}{ml} < 0$ and so there are no periodic solutions.

(b) If (x,y) is a critical point, $y = 0$ and so $\sin x(\omega^2 \cos x - g/l) = 0$. Either $\sin x = 0$ (in which case $x = 0$) of $\cos x = g/\omega^2 l$. But if $\omega^2 < g/l$, $g/\omega^2 l > 1$ and so the latter equation has no real solutions. Therefore $(0,0)$ is the only critical point if $\omega^2 < g/l$. The Jacobian matrix is

$$\mathbf{g}'(\mathbf{X}) = \begin{pmatrix} 0 & 1 \\ \omega^2 \cos 2x - \dfrac{g}{l}\cos x & -\dfrac{\beta}{ml} \end{pmatrix}$$

and so $\tau = -\beta/ml < 0$ and $\Delta = g/l - \omega^2 > 0$ for $\mathbf{X} = (0,0)$. It follows that $(0,0)$ is asymptotically stable and so after a small displacement, the pendulum will return to $\theta = 0$, $\theta' = 0$.

(c) If $\omega^2 > g/l$, $\cos x = g/\omega^2 l$ will have two solutions $x = \pm\hat{x}$ that satisfy $-\pi < x < \pi$. Therefore $(\pm\hat{x}, 0)$ are two additional critical points. If $\mathbf{X}_1 = (0,0)$, $\Delta = g/l - \omega^2 < 0$ and so $(0,0)$ is a saddle point. If $\mathbf{X}_1 = (\pm\hat{x}, 0)$, $\tau = -\beta/ml < 0$ and

$$\Delta = \frac{g}{l}\cos\hat{x} - \omega^2\cos 2\hat{x} = \frac{g^2}{\omega^2 l^2} - \omega^2\left(2\frac{g^2}{\omega^4 l^2} - 1\right) = \omega^2 - \frac{g^2}{\omega^2 l^2} > 0.$$

Therefore $(\hat{x}, 0)$ and $(-\hat{x}, 0)$ are each stable. When $\theta(0) = \theta_0$, $\theta'(0) = 0$ and θ_0 is small we expect the pendulum to reach one of these two stable equilibrium positions.

(d) In **(b)**, $(0,0)$ is a stable spiral point provided

$$\tau^2 - 4\Delta = \frac{\beta^2}{m^2 l^2} - 4\left(\frac{g}{l} - \omega^2\right) < 0.$$

This condition is equivalent to $\beta < 2ml\sqrt{g/l - \omega^2}$. In **(c)**, $(\pm\hat{x}, 0)$ are stable spiral points provided that

$$\tau^2 - 4\Delta = \frac{\beta^2}{m^2 l^2} - 4\left(\omega^2 - \frac{g^2}{\omega^2 l^2}\right) < 0.$$

This condition is equivalent to $\beta < 2ml\sqrt{\omega^2 - g^2/(\omega^2 l^2)}$.

Chapter 12

Fourier Series

12.1 Orthogonal Functions

3. $\displaystyle\int_0^2 e^x(xe^{-x} - e^{-x})\,dx = \int_0^2 (x - 1)\,dx = \left(\frac{1}{2}x^2 - x\right)\bigg|_0^2 = 0$

6. $\displaystyle\int_{\pi/4}^{5\pi/4} e^x \sin x\,dx = \left(\frac{1}{2}e^x \sin x - \frac{1}{2}e^x \cos x\right)\bigg|_{\pi/4}^{5\pi/4} = 0$

9. For $m \neq n$ $\displaystyle\int_0^\pi \cos nx \cos mx\,dx = \frac{1}{2}\int_0^\pi (\cos(n+m)x + \cos(n-m)x)\,dx$

$$= \frac{1}{2(n+m)}\sin(n+m)x\bigg|_0^\pi + \frac{1}{2(n-m)}\sin(n-m)x\bigg|_0^\pi$$

$$\int_0^\pi \sin nx \sin mx\,dx = \frac{1}{2}\int_0^\pi (\cos(n-m)x - \cos(n+m)x)\,dx$$

$$= \frac{1}{2(n-m)}\sin(n-m)x\bigg|_0^\pi - \frac{1}{2(n+m)}\sin(n+m)x\bigg|_0^\pi = 0.$$

For $m = n$

$$\int_0^\pi \sin^2 nx\,dx = \int_0^\pi \left(\frac{1}{2} - \frac{1}{2}\cos 2nx\right)\,dx = \frac{1}{2}x\bigg|_0^\pi - \frac{1}{4n}\sin 2nx\bigg|_0^\pi = \frac{\pi}{2}$$

so that

$$\|\sin nx\| = \sqrt{\frac{\pi}{2}}.$$

12. For $m \neq n$, we use Problems 11 and 10:

$$\int_{-p}^p \cos\frac{n\pi}{p}x \cos\frac{m\pi}{p}x\,dx = 2\int_0^p \cos\frac{n\pi}{p}x \cos\frac{m\pi}{p}x\,dx = 0,$$

$$\int_{-p}^p \sin\frac{n\pi}{p}x \sin\frac{m\pi}{p}x\,dx = 2\int_0^p \sin\frac{n\pi}{p}x \sin\frac{m\pi}{p}x\,dx = 0.$$

Also

$$\int_{-p}^{p} \sin\frac{n\pi}{p}x \cos\frac{m\pi}{p}x\,dx = \frac{1}{2}\int_{-p}^{p}\left(\sin\frac{(n-m)\pi}{p}x + \sin\frac{(n+m)\pi}{p}x\right)dx = 0,$$

$$\int_{-p}^{p} 1\cdot\cos\frac{n\pi}{p}x\,dx = \frac{p}{n\pi}\sin\frac{n\pi}{p}x\bigg|_{-p}^{p} = 0,$$

$$\int_{-p}^{p} 1\cdot\sin\frac{n\pi}{p}x\,dx = -\frac{p}{n\pi}\cos\frac{n\pi}{p}x\bigg|_{-p}^{p} = 0,$$

and

$$\int_{-p}^{p} \sin\frac{n\pi}{p}x \cos\frac{n\pi}{p}x\,dx = \int_{-p}^{p}\frac{1}{2}\sin\frac{2n\pi}{p}x\,dx = \frac{-p}{4n\pi}\cos\frac{2n\pi}{p}x\bigg|_{-p}^{p} = 0.$$

For $m = n$

$$\int_{-p}^{p} \cos^2\frac{n\pi}{p}x\,dx = \int_{-p}^{p}\left(\frac{1}{2} + \frac{1}{2}\cos\frac{2n\pi}{p}x\right)dx = p,$$

$$\int_{-p}^{p} \sin^2\frac{n\pi}{p}x\,dx = \int_{-p}^{p}\left(\frac{1}{2} - \frac{1}{2}\cos\frac{2n\pi}{p}x\right)dx = p,$$

and

$$\int_{-p}^{p} 1^2\,dx = 2p$$

so that

$$\|1\| = \sqrt{2p}, \quad \left\|\cos\frac{n\pi}{p}x\right\| = \sqrt{p}, \quad \text{and} \quad \left\|\sin\frac{n\pi}{p}x\right\| = \sqrt{p}.$$

15. By orthogonality $\int_a^b \phi_0(x)\phi_n(x)dx = 0$ for $n = 1, 2, 3, \dots$; that is, $\int_a^b \phi_n(x)\,dx = 0$ for $n = 1, 2, 3, \dots$.

18. Setting

$$0 = \int_{-2}^{2} f_3(x)f_1(x)\,dx = \int_{-2}^{2}\left(x^2 + c_1 x^3 + c_2 x^4\right)dx = \frac{16}{3} + \frac{64}{5}c_2$$

and

$$0 = \int_{-2}^{2} f_3(x)f_2(x)\,dx = \int_{-2}^{2}\left(x^3 + c_1 x^4 + c_2 x^5\right)dx = \frac{64}{5}c_1$$

we obtain $c_1 = 0$ and $c_2 = -5/12$.

21. The fundamental period is $2\pi/2\pi = 1$.

24. The fundamental period of $\sin 2x + \cos 4x$ is $2\pi/2 = \pi$.

12.2 | Fourier Series

3. $a_0 = \int_{-1}^{1} f(x)\,dx = \int_{-1}^{0} 1\,dx + \int_{0}^{1} x\,dx = \dfrac{3}{2}$

$a_n = \int_{-1}^{1} f(x)\cos n\pi x\,dx = \int_{-1}^{0} \cos n\pi x\,dx + \int_{0}^{1} x\cos n\pi x\,dx = \dfrac{1}{n^2\pi^2}[(-1)^n - 1]$

$b_n = \int_{-1}^{1} f(x)\sin n\pi x\,dx = \int_{-1}^{0} \sin n\pi x\,dx + \int_{0}^{1} x\sin n\pi x\,dx = -\dfrac{1}{n\pi}$

$f(x) = \dfrac{3}{4} + \sum_{n=1}^{\infty}\left[\dfrac{(-1)^n - 1}{n^2\pi^2}\cos n\pi x - \dfrac{1}{n\pi}\sin n\pi x\right]$

Converges to $\frac{1}{2}$ at $x = 0$.

6. $a_0 = \dfrac{1}{\pi}\int_{-\pi}^{\pi} f(x)\,dx = \dfrac{1}{\pi}\int_{-\pi}^{0} \pi^2\,dx + \dfrac{1}{\pi}\int_{0}^{\pi} \left(\pi^2 - x^2\right)\,dx = \dfrac{5}{3}\pi^2$

$a_n = \dfrac{1}{\pi}\int_{-\pi}^{\pi} f(x)\cos nx\,dx = \dfrac{1}{\pi}\int_{-\pi}^{0} \pi^2\cos nx\,dx + \dfrac{1}{\pi}\int_{0}^{\pi} \left(\pi^2 - x^2\right)\cos nx\,dx$

$= \dfrac{1}{\pi}\left(\dfrac{\pi^2 - x^2}{n}\sin nx\bigg|_0^\pi + \dfrac{2}{n}\int_0^\pi x\sin nx\,dx\right) = \dfrac{2}{n^2}(-1)^{n+1}$

$b_n = \dfrac{1}{\pi}\int_{-\pi}^{\pi} f(x)\sin nx\,dx = \dfrac{1}{\pi}\int_{-\pi}^{0} \pi^2\sin nx\,dx + \dfrac{1}{\pi}\int_{0}^{\pi} \left(\pi^2 - x^2\right)\sin nx\,dx$

$= \dfrac{\pi}{n}[(-1)^n - 1] + \dfrac{1}{\pi}\left(\dfrac{x^2 - \pi^2}{n}\cos nx\bigg|_0^\pi - \dfrac{2}{n}\int_0^\pi x\cos nx\,dx\right) = \dfrac{\pi}{n}(-1)^n + \dfrac{2}{n^3\pi}[1 - (-1)^n]$

$f(x) = \dfrac{5\pi^2}{6} + \sum_{n=1}^{\infty}\left[\dfrac{2}{n^2}(-1)^{n+1}\cos nx + \left(\dfrac{\pi}{n}(-1)^n + \dfrac{2[1 - (-1)^n]}{n^3\pi}\right)\sin nx\right]$

f is continous on the interval.

9. $a_0 = \dfrac{1}{\pi}\int_{-\pi}^{\pi} f(x)\,dx = \dfrac{1}{\pi}\int_{0}^{\pi} \sin x\,dx = \dfrac{2}{\pi}$

$a_n = \dfrac{1}{\pi}\int_{-\pi}^{\pi} f(x)\cos nx\,dx = \dfrac{1}{\pi}\int_{0}^{\pi} \sin x\,\cos nx\,dx = \dfrac{1}{2\pi}\int_{0}^{\pi} \left(\sin(1+n)x + \sin(1-n)x\right)\,dx$

$= \dfrac{1 + (-1)^n}{\pi(1 - n^2)}\quad$ for $n = 2, 3, 4, \ldots$

$a_1 = \dfrac{1}{2\pi}\int_{0}^{\pi} \sin 2x\,dx = 0$

$b_n = \dfrac{1}{\pi}\int_{-\pi}^{\pi} f(x)\sin nx\,dx = \dfrac{1}{\pi}\int_{0}^{\pi} \sin x\,\sin nx\,dx$

$= \dfrac{1}{2\pi}\int_{0}^{\pi} \left(\cos(1-n)x - \cos(1+n)x\right)\,dx = 0\quad$ for $n = 2, 3, 4, \ldots$

$$b_1 = \frac{1}{2\pi} \int_0^\pi (1 - \cos 2x)\, dx = \frac{1}{2}$$

$$f(x) = \frac{1}{\pi} + \frac{1}{2}\sin x + \sum_{n=2}^\infty \frac{1 + (-1)^n}{\pi(1 - n^2)}\cos nx$$

f is continous on the interval.

12. $a_0 = \dfrac{1}{2}\displaystyle\int_{-2}^{2} f(x)\, dx = \dfrac{1}{2}\left(\int_0^1 x\, dx + \int_1^2 1\, dx\right) = \dfrac{3}{4}$

$a_n = \dfrac{1}{2}\displaystyle\int_{-2}^{2} f(x)\cos\frac{n\pi}{2}x\, dx = \dfrac{1}{2}\left(\int_0^1 x\cos\frac{n\pi}{2}x\, dx + \int_1^2 \cos\frac{n\pi}{2}x\, dx\right) = \dfrac{2}{n^2\pi^2}\left(\cos\frac{n\pi}{2} - 1\right)$

$b_n = \dfrac{1}{2}\displaystyle\int_{-2}^{2} f(x)\sin\frac{n\pi}{2}x\, dx = \dfrac{1}{2}\left(\int_0^1 x\sin\frac{n\pi}{2}x\, dx + \int_1^2 \sin\frac{n\pi}{2}x\, dx\right)$

$\qquad = \dfrac{2}{n^2\pi^2}\left(\sin\frac{n\pi}{2} + \frac{n\pi}{2}(-1)^{n+1}\right)$

$f(x) = \dfrac{3}{8} + \displaystyle\sum_{n=1}^\infty \left[\frac{2}{n^2\pi^2}\left(\cos\frac{n\pi}{2} - 1\right)\cos\frac{n\pi}{2}x + \frac{2}{n^2\pi^2}\left(\sin\frac{n\pi}{2} + \frac{n\pi}{2}(-1)^{n+1}\right)\sin\frac{n\pi}{2}x\right]$

f is continous on the interval.

15. $a_0 = \dfrac{1}{\pi}\displaystyle\int_{-\pi}^{\pi} f(x)\, dx = \dfrac{1}{\pi}\int_{-\pi}^{\pi} e^x\, dx = \dfrac{1}{\pi}(e^\pi - e^{-\pi})$

$a_n = \dfrac{1}{\pi}\displaystyle\int_{-\pi}^{\pi} f(x)\cos nx\, dx = \dfrac{(-1)^n(e^\pi - e^{-\pi})}{\pi(1 + n^2)}$

$b_n = \dfrac{1}{\pi}\displaystyle\int_{-\pi}^{\pi} f(x)\sin nx\, dx = \dfrac{1}{\pi}\int_{-\pi}^{\pi} e^x \sin nx\, dx = \dfrac{(-1)^n n(e^{-\pi} - e^\pi)}{\pi(1 + n^2)}$

$f(x) = \dfrac{e^\pi - e^{-\pi}}{2\pi} + \displaystyle\sum_{n=1}^\infty \left[\frac{(-1)^n(e^\pi - e^{-\pi})}{\pi(1 + n^2)}\cos nx + \frac{(-1)^n n(e^{-\pi} - e^\pi)}{\pi(1 + n^2)}\sin nx\right]$

f is continous on the interval.

18. $a_0 = \dfrac{1}{\pi}\displaystyle\int_{-\pi}^{\pi} (\pi^2 - x^2)\, dx = \dfrac{1}{\pi}\left[\pi^2 x - \frac{1}{3}x^3\right]_{-\pi}^{\pi} = \dfrac{4\pi^2}{3}$

$a_n = \dfrac{1}{\pi}\displaystyle\int_{-\pi}^{\pi} (\pi^2 - x^2)\cos nx\, dx = \dfrac{1}{\pi}\left[(\pi^2 - x^2)\frac{\sin nx}{n}\Big|_{-\pi}^{\pi} + \frac{2}{n}\int_{\pi}^{\pi} x\sin nx\, dx\right]$

$\qquad = \dfrac{1}{\pi}\left[(\pi^2 - x^2)\frac{\sin nx}{n} + \frac{2}{n^2}\left(-x\cos nx + \frac{\sin nx}{n}\right)\right]_{-\pi}^{\pi} = \dfrac{4(-1)^{n+1}}{n^2}$

$b_n = \dfrac{1}{\pi}\displaystyle\int_{-\pi}^{\pi} (\pi^2 - x^2)\sin nx\, dx = \dfrac{1}{\pi}\left[\pi^2\int_{-\pi}^{\pi} \sin nx\, dx - \int_{\pi}^{\pi} x^2 \sin nx\right]$

$\qquad = \dfrac{1}{\pi}\left[-\frac{\pi^2\cos nx}{n} + x^2\frac{\cos nx}{n}\Big|_{\pi}^{\pi} - \frac{2}{n}\int_{\pi}^{\pi} x\cos nx\, dx\right]$

$\qquad = \dfrac{1}{\pi}\left[-\frac{\pi^2\cos nx}{n} + x^2\frac{\cos nx}{n} - \frac{2}{n^2}\left(x\sin nx + \frac{\cos nx}{n}\right)\right]_{-\pi}^{\pi} = 0$

$$f(x) = \frac{2\pi^2}{3} + 4\sum_{n=1}^{\infty} \frac{(-1)^{n+1}}{n^2} \cos nx$$

f is continuous on the interval.

21. The function in Problem 5 is discontinuous at $x = \pi$, so the corresponding Fourier series converges to $\pi^2/2$ at $x = \pi$. That is,

$$\frac{\pi^2}{2} = \frac{\pi^2}{6} + \sum_{n=1}^{\infty} \left[\frac{2(-1)^n}{n^2} \cos n\pi + \left(\frac{\pi}{n}(-1)^{n+1} + \frac{2[(-1)^n - 1]}{n^3\pi} \right) \sin n\pi \right]$$

$$= \frac{\pi^2}{6} + \sum_{n=1}^{\infty} \frac{2(-1)^n}{n^2}(-1)^n = \frac{\pi^2}{6} + \sum_{n=1}^{\infty} \frac{2}{n^2} = \frac{\pi^2}{6} + 2\left(1 + \frac{1}{2^2} + \frac{1}{3^2} + \cdots\right)$$

and

$$\frac{\pi^2}{6} = \frac{1}{2}\left(\frac{\pi^2}{2} - \frac{\pi^2}{6}\right) = 1 + \frac{1}{2^2} + \frac{1}{3^2} + \cdots .$$

At $x = 0$ the series converges to 0 and

$$0 = \frac{\pi^2}{6} + \sum_{n=1}^{\infty} \frac{2(-1)^n}{n^2} = \frac{\pi^2}{6} + 2\left(-1 + \frac{1}{2^2} - \frac{1}{3^2} + \frac{1}{4^2} - \cdots\right)$$

so

$$\frac{\pi^2}{12} = 1 - \frac{1}{2^2} + \frac{1}{3^2} - \frac{1}{4^2} + \cdots .$$

24. The function in Problem 9 is continuous at $x = \pi/2$ so

$$1 = f\left(\frac{\pi}{2}\right) = \frac{1}{\pi} + \frac{1}{2} + \sum_{n=2}^{\infty} \frac{1 + (-1)^n}{\pi(1 - n^2)} \cos \frac{n\pi}{2}$$

$$1 = \frac{1}{\pi} + \frac{1}{2} + \frac{2}{3\pi} - \frac{2}{3 \cdot 5\pi} + \frac{2}{5 \cdot 7\pi} - \cdots$$

and

$$\pi = 1 + \frac{\pi}{2} + \frac{2}{3} - \frac{2}{3 \cdot 5} + \frac{2}{5 \cdot 7} - \cdots$$

or

$$\frac{\pi}{4} = \frac{1}{2} + \frac{1}{1 \cdot 3} - \frac{1}{3 \cdot 5} + \frac{1}{5 \cdot 7} - \cdots .$$

12.3   Fourier Cosine and Sine Series

3. Since $f(-x) = (-x)^2 - x = x^2 - x$, $f(x)$ is neither even nor odd.

6. Since $f(-x) = e^{-x} - e^x = -f(x)$, $f(x)$ is an odd function.

9. Since $f(x)$ is not defined for $x < 0$, it is neither even nor odd.

12. Since $f(x)$ is an even function, we expand in a cosine series:

$$a_0 = \int_1^2 1\, dx = 1 \qquad\qquad a_n = \int_1^2 \cos\frac{n\pi}{2}x\, dx = -\frac{2}{n\pi}\sin\frac{n\pi}{2}.$$

Thus

$$f(x) = \frac{1}{2} + \sum_{n=1}^{\infty} \frac{-2}{n\pi}\sin\frac{n\pi}{2}\cos\frac{n\pi}{2}x.$$

15. Since $f(x)$ is an even function, we expand in a cosine series:

$$a_0 = 2\int_0^1 x^2\, dx = \frac{2}{3}$$

$$a_n = 2\int_0^1 x^2\cos n\pi x\, dx = 2\left(\left.\frac{x^2}{n\pi}\sin n\pi x\right|_0^1 - \frac{2}{n\pi}\int_0^1 x\sin n\pi x\, dx\right) = \frac{4}{n^2\pi^2}(-1)^n.$$

Thus

$$f(x) = \frac{1}{3} + \sum_{n=1}^{\infty} \frac{4}{n^2\pi^2}(-1)^n\cos n\pi x.$$

18. Since $f(x)$ is an odd function, we expand in a sine series:

$$b_n = \frac{2}{\pi}\int_0^\pi x^3\sin nx\, dx = \frac{2}{\pi}\left(\left.-\frac{x^3}{n}\cos nx\right|_0^\pi + \frac{3}{n}\int_0^\pi x^2\cos nx\, dx\right)$$

$$= \frac{2\pi^2}{n}(-1)^{n+1} - \frac{12}{n^2\pi}\int_0^\pi x\sin nx\, dx$$

$$= \frac{2\pi^2}{n}(-1)^{n+1} - \frac{12}{n^2\pi}\left(\left.-\frac{x}{n}\cos nx\right|_0^\pi + \frac{1}{n}\int_0^\pi\cos nx\, dx\right) = \frac{2\pi^2}{n}(-1)^{n+1} + \frac{12}{n^3}(-1)^n.$$

Thus

$$f(x) = \sum_{n=1}^{\infty}\left(\frac{2\pi^2}{n}(-1)^{n+1} + \frac{12}{n^3}(-1)^n\right)\sin nx.$$

21. Since $f(x)$ is an even function, we expand in a cosine series:

$$a_0 = \int_0^1 x\, dx + \int_1^2 1\, dx = \frac{3}{2}$$

$$a_n = \int_0^1 x\cos\frac{n\pi}{2}x\, dx + \int_1^2 \cos\frac{n\pi}{2}x\, dx = \frac{4}{n^2\pi^2}\left(\cos\frac{n\pi}{2} - 1\right).$$

Thus

$$f(x) = \frac{3}{4} + \sum_{n=1}^{\infty} \frac{4}{n^2\pi^2}\left(\cos\frac{n\pi}{2} - 1\right)\cos\frac{n\pi}{2}x.$$

24. Since $f(x)$ is an even function, we expand in a cosine series. [See the solution of Problem 10 in Exercises 12.2 for the computation of the integrals.]

$$a_0 = \frac{2}{\pi/2} \int_0^{\pi/2} \cos x \, dx = \frac{4}{\pi} \qquad a_n = \frac{2}{\pi/2} \int_0^{\pi/2} \cos x \cos \frac{n\pi}{\pi/2} x \, dx = \frac{4(-1)^{n+1}}{\pi(4n^2 - 1)}$$

Thus

$$f(x) = \frac{2}{\pi} + \sum_{n=1}^{\infty} \frac{4(-1)^{n+1}}{\pi(4n^2 - 1)} \cos 2nx.$$

27. $a_0 = \dfrac{4}{\pi} \displaystyle\int_0^{\pi/2} \cos x \, dx = \dfrac{4}{\pi}$

$$a_n = \frac{4}{\pi} \int_0^{\pi/2} \cos x \cos 2nx \, dx = \frac{2}{\pi} \int_0^{\pi/2} [\cos(2n+1)x + \cos(2n-1)x] \, dx = \frac{4(-1)^n}{\pi(1 - 4n^2)}$$

$$b_n = \frac{4}{\pi} \int_0^{\pi/2} \cos x \sin 2nx \, dx = \frac{2}{\pi} \int_0^{\pi/2} [\sin(2n+1)x + \sin(2n-1)x] \, dx = \frac{8n}{\pi(4n^2 - 1)}$$

$$f(x) = \frac{2}{\pi} + \sum_{n=1}^{\infty} \frac{4(-1)^n}{\pi(1 - 4n^2)} \cos 2nx$$

$$f(x) = \sum_{n=1}^{\infty} \frac{8n}{\pi(4n^2 - 1)} \sin 2nx$$

30. $a_0 = \dfrac{1}{\pi} \displaystyle\int_{\pi}^{2\pi} (x - \pi) \, dx = \dfrac{\pi}{2}$

$$a_n = \frac{1}{\pi} \int_{\pi}^{2\pi} (x - \pi) \cos \frac{n}{2} x \, dx = \frac{4}{n^2\pi} \left((-1)^n - \cos \frac{n\pi}{2} \right)$$

$$b_n = \frac{1}{\pi} \int_{\pi}^{2\pi} (x - \pi) \sin \frac{n}{2} x \, dx = \frac{2}{n}(-1)^{n+1} - \frac{4}{n^2\pi} \sin \frac{n\pi}{2}$$

$$f(x) = \frac{\pi}{4} + \sum_{n=1}^{\infty} \frac{4}{n^2\pi} \left((-1)^n - \cos \frac{n\pi}{2} \right) \cos \frac{n}{2} x$$

$$f(x) = \sum_{n=1}^{\infty} \left(\frac{2}{n}(-1)^{n+1} - \frac{4}{n^2\pi} \sin \frac{n\pi}{2} \right) \sin \frac{n}{2} x$$

33. $a_0 = 2 \displaystyle\int_0^1 (x^2 + x) \, dx = \dfrac{5}{3}$

$$a_n = 2 \int_0^1 (x^2 + x) \cos n\pi x \, dx = \frac{2(x^2 + x)}{n\pi} \sin n\pi x \Big|_0^1 - \frac{2}{n\pi} \int_0^1 (2x + 1) \sin n\pi x \, dx$$

$$= \frac{2}{n^2\pi^2} [3(-1)^n - 1]$$

$$b_n = 2 \int_0^1 (x^2 + x) \sin n\pi x \, dx = -\frac{2(x^2 + x)}{n\pi} \cos n\pi x \Big|_0^1 + \frac{2}{n\pi} \int_0^1 (2x + 1) \cos n\pi x \, dx$$

$$= \frac{4}{n\pi}(-1)^{n+1} + \frac{4}{n^3\pi^3}[(-1)^n - 1]$$

$$f(x) = \frac{5}{6} + \sum_{n=1}^{\infty} \frac{2}{n^2\pi^2}[3(-1)^n - 1] \cos n\pi x$$

$$f(x) = \sum_{n=1}^{\infty} \left(\frac{4}{n\pi}(-1)^{n+1} + \frac{4}{n^3\pi^3}[(-1)^n - 1] \right) \sin n\pi x$$

36. $a_0 = \dfrac{2}{\pi} \displaystyle\int_0^\pi x \, dx = \pi$ $a_n = \dfrac{2}{\pi} \displaystyle\int_0^\pi x \cos 2nx \, dx = 0$

$$b_n = \frac{2}{\pi} \int_0^\pi x \sin 2nx \, dx = -\frac{1}{n}$$

$$f(x) = \frac{\pi}{2} - \sum_{n=1}^{\infty} \frac{1}{n} \sin 2nx$$

39. The periodic extensions for the cosine, sine, and Fourier series are shown below:

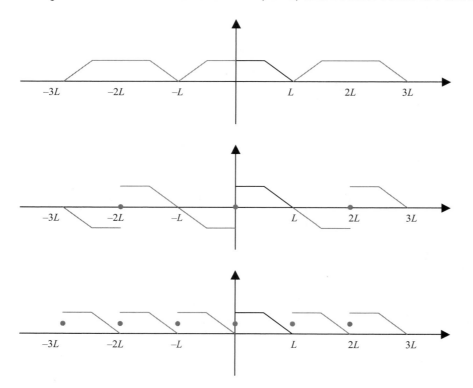

42. The periodic extensions for the cosine, sine, and Fourier series are shown below:

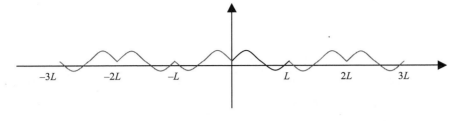

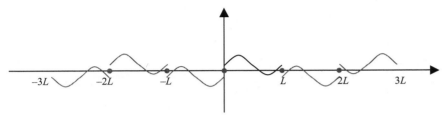

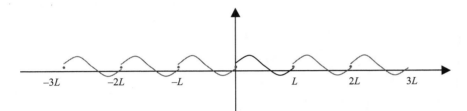

45. We have

$$a_0 = \frac{2}{\pi} \int_0^\pi (2\pi t - t^2)\, dt = \frac{4}{3}\pi^2 \qquad a_n = \frac{2}{\pi} \int_0^\pi (2\pi t - t^2) \cos nt\, dt = -\frac{4}{n^2}$$

so that

$$f(t) = \frac{2\pi^2}{3} - \sum_{n=1}^\infty \frac{4}{n^2} \cos nt.$$

Substituting the assumption

$$x_p(t) = \frac{A_0}{2} + \sum_{n=1}^\infty A_n \cos nt$$

into the differential equation then gives

$$\frac{1}{4}\, x_p'' + 12 x_p = 6 A_0 + \sum_{n=1}^\infty A_n \left(-\frac{1}{4} n^2 + 12 \right) \cos nt = \frac{2\pi^2}{3} - \sum_{n=1}^\infty \frac{4}{n^2} \cos nt$$

and $A_0 = \pi^2/9$, $A_n = 16/n^2(n^2 - 48)$. Thus

$$x_p(t) = \frac{\pi^2}{18} + 16 \sum_{n=1}^\infty \frac{1}{n^2(n^2 - 48)} \cos nt.$$

48. (a) The general solution is $x(t) = c_1 \cos 4\sqrt{3}\, t + c_2 \sin 4\sqrt{3} t + x_p(t)$, where

$$x_p(t) = \frac{\pi^2}{18} + 16 \sum_{n=1}^\infty \frac{1}{n^2(n^2 - 48)} \cos nt.$$

The initial condition $x(0) = 0$ implies $c_1 + x_p(0) = 1$ or

$$c_1 = 1 - x_p(0) = 1 - \frac{\pi^2}{18} - 16 \sum_{n=1}^{\infty} \frac{1}{n^2(n^2 - 48)}.$$

Now $x'(t) = -4\sqrt{3}c_1 \sin 4\sqrt{3}\,t + 4\sqrt{3}c_2 \cos 4\sqrt{3}\,t + x_p'(t)$, so $x'(0) = 0$ implies $4\sqrt{3}c_2 + x_p'(0) = 0$. Since $x_p'(0) = 0$, we have $c_2 = 0$ and

$$x(t) = \left(1 - \frac{\pi^2}{18} - 16 \sum_{n=1}^{\infty} \frac{1}{n^2(n^2 - 48)}\right) \cos 4\sqrt{3}\,t + \frac{\pi^2}{18} + 16 \sum_{n=1}^{\infty} \frac{1}{n^2(n^2 - 48)} \cos nt$$

$$= \frac{\pi^2}{18} + \left(1 - \frac{\pi^2}{18}\right) \cos 4\sqrt{3}\,t + 16 \sum_{n=1}^{\infty} \frac{1}{n^2(n^2 - 48)} \left[\cos nt - \cos 4\sqrt{3}\,t\right].$$

(b) The graph is plotted using five nonzero terms in the series expansion of $x(t)$.

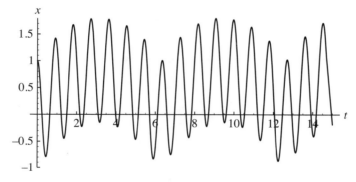

12.4 | Complex Fourier Series

In this section we make use of the following identities due to Euler's formula:
$$e^{in\pi} = e^{-in\pi} = (-1)^n, \qquad e^{-2in\pi} = 1, \qquad e^{-in\pi/2} = (-i)^n.$$

3. Identifying $p = 1/2$ we have

$$c_n = \int_{-1/2}^{1/2} f(x) e^{-2in\pi x}\, dx = \int_{0}^{1/4} e^{-2in\pi x}\, dx = -\frac{1}{2in\pi} e^{-2in\pi x} \Big|_{0}^{1/4}$$

$$= -\frac{1}{2in\pi} \left[e^{-in\pi/2} - 1\right] = -\frac{1}{2in\pi}[(-i)^n - 1] = \frac{i}{2n\pi}[(-i)^n - 1]$$

and

$$c_0 = \int_{0}^{1/4} dx = \frac{1}{4}.$$

Thus

$$f(x) = \frac{1}{4} + \frac{i}{2\pi} \sum_{\substack{n=-\infty \\ n \neq 0}}^{\infty} \frac{(-i)^n - 1}{n} e^{2in\pi x}.$$

6. Identifying $p = 1$ we have

$$c_n = \frac{1}{2} \int_{-1}^{1} f(x)e^{-in\pi x}\, dx = \frac{1}{2}\left[\int_{-1}^{0} e^x e^{-in\pi x}\, dx + \int_{0}^{1} e^{-x}e^{-in\pi x}\, dx\right]$$

$$= \frac{1}{2}\left[-\frac{1}{1-in\pi}e^{(1-in\pi)x}\Big|_{-1}^{0} - \frac{1}{1+in\pi}e^{-(1+in\pi)x}\Big|_{0}^{1}\right]$$

$$= \frac{e-(-1)^n}{e(1-in\pi)} + \frac{1-e^{-1}(-1)^n}{1+in\pi} = \frac{2[e-(-1)^n]}{e(1+n^2\pi^2)}.$$

Thus

$$f(x) = \sum_{n=-\infty}^{\infty} \frac{2[e-(-1)^n]}{e(1+n^2\pi^2)}e^{in\pi x}.$$

9. Identifying $2p = \pi$ or $p = \pi/2$, and using $\sin x = (e^{ix} - e^{-ix})/2i$, we have

$$c_n = \frac{1}{\pi}\int_{0}^{\pi} f(x)e^{-2inx/\pi}\, dx = \frac{1}{\pi}\int_{0}^{\pi}(\sin x)e^{-2inx/\pi}\, dx$$

$$= \frac{1}{\pi}\int_{0}^{\pi}\frac{1}{2i}(e^{ix}-e^{-ix})e^{-2inx/\pi}\, dx$$

$$= \frac{1}{2\pi i}\int_{0}^{\pi}\left(e^{(1-2n/\pi)ix} - e^{-(1+2n/\pi)ix}\right)dx$$

$$= \frac{1}{2\pi i}\left[\frac{1}{i(1-2n/\pi)}e^{(1-2n/\pi)ix}\right.$$

$$\left. + \frac{1}{i(1+2n/\pi)}e^{-(1+2n/\pi)ix}\right]_{0}^{\pi}$$

$$= \frac{\pi(1+e^{-2in})}{\pi^2-4n^2}.$$

The fundamental period is $T = \pi$, so $\omega = 2\pi/\pi = 2$ and the values of $n\omega$ are
$0, \pm 2, \pm 4, \pm 6, \ldots$. Values of $|c_n|$ for $n = 0, \pm 1, \pm 2, \pm 3, \pm 4$, and ± 5 are shown in the table.
The bottom graph is a portion of the frequency spectrum.

n	−5	−4	−3	−2	−1	0	1	2	3	4	5
c_n	0.0198	0.0759	0.2380	0.4265	0.5784	0.6366	0.5784	0.4285	0.2380	0.0759	0.0198

12. From Problem 11 and the fact that f is odd, $c_n + c_{-n} = a_n = 0$, so $c_{-n} = -c_n$. Then
$b_n = i(c_n - c_{-n}) = 2ic_n$. From Problem 1, $b_n = 2i[1-(-1)^n]/n\pi i = 2[1-(-1)^n]/n\pi$, and
the Fourier sine series of f is

$$f(x) = \sum_{i=1}^{\infty} \frac{2[1-(-1)^n]}{n\pi}\sin\frac{n\pi x}{2}.$$

12.5   Sturm–Liouville Problem

3. For $\lambda = 0$ the solution of $y'' = 0$ is $y = c_1 x + c_2$. The condition $y'(0) = 0$ implies $c_1 = 0$, so $\lambda = 0$ is an eigenvalue with corresponding eigenfunction 1.

For $\lambda = -\alpha^2 < 0$ we have $y = c_1 \cosh \alpha x + c_2 \sinh \alpha x$ and $y' = c_1 \alpha \sinh \alpha x + c_2 \alpha \cosh \alpha x$. The condition $y'(0) = 0$ implies $c_2 = 0$ and so $y = c_1 \cosh \alpha x$. Now the condition $y'(L) = 0$ implies $c_1 = 0$. Thus $y = 0$ and there are no negative eigenvalues.

For $\lambda = \alpha^2 > 0$ we have $y = c_1 \cos \alpha x + c_2 \sin \alpha x$ and $y' = -c_1 \alpha \sin \alpha x + c_2 \alpha \cos \alpha x$. The condition $y'(0) = 0$ implies $c_2 = 0$ and so $y = c_1 \cos \alpha x$. Now the condition $y'(L) = 0$ implies $-c_1 \alpha \sin \alpha L = 0$. For $c_1 \neq 0$ this condition will hold when $\alpha L = n\pi$ or $\lambda = \alpha^2 = n^2 \pi^2 / L^2$, where $n = 1, 2, 3, \dots$. These are the positive eigenvalues with corresponding eigenfunctions $\cos(n\pi x / L)$, $n = 1, 2, 3, \dots$.

6. The eigenfunctions are $\sin \alpha_n x$ where $\tan \alpha_n = -\alpha_n$. Thus

$$\| \sin \alpha_n x \|^2 = \int_0^1 \sin^2 \alpha_n x \, dx = \frac{1}{2} \int_0^1 (1 - \cos 2\alpha_n x) \, dx$$

$$= \frac{1}{2} \left(x - \frac{1}{2\alpha_n} \sin 2\alpha_n x \right) \Bigg|_0^1 = \frac{1}{2} \left(1 - \frac{1}{2\alpha_n} \sin 2\alpha_n \right)$$

$$= \frac{1}{2} \left[1 - \frac{1}{2\alpha_n} (2 \sin \alpha_n \cos \alpha_n) \right] = \frac{1}{2} \left[1 - \frac{1}{\alpha_n} \tan \alpha_n \cos \alpha_n \cos \alpha_n \right]$$

$$= \frac{1}{2} \left[1 - \frac{1}{\alpha_n} (-\alpha_n \cos^2 \alpha_n) \right] = \frac{1}{2} \left(1 + \cos^2 \alpha_n \right).$$

9. We divide by lead coefficient of the differential equation to obtain the form $y'' + \left(-1 + \dfrac{1}{x} \right) y' + \dfrac{n}{x} y = 0$. The integrating factor is then

$$e^{\int (-1 + 1/x) \, dx} = e^{-x + \ln x} = e^{-x} e^{\ln x} = x e^{-x}.$$

Thus, the differential equation is

$$x e^{-x} y'' + (1 - x) e^{-x} y' + n e^{-x} y = 0$$

and the self-adjoint form is

$$\frac{d}{dx} \left[x e^{-x} y' \right] + n e^{-x} y = 0.$$

Identifying the weight function $p(x) = e^{-x}$ and noting that since $r(x) = x e^{-x}$, $r(0) = 0$ and $\lim_{x \to \infty} r(x) = 0$, we have the orthogonality relation

$$\int_0^\infty e^{-x} L_m(x) L_n(x) \, dx = 0, \ m \neq n.$$

12. (a) Letting $\lambda = \alpha^2$ the differential equation becomes $x^2 y'' + x y' + (\alpha^2 x^2 - 1) y = 0$. This is the parametric Bessel equation with $\nu = 1$. The general solution is

$$y = c_1 J_1(\alpha x) + c_2 Y_1(\alpha x).$$

Since Y is unbounded at 0 we must have $c_2 = 0$, so that $y = c_1 J_1(\alpha x)$. The condition $J_1(3\alpha) = 0$ defines the eigenvalues $\lambda_n = \alpha_n^2$ for $n = 1, 2, 3, \ldots$. The corresponding eigenfunctions are $J_1(\alpha_n x)$.

(b) Using a CAS or Table 5.3.1 in the text to solve $J_1(3\alpha) = 0$ we find $3\alpha_1 = 3.8317$, $3\alpha_2 = 7.0156$, $3\alpha_3 = 10.1735$, and $3\alpha_4 = 13.3237$. The corresponding eigenvalues are $\lambda_1 = \alpha_1^2 = 1.6313$, $\lambda_2 = \alpha_2^2 = 5.4687$, $\lambda_3 = \alpha_3^2 = 11.4999$, and $\lambda_4 = \alpha_4^2 = 19.7245$.

12.6 Bessel and Legendre Series

3. The boundary condition indicates that we use (15) and (16) in the text. With $b = 2$ we obtain

$$c_i = \frac{2}{4 J_1^2(2\alpha_i)} \int_0^2 x J_0(\alpha_i x)\, dx \qquad \boxed{t = \alpha_i x, \quad dt = \alpha_i\, dx}$$

$$= \frac{1}{2 J_1^2(2\alpha_i)} \cdot \frac{1}{\alpha_i^2} \int_0^{2\alpha_i} t J_0(t)\, dt = \frac{1}{2\alpha_i^2 J_1^2(2\alpha_i)} \int_0^{2\alpha_i} \frac{d}{dt}[t J_1(t)]\, dt \qquad \text{[From (5) in the text]}$$

$$= \frac{1}{2\alpha_i^2 J_1^2(2\alpha_i)} t J_1(t) \Big|_0^{2\alpha_i} = \frac{1}{\alpha_i J_1(2\alpha_i)}.$$

Thus

$$f(x) = \sum_{i=1}^{\infty} \frac{1}{\alpha_i J_1(2\alpha_i)} J_0(\alpha_i x).$$

6. Writing the boundary condition in the form

$$2 J_0(2\alpha) + 2\alpha J_0'(2\alpha) = 0$$

we identify $b = 2$ and $h = 2$. Using (17) and (18) in the text we obtain

$$c_i = \frac{2\alpha_i^2}{(4\alpha_i^2 + 4) J_0^2(2\alpha_i)} \int_0^2 x J_0(\alpha_i x)\, dx \qquad \boxed{t = \alpha_i x, \quad dt = \alpha_i\, dx}$$

$$= \frac{\alpha_i^2}{2(\alpha_i^2 + 1) J_0^2(2\alpha_i)} \cdot \frac{1}{\alpha_i^2} \int_0^{2\alpha_i} t J_0(t)\, dt$$

$$= \frac{1}{2(\alpha_i^2 + 1) J_0^2(2\alpha_i)} \int_0^{2\alpha_i} \frac{d}{dt}[t J_1(t)]\, dt \qquad \text{[From (5) in the text]}$$

$$= \frac{1}{2(\alpha_i^2 + 1) J_0^2(2\alpha_i)} t J_1(t) \Big|_0^{2\alpha_i} = \frac{\alpha_i J_1(2\alpha_i)}{(\alpha_i^2 + 1) J_0^2(2\alpha_i)}.$$

Thus

$$f(x) = \sum_{i=1}^{\infty} \frac{\alpha_i J_1(2\alpha_i)}{(\alpha_i^2 + 1)J_0^2(2\alpha_i)} J_0(\alpha_i x).$$

9. The boundary condition indicates that we use (19) and (20) in the text. With $b = 3$ we obtain

$$c_1 = \frac{2}{9} \int_0^3 xx^2 \, dx = \frac{2}{9} \frac{x^4}{4} \Big|_0^3 = \frac{9}{2},$$

$$c_i = \frac{2}{9J_0^2(3\alpha_i)} \int_0^3 xJ_0(\alpha_i x)x^2 \, dx \qquad \boxed{t = \alpha_i x, \quad dt = \alpha_i \, dx}$$

$$= \frac{2}{9J_0^2(3\alpha_i)} \cdot \frac{1}{\alpha_i^4} \int_0^{3\alpha_i} t^3 J_0(t) \, dt$$

$$= \frac{2}{9\alpha_i^4 J_0^2(3\alpha_i)} \int_0^{3\alpha_i} t^2 \frac{d}{dt}[tJ_1(t)] \, dt \qquad \boxed{\begin{array}{ll} u = t^2 & dv = \frac{d}{dt}[tJ_1(t)] \, dt \\ du = 2t \, dt & v = tJ_1(t) \end{array}}$$

$$= \frac{2}{9\alpha_i^4 J_0^2(3\alpha_i)} \left(t^3 J_1(t) \Big|_0^{3\alpha_i} - 2 \int_0^{3\alpha_i} t^2 J_1(t) \, dt \right).$$

With $n = 0$ in equation (6) in the text we have $J_0'(x) = -J_1(x)$, so the boundary condition $J_0'(3\alpha_i) = 0$ implies $J_1(3\alpha_i) = 0$. Then

$$c_i = \frac{2}{9\alpha_i^4 J_0^2(3\alpha_i)} \left(-2 \int_0^{3\alpha_i} \frac{d}{dt}[t^2 J_2(t)] \, dt \right) = \frac{2}{9\alpha_i^4 J_0^2(3\alpha_i)} \left(-2t^2 J_2(t) \Big|_0^{3\alpha_i} \right)$$

$$= \frac{2}{9\alpha_i^4 J_0^2(3\alpha_i)} \left[-18\alpha_i^2 J_2(3\alpha_i) \right] = \frac{-4J_2(3\alpha_i)}{\alpha_i^2 J_0^2(3\alpha_i)}.$$

Thus

$$f(x) = \frac{9}{2} - 4 \sum_{i=1}^{\infty} \frac{J_2(3\alpha_i)}{\alpha_i^2 J_0^2(3\alpha_i)} J_0(\alpha_i x).$$

15. We compute

$$c_0 = \frac{1}{2} \int_0^1 xP_0(x) \, dx = \frac{1}{2} \int_0^1 x \, dx = \frac{1}{4}$$

$$c_1 = \frac{3}{2} \int_0^1 xP_1(x) \, dx = \frac{3}{2} \int_0^1 x^2 \, dx = \frac{1}{2}$$

$$c_2 = \frac{5}{2} \int_0^1 xP_2(x) \, dx = \frac{5}{2} \int_0^1 \frac{1}{2}(3x^3 - x) \, dx = \frac{5}{16}$$

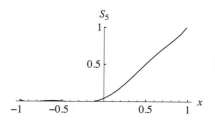

$$c_3 = \frac{7}{2} \int_0^1 x P_3(x)\, dx = \frac{7}{2} \int_0^1 \frac{1}{2}(5x^4 - 3x^2)\, dx = 0$$

$$c_4 = \frac{9}{2} \int_0^1 x P_4(x)\, dx = \frac{9}{2} \int_0^1 \frac{1}{8}(35x^5 - 30x^3 + 3x)\, dx = \frac{3}{32}$$

$$c_5 = \frac{11}{2} \int_0^1 x P_5(x)\, dx = \frac{11}{2} \int_0^1 \frac{1}{8}(63x^6 - 70x^4 + 15x^2)\, dx = 0$$

$$c_6 = \frac{13}{2} \int_0^1 x P_6(x)\, dx = \frac{13}{2} \int_0^1 \frac{1}{16}(231x^7 - 315x^5 + 105x^3 - 5x)\, dx = \frac{13}{256}.$$

Thus

$$f(x) = \frac{1}{4} P_0(x) + \frac{1}{2} P_1(x) + \frac{5}{16} P_2(x) - \frac{3}{32} P_4(x) + \frac{13}{256} P_6(x) + \cdots.$$

The figure above is the graph of $S_5(x) = \frac{1}{4} P_0(x) + \frac{1}{2} P_1(x) + \frac{5}{16} P_2(x) - \frac{3}{32} P_4(x) + \frac{13}{256} P_6(x)$.

18. From Problem 17 we have

$$P_2(\cos\theta) = \frac{1}{4}(3\cos 2\theta + 1) \qquad \text{or} \qquad \cos 2\theta = \frac{4}{3} P_2(\cos\theta) - \frac{1}{3}.$$

Then, using $P_0(\cos\theta) = 1$,

$$F(\theta) = 1 - \cos 2\theta = 1 - \left[\frac{4}{3} P_2(\cos\theta) - \frac{1}{3} \right] = \frac{4}{3} - \frac{4}{3} P_2(\cos\theta) = \frac{4}{3} P_0(\cos\theta) - \frac{4}{3} P_2(\cos\theta).$$

21. From (26) in Problem 19 in the text we find

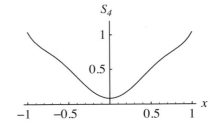

S_4

$$c_0 = \int_0^1 x P_0(x)\, dx = \int_0^1 x\, dx = \frac{1}{2}$$

$$c_2 = 5 \int_0^1 x P_2(x)\, dx = 5 \int_0^1 \frac{1}{2}(3x^3 - x)\, dx = \frac{5}{8}$$

$$c_4 = 9 \int_0^1 x P_4(x)\, dx = 9 \int_0^1 \frac{1}{8}(35x^5 - 30x^3 + 3x)\, dx = -\frac{3}{16}$$

and

$$c_6 = 13 \int_0^1 x P_6(x)\, dx = 13 \int_0^1 \frac{1}{16}(231x^7 - 315x^5 + 105x^3 - 5x)\, dx = \frac{13}{128}.$$

Hence, from (25) in the text,

$$f(x) = \frac{1}{2} P_0(x) + \frac{5}{8} P_2(x) - \frac{3}{16} P_4(x) + \frac{13}{128} P_6 + \cdots.$$

On the interval $-1 < x < 1$ this series represents the function $f(x) = |x|$.

Chapter 12 in Review

3. cosine, since f is even

6. Periodically extending the function we see that at $x = -1$ the function converges to $\frac{1}{2}(-1 + 0) = -\frac{1}{2}$; at $x = 0$ it converges to $\frac{1}{2}(0+1) = \frac{1}{2}$, and at $x = 1$ it converges to $\frac{1}{2}(-1+0) = -\frac{1}{2}$.

9. True, since $\int_0^1 P_{2m}(x)P_{2n}(x)\,dx = \frac{1}{2}\int_{-1}^1 P_{2m}(x)P_{2n}(x)\,dx = 0$ when $m \neq n$.

12. (a) For $m \neq n$

$$\int_0^L \sin\frac{(2n+1)\pi}{2L}x \, \sin\frac{(2m+1)\pi}{2L}x\,dx = \frac{1}{2}\int_0^L \left(\cos\frac{n-m}{L}\pi x - \cos\frac{n+m+1}{L}\pi x\right)dx$$

$$= 0.$$

(b) From

$$\int_0^L \sin^2\frac{(2n+1)\pi}{2L}x\,dx = \int_0^L \left(\frac{1}{2} - \frac{1}{2}\cos\frac{(2n+1)\pi}{L}x\right)dx = \frac{L}{2}$$

we see that

$$\left\|\sin\frac{(2n+1)\pi}{2L}x\right\| = \sqrt{\frac{L}{2}}.$$

15. Since

$$a_0 = \frac{2}{1}\int_0^1 e^{-x}\,dx = 2(1 - e^{-1})$$

and

$$a_n = \frac{2}{1}\int_0^1 e^{-x}\cos n\pi x\,dx = \frac{2}{1+n^2\pi^2}[1 - (-1)^n e^{-1}],$$

for $n = 1, 2, 3, \ldots$, we have the cosine series

$$f(x) = 1 - e^{-1} + 2\sum_{n=1}^{\infty}\frac{1 - (-1)^n e^{-1}}{1+n^2\pi^2}\cos n\pi x.$$

Since

$$b_n = \frac{2}{1}\int_0^1 e^{-x}\sin n\pi x\,dx = \frac{2n\pi}{1+n^2\pi^2}[1 - (-1)^n e^{-1}],$$

for $n = 1, 2, 3, \ldots$, we have the sine series

$$f(x) = 2\pi\sum_{n=1}^{\infty}\frac{n[1 - (-1)^n e^{-1}]}{1+n^2\pi^2}\sin n\pi x.$$

18. To obtain the self-adjoint form of the differential equation in Problem 17 we note that an integrating factor is $(1/x^2)e^{\int dx/x} = 1/x$. Thus the weight function is $1/x$ and an orthogonality relation is

$$\int_1^e \frac{1}{x}\cos\left(\frac{2n-1}{2}\pi \ln x\right)\cos\left(\frac{2m-1}{2}\pi \ln x\right)dx = 0, \ m \neq n.$$

21. The boundary condition indicates that we use (15) and (16) of Section 12.6 in the text. With $b = 4$ we obtain

$$c_i = \frac{2}{16J_1^2(4\alpha_i)} \int_0^4 x J_0(\alpha_i x) f(x)\, dx$$

$$= \frac{1}{8J_1^2(4\alpha_i)} \int_0^2 x J_0(\alpha_i x)\, dx \qquad \boxed{t = \alpha_i x, \quad dt = \alpha_i\, dx}$$

$$= \frac{1}{8J_1^2(4\alpha_i)} \cdot \frac{1}{\alpha_i^2} \int_0^{2\alpha_i} t J_0(t)\, dt$$

$$= \frac{1}{8J_1^2(4\alpha_i)} \int_0^{2\alpha_i} \frac{d}{dt}[t J_1(t)]\, dt \qquad \text{[From (5) in 12.6 in the text]}$$

$$= \frac{1}{8J_1^2(4\alpha_i)} t J_1(t)\Big|_0^{2\alpha_i} = \frac{J_1(2\alpha_i)}{4\alpha_i J_1^2(4\alpha_i)}\,.$$

Thus

$$f(x) = \frac{1}{4} \sum_{i=1}^{\infty} \frac{J_1(2\alpha_i)}{\alpha_i J_1^2(4\alpha_i)} J_0(\alpha_i x).$$

24. $f_e(x) = \dfrac{f(x) + f(-x)}{2} = \dfrac{e^x + e^{-x}}{2} = \cosh x$

$f_o(x) = \dfrac{f(x) - f(-x)}{2} = \dfrac{e^x - e^{-x}}{2} = \sinh x$

Chapter 13

Boundary-Value Problems in Rectangular Coordinates

13.1 | Separable Partial Differential Equations

3. Substituting $u(x, y) = X(x)Y(y)$ into the partial differential equation yields $X'Y + XY' = XY$. Separating variables and using the separation constant $-\lambda$ we obtain

$$\frac{X'}{X} = \frac{Y - Y'}{Y} = -\lambda.$$

Then

$$X' + \lambda X = 0 \quad \text{and} \quad Y' - (1 + \lambda)Y = 0$$

so that

$$X = c_1 e^{-\lambda x} \quad \text{and} \quad Y = c_2 e^{(1+\lambda)y}.$$

A particular product solution of the partial differential equation is

$$u = XY = c_3 e^{y + \lambda(y - x)}.$$

6. Substituting $u(x, y) = X(x)Y(y)$ into the partial differential equation yields $yX'Y + xXY' = 0$. Separating variables and using the separation constant $-\lambda$ we obtain

$$\frac{X'}{xX} = -\frac{Y'}{yY} = -\lambda.$$

When $\lambda \neq 0$

$$X' + \lambda x X = 0 \quad \text{and} \quad Y' - \lambda y Y = 0$$

so that

$$X = c_1 e^{\lambda x^2/2} \quad \text{and} \quad Y = c_2 e^{-\lambda y^2/2}.$$

A particular product solution of the partial differential equation is

$$u = XY = c_3 e^{\lambda(x^2 - y^2)/2}.$$

When $\lambda = 0$ the differential equations become $X' = 0$ and $Y' = 0$, so in this case $X = c_4$, $Y = c_5$, and $u = XY = c_6$.

9. Substituting $u(x,t) = X(x)T(t)$ into the partial differential equation yields $kX''T - XT = XT'$. Separating variables and using the separation constant $-\lambda$ we obtain

$$\frac{kX'' - X}{X} = \frac{T'}{T} = -\lambda.$$

Then

$$X'' + \frac{\lambda - 1}{k}X = 0 \qquad \text{and} \qquad T' + \lambda T = 0.$$

The second differential equation implies $T(t) = c_1 e^{-\lambda t}$. For the first differential equation we consider three cases:

I. If $(\lambda - 1)/k = 0$ then $\lambda = 1$, $X'' = 0$, and $X(x) = c_2 x + c_3$, so

$$u = XT = e^{-t}(A_1 x + A_2).$$

II. If $(\lambda - 1)/k = -\alpha^2 < 0$, then $\lambda = 1 - k\alpha^2$, $X'' - \alpha^2 X = 0$, and $X(x) = c_4 \cosh \alpha x + c_5 \sinh \alpha x$, so

$$u = XT = (A_3 \cosh \alpha x + A_4 \sinh \alpha x)e^{-(1-k\alpha^2)t}.$$

III. If $(\lambda - 1)/k = \alpha^2 > 0$, then $\lambda = 1 + \lambda\alpha^2$, $X'' + \alpha^2 X = 0$, and $X(x) = c_6 \cos \alpha x + c_7 \sin \alpha x$, so

$$u = XT = (A_5 \cos \alpha x + A_6 \sin \alpha x)e^{-(1+\lambda\alpha^2)t}.$$

12. Substituting $u(x,t) = X(x)T(t)$ into the partial differential equation yields $a^2 X''T = XT'' + 2kXT'$. Separating variables and using the separation constant $-\lambda$ we obtain

$$\frac{X''}{X} = \frac{T'' + 2kT'}{a^2 T} = -\lambda.$$

Then

$$X'' + \lambda X = 0 \qquad \text{and} \qquad T'' + 2kT' + a^2\lambda T = 0.$$

We consider three cases:

I. If $\lambda = 0$ then $X'' = 0$ and $X(x) = c_1 x + c_2$. Also, $T'' + 2kT' = 0$ and $T(t) = c_3 + c_4 e^{-2kt}$, so

$$u = XT = (c_1 x + c_2)(c_3 + c_4 e^{-2kt}).$$

II. If $\lambda = -\alpha^2 < 0$, then $X'' - \alpha^2 X = 0$, and $X(x) = c_5 \cosh \alpha x + c_6 \sinh \alpha x$. The auxiliary equation of $T'' + 2kT' - \alpha^2 a^2 T = 0$ is $m^2 + 2km - \alpha^2 a^2 = 0$. Solving for m we obtain $m = -k \pm \sqrt{k^2 + \alpha^2 a^2}$, so $T(t) = c_7 e^{(-k+\sqrt{k^2+\alpha^2 a^2})t} + c_8 e^{(-k-\sqrt{k^2+\alpha^2 a^2})t}$. Then

$$u = XT = (c_5 \cosh \alpha x + c_6 \sinh \alpha x)\left(c_7 e^{(-k+\sqrt{k^2+\alpha^2 a^2})t} + c_8 e^{(-k-\sqrt{k^2+\alpha^2 a^2})t}\right).$$

III. If $\lambda = \alpha^2 > 0$, then $X'' + \alpha^2 X = 0$, and $X(x) = c_9 \cos \alpha x + c_{10} \sin \alpha x$. The auxiliary equation of $T'' + 2kT' + \alpha^2 a^2 T = 0$ is $m^2 + 2km + \alpha^2 a^2 = 0$. Solving for m we obtain $m = -k \pm \sqrt{k^2 - \alpha^2 a^2}$. We consider three possibilities for the discriminant $k^2 - \alpha^2 a^2$:

(i) If $k^2 - \alpha^2 a^2 = 0$ then $T(t) = c_{11}e^{-kt} + c_{12}te^{-kt}$ and

$$u = XT = (c_9 \cos \alpha x + c_{10} \sin \alpha x)(c_{11}e^{-kt} + c_{12}te^{-kt}).$$

From $k^2 - \alpha^2 a^2 = 0$ we have $\alpha = k/a$ so the solution can be written

$$u = XT = (c_9 \cos kx/a + c_{10} \sin kx/a)(c_{11}e^{-kt} + c_{12}te^{-kt}).$$

(ii) If $k^2 - \alpha^2 a^2 < 0$ then $T(t) = e^{-kt}\left(c_{13} \cos \sqrt{\alpha^2 a^2 - k^2}\, t + c_{14} \sin \sqrt{\alpha^2 a^2 - k^2}\, t\right)$ and

$$u = XT = (c_9 \cos \alpha x + c_{10} \sin \alpha x)e^{-kt}\left(c_{13} \cos \sqrt{\alpha^2 a^2 - k^2}\, t + c_{14} \sin \sqrt{\alpha^2 a^2 - k^2}\, t\right).$$

(iii) If $k^2 - \alpha^2 a^2 > 0$ then $T(t) = c_{15}e^{(-k+\sqrt{k^2 - \alpha^2 a^2})t} + c_{16}e^{(-k-\sqrt{k^2 - \alpha^2 a^2})t}$ and

$$u = XT = (c_9 \cos \alpha x + c_{10} \sin \alpha x)\left(c_{15}e^{(-k+\sqrt{k^2 - \alpha^2 a^2})t} + c_{16}e^{(-k-\sqrt{k^2 - \alpha^2 a^2})t}\right).$$

15. Substituting $u(x, y) = X(x)Y(y)$ into the partial differential equation yields $X''Y + XY'' = XY$. Separating variables and using the separation constant $-\lambda$ we obtain

$$\frac{X''}{X} = \frac{Y - Y''}{Y} = -\lambda.$$

Then

$$X'' + \lambda X = 0 \qquad \text{and} \qquad Y'' - (1 + \lambda)Y = 0.$$

We consider three cases:

I. If $\lambda = 0$ then $X'' = 0$ and $X(x) = c_1 x + c_2$. Also $Y'' - Y = 0$ and $Y(y) = c_3 \cosh y + c_4 \sinh y$ so

$$u = XY = (c_1 x + c_2)(c_3 \cosh y + c_4 \sinh y).$$

II. If $\lambda = -\alpha^2 < 0$ then $X'' - \alpha^2 X = 0$ and $Y'' + (\alpha^2 - 1)Y = 0$. The solution of the first differential equation is $X(x) = c_5 \cosh \alpha x + c_6 \sinh \alpha x$. The solution of the second differential equation depends on the nature of $\alpha^2 - 1$. We consider three cases:

(i) If $\alpha^2 - 1 = 0$, or $\alpha^2 = 1$, then $Y(y) = c_7 y + c_8$ and

$$u = XY = (c_5 \cosh \alpha x + c_6 \sinh \alpha x)(c_7 y + c_8).$$

(ii) If $\alpha^2 - 1 < 0$, or $0 < \alpha^2 < 1$, then $Y(y) = c_9 \cosh \sqrt{1 - \alpha^2}\, y + c_{10} \sinh \sqrt{1 - \alpha^2}\, y$ and

$$u = XY = (c_5 \cosh \alpha x + c_6 \sinh \alpha x)\left(c_9 \cosh \sqrt{1 - \alpha^2}\, y + c_{10} \sinh \sqrt{1 - \alpha^2}\, y\right).$$

(iii) If $\alpha^2 - 1 > 0$, or $\alpha^2 > 1$, then $Y(y) = c_{11} \cos \sqrt{\alpha^2 - 1}\, y + c_{12} \sin \sqrt{\alpha^2 - 1}\, y$ and

$$u = XY = (c_5 \cosh \alpha x + c_6 \sinh \alpha x)\left(c_{11} \cos \sqrt{\alpha^2 - 1}\, y + c_{12} \sin \sqrt{\alpha^2 - 1}\, y\right).$$

III. If $\lambda = \alpha^2 > 0$, then $X'' + \alpha^2 X = 0$ and $X(x) = c_{13} \cos \alpha x + c_{14} \sin \alpha x$. Also,

$$Y'' - (1 + \alpha^2)Y = 0 \text{ and } Y(y) = c_{15} \cosh \sqrt{1 + \alpha^2}\, y + c_{16} \sinh \sqrt{1 + \alpha^2}\, y \text{ so}$$

$$u = XY = (c_{13} \cos \alpha x + c_{14} \sin \alpha x)\left(c_{15} \cosh \sqrt{1 + \alpha^2}\, y + c_{16} \sinh \sqrt{1 + \alpha^2}\, y\right).$$

18. Identifying $A = 3$, $B = 5$, and $C = 1$, we compute $B^2 - 4AC = 13 > 0$. The equation is hyperbolic.

21. Identifying $A = 1$, $B = -9$, and $C = 0$, we compute $B^2 - 4AC = 81 > 0$. The equation is hyperbolic.

24. Identifying $A = 1$, $B = 0$, and $C = 1$, we compute $B^2 - 4AC = -4 < 0$. The equation is elliptic.

27. Substituting $u(r, t) = R(r)T(t)$ into the partial differential equation yields

$$k\left(R''T + \frac{1}{r}R'T\right) = RT'.$$

Separating variables and using the separation constant $-\lambda$ we obtain

$$\frac{rR'' + R'}{rR} = \frac{T'}{kT} = -\lambda.$$

Then

$$rR'' + R' + \lambda rR = 0 \qquad \text{and} \qquad T' + \lambda kT = 0.$$

Letting $\lambda = \alpha^2$ and writing the first equation as $r^2 R'' + rR' = \alpha^2 r^2 R = 0$ we see that it is a parametric Bessel equation of order 0. As discussed in Chapter 5 of the text, it has solution $R(r) = c_1 J_0(\alpha r) + c_2 Y_0(\alpha r)$. Since a solution of $T' + \alpha^2 kT$ is $T(t) = e^{-k\alpha^2 t}$, we see that a solution of the partial differential equation is

$$u = RT = e^{-k\alpha^2 t}[c_1 J_0(\alpha r) + c_2 Y_0(\alpha r)].$$

30. We identify $A = xy + 1$, $B = x + 2y$, and $C = 1$. Then $B^2 - 4AC = x^2 + 4y^2 - 4$. The equation $x^2 + 4y^2 = 4$ defines an ellipse. The partial differential equation is hyperbolic outside the ellipse, parabolic on the ellipse, and elliptic inside the ellipse.

13.2   Classical PDEs and Boundary-Value Problems

3. $k\dfrac{\partial^2 u}{\partial x^2} = \dfrac{\partial u}{\partial t}$, $0 < x < L,\ t > 0$

$u(0,t) = 100$, $\left.\dfrac{\partial u}{\partial x}\right|_{x=L} = -hu(L,t)$, $t > 0$

$u(x,0) = f(x)$, $0 < x < L$

6. $k\dfrac{\partial^2 u}{\partial x^2} + h(u - 50) = \dfrac{\partial u}{\partial t}$, $0 < x < L,\ t > 0$

$\left.\dfrac{\partial u}{\partial x}\right|_{x=0} = 0$, $\left.\dfrac{\partial u}{\partial x}\right|_{x=L} = 0$, $t > 0$

$u(x,0) = 100$, $0 < x < L$

9. $a^2\dfrac{\partial^2 u}{\partial x^2} - 2\beta\dfrac{\partial u}{\partial t} = \dfrac{\partial^2 u}{\partial t^2}$, $0 < x < L,\ t > 0$

$u(0,t) = 0$, $u(L,t) = \sin \pi t$, $t > 0$

$u(x,0) = f(x)$, $\left.\dfrac{\partial u}{\partial t}\right|_{t=0} = 0$, $0 < x < L$

12. $\dfrac{\partial^2 u}{\partial x^2} + \dfrac{\partial^2 u}{\partial y^2} = 0$, $0 < x < \pi,\ y > 0$

$u(0,y) = e^{-y}$, $u(\pi,y) = \begin{cases} 100, & 0 < y \le 1 \\ 0, & y > 1 \end{cases}$

$u(x,0) = f(x)$, $0 < x < \pi$

13.3   Heat Equation

3. Using $u = XT$ and $-\lambda$ as a separation constant we obtain

$$X'' + \lambda X = 0,$$
$$X(0) = 0,$$
$$X(L) = 0,$$

and

$$T' + k\lambda T = 0.$$

This leads to

$$X = c_1 \sin \frac{n\pi}{L} x \quad \text{and} \quad T = c_2 e^{-kn^2\pi^2 t/L^2}$$

for $n = 1, 2, 3, \ldots$ so that

$$u = \sum_{n=1}^{\infty} A_n e^{-kn^2\pi^2 t/L^2} \sin \frac{n\pi}{L} x.$$

Imposing

$$u(x, 0) = \sum_{n=1}^{\infty} A_n \sin \frac{n\pi}{L} x$$

gives

$$A_n = \frac{2}{L} \int_0^{L/2} \sin \frac{n\pi}{L} x \, dx = \frac{2}{n\pi} \left(1 - \cos \frac{n\pi}{2}\right)$$

for $n = 1, 2, 3, \ldots$ so that

$$u(x, t) = \frac{2}{\pi} \sum_{n=1}^{\infty} \frac{1 - \cos \frac{n\pi}{2} n}{n} e^{-kn^2\pi^2 t/L^2} \sin \frac{n\pi}{L} x.$$

6. If $L = 2$ and $f(x)$ is x for $0 < x < 1$ and $f(x)$ is 0 for $1 < x < 2$ then

$$u(x, t) = \frac{1}{4} + 4 \sum_{n=1}^{\infty} \left[\frac{1}{2n\pi} \sin \frac{n\pi}{2} + \frac{1}{n^2\pi^2} \left(\cos \frac{n\pi}{2} - 1\right) \right] e^{-kn^2\pi^2 t/4} \cos \frac{n\pi}{2} x.$$

9. Using $u = XT$ and $-\lambda$ as a separation constant leads to

$$X'' + \lambda X = 0,$$
$$X'(0) = 0,$$
$$X'(L) = 0,$$

and

$$T' + (h + k\lambda)T = 0.$$

Then

$$X = c_1 \cos \frac{n\pi}{L} x \quad \text{and} \quad T = c_2 e^{-ht - kn^2\pi^2 t/L^2}$$

for $n = 0, 1, 2, \ldots$ ($\lambda = 0$ is an eigenvalue in this case) so that

$$u = A_0 e^{-ht} + e^{-ht} \sum_{n=1}^{\infty} A_n e^{-kn^2\pi^2 t/L^2} \cos \frac{n\pi}{L} x.$$

Imposing

$$u(x, 0) = f(x) = \sum_{n=0}^{\infty} A_n \cos \frac{n\pi}{L} x$$

gives

$$u(x, t) = \frac{e^{-ht}}{L} \int_0^L f(x) \, dx + \frac{2e^{-ht}}{L} \sum_{n=1}^{\infty} \left(\int_0^L f(x) \cos \frac{n\pi}{L} x \, dx \right) e^{-kn^2\pi^2 t/L^2} \cos \frac{n\pi}{L} x.$$

13.4 | Wave Equation

For Problems 1–10, recall that the solution to the wave equation is given by

$$u(x,t) = \sum_{n=1}^{\infty} \left(A_n \cos \frac{n\pi a}{L} t + B_n \sin \frac{n\pi a}{L} t \right) \sin \frac{n\pi}{L} x$$

where $A_n = \dfrac{2}{L} \displaystyle\int_0^L f(x) \sin \dfrac{n\pi}{L} x \, dx$ *and* $B_n = \dfrac{2}{n\pi a} \displaystyle\int_0^L g(x) \sin \dfrac{n\pi}{L} x \, dx.$

Here, $f(x) = u(x,0)$ *and* $g(x) = \dfrac{\partial u}{\partial t}\Big|_{t=0}.$

3. By the discussion in Section 13.4, pages 728–730,

$$u(x,t) = \sum_{n=1}^{\infty} \left(A_n \cos nat + B_n \sin nat \right) \sin nx.$$

The coefficients are given by

$$A_n = \frac{2}{\pi} \int_0^{\pi} f(x) \sin nx \, dx = 0$$

$$B_n = \frac{2}{n\pi a} \int_0^{\pi} g(x) \sin nx \, dx = \frac{1}{n\pi a} \int_0^{\pi} \sin x \sin nx \, dx = 0 \quad \text{for } n = 2,3,4,\dots$$

$$B_1 = \frac{2}{\pi a} \int_0^{\pi} \sin x \sin x \, dx = \frac{1}{a}.$$

Therefore the solution to the problem is

$$u(x,t) = B_1 \sin at \sin x = \frac{1}{a} \sin at \sin x.$$

6. By the discussion in Section 13.4, pages 728–730,

$$u(x,t) = \sum_{n=1}^{\infty} \left(A_n \cos n\pi at + B_n \sin n\pi at \right) \sin n\pi x.$$

The coefficients are

$$A_n = \frac{2}{1} \int_0^1 f(x) \sin n\pi x \, dx = \frac{2}{1} \int_0^1 0.01 \sin 3\pi x \sin n\pi x \, dx = 0 \quad \text{for } n = 1,2,4,5,6,\dots$$

$$A_3 = \frac{2}{1} \int_0^1 0.01 \sin 3\pi x \sin n\pi x \, dx = 0.01$$

$$B_n = \frac{2}{n\pi a} \int_0^1 g(x) \sin n\pi x \, dx = 0.$$

Therefore the solution to the problem is

$$u(x,t) = (A_3 \cos 3\pi at) \sin 3\pi x = 0.01 \cos 3\pi at \sin 3\pi x.$$

For Problems 7–10, we have $g(x) = 0$ because the string is released from the rest, so $B_n = 0$. So, our general solution is of the form

$$u(x,t) = \sum_{n=1}^{\infty} A_n \cos \frac{n\pi at}{L} \sin \frac{n\pi x}{L}$$

where $A_n = \dfrac{2}{L} \displaystyle\int_0^L f(x) \sin \dfrac{n\pi}{L} x\, dx$ as before.

9. By the discussion in Section 13.4, pages 728–730,

$$u(x,t) = \sum_{n=1}^{\infty} \left(A_n \cos \frac{n\pi a}{L} t + B_n \sin \frac{n\pi a}{L} t \right) \sin \frac{n\pi}{L} x.$$

The coefficients are

$$A_n = \frac{2}{L} \int_0^L f(x) \sin \frac{n\pi}{L} x\, dx$$

$$= \frac{2}{L} \left[\int_0^{L/3} \frac{3hx}{L} \sin \frac{n\pi}{L} x\, dx + \int_{L/3}^{2L/3} h \sin \frac{n\pi}{L} x\, dx + \int_{2L/3}^{L} 3h \left(1 - \frac{x}{L} \right) \sin \frac{n\pi}{L} x\, dx \right]$$

$$= \frac{6h \left[\sin \left(\frac{2n\pi}{3} \right) + \sin \left(\frac{n\pi}{3} \right) \right]}{n^2 \pi^2}$$

$$B_n = \frac{2}{n\pi a} \int_0^L g(x) \sin \frac{n\pi}{L} x\, dx = 0.$$

Therefore the solution to the problem is

$$u(x,t) = \sum_{n=1}^{\infty} \left(\frac{6h \left[\sin \left(\frac{2n\pi}{3} \right) + \sin \left(\frac{n\pi}{3} \right) \right]}{n^2 \pi^2} \cos \frac{n\pi a}{L} t \right) \sin \frac{n\pi}{L} x.$$

Using the trigonometric identity

$$\sin \frac{2n\pi}{3} = \sin \left(n\pi - \frac{n\pi}{3} \right) = \sin n\pi \cos n\pi 3 - \cos n\pi \sin \frac{n\pi}{3} = -(-1)^n \sin \frac{n\pi}{3}$$

we have

$$u(x,t) = \frac{6h}{\pi^2} \sum_{n=1}^{\infty} \frac{1 - (-1)^n}{n^2} \sin \frac{n\pi}{3} \cos \frac{n\pi a}{L} t \sin \frac{n\pi}{L} x.$$

12. By the discussion in Section 13.4, pages 728–730, we get

$$\begin{cases} X'' + \lambda X = 0, \\ X'(0) = 0, \\ X'(L) = 0. \end{cases}$$

For $\lambda = 0$,

$$\{ X'' = 0, X'(0) = 0, X'(L) = 0,$$

Thus $X = c_1 + c_2 x$ and therefore $X = c_1$. Morevoer $\{ T'' = 0$ implies $T = c_3 + c_4 t$.

Therefore we have $u_0(x, t) = c_1(c_3 + c_4 t) = A_0 + B_0 t$. Using the given conditions and the results from Problem 3 in Section 12.5, the rest of the eigenvalues are $\lambda_n = n^2 \pi^2 / L^2$ with corresponding eigenfunctions $X_n = \cos \frac{n\pi}{L} x$, $n = 1, 2, 3, \ldots$ therefore $T = c_3 \cos \frac{n\pi a}{L} t + c_4 \sin \frac{n\pi a}{L} t$ and we now have

$$u(x, t) = (A_0 + B_0 t) + \sum_{n=1}^{\infty} \left(A_n \cos \frac{n\pi a}{L} t + B_n \sin \frac{n\pi a}{L} t \right) \cos \frac{n\pi}{L} x.$$

To complete the problem we need only to find the coefficients. At $t = 0$ we have

$$u(x, 0) = f(x) = A_0 + \sum_{n=1}^{\infty} A_n \cos \frac{n\pi}{L} x \qquad \longleftarrow \text{A Fourier cosine series for } f(x)$$

where

$$A_0 = \frac{1}{L} \int_0^L f(x) \, dx \qquad \text{and} \qquad A_n = \frac{2}{L} \int_0^L f(x) \cos \frac{n\pi}{L} x \, dx.$$

Similarly at $t = 0$,

$$u_t(x, 0) = g(x) = B_0 + \sum_{n=1}^{\infty} \left(B_n \cdot \frac{n\pi a}{L} \right) \cos \frac{n\pi}{L} x$$

and so we get

$$B_0 = \frac{1}{L} \int_0^L g(x) \, dx \qquad \text{and} \qquad B_n = \frac{2}{n\pi a} \int_0^L g(x) \cos \frac{n\pi}{L} x \, dx.$$

15. Using $u = XT$ and $-\lambda$ as a separation constant we obtain

$$X'' + \lambda X = 0,$$
$$X(0) = 0,$$
$$X(\pi) = 0,$$

and

$$T'' + 2\beta T' + \lambda T = 0,$$
$$T'(0) = 0.$$

Solving the differential equations we get

$$X = c_1 \sin nx + c_2 \cos nx \qquad \text{and} \qquad T = e^{-\beta t} \left(c_3 \cos \sqrt{n^2 - \beta^2} \, t + c_4 \sin \sqrt{n^2 - \beta^2} \, t \right).$$

The boundary conditions on X imply $c_2 = 0$ so

$$X = c_1 \sin nx \qquad \text{and} \qquad T = e^{-\beta t}\left(c_3 \cos \sqrt{n^2 - \beta^2}\, t + c_4 \sin \sqrt{n^2 - \beta^2}\, t\right)$$

and

$$u = \sum_{n=1}^{\infty} e^{-\beta t}\left(A_n \cos \sqrt{n^2 - \beta^2}\, t + B_n \sin \sqrt{n^2 - \beta^2}\, t\right) \sin nx.$$

Imposing

$$u(x,0) = f(x) = \sum_{n=1}^{\infty} A_n \sin nx$$

and

$$u_t(x,0) = 0 = \sum_{n=1}^{\infty}\left(B_n \sqrt{n^2 - \beta^2} - \beta A_n\right) \sin nx$$

gives

$$u(x,t) = e^{-\beta t} \sum_{n=1}^{\infty} A_n \left(\cos \sqrt{n^2 - \beta^2}\, t + \frac{\beta}{\sqrt{n^2 - \beta^2}} \sin \sqrt{n^2 - \beta^2}\, t\right) \sin nx,$$

where

$$A_n = \frac{2}{\pi} \int_0^{\pi} f(x) \sin nx \, dx.$$

18. (a) We note that $\xi_x = \eta_x = 1$, $\xi_t = a$, and $\eta_t = -a$. Then

$$\frac{\partial u}{\partial x} = \frac{\partial u}{\partial \xi}\frac{\partial \xi}{\partial x} + \frac{\partial u}{\partial \eta}\frac{\partial \eta}{\partial x} = u_\xi + u_\eta$$

and

$$\frac{\partial^2 u}{\partial x^2} = \frac{\partial}{\partial x}(u_\xi + u_\eta) = \frac{\partial u_\xi}{\partial \xi}\frac{\partial \xi}{\partial x} + \frac{\partial u_\xi}{\partial \eta}\frac{\partial \eta}{\partial x} + \frac{\partial u_\eta}{\partial \xi}\frac{\partial \xi}{\partial x} + \frac{\partial u_\eta}{\partial \eta}\frac{\partial \eta}{\partial x}$$

$$= u_{\xi\xi} + 2u_{\xi\eta} + u_{\eta\eta}.$$

Similarly

$$\frac{\partial^2 u}{\partial t^2} = a^2\left(u_{\xi\xi} - 2u_{\xi\eta} + u_{\eta\eta}\right).$$

Thus

$$a^2 \frac{\partial^2 u}{\partial x^2} = \frac{\partial^2 u}{\partial t^2} \qquad \text{becomes} \qquad \frac{\partial^2 u}{\partial \xi \partial \eta} = 0.$$

(b) Integrating

$$\frac{\partial^2 u}{\partial \xi \partial \eta} = \frac{\partial}{\partial \eta} u_\xi = 0$$

we obtain

$$\int \frac{\partial}{\partial \eta} u_\xi \, d\eta = \int 0 \, d\eta$$

$$u_\xi = f(\xi).$$

Integrating this result with respect to ξ we obtain

$$\int \frac{\partial u}{\partial \xi}\, d\xi = \int f(\xi)\, d\xi$$

$$u = F(\xi) + G(\eta).$$

Since $\xi = x + at$ and $\eta = x - at$, we then have

$$u = F(\xi) + G(\eta) = F(x + at) + G(x - at).$$

Next, we have

$$u(x, t) = F(x + at) + G(x - at)$$

$$u(x, 0) = F(x) + G(x) = f(x)$$

$$u_t(x, 0) = aF'(x) - aG'(x) = g(x).$$

Integrating the last equation with respect to x gives

$$F(x) - G(x) = \frac{1}{a} \int_{x_0}^{x} g(s)\, ds + c_1.$$

Substituting $G(x) = f(x) - F(x)$ we obtain

$$F(x) = \frac{1}{2} f(x) + \frac{1}{2a} \int_{x_0}^{x} g(s)\, ds + c$$

where $c = c_1/2$. Thus

$$G(x) = \frac{1}{2} f(x) - \frac{1}{2a} \int_{x_0}^{x} g(s)\, ds - c.$$

(c) From the expressions for F and G,

$$F(x + at) = \frac{1}{2} f(x + at) + \frac{1}{2a} \int_{x_0}^{x+at} g(s)\, ds + c$$

$$G(x - at) = \frac{1}{2} f(x - at) - \frac{1}{2a} \int_{x_0}^{x-at} g(s)\, ds - c.$$

Thus,

$$u(x, t) = F(x + at) + G(x - at) = \frac{1}{2}[f(x + at) + f(x - at)] + \frac{1}{2a} \int_{x-at}^{x+at} g(s)\, ds.$$

Here we have used $-\int_{x_0}^{x-at} g(s)\, ds = \int_{x-at}^{x_0} g(s)\, ds.$

21. $u(x, t) = 0 + \dfrac{1}{2a} \displaystyle\int_{x-at}^{x+at} \sin 2s\, ds = \dfrac{1}{2a} \left[\dfrac{-\cos(2x + 2at) + \cos(2x - 2at)}{2} \right]$

$$= \frac{1}{4a} \left[-\cos 2x \cos 2at + \sin 2x \sin 2at + \cos 2x \cos 2at + \sin 2x \sin 2at \right]$$

$$= \frac{1}{2a} \sin 2x \sin 2at$$

13.5 Laplace's Equation

3. Using $u = XY$ and $-\lambda$ as a separation constant we obtain

$$X'' + \lambda X = 0,$$
$$X(0) = 0,$$
$$X(a) = 0,$$

and

$$Y'' - \lambda Y = 0,$$
$$Y(b) = 0.$$

With $\lambda = \alpha^2 > 0$ the solutions of the differential equations are

$$X = c_1 \cos \alpha x + c_2 \sin \alpha x \quad \text{and} \quad Y = c_3 \cosh \alpha y + c_4 \sinh \alpha y$$

The boundary and initial conditions imply

$$X = c_2 \sin \frac{n\pi}{a} x \quad \text{and} \quad Y = c_2 \cosh \frac{n\pi}{a} y - c_2 \frac{\cosh \frac{n\pi b}{a}}{\sinh \frac{n\pi b}{a}} \sinh \frac{n\pi}{a} y$$

for $n = 1, 2, 3, \ldots$ so that

$$u = \sum_{n=1}^{\infty} A_n \left(\cosh \frac{n\pi}{a} y - \frac{\cosh \frac{n\pi b}{a}}{\sinh \frac{n\pi b}{a}} \sinh \frac{n\pi}{a} y \right) \sin \frac{n\pi}{a} x.$$

Imposing

$$u(x, 0) = f(x) = \sum_{n=1}^{\infty} A_n \sin \frac{n\pi}{a} x$$

gives

$$A_n = \frac{2}{a} \int_0^a f(x) \sin \frac{n\pi}{a} x \, dx$$

so that

$$u(x, y) = \frac{2}{a} \sum_{n=1}^{\infty} \left(\int_0^a f(x) \sin \frac{n\pi}{a} x \, dx \right) \left(\cosh \frac{n\pi}{a} y - \frac{\cosh \frac{n\pi b}{a}}{\sinh \frac{n\pi b}{a}} \sinh \frac{n\pi}{a} y \right) \sin \frac{n\pi}{a} x.$$

6. Using $u = XY$ and $-\lambda$ as a separation constant we obtain

$$X'' + \lambda X = 0,$$
$$X'(1) = 0$$

and

$$Y'' - \lambda Y = 0,$$
$$Y'(0) = 0,$$
$$Y'(\pi) = 0.$$

With $\lambda = \alpha^2 < 0$ the solutions of the differential equations are

$$X = c_1 \cosh \alpha x + c_2 \sinh \alpha x \quad \text{and} \quad Y = c_3 \cos \alpha y + c_4 \sin \alpha y.$$

The boundary and initial conditions imply

$$X = c_1 \cosh nx - c_1 \frac{\sinh n}{\cosh n} \sinh nx \quad \text{and} \quad Y = c_3 \cos ny$$

for $n = 1, 2, 3, \ldots$. Since $\lambda = 0$ is an eigenvalue for both differential equations with corresponding eigenfunctions 1 and 1 we have

$$u = A_0 + \sum_{n=1}^{\infty} A_n \left(\cosh nx - \frac{\sinh n}{\cosh n} \sinh nx \right) \cos ny.$$

Imposing

$$u(0, y) = g(y) = A_0 + \sum_{n=1}^{\infty} A_n \cos ny$$

gives

$$A_0 = \frac{1}{\pi} \int_0^\pi g(y) \, dy \quad \text{and} \quad A_n = \frac{2}{\pi} \int_0^\pi g(y) \cos ny \, dy$$

for $n = 1, 2, 3, \ldots$ so that

$$u(x, y) = \frac{1}{\pi} \int_0^\pi g(y) \, dy + \sum_{n=1}^{\infty} \left(\frac{2}{\pi} \int_0^\pi g(y) \cos ny \, dy \right) \left(\cosh nx - \frac{\sinh n}{\cosh n} \sinh nx \right) \cos ny.$$

9. This boundary-value problem has the form of Problem 1 from the text of this section, with $a = b = 1$, $f(x) = 100$, and $g(x) = 200$. The solution, then, is

$$u(x, y) = \sum_{n=1}^{\infty} (A_n \cosh n\pi y + B_n \sinh n\pi y) \sin n\pi x,$$

where

$$A_n = 2 \int_0^1 100 \sin n\pi x \, dx = 200 \left(\frac{1 - (-1)^n}{n\pi} \right)$$

and

$$B_n = \frac{1}{\sinh n\pi} \left[2 \int_0^1 200 \sin n\pi x \, dx - A_n \cosh n\pi \right]$$

$$= \frac{1}{\sinh n\pi} \left[400 \left(\frac{1 - (-1)^n}{n\pi} \right) - 200 \left(\frac{1 - (-1)^n}{n\pi} \right) \cosh n\pi \right]$$

$$= 200 \left[\frac{1 - (-1)^n}{n\pi} \right] [2 \operatorname{csch} n\pi - \coth n\pi].$$

12. Using $u = XY$ and $-\lambda$ as a separation constant we obtain

$$X'' + \lambda X = 0,$$
$$X'(0) = 0,$$
$$X'(\pi) = 0,$$

and

$$Y'' - \lambda Y = 0.$$

With $\lambda = \alpha^2 > 0$ the solutions of the differential equations are

$$X = c_1 \cos \alpha x + c_2 \sin \alpha x \qquad \text{and} \qquad Y = c_3 e^{\alpha y} + c_4 e^{-\alpha y}.$$

The boundary conditions at $x = 0$ and $x = \pi$ imply $c_2 = 0$ so $X = c_1 \cos nx$ for $n = 1, 2, 3, \ldots$. Now the boundedness of u as $y \to \infty$ implies $c_3 = 0$, so $Y = c_4 e^{-ny}$. In this problem $\lambda = 0$ is also an eigenvalue with corresponding eigenfunction 1 so that

$$u = A_0 + \sum_{n=1}^{\infty} A_n e^{-ny} \cos nx.$$

Imposing

$$u(x, 0) = f(x) = A_0 + \sum_{n=1}^{\infty} A_n \cos nx$$

gives

$$A_0 = \frac{1}{\pi} \int_0^{\pi} f(x)\, dx \qquad \text{and} \qquad A_n = \frac{2}{\pi} \int_0^{\pi} f(x) \cos nx\, dx$$

so that

$$u(x, y) = \frac{1}{\pi} \int_0^{\pi} f(x)\, dx + \sum_{n=1}^{\infty} \left(\frac{2}{\pi} \int_0^{\pi} f(x) \cos nx\, dx \right) e^{-ny} \cos nx.$$

15. Referring to the discussion in this section of the text we identify $a = b = \pi$, $f(x) = 0$, $g(x) = 1$, $F(y) = 1$, and $G(y) = 1$. Then $A_n = 0$ and

$$u_1(x, y) = \sum_{n=1}^{\infty} B_n \sinh ny \sin nx$$

where

$$B_n = \frac{2}{\pi \sinh n\pi} \int_0^{\pi} \sin nx\, dx = \frac{2[1 - (-1)^n]}{n\pi \sinh n\pi}.$$

Next

$$u_2(x, y) = \sum_{n=1}^{\infty} (A_n \cosh nx + B_n \sinh nx) \sin ny$$

where

$$A_n = \frac{2}{\pi} \int_0^{\pi} \sin ny\, dy = \frac{2[1 - (-1)^n]}{n\pi}$$

and

$$B_n = \frac{1}{\sinh n\pi} \left(\frac{2}{\pi} \int_0^\pi \sin ny \, dy - A_n \cosh n\pi \right)$$

$$= \frac{1}{\sinh n\pi} \left(\frac{2[1-(-1)^n]}{n\pi} - \frac{2[1-(-1)^n]}{n\pi} \cosh n\pi \right)$$

$$= \frac{2[1-(-1)^n]}{n\pi \sinh n\pi} (1 - \cosh n\pi).$$

Now

$$A_n \cosh nx + B_n \sinh nx = \frac{2[1-(-1)^n]}{n\pi} \left[\cosh nx + \frac{\sinh nx}{\sinh n\pi} (1 - \cosh n\pi) \right]$$

$$= \frac{2[1-(-1)^n]}{n\pi \sinh n\pi} [\cosh nx \sinh n\pi + \sinh nx - \sinh nx \cosh n\pi]$$

$$= \frac{2[1-(-1)^n]}{n\pi \sinh n\pi} [\sinh nx + \sinh n(\pi - x)]$$

and

$$u(x,y) = u_1 + u_2 = \frac{2}{\pi} \sum_{n=1}^\infty \frac{1-(-1)^n}{n \sinh n\pi} \sinh ny \sin nx$$

$$+ \frac{2}{\pi} \sum_{n=1}^\infty \frac{[1-(-1)^n][\sinh nx + \sinh n(\pi - x)]}{n \sinh n\pi} \sin ny.$$

13.6   Nonhomogeneous Boundary-Value Problems

3. If we let $u(x,t) = v(x,t) + \psi(x)$, then we obtain as in Example 1 in the text

$$k\psi'' + r = 0$$

or

$$\psi(x) = -\frac{r}{2k}x^2 + c_1 x + c_2.$$

The boundary conditions become

$$u(0,t) = v(0,t) + \psi(0) = u_0$$

$$u(1,t) = v(1,t) + \psi(1) = u_0.$$

Letting $\psi(0) = \psi(1) = u_0$ we obtain homogeneous boundary conditions in v:

$$v(0,t) = 0 \quad \text{and} \quad v(1,t) = 0.$$

Now $\psi(0) = \psi(1) = u_0$ implies $c_2 = u_0$ and $c_1 = r/2k$. Thus

$$\psi(x) = -\frac{r}{2k}x^2 + \frac{r}{2k}x + u_0 = u_0 - \frac{r}{2k}x(x - 1).$$

To determine $v(x, t)$ we solve

$$k\frac{\partial^2 v}{\partial x^2} = \frac{\partial v}{dt}, \quad 0 < x < 1, \ t > 0$$

$$v(0, t) = 0, \quad v(1, t) = 0,$$

$$v(x, 0) = \frac{r}{2k}x(x - 1) - u_0.$$

Separating variables, we find

$$v(x, t) = \sum_{n=1}^{\infty} A_n e^{-kn^2\pi^2 t} \sin n\pi x,$$

where

$$A_n = 2\int_0^1 \left[\frac{r}{2k}x(x - 1) - u_0\right] \sin n\pi x \, dx = 2\left[\frac{u_0}{n\pi} + \frac{r}{kn^3\pi^3}\right][(-1)^n - 1]. \quad \textbf{(1)}$$

Hence, a solution of the original problem is

$$u(x, t) = \psi(x) + v(x, t) = u_0 - \frac{r}{2k}x(x - 1) + \sum_{n=1}^{\infty} A_n e^{-kn^2\pi^2 t} \sin n\pi x,$$

where A_n is defined in (**1**).

6. Substituting $u(x, t) = v(x, t) + \psi(x)$ into the partial differential equation gives

$$k\frac{\partial^2 v}{\partial x^2} + k\psi'' - hv - h\psi = \frac{\partial v}{\partial t}.$$

This equation will be homogeneous provided ψ satisfies

$$k\psi'' - h\psi = 0.$$

Since k and h are positive, the general solution of this latter linear second-order equation is

$$\psi(x) = c_1 \cosh\sqrt{\frac{h}{k}}\,x + c_2 \sinh\sqrt{\frac{h}{k}}\,x.$$

From $\psi(0) = 0$ and $\psi(\pi) = u_0$ we find $c_1 = 0$ and $c_2 = u_0/\sinh\sqrt{h/k}\,\pi$. Hence

$$\psi(x) = u_0 \frac{\sinh\sqrt{h/k}\,x}{\sinh\sqrt{h/k}\,\pi}.$$

Now the new problem is

$$k\frac{\partial^2 v}{\partial x^2} - hv = \frac{\partial v}{\partial t}, \quad 0 < x < \pi, \ t > 0$$

$$v(0, t) = 0, \quad v(\pi, t) = 0, \quad t > 0$$

$$v(x, 0) = -\psi(x), \quad 0 < x < \pi.$$

If we let $v = XT$ then

$$\frac{X''}{X} = \frac{T' + hT}{kT} = -\lambda.$$

With $\lambda = \alpha^2 > 0$, the separated differential equations

$$X'' + \alpha^2 X = 0 \quad \text{and} \quad T' + \left(h + k\alpha^2\right)T = 0$$

have the respective solutions

$$X(x) = c_3 \cos \alpha x + c_4 \sin \alpha x$$

$$T(t) = c_5 e^{-\left(h + k\alpha^2\right)t}.$$

From $X(0) = 0$ we get $c_3 = 0$ and from $X(\pi) = 0$ we find $\alpha = n$ for $n = 1, 2, 3, \ldots$. Consequently, it follows that

$$v(x, t) = \sum_{n=1}^{\infty} A_n e^{-\left(h + kn^2\right)t} \sin nx$$

where

$$A_n = -\frac{2}{\pi} \int_0^\pi \psi(x) \sin nx \, dx.$$

Hence a solution of the original problem is

$$u(x, t) = u_0 \frac{\sinh \sqrt{h/k}\, x}{\sinh \sqrt{h/k}\, \pi} + e^{-ht} \sum_{n=1}^{\infty} A_n e^{-kn^2 t} \sin nx$$

where

$$A_n = -\frac{2}{\pi} \int_0^\pi u_0 \frac{\sinh \sqrt{h/k}\, x}{\sinh \sqrt{h/k}\, \pi} \sin nx \, dx.$$

Using the exponential definition of the hyperbolic sine and integration by parts we find

$$A_n = \frac{2u_0 nk(-1)^n}{\pi \left(h + kn^2\right)}.$$

9. Substituting $u(x, t) = v(x, t) + \psi(x)$ into the partial differential equation gives

$$a^2 \frac{\partial^2 v}{\partial x^2} + a^2 \psi'' + Ax = \frac{\partial^2 v}{\partial t^2}.$$

This equation will be homogeneous provided ψ satisfies

$$a^2 \psi'' + Ax = 0.$$

The solution of this differential equation is

$$\psi(x) = -\frac{A}{6a^2} x^3 + c_1 x + c_2.$$

From $\psi(0) = 0$ we obtain $c_2 = 0$, and from $\psi(1) = 0$ we obtain $c_1 = A/6a^2$. Hence

$$\psi(x) = \frac{A}{6a^2}(x - x^3).$$

Now the new problem is

$$a^2 \frac{\partial^2 v}{\partial x^2} = \frac{\partial^2 v}{\partial t^2}$$

$$v(0, t) = 0, \quad v(1, t) = 0, \quad t > 0,$$

$$v(x, 0) = -\psi(x), \quad v_t(x, 0) = 0, \quad 0 < x < 1.$$

Identifying this as the wave equation solved in Section 13.4 in the text with $L = 1$, $f(x) = -\psi(x)$, and $g(x) = 0$ we obtain

$$v(x, t) = \sum_{n=1}^{\infty} A_n \cos n\pi a t \sin n\pi x$$

where

$$A_n = 2 \int_0^1 [-\psi(x)] \sin n\pi x \, dx = \frac{A}{3a^2} \int_0^1 (x^3 - x) \sin n\pi x \, dx = \frac{2A(-1)^n}{a^2 \pi^3 n^3}.$$

Thus

$$u(x, t) = \frac{A}{6a^2}(x - x^3) + \frac{2A}{a^2 \pi^3} \sum_{n=1}^{\infty} \frac{(-1)^n}{n^3} \cos n\pi a t \sin n\pi x.$$

12. Substituting $u(x, y) = v(x, y) + \psi(x)$ into Poisson's equation we obtain

$$\frac{\partial^2 v}{\partial x^2} + \psi''(x) + h + \frac{\partial^2 v}{\partial y^2} = 0.$$

The equation will be homogeneous provided ψ satisfies $\psi''(x) + h = 0$ or $\psi(x) = -\frac{h}{2}x^2 + c_1 x + c_2$. From $\psi(0) = 0$ we obtain $c_2 = 0$. From $\psi(\pi) = 1$ we obtain

$$c_1 = \frac{1}{\pi} + \frac{h\pi}{2}.$$

Then

$$\psi(x) = \left(\frac{1}{\pi} + \frac{h\pi}{2}\right)x - \frac{h}{2}x^2.$$

The new boundary-value problem is

$$\frac{\partial^2 v}{\partial x^2} + \frac{\partial^2 v}{\partial y^2} = 0$$

$$v(0, y) = 0, \quad v(\pi, y) = 0,$$

$$v(x, 0) = -\psi(x), \quad 0 < x < \pi.$$

This is Problem 11 in Section 13.5. The solution is

$$v(x, y) = \sum_{n=1}^{\infty} A_n e^{-ny} \sin nx$$

where

$$A_n = \frac{2}{\pi} \int_0^{\pi} [-\psi(x) \sin nx] \, dx = \frac{2(-1)^n}{n} \left(\frac{1}{\pi} + \frac{h\pi}{2} \right) - h(-1)^n \left(\frac{\pi}{n} + \frac{2}{n^2} \right).$$

Thus

$$u(x, y) = v(x, y) + \psi(x) = \left(\frac{1}{\pi} + \frac{h\pi}{2} \right) x - \frac{h}{2} x^2 + \sum_{n=1}^{\infty} A_n e^{-ny} \sin nx.$$

15. From (13) and (14) we have

$$\psi(x, t) = u_0(t) + \frac{X}{L} [u_1(t) - u_0(t)] = x \sin t.$$

Then the substitution

$$u(x, t) = v(x, t) + \psi(x, t) = x \sin t$$

leads to the boundary-value problem

$$\frac{\partial^2 v}{\partial x^2} + x \sin t = \frac{\partial^2 v}{\partial t^2}, \quad 0 < x < 1, \quad t > 0$$

$$v(0, t) = 0, \quad v(1, t) = 0, \quad t > 0$$

$$v(x, 0) = 0, \quad \left. \frac{\partial v}{\partial t} \right|_{t=0} = -x, \quad 0 < x < 1.$$

The eigenvalues and eigenfunctions of the Sturm–Liouville problem

$$X'' + \lambda X = 0, \quad X(0) = 0, \quad X(1) = 0$$

are $\lambda_n = \alpha_n^2 = n^2 \pi^2$ and $\sin n\pi x$, $n = 1, 2, 3, \ldots$. With $G(x, t) = x \sin t$ we assume for fixed t that v and G can be written as Fourier sine series:

$$v(x, t) = \sum_{n=1}^{\infty} v_n(t) \sin n\pi x \quad \text{and} \quad G(x, t) = \sum_{n=1}^{\infty} G_n(t) \sin n\pi x.$$

By treating t as a parameter, the coefficients G_n can be computed:

$$G_n(t) = \frac{2}{1} \int_0^1 x \sin t \sin n\pi x \, dx = \frac{2}{1} \sin t \int_0^1 x \sin n\pi x \, dx = -2 \frac{(-1)^n \sin t}{n\pi}.$$

Hence

$$x \sin t = \sum_{n=1}^{\infty} -2 \frac{(-1)^n \sin t}{n\pi} \sin n\pi x.$$

Now, using the series representation for $v(x,t)$, we have

$$\frac{\partial^2 v}{\partial x^2} = \sum_{n=1}^{\infty} v_n(t)(-n^2\pi^2)\sin n\pi x \quad \text{and} \quad \frac{\partial v}{\partial t} = \sum_{n=1}^{\infty} v_n'(t)\sin n\pi x.$$

Writing the partial differential equation as $v_t - v_{xx} = x\sin t$ and using the above results we have

$$\sum_{n=1}^{\infty} [v_n'(t) + n^2\pi^2 v_n(t)]\sin n\pi x = \sum_{n=1}^{\infty} -2\frac{(-1)^n \sin t}{n\pi}\sin n\pi x.$$

Equating coefficients we get

$$v_n''(t) + n^2\pi^2 v_n(t) = -2\frac{(-1)^n \sin t}{n\pi}.$$

For each n the general solution is

$$v_n(t) = A_n\cos n\pi t + B_n\sin n\pi t - 2\frac{(-1)^n}{n\pi\,(n^2\pi^2 - 1)}\sin t.$$

Thus

$$v(x,t) = \sum_{n=1}^{\infty}\left[A_n\cos n\pi t + B_n\sin n\pi t - 2\frac{(-1)^n}{n\pi\,(n^2\pi^2 - 1)}\sin t\right]\sin n\pi x.$$

The initial condition $v(x,0) = 0$ implies

$$\sum_{n=1}^{\infty} A_n\sin n\pi x = 0$$

or $A_n = 0$ for $n = 1,\,2,\,3,\,\dots$. So

$$v(x,t) = \sum_{n=1}^{\infty}\left[B_n\sin n\pi t - 2\frac{(-1)^n}{n\pi\,(n^2\pi^2 - 1)}\sin t\right]\sin n\pi x$$

and

$$\left.\frac{\partial v}{\partial t}\right|_{t=0} = -x = \sum_{n=1}^{\infty}\left[n\pi B_n - 2\frac{(-1)^n}{n\pi\,(n^2\pi^2 - 1)}\right]\sin n\pi x.$$

Thinking of $-x$ as a Fourier sine series with coefficients

$$-2\int_0^1 x\sin n\pi x\,dx = 2\frac{(-1)^n}{n\pi}$$

we equate coefficients to obtain

$$n\pi B_n - 2\frac{(-1)^n}{n\pi\,(n^2\pi^2 - 1)} = 2\frac{(-1)^n}{n\pi}$$

so

$$B_n = 2\frac{(-1)^n}{n\pi} + 2\frac{(-1)^n}{n\pi\,(n^2\pi^2 - 1)}.$$

Therefore

$$v(x,t) = \sum_{n=1}^{\infty} \left[\left(2\frac{(-1)^n}{n\pi} + 2\frac{(-1)^n}{n\pi\,(n^2\pi^2 - 1)} \right) \sin n\pi t - 2\frac{(-1)^n}{n\pi\,(n^2\pi^2 - 1)} \sin t \right] \sin n\pi x$$

and

$$u(x,t) = v(x,t) + \psi(x,t)$$

$$= x\sin t + 2\sum_{n=1}^{\infty} \left[\left(\frac{(-1)^n}{n\pi} + \frac{(-1)^n}{n\pi\,(n^2\pi^2 - 1)} \right) \sin n\pi t - \frac{(-1)^n}{n\pi\,(n^2\pi^2 - 1)} \sin t \right] \sin n\pi x.$$

18. Identifying $k = 1$ and $L = \pi$ we see that the eigenfunctions of $X'' + \lambda X = 0$, $X(0) = 0$, $X'(\pi) = 0$ are 1, $\cos nx$, $n = 1, 2, 3, \dots$. Assuming that $u(x,t) = \frac{1}{2}u_0(t) + \sum_{n=1}^{\infty} u_n(t)\cos nx$, the formal partial derivatives of u are

$$\frac{\partial^2 u}{\partial x^2} = \sum_{n=1}^{\infty} u_n(t)(-n^2)\cos nx \qquad \text{and} \qquad \frac{\partial u}{\partial t} = \frac{1}{2}u_0' + \sum_{n=1}^{\infty} u_n'(t)\cos nx.$$

Assuming that $xe^{-3t} = \frac{1}{2}F_0(t) + \sum_{n=1}^{\infty} F_n(t)\cos nx$ we have

$$F_0(t) = \frac{2e^{-3t}}{\pi} \int_0^{\pi} x\,dx = \pi e^{-3t}$$

and

$$F_n(t) = \frac{2e^{-3t}}{\pi} \int_0^{\pi} x\cos nx\,dx = \frac{2e^{-3t}[(-1)^n - 1]}{\pi n^2}.$$

Then

$$xe^{-3t} = \frac{\pi}{2}e^{-3t} + \sum_{n=1}^{\infty} \frac{2e^{-3t}[(-1)^n - 1]}{\pi n^2}\cos nx$$

and

$$u_t - u_{xx} = \frac{1}{2}u_0'(t) + \sum_{n=1}^{\infty} [u_n'(t) + n^2 u_n(t)]\cos nx$$

$$= xe^{-3t} = \frac{\pi}{2}e^{-3t} + \sum_{n=1}^{\infty} \frac{2e^{-3t}[(-1)^n - 1]}{\pi n^2}\cos nx.$$

Equating coefficients, we obtain

$$u_0'(t) = \pi e^{-3t} \qquad \text{and} \qquad u_n'(t) + n^2 u_n(t) = \frac{2e^{-3t}[(-1)^n - 1]}{\pi n^2}\cos nx.$$

The first equation yields $u_0(t) = -(\pi/3)e^{-3t} + C_0$ and the second equation, which is a linear first-order differential equation, yields

$$u_n(t) = \frac{2[(-1)^n - 1]}{\pi n^2(n^2 - 3)}e^{-3t} + C_n e^{-n^2 t}.$$

Thus

$$u(x,t) = -\frac{\pi}{3}e^{-3t} + C_0 + \sum_{n=1}^{\infty} \frac{2[(-1)^n - 1]}{\pi n^2(n^2 - 3)} e^{-3t} \cos nx + \sum_{n=1}^{\infty} C_n e^{-n^2 t} \cos nx$$

and $u(x,0) = 0$ implies

$$-\frac{\pi}{3} + C_0 + \sum_{n=1}^{\infty} \frac{2[(-1)^n - 1]}{\pi n^2(n^2 - 3)} \cos nx + \sum_{n=1}^{\infty} C_n \cos nx = 0$$

so that $C_0 = \pi/3$ and $C_n = 2[(-1)^n - 1]/\pi n^2(n^2 - 3)$. Therefore

$$u(x,t) = \frac{\pi}{3}(1 - e^{-3t}) + \frac{2}{\pi} \sum_{n=1}^{\infty} \frac{(-1)^n - 1}{n^2(n^2 - 3)} e^{-3t} \cos nx + \frac{2}{\pi} \sum_{n=1}^{\infty} \frac{1 - (-1)^n}{n^2(n^2 - 3)} e^{-n^2 t} \cos nx.$$

13.7 | Orthogonal Series Expansions

3. Separating variables in Laplace's equation gives

$$X'' + \alpha^2 X = 0$$

$$Y'' - \alpha^2 Y = 0$$

and

$$X(x) = c_1 \cos \alpha x + c_2 \sin \alpha x$$

$$Y(y) = c_3 \cosh \alpha y + c_4 \sinh \alpha y.$$

From $u(0,y) = 0$ we obtain $X(0) = 0$ and $c_1 = 0$. From $u_x(a,y) = -hu(a,y)$ we obtain $X'(a) = -hX(a)$ and

$$\alpha \cos \alpha a = -h \sin \alpha a \qquad \text{or} \qquad \tan \alpha a = -\frac{\alpha}{h}.$$

Let α_n, where $n = 1, 2, 3, \ldots$, be the consecutive positive roots of this equation. From $u(x,0) = 0$ we obtain $Y(0) = 0$ and $c_3 = 0$. Thus

$$u(x,y) = \sum_{n=1}^{\infty} A_n \sinh \alpha_n y \sin \alpha_n x.$$

Now

$$f(x) = \sum_{n=1}^{\infty} A_n \sinh \alpha_n b \sin \alpha_n x$$

and

$$A_n \sinh \alpha_n b = \frac{\int_0^a f(x) \sin \alpha_n x \, dx}{\int_0^a \sin^2 \alpha_n x \, dx}.$$

Since

$$\int_0^a \sin^2 \alpha_n x \, dx = \frac{1}{2}\left[a - \frac{1}{2\alpha_n}\sin 2\alpha_n a\right] = \frac{1}{2}\left[a - \frac{1}{\alpha_n}\sin \alpha_n a \cos \alpha_n a\right]$$

$$= \frac{1}{2}\left[a - \frac{1}{h\alpha_n}(h\sin \alpha_n a)\cos \alpha_n a\right]$$

$$= \frac{1}{2}\left[a - \frac{1}{h\alpha_n}(-\alpha_n \cos \alpha_n a)\cos \alpha_n a\right] = \frac{1}{2h}\left[ah + \cos^2 \alpha_n a\right],$$

we have

$$A_n = \frac{2h}{\sinh \alpha_n b[ah + \cos^2 \alpha_n a]}\int_0^a f(x)\sin \alpha_n x \, dx.$$

6. Substituting $u(x,t) = v(x,t) + \psi(x)$ into the partial differential equation gives

$$a^2 \frac{\partial^2 v}{\partial x^2} + \psi''(x) = \frac{\partial^2 v}{\partial t^2}.$$

This equation will be homogeneous if $\psi''(x) = 0$ or $\psi(x) = c_1 x + c_2$. The boundary condition $u(0,t) = 0$ implies $\psi(0) = 0$ which implies $c_2 = 0$. Thus $\psi(x) = c_1 x$. Using the second boundary condition, we obtain

$$E\left(\frac{\partial v}{\partial x} + \psi'\right)\bigg|_{x=L} = F_0,$$

which will be homogeneous when

$$E\psi'(L) = F_0.$$

Since $\psi'(x) = c_1$ we conclude that $c_1 = F_0/E$ and

$$\psi(x) = \frac{F_0}{E}x.$$

The new boundary-value problem is

$$a^2 \frac{\partial^2 v}{\partial x^2} = \frac{\partial^2 v}{\partial t^2}, \quad 0 < x < L, \quad t > 0$$

$$v(0,t) = 0, \quad \frac{\partial v}{\partial x}\bigg|_{x=L} = 0, \quad t > 0,$$

$$v(x,0) = -\frac{F_0}{E}x, \quad \frac{\partial v}{\partial t}\bigg|_{t=0} = 0, \quad 0 < x < L.$$

Referring to Example 2 in the text we see that

$$v(x,t) = \sum_{n=1}^{\infty} A_n \cos a\left(\frac{2n-1}{2L}\right)\pi t \sin\left(\frac{2n-1}{2L}\right)\pi x$$

where

$$-\frac{F_0}{E}x = \sum_{n=1}^{\infty} A_n \sin\left(\frac{2n-1}{2L}\right)\pi x$$

and

$$A_n = \frac{-F_0 \int_0^L x \sin\left(\frac{2n-1}{2L}\right)\pi x \, dx}{E \int_0^L \sin^2\left(\frac{2n-1}{2L}\right)\pi x \, dx} = \frac{8F_0 L(-1)^n}{E\pi^2(2n-1)^2}.$$

Thus

$$u(x,t) = v(x,t) + \psi(x) = \frac{F_0}{E}x + \frac{8F_0 L}{E\pi^2}\sum_{n=1}^{\infty}\frac{(-1)^n}{(2n-1)^2}\cos a\left(\frac{2n-1}{2L}\right)\pi t \sin\left(\frac{2n-1}{2L}\right)\pi x.$$

9. The boundary-value problem is

$$k\frac{\partial^2 u}{\partial x^2} = \frac{\partial u}{\partial t}, \quad 0 < x < 1, \quad t > 0$$

$$\left.\frac{\partial u}{\partial x}\right|_{x=0} = hu(0,t), \quad \left.\frac{\partial u}{\partial x}\right|_{x=1} = -hu(1,t), \quad h > 0, \quad t > 0,$$

$$u(x,0) = f(x), \quad 0 < x < 1.$$

Referring to Example 1 in the text we have

$$X(x) = c_1 \cos \alpha x + c_2 \sin \alpha x \quad \text{and} \quad T(t) = c_3 e^{-k\alpha^2 t}.$$

Applying the boundary conditions, we obtain

$$X'(0) = hX(0)$$

$$X'(1) = -hX(1)$$

or

$$\alpha c_2 = hc_1$$

$$-\alpha c_1 \sin \alpha + \alpha c_2 \cos \alpha = -hc_1 \cos \alpha - hc_2 \sin \alpha.$$

Choosing $c_1 = \alpha$ and $c_2 = h$ (to satisfy the first equation above) we obtain

$$-\alpha^2 \sin \alpha + h\alpha \cos \alpha = -h\alpha \cos \alpha - h^2 \sin \alpha$$

$$2h\alpha \cos \alpha = (\alpha^2 - h^2)\sin \alpha.$$

The eigenvalues α_n are the consecutive positive roots of

$$\tan \alpha = \frac{2h\alpha}{\alpha^2 - h^2}.$$

Then

$$u(x,t) = \sum_{n=1}^{\infty} A_n e^{-k\alpha_n^2 t}(\alpha_n \cos \alpha_n x + h \sin \alpha_n x)$$

where

$$f(x) = u(x,0) = \sum_{n=1}^{\infty} A_n(\alpha_n \cos \alpha_n x + h \sin \alpha_n x)$$

and

$$A_n = \frac{\displaystyle\int_0^1 f(x)(\alpha_n \cos \alpha_n x + h \sin \alpha_n x)\, dx}{\displaystyle\int_0^1 (\alpha_n \cos \alpha_n x + h \sin \alpha_n x)^2\, dx}$$

$$= \frac{2}{\alpha_n^2 + 2h + h^2} \int_0^1 f(x)(\alpha_n \cos \alpha_n x + h \sin \alpha_n x)\, dx.$$

[Note: the evaluation and simplification of the integral in the denominator requires the use of the relationship $(\alpha^2 - h^2)\sin \alpha = 2h\alpha \cos \alpha$.]

13.8 | Higher-Dimensional Problems

3. The conditions $X(0) = 0$ and $Y(0) = 0$ give $c_1 = 0$ and $c_3 = 0$. The conditions $X(\pi) = 0$ and $Y(\pi) = 0$ yield two sets of eigenvalues:

$$\alpha = m, \ m = 1, 2, 3, \ldots \qquad \text{and} \qquad \beta = n, \ n = 1, 2, 3, \ldots .$$

A product solution of the partial differential equation that satisfies the boundary conditions is

$$u_{mn}(x, y, t) = \left(A_{mn} \cos a\sqrt{m^2 + n^2}\, t + B_{mn} \sin a\sqrt{m^2 + n^2}\, t\right) \sin mx \sin ny.$$

To satisfy the initial conditions we use the superposition principle:

$$u(x, y, t) = \sum_{m=1}^{\infty}\sum_{n=1}^{\infty} \left(A_{mn} \cos a\sqrt{m^2 + n^2}\, t + B_{mn} \sin a\sqrt{m^2 + n^2}\, t\right) \sin mx \sin ny.$$

The initial condition $u_t(x, y, 0) = 0$ implies $B_{mn} = 0$ and

$$u(x, y, t) = \sum_{m=1}^{\infty}\sum_{n=1}^{\infty} A_{mn} \cos a\sqrt{m^2 + n^2}\, t \sin mx \sin ny.$$

At $t = 0$ we have

$$xy(x - \pi)(y - \pi) = \sum_{m=1}^{\infty}\sum_{n=1}^{\infty} A_{mn} \sin mx \sin ny.$$

Using (12) and (13) in the text, it follows that

$$A_{mn} = \frac{4}{\pi^2} \int_0^\pi \int_0^\pi xy(x-\pi)(y-\pi) \sin mx \sin ny \, dx \, dy$$

$$= \frac{4}{\pi^2} \int_0^\pi x(x-\pi) \sin mx \, dx \int_0^\pi y(y-\pi) \sin ny \, dy$$

$$= \frac{16}{m^3 n^3 \pi^2} [(-1)^m - 1][(-1)^n - 1].$$

6. The boundary and initial conditions are

$$u(0,y,z) = 0, \qquad u(a,y,z) = 0,$$
$$u(x,0,z) = 0, \qquad u(x,b,z) = 0,$$
$$u(x,y,0) = f(x,y), \qquad u(x,y,c) = 0.$$

The conditions $X(0) = Y(0) = 0$ give $c_1 = c_3 = 0$. The conditions $X(a) = Y(b) = 0$ yield two sets of eigenvalues:

$$\alpha = \frac{m\pi}{a}, \; m = 1,2,3,\ldots \qquad \text{and} \qquad \beta = \frac{n\pi}{b}, \; n = 1,2,3,\ldots.$$

Let

$$\omega_{mn}^2 = \frac{m^2 \pi^2}{a^2} + \frac{n^2 \pi^2}{b^2}.$$

Then the boundary condition $Z(c) = 0$ gives

$$c_5 \cosh c\,\omega_{mn} + c_6 \sinh c\,\omega_{mn} = 0$$

from which we obtain

$$Z(z) = c_5 \left(\cosh w_{mn}z - \frac{\cosh c\,\omega_{mn}}{\sinh c\,\omega_{mn}} \sinh \omega z \right)$$

$$= \frac{c_5}{\sinh c\,\omega_{mn}} (\sinh c\,\omega_{mn} \cosh \omega_{mn}z - \cosh c\,\omega_{mn} \sinh \omega_{mn}z) = c_{mn} \sinh \omega_{mn}(c - z).$$

By the superposition principle

$$u(x,y,t) = \sum_{m=1}^\infty \sum_{n=1}^\infty A_{mn} \sinh \omega_{mn}(c-z) \sin \frac{m\pi}{a} x \sin \frac{n\pi}{b} y$$

where

$$A_{mn} = \frac{4}{ab \sinh c\,\omega_{mn}} \int_0^b \int_0^a f(x,y) \sin \frac{m\pi}{a} x \sin \frac{n\pi}{b} y \, dx \, dy.$$

> **Chapter 13 in Review**

3. Substituting $u(x, t) = v(x, t) + \psi(x)$ into the partial differential equation we obtain

$$k\frac{\partial^2 v}{\partial x^2} + k\psi''(x) = \frac{\partial v}{\partial t}.$$

This equation will be homogeneous provided ψ satisfies

$$k\psi'' = 0 \quad \text{or} \quad \psi = c_1 x + c_2.$$

Considering

$$u(0, t) = v(0, t) + \psi(0) = u_0$$

we set $\psi(0) = u_0$ so that $\psi(x) = c_1 x + u_0$. Now

$$-\frac{\partial u}{\partial x}\bigg|_{x=\pi} = -\frac{\partial v}{\partial x}\bigg|_{x=\pi} - \psi'(x) = v(\pi, t) + \psi(\pi) - u_1$$

is equivalent to

$$\frac{\partial v}{\partial x}\bigg|_{x=\pi} + v(\pi, t) = u_1 - \psi'(x) - \psi(\pi) = u_1 - c_1 - (c_1\pi + u_0),$$

which will be homogeneous when

$$u_1 - c_1 - c_1\pi - u_0 = 0 \quad \text{or} \quad c_1 = \frac{u_1 - u_0}{1 + \pi}.$$

The steady-state solution is

$$\psi(x) = \left(\frac{u_1 - u_0}{1 + \pi}\right)x + u_0.$$

6. The boundary-value problem is

$$\frac{\partial^2 u}{\partial x^2} + x^2 = \frac{\partial^2 u}{\partial t^2}, \quad 0 < x < 1, \quad t > 0,$$

$$u(0, t) = 1, \quad u(1, t) = 0, \quad t > 0,$$

$$u(x, 0) = f(x), \quad u_t(x, 0) = 0, \quad 0 < x < 1.$$

Substituting $u(x, t) = v(x, t) + \psi(x)$ into the partial differential equation gives

$$\frac{\partial^2 v}{\partial x^2} + \psi''(x) + x^2 = \frac{\partial^2 v}{\partial t^2}.$$

This equation will be homogeneous provided $\psi''(x) + x^2 = 0$ or

$$\psi(x) = -\frac{1}{12}x^4 + c_1 x + c_2.$$

From $\psi(0) = 1$ and $\psi(1) = 0$ we obtain $c_1 = -11/12$ and $c_2 = 1$. The new problem is

$$\frac{\partial^2 v}{\partial x^2} = \frac{\partial^2 v}{\partial t^2}, \quad 0 < x < 1, \quad t > 0,$$

$$v(0, t) = 0, \quad v(1, t) = 0, \quad t > 0,$$

$$v(x, 0) = f(x) - \psi(x), \quad v_t(x, 0) = 0, \quad 0 < x < 1.$$

From Section 13.4 in the text we see that $B_n = 0$,

$$A_n = 2 \int_0^1 [f(x) - \psi(x)] \sin n\pi x \, dx = 2 \int_0^1 \left[f(x) + \frac{1}{12} x^4 + \frac{11}{12} x - 1 \right] \sin n\pi x \, dx,$$

and

$$v(x, t) = \sum_{n=1}^{\infty} A_n \cos n\pi t \sin n\pi x.$$

Thus

$$u(x, t) = v(x, t) + \psi(x) = -\frac{1}{12} x^4 - \frac{11}{12} x + 1 + \sum_{n=1}^{\infty} A_n \cos n\pi t \sin n\pi x.$$

9. Using $u = XY$ and $-\lambda$ as a separation constant leads to

$$X'' - \lambda X = 0,$$

and

$$Y'' + \lambda Y = 0,$$
$$Y(0) = 0,$$
$$Y(\pi) = 0.$$

Then

$$X = c_1 e^{nx} + c_2 e^{-nx} \quad \text{and} \quad Y = c_3 \cos ny + c_4 \sin ny$$

for $n = 1, 2, 3, \ldots$. Since u must be bounded as $x \to \infty$, we define $c_1 = 0$. Also $Y(0) = 0$ implies $c_3 = 0$ so

$$u = \sum_{n=1}^{\infty} A_n e^{-nx} \sin ny.$$

Imposing

$$u(0, y) = 50 = \sum_{n=1}^{\infty} A_n \sin ny$$

gives

$$A_n = \frac{2}{\pi} \int_0^{\pi} 50 \sin ny \, dy = \frac{100}{n\pi} [1 - (-1)^n]$$

so that

$$u(x, y) = \sum_{n=1}^{\infty} \frac{100}{n\pi} [1 - (-1)^n] e^{-nx} \sin ny.$$

12. Substituting $u(x,t) = v(x,t) + \psi(x)$ into the partial differential equation results in $\psi'' = -\sin x$ and $\psi(x) = c_1 x + c_2 + \sin x$. The boundary conditions $\psi(0) = 400$ and $\psi(\pi) = 200$ imply $c_1 = -200/\pi$ and $c_2 = 400$ so

$$\psi(x) = -\frac{200}{\pi}x + 400 + \sin x.$$

Solving

$$\frac{\partial^2 v}{\partial x^2} = \frac{\partial v}{\partial t}, \qquad 0 < x < \pi, \quad t > 0$$

$$v(0,t) = 0, \qquad v(\pi,t) = 0, \quad t > 0$$

$$v(x,0) = 400 + \sin x - \left(-\frac{200}{\pi}x + 400 + \sin x\right) = \frac{200}{\pi}x, \qquad 0 < x < \pi$$

using separation of variables with separation constant $-\lambda$, where $\lambda = \alpha^2$, gives

$$X'' + \alpha^2 X = 0 \qquad \text{and} \qquad T' + \alpha^2 T = 0.$$

Using $X(0) = 0$ and $X(\pi) = 0$ we determine $\alpha^2 = n^2$, $X(x) = c_2 \sin nx$, and $T(t) = c_3 e^{-n^2 t}$. Then

$$v(x,t) = \sum_{n=1}^{\infty} A_n e^{-n^2 t} \sin nx$$

and

$$v(x,0) = \frac{200}{\pi}x = \sum_{n=1}^{\infty} A_n \sin nx$$

so

$$A_n = \frac{400}{\pi^2} \int_0^\pi x \sin nx \, dx = \frac{400}{n\pi}(-1)^{n+1}.$$

Thus

$$u(x,t) = -\frac{200}{\pi}x + 400 + \sin x + \frac{400}{\pi}\sum_{n=1}^{\infty}\frac{(-1)^{n+1}}{n}e^{-n^2 t}\sin nx.$$

15. By the discussion in Section 13.6, we assume a solution in the form $u(x,t) = v(x,t) + \psi(x)$. Force this into the differential equation to get $\psi''(x) = 0$ together with the conditions $\psi(0) = 0$ and $\psi'(1) + \psi(1) = u_1$.

Integrating twice and imposing the two conditions leads to $\psi(x) = \frac{1}{2}(u_1 - u_0)x + u_0$. Now

$$u(x,t) = v(x,t) + \psi(x)$$

$$u(0,t) = v(0,t) + \psi(0) = u_0$$

$$v(0,t) + \left[\frac{1}{2}(u_1 - u_0)(0) + u_0\right] = u_0$$

$$v(0,t) = 0.$$

Similarly we find that

$$u_x(x,t) = v_x(x,t) + \psi'(x)$$

$$u_x(1,t) = v_x(1,t) + \psi'(1) = u_1 - u(1,t)$$

$$v_x(1,t) + \left[\frac{1}{2}(u_1 - u_0)\right] = u_1 - [v(1,t) + \psi(1)]$$

$$v_x(1,t) = u_1 - [v(1,t) + \psi(1)] - \frac{1}{2}(u_1 - u_0)$$

$$v_x(1,t) = -v(1,t).$$

So now we solve

$$\frac{\partial^2 v}{\partial x^2} = \frac{\partial v}{\partial t}, \quad 0 < x < 1, \ t > 0$$

$$v(0,t) = 0, \quad \left.\frac{\partial v}{\partial x}\right|_{x=1} = -v(1,t), \ t > 0$$

$$v(x,0) = \frac{1}{2}(u_0 - u_1)x, \ 0 < x < 1.$$

Using the results of Example 1 from Section 13.7, $v(x,t) = \sum_{n=1}^{\infty} A_n e^{-\alpha_n^2 t} \sin \alpha_n x$ and the coefficients are given by

$$A_n = \frac{\frac{1}{2}(u_0 - u_1) \int_0^1 x \sin \alpha_n x \, dx}{\int_0^1 \sin^2 \alpha_n x \, dx} = \frac{1}{2}(u_0 - u_1)\frac{\frac{\sin \alpha_n - \alpha_n \cos \alpha_n}{\alpha_n^2}}{\frac{1}{2}(1 + \cos^2 \alpha_n)}$$

$$= (u_0 - u_1)\frac{\sin \alpha_n - \alpha_n \cos \alpha_n}{\alpha_n^2 (1 + \cos^2 \alpha_n)}.$$

Since $\tan \alpha_n = -\alpha_n$ we have $\sin \alpha_n = -\alpha_n \cos \alpha_n$ and so

$$A_n = 2(u_1 - u_0) \sum_{n=1}^{\infty} \frac{\cos \alpha_n}{\alpha_n (a + \cos^2 \alpha_n)} e^{-\alpha_n^2 t} \sin (\alpha_n x)$$

and

$$u(x,t) = u_0 + \frac{1}{2}(u_1 - u_0)x + 2(u_1 - u_0) \sum_{n=1}^{\infty} \frac{\cos \alpha_n}{\alpha_n (a + \cos^2 \alpha_n)} e^{-\alpha_n^2 t} \sin (\alpha_n x).$$

18. Using the substitution $u(x,t) = v(x,t) + \psi(x)$ into the boundary-value problem yields

$$\frac{\partial^2 v}{\partial x^2} + \psi'' + e^{-x} = \frac{\partial v}{\partial t}.$$

Now $u(0,t) = v(0,t) + \psi(0) = 0$ and $u_x(\pi,t) = v_x(\pi,t) + \psi'(\pi) = 0$. The condition $\psi(0) = 0$ gives $-1 + c_1 = 0$ or $c_1 = 1$. The condition $\psi'(\pi) = 0$ gives $e^{-\pi} + c_2 = 0$ or $c_2 = -e^{-\pi}$. Thus,

$$\psi(x) = -e^{-x} + 1 - e^{-\pi}x.$$

Now the homogeneous boundary-value problem for $v(x, t)$ is

$$\frac{\partial^2 v}{\partial x^2} = \frac{\partial v}{\partial t}, \quad 0 < x < \pi, \quad t > 0$$

$$v(0, t) = 0, \quad \left.\frac{\partial v}{\partial x}\right|_{x=\pi}, \quad t > 0$$

$$v(x, 0) = f(x) + e^{-x} - 1 + e^{-\pi} x, \quad 0 < x < \pi.$$

Using $v = XT$ and separation of variables leads to

$$X'' + \lambda X = 0$$

$$T' + \lambda T = 0$$

with

$$X(0) = 0 \quad \text{and} \quad X'(\pi) = 0.$$

This is a regular Sturm–Liouville problem. For $\lambda = \alpha^2$ the general solution is $X = c_3 \cos \alpha x + c_4 \sin \alpha x$. From $X(0) = 0$ we find $c_3 = 0$, so $X = c_4 \sin \alpha x$. From $X'(\pi) = 0$ we find $\cos \alpha \pi = 0$ or $\alpha \pi = \left(\dfrac{2n-1}{2}\right) \pi$ for $n = 1, 2, 3, \ldots$. Therefore

$$X = c_1 \sin \left(\frac{2n-1}{2}\right) x \quad \text{and} \quad T = c_5 e^{-(2n-1)^2 t/4}.$$

Thus

$$v(x, t) = \sum_{n=1}^{\infty} A_n e^{-(2n-1)^2 t/4} \sin \left(\frac{2n-1}{2}\right) x.$$

When $t = 0$,

$$f(x) + e^{-x} - 1 + e^{-\pi} x = \sum_{n=1}^{\infty} A_n \sin \left(\frac{2n-1}{2}\right) x$$

where

$$A_n = \frac{\displaystyle\int_0^\pi \left(f(x) + e^{-x} - 1 + e^{-\pi} x\right) \sin \left(\frac{2n-1}{2}\right) x \, dx}{\displaystyle\int_0^\pi \sin^2 \left(\frac{2n-1}{2}\right) x}$$

$$= \frac{2}{\pi} \int_0^\pi \left(f(x) + e^{-x} - 1 + e^{-\pi} x\right) \sin \left(\frac{2n-1}{2}\right) x \, dx.$$

The solution is then

$$u(x, t) = -e^{-x} + 1 - e^{-\pi} x + \sum_{n=1}^{\infty} A_n e^{-(2n-1)^2 t/4} \sin \left(\frac{2n-1}{2}\right) x$$

where the coefficients A_n are defined above.

Chapter 14

Boundary-Value Problems in Other Coordinate Systems

14.1 | Polar Coordinates

3. We have

$$A_0 = \frac{1}{2\pi} \int_0^{2\pi} (2\pi\theta - \theta^2)\, d\theta = \frac{2\pi^2}{3}$$

$$A_n = \frac{1}{\pi} \int_0^{2\pi} (2\pi\theta - \theta^2) \cos n\theta\, d\theta = -\frac{4}{n^2}$$

$$B_n = \frac{1}{\pi} \int_0^{2\pi} (2\pi\theta - \theta^2) \sin n\theta\, d\theta = 0$$

and so

$$u(r, \theta) = \frac{2\pi^2}{3} - 4 \sum_{n=1}^{\infty} \frac{r^n}{n^2} \cos n\theta.$$

6. The nature of the given conditions suggest we let $u(r, \theta) = v(r, \theta) + \psi(\theta)$. From this we get $\psi''(\theta) = 0$ so that $\psi(\theta) = c_1\theta + c_2$. Now $\psi(0) = 0$ gives $c_2 = 0$ and $\psi(\pi) = u_0$ gives $c_1 = u_0/\pi$. Therefore $\psi(\theta) = (u_0/\pi)\,\theta$. Next we have

$$u(1, \theta) = v(1, \theta) + \psi(\theta)$$

$$u_0 = v(1, \theta) + \frac{u_0}{\pi}\theta \quad \text{or} \quad v(1, \theta) = u_0 - \frac{u_0}{\pi}\theta$$

From

$$v(r, \theta) = \sum_{n=1}^{\infty} A_n r^n \sin n\theta \qquad \text{and} \qquad v(1, \theta) = \sum_{n=1}^{\infty} A_n \sin n\theta$$

we get the coefficients

$$A_n = \frac{2}{\pi} \int_0^{\pi} \left(u_0 - \frac{u_0}{\pi}\theta \right) \sin n\theta\, d\theta = \frac{2u_0}{n\pi}.$$

The final solution is therefore

$$u(r, \theta) = \frac{u_0}{\pi}\theta + \sum_{n=1}^{\infty}\left(\frac{2u_0}{n\pi}\right)r^n \sin n\theta.$$

9. Using $u = R(r)\Theta(\theta)$, the boundary-value problem

$$\frac{\partial^2 u}{\partial r^2} + \frac{1}{r}\frac{\partial u}{\partial r} + \frac{1}{r^2}\frac{\partial^2 u}{\partial \theta^2} = 0, \quad 0 < r < c, \quad 0 < \theta < \beta$$

$$u(r, 0) = 0, \quad u(r, \beta) = 0, \quad 0 < r < c$$

$$u(c, \theta) = f(\theta), \quad 0 < \theta < \beta$$

yields $\Theta(\theta) = c_1 \cos \alpha\theta + c_2 \sin \alpha\theta$. The condition $\Theta(0) = 0$ gives $c_1 = 0$. So $\Theta(\theta) = c_2 \sin \alpha\theta$. Then,

$$\Theta(\beta) = c_2 \sin \alpha\beta = 0$$

$$\alpha = \frac{n\pi}{\beta}.$$

Now $R(r) = c_3 r^{n\pi/\beta} + c_4 r^{-n\pi/\beta}$ and boundedness at $r = 0$ gives $c_4 = 0$. So, $R(r) = c_3 r^{n\pi/\beta}$ and

$$u(r, \theta) = \sum_{n=1}^{\infty} A_n r^{n\pi/\beta} \sin \frac{n\pi}{\beta}\theta.$$

For $r = c$ we have

$$f(\theta) = \sum_{n=1}^{\infty} A_n c^{n\pi/\beta} \sin \frac{n\pi}{\beta}\theta$$

where

$$A_n = \frac{2}{\beta c^{n\pi/\beta}} \int_0^\beta f(\theta) \sin \frac{n\pi}{\beta}\theta\, d\theta.$$

12. Taking the hint and substituting $u(r, \theta) = v(r, \theta) + \psi(r)$ into the PDE leads to $r^2\psi''(r) + r\psi'(r) = 0$ whose general solution is $\psi(r) = c_1 + c_2 \ln r$. Now from the given conditions we have

$$u(a, \theta) = v(a, \theta) + \psi(a) = u_0$$

$$u(b, \theta) = v(b, \theta) + \psi(b) = u_1.$$

In order for $v(a, \theta) = v(b, \theta) = 0$, we must take $\psi(a) = u_0$ and $\psi(b) = u_1$. Therefore we have

$$\begin{cases} \psi(a) = c_1 + c_2 \ln a = u_0 \\ \psi(b) = c_1 + c_2 \ln b = u_1. \end{cases}$$

So

$$c_1 = \frac{u_1 \ln a - u_0 \ln b}{\ln(a/b)} \quad \text{and} \quad c_2 = \frac{u_0 - u_1}{\ln(a/b)}.$$

Therefore

$$\psi(r) = c_1 + c_2 \ln r = \frac{u_1 \ln a - u_0 \ln b}{\ln (a/b)} + \frac{u_0 - u_1}{\ln (a/b)} \ln r$$

$$= \frac{u_1 \ln a - u_0 \ln b + u_0 \ln r - u_1 \ln r}{\ln (a/b)} = \frac{u_0 \ln r/b - u_1 \ln r/a}{\ln (a/b)}.$$

From the results of Problem 11, we know that $v(r, \theta) = 0$, and so the steady-state temperature is

$$u(r, \theta) = v(r, \theta) + \psi(r) = 0 + \frac{u_1 \ln a - u_0 \ln b + u_0 \ln r - u_1 \ln r}{\ln (a/b)} = \frac{u_0 \ln r/b - u_1 \ln r/a}{\ln (a/b)}.$$

15. Proceeding in the usual way by letting $u(r, \theta) = R(r)\Theta(\theta)$ and forcing it into the PDE we get

$$r^2 R'' + rR' - \lambda R = 0 \quad \text{and} \quad \Theta'' + \lambda\Theta = 0.$$

Using the results from Section 12.5, the eigenvalues and eigenfunctions for the boundary-value problem $\Theta'' + \lambda\Theta = 0$, $\Theta(0) = \Theta(\pi) = 0$ are $\lambda_n = n^2$ for $n = 1, 2, 3, \ldots$ and $\Theta = c_2 \sin (n\theta)$. When $n = 1, 2, 3, \ldots$, the equation $r^2 R'' + rR' - \lambda R = 0$ has solution $R(r) = c_3 r^n + c_4 r^{-n}$ so that the boundary condition $R(b) = 0$ gives us $c_4 = -c_3 b^{2n}$ therefore

$$R(r) = c_3 r^n + c_4 r^{-n} = R(r) = c_3 r^n - c_3 b^{2n} r^{-n} = c_3 \left(\frac{r^{2n} - b^{2n}}{r^n} \right).$$

We now have

$$u(r, \theta) = \sum_{n=1}^{\infty} A_n \left(\frac{r^{2n} - b^{2n}}{r^n} \right) \sin n\theta.$$

From the given condition at $r = a$, we obtain the coefficients

$$A_n = \frac{a^n}{a^{2n} - b^{2n}} \cdot \frac{2}{\pi} \int_0^\pi \theta(\pi - \theta) \sin n\theta \, d\theta = \frac{a^n}{a^{2n} - b^{2n}} \cdot \frac{4[1 - (-1)^n]}{n^3 \pi}.$$

The final solution is therefore

$$u(r, \theta) = \sum_{n=1}^{\infty} \left(\frac{a^n}{a^{2n} - b^{2n}} \cdot \frac{4[1 - (-1)^n]}{n^3 \pi} \right) \left(\frac{r^{2n} - b^{2n}}{r^n} \right) \sin n\theta.$$

18. Proceeding in the usual way by letting $u(r, \theta) = R(r)\Theta(\theta)$ and forcing it into the PDE we get

$$r^2 R'' + rR' - \lambda R = 0 \quad \text{and} \quad \Theta'' + \lambda\Theta = 0.$$

Using the results from Section 12.5, the eigenvalues and eigenfunctions for the boundary-value problem $\Theta'' + \lambda\Theta = 0$, $\Theta(0) = \Theta(\pi/4) = 0$ are $\lambda_n = (4n)^2$ for $n = 1, 2, 3, \ldots$ and $\Theta = c_2 \sin (4n\theta)$. When $n = 1, 2, 3, \ldots$, the equation $r^2 R'' + rR' - \lambda R = 0$ has solution

$R(r) = c_3 r^{4n} + c_4 r^{-4n}$ so that the boundary condition $R(a) = 0$ gives us $c_4 = -c_3 \left(a^{4n} / a^{-4n} \right)$ therefore

$$R(r) = c_3 r^n + c_4 r^{-n} = R(r) = c_3 \left(\frac{(r/a)^{4n} - (a/r)^{4n}}{a^{4n}} \right).$$

We now have

$$u(r, \theta) = \sum_{n=1}^{\infty} A_n \left(\frac{(r/a)^{4n} - (a/r)^{4n}}{a^{4n}} \right) \sin 4n\theta.$$

From the given condition at $r = b$, we obtain the coefficients

$$A_n = \frac{a^{4n}}{(b/a)^{4n} - (a/b)^{4n}} \cdot \int_0^{\pi/4} 100 \sin 4n\theta \, d\theta = \frac{a^{4n}}{(b/a)^{4n} - (a/b)^{4n}} \cdot \frac{200 \left[1 - (-1)^n \right]}{n\pi}.$$

The final solution is therefore

$$u(r, \theta) = \sum_{n=1}^{\infty} \frac{200 \left[1 - (-1)^n \right]}{n\pi} \left(\frac{\left(\frac{r}{a} \right)^{4n} - \left(\frac{a}{r} \right)^{4n}}{\left(\frac{b}{a} \right)^{4n} - \left(\frac{a}{b} \right)^{4n}} \right) \sin 4n\theta.$$

14.2 Cylindrical Coordinates

3. Referring to Example 2 in the text we have

$$R(r) = c_1 J_0(\alpha r) + c_2 Y_0(\alpha r)$$
$$Z(z) = c_3 \cosh \alpha z + c_4 \sinh \alpha z$$

where $c_2 = 0$ and $J_0(2\alpha) = 0$ defines the positive eigenvalues $\lambda_n = \alpha_n^2$. From $Z(4) = 0$ we obtain

$$c_3 \cosh 4\alpha_n + c_4 \sinh 4\alpha_n = 0 \qquad \text{or} \qquad c_4 = -c_3 \frac{\cosh 4\alpha_n}{\sinh 4\alpha_n}.$$

Then

$$Z(z) = c_3 \left[\cosh \alpha_n z - \frac{\cosh 4\alpha_n}{\sinh 4\alpha_n} \sinh \alpha_n z \right] = c_3 \frac{\sinh 4\alpha_n \cosh \alpha_n z - \cosh 4\alpha_n \sinh \alpha_n z}{\sinh 4\alpha_n}$$

$$= c_3 \frac{\sinh \alpha_n (4 - z)}{\sinh 4\alpha_n}$$

and

$$u(r, z) = \sum_{n=1}^{\infty} A_n \frac{\sinh \alpha_n (4 - z)}{\sinh 4\alpha_n} J_0(\alpha_n r).$$

From

$$u(r, 0) = u_0 = \sum_{n=1}^{\infty} A_n J_0(\alpha_n r)$$

we obtain

$$A_n = \frac{2u_0}{4 J_1^2(2\alpha_n)} \int_0^2 r J_0(\alpha_n r) \, dr = \frac{u_0}{\alpha_n J_1(2\alpha_n)}.$$

Thus the temperature in the cylinder is

$$u(r, z) = u_0 \sum_{n=1}^{\infty} \frac{\sinh \alpha_n (4 - z) J_0(\alpha_n r)}{\alpha_n \sinh 4\alpha_n J_1(2\alpha_n)}.$$

6. Using λ as the separation constant the separated equations are

$$rR'' + R' - r\lambda R = 0 \quad \text{and} \quad Z'' + \lambda Z = 0.$$

The boundary conditions are $Z'(0) = 0$ and $Z(1) = 0$.

If $\lambda = 0$ the solutions of the ordinary differential equations are

$$R = c_1 + c_2 \ln r \quad \text{and} \quad Z = c_3 + c_4 z.$$

Since $Z'(0) = 0$, $c_4 = 0$. Therefore $Z = c_3$ and $Z(1) = 0$ so $Z(1) = c_3 = 0$. The product solution is $u = R(r)Z(z) = 0$. Thus $\lambda = 0$ is not an eigenvalue.

If $\lambda = -\alpha^2 < 0$, the solutions of the ordinary differential equations are

$$R = c_1 J_0(\alpha r) + c_2 Y_0(\alpha r) \quad \text{and} \quad Z = c_3 \cosh \alpha z + c_4 \sinh \alpha z.$$

Since $Z'(0) = 0$, $c_4 = 0$. Therefore $Z = c_3 \cosh \alpha z$ and $Z(1) = 0$ so $c_4 \cosh \alpha = 0$, which implies $c_4 = 0$. Thus $Z = 0$ and therefore $u = 0$.

If $\lambda = \alpha^2 > 0$, the solutions of the ordinary differential equations are

$$R = c_1 I_0(\alpha r) + c_2 K_0(\alpha r) \quad \text{and} \quad Z = c_3 \cos \alpha z + c_4 \sin \alpha z.$$

Since $Z'(0) = 0$, $c_4 = 0$ and so $Z = c_3 \cos \alpha z$. Now $Z(1) = 0$ so $c_4 \cos \alpha = 0$ and $\alpha = (2n-1)\pi/2$, $n = 1, 2, 3, \ldots$. The eigenvalues are $\lambda_n = (2n - 1)^2 \pi^2 / 4$ and corresponding eigenfunctions are $Z = c_3 \cos (2n - 1)\pi z/2$. Now, the usual implicit requirement that u be bounded at $r = 0$ implies $c_2 = 0$. (See Figure 5.3.4 in the text.) Therefore $R = c_1 I_0(\alpha r)$ or $R = c_1 I_0 ((2n - 1)\pi r/2)$. The superposition principle then yields

$$u(r, z) = \sum_{n=1}^{\infty} A_n I_0 \left(\frac{2n - 1}{2} \pi r \right) \cos \frac{2n - 1}{2} \pi z.$$

At $r = 1$

$$u(1, z) = z = \sum_{n=1}^{\infty} A_n I_0 \left(\frac{2n - 1}{2} \pi \right) \cos \frac{2n - 1}{2} \pi z,$$

which is not a Fourier series. Thus

$$A_n I_0 \left(\frac{2n - 1}{2} \pi \right) = \frac{\int_0^1 z \cos \frac{2n - 1}{2} \pi z \, dz}{\int_0^1 \cos^2 \frac{2n - 1}{2} \pi z \, dz} = \frac{-8 - 4(2n - 1)\pi(-1)^n}{(2n - 1)^2 \pi^2}.$$

Therefore

$$u(r,z) = 4 \sum_{n=1}^{\infty} \frac{(2n-1)\pi(-1)^{n+1} - 2}{(2n-1)^2\pi^2 I_0\left(\frac{2n-1}{2}\pi\right)} I_0\left(\frac{2n-1}{2}\pi r\right) \cos\frac{2n-1}{2}\pi z.$$

9. Letting $u(r,t) = R(r)T(t)$ and separating variables we obtain

$$\frac{R'' + \frac{1}{r}R'}{R} = \frac{T'}{kT} = -\lambda \qquad \text{and} \qquad R'' + \frac{1}{r}R' + \lambda R = 0, \quad T' + \lambda kT = 0.$$

From the last equation we find $T(t) = e^{-\lambda kt}$. If $\lambda < 0$, $T(t)$ increases without bound as $t \to \infty$. Thus we assume $\lambda = \alpha^2 > 0$. Now

$$R'' + \frac{1}{r}R' + \alpha^2 R = 0$$

is a parametric Bessel equation with solution

$$R(r) = c_1 J_0(\alpha r) + c_2 Y_0(\alpha r).$$

Since Y_0 is unbounded as $r \to 0$ we take $c_2 = 0$. Then $R(r) = c_1 J_0(\alpha r)$ and the boundary condition $u(c,t) = R(c)T(t) = 0$ implies $J_0(\alpha c) = 0$. This latter equation defines the positive eigenvalues $\lambda_n = \alpha_n^2$. Thus

$$u(r,t) = \sum_{n=1}^{\infty} A_n J_0(\alpha_n r)e^{-\alpha_n^2 kt}.$$

From

$$u(r,0) = f(r) = \sum_{n=1}^{\infty} A_n J_0(\alpha_n r)$$

we find

$$A_n = \frac{2}{c^2 J_1^2(\alpha_n c)} \int_0^c r J_0(\alpha_n r) f(r)\, dr, \ n = 1, 2, 3, \ldots.$$

12. We solve

$$\frac{\partial^2 u}{\partial r^2} + \frac{1}{r}\frac{\partial u}{\partial r} + \frac{\partial^2 u}{\partial z^2} = 0, \quad 0 < r < 1, \quad z > 0$$

$$\left.\frac{\partial u}{\partial r}\right|_{r=1} = -hu(1,z), \quad z > 0$$

$$u(r,0) = u_0, \quad 0 < r < 1.$$

Assuming $u = RZ$ we get

$$\frac{R'' + \frac{1}{r}R'}{R} = -\frac{Z''}{Z} = -\lambda$$

and so

$$rR'' + R' + \lambda^2 rR = 0 \qquad \text{and} \qquad Z'' - \lambda Z = 0.$$

Letting $\lambda = \alpha^2$ we then have

$$R(r) = c_1 J_0(\alpha r) + c_2 Y_0(\alpha r) \qquad \text{and} \qquad Z(z) = c_3 e^{-\alpha z} + c_4 e^{\alpha z}.$$

We use the exponential form of the solution of $Z'' - \alpha^2 Z = 0$ since the domain of the variable z is a semi-infinite interval. As usual we define $c_2 = 0$ since the temperature is surely bounded as $r \to 0$. Hence $R(r) = c_1 J_0(\alpha r)$. Now the boundary condition $u_r(1, z) + hu(1, z) = 0$ is equivalent to

$$\alpha J_0'(\alpha) + h J_0(\alpha) = 0. \tag{4}$$

The eigenvalues α_n are the positive roots of (**4**) above. Finally, we must now define $c_4 = 0$ since the temperature is also expected to be bounded as $z \to \infty$. A product solution of the partial differential equation that satisfies the first boundary condition is given by

$$u_n(r, z) = A_n e^{-\alpha_n z} J_0(\alpha_n r).$$

Therefore

$$u(r, z) = \sum_{n=1}^{\infty} A_n e^{-\alpha_n z} J_0(\alpha_n r)$$

is another formal solution. At $z = 0$ we have $u_0 = A_n J_0(\alpha_n r)$. In view of (**4**) above we use equations (17) and (18) of Section 12.6 in the text with the identification $b = 1$:

$$A_n = \frac{2\alpha_n^2}{(\alpha_n^2 + h^2) J_0^2(\alpha_n)} \int_0^1 r J_0(\alpha_n r) u_0 \, dr$$

$$= \frac{2\alpha_n^2 u_0}{(\alpha_n^2 + h^2) J_0^2(\alpha_n)\alpha_n^2} \, t J_1(t) \Big|_0^{\alpha_n} = \frac{2\alpha_n u_0 J_1(\alpha_n)}{(\alpha_n^2 + h^2) J_0^2(\alpha_n)}. \tag{5}$$

Since $J_0' = -J_1$ [see equation (6) of Section 12.6 in the text] it follows from (**4**) above that $\alpha_n J_1(\alpha_n) = h J_0(\alpha_n)$. Thus (**5**) from this section of the manual simplifies to

$$A_n = \frac{2u_0 h}{(\alpha_n^2 + h^2) J_0(\alpha_n)}.$$

A solution to the boundary-value problem is then

$$u(r, z) = 2u_0 h \sum_{n=1}^{\infty} \frac{e^{-\alpha_n z}}{(\alpha_n^2 + h^2) J_0(\alpha_n)} J_0(\alpha_n r).$$

15. (a) Writing the partial differential equation in the form

$$g\left(x \frac{\partial^2 u}{\partial x^2} + \frac{\partial u}{\partial x}\right) = \frac{\partial^2 u}{\partial t^2}$$

and separating variables we obtain

$$\frac{xX'' + X'}{X} = \frac{T''}{gT} = -\lambda.$$

Letting $\lambda = \alpha^2$ we obtain

$$xX'' + X' + \alpha^2 X = 0 \qquad \text{and} \qquad T'' + g\alpha^2 T = 0.$$

Letting $x = \tau^2/4$ in the first equation we obtain $dx/d\tau = \tau/2$ or $d\tau/dx = 2\tau$. Then

$$\frac{dX}{dx} = \frac{dX}{d\tau}\frac{d\tau}{dx} = \frac{2}{\tau}\frac{dX}{d\tau}$$

and

$$\frac{d^2X}{dx^2} = \frac{d}{dx}\left(\frac{2}{\tau}\frac{dX}{d\tau}\right) = \frac{2}{\tau}\frac{d}{dx}\left(\frac{dX}{d\tau}\right) + \frac{dX}{d\tau}\frac{d}{dx}\left(\frac{2}{\tau}\right)$$

$$= \frac{2}{\tau}\frac{d}{d\tau}\left(\frac{dX}{d\tau}\right)\frac{d\tau}{dx} + \frac{dX}{d\tau}\frac{d}{d\tau}\left(\frac{2}{\tau}\right)\frac{d\tau}{dx} = \frac{4}{\tau^2}\frac{d^2X}{d\tau^2} - \frac{4}{\tau^3}\frac{dX}{d\tau}.$$

Thus

$$xX'' + X' + \alpha^2 X = \frac{\tau^2}{4}\left(\frac{4}{\tau^2}\frac{d^2X}{d\tau^2} - \frac{4}{\tau^3}\frac{dX}{d\tau}\right) + \frac{2}{\tau}\frac{dX}{d\tau} + \alpha^2 X = \frac{d^2X}{d\tau^2} + \frac{1}{\tau}\frac{dX}{d\tau} + \alpha^2 X = 0.$$

This is a parametric Bessel equation with solution

$$X(\tau) = c_1 J_0(\alpha\tau) + c_2 Y_0(\alpha\tau).$$

(b) To ensure a finite solution at $x = 0$ (and thus $\tau = 0$) we set $c_2 = 0$. The condition $u(L, t) = X(L)T(t) = 0$ implies $X\big|_{x=L} = X\big|_{\tau=2\sqrt{L}} = c_1 J_0(2\alpha\sqrt{L}) = 0$, which defines positive eigenvalues $\lambda_n = \alpha_n^2$. The solution of $T'' + g\alpha^2 T = 0$ is

$$T(t) = c_3 \cos\left(\alpha_n \sqrt{g}\,t\right) + c_4 \sin\left(\alpha_n \sqrt{g}\,t\right).$$

The boundary condition $u_t(x, 0) = X(x)T'(0) = 0$ implies $c_4 = 0$. Thus

$$u(\tau, t) = \sum_{n=1}^{\infty} A_n \cos\left(\alpha_n \sqrt{g}\,t\right) J_0(\alpha_n \tau).$$

From

$$u(\tau, 0) = f(\tau^2/4) = \sum_{n=1}^{\infty} A_n J_0(\alpha_n \tau)$$

we find

$$A_n = \frac{2}{(2\sqrt{L})^2 J_1^2(2\alpha_n \sqrt{L})} \int_0^{2\sqrt{L}} \tau J_0(\alpha_n \tau) f(\tau^2/4)\, d\tau \qquad \boxed{v = \tau/2, \ dv = d\tau/2}$$

$$= \frac{1}{2L J_1^2(2\alpha_n \sqrt{L})} \int_0^{\sqrt{L}} 2v J_0(2\alpha_n v) f(v^2) 2\, dv$$

$$= \frac{2}{L J_1^2(2\alpha_n \sqrt{L})} \int_0^{\sqrt{L}} v J_0(2\alpha_n v) f(v^2)\, dv.$$

The solution of the boundary-value problem is

$$u(x,t) = \sum_{n=1}^{\infty} A_n \cos\left(\alpha_n \sqrt{g}\, t\right) J_0(2\alpha_n \sqrt{x}\,).$$

18. (a) First we see that

$$\frac{R''\Theta + \dfrac{1}{r} R'\Theta + \dfrac{1}{r^2} R\Theta''}{R\Theta} = \frac{T''}{a^2 T} = -\lambda.$$

This gives $T'' + a^2\lambda T = 0$ and from

$$\frac{R'' + \dfrac{1}{r} R' + \lambda R}{-R/r^2} = \frac{\Theta''}{\Theta} = -\nu$$

we get $\Theta'' + \nu\Theta = 0$ and $r^2 R'' + r R' + (\lambda r^2 - \nu)R = 0$.

(b) With $\lambda = \alpha^2$ and $\nu = \beta^2$ the general solutions of the differential equations in part **(a)** are

$$T = c_1 \cos a\alpha t + c_2 \sin a\alpha t$$
$$\Theta = c_3 \cos \beta\theta + c_4 \cos \beta\theta$$
$$R = c_5 J_\beta(\alpha r) + c_6 Y_\beta(\alpha r).$$

(c) Implicitly we expect $u(r,\theta,t) = u(r,\theta + 2\pi, t)$ and so Θ must be 2π-periodic. Therefore $\beta = n$, $n = 0, 1, 2, \ldots$. The corresponding eigenfunctions are 1, $\cos\theta$, $\cos 2\theta$, $\ldots$, $\sin\theta$, $\sin 2\theta$, $\ldots$. Arguing that $u(r,\theta,t)$ is bounded as $r \to 0$ we then define $c_6 = 0$ and so $R = c_3 J_n(\alpha r)$. But $R(c) = 0$ gives $J_n(\alpha c) = 0$; this equation defines the eigenvalues $\lambda_n = \alpha_n^2$. For each n, $\alpha_{ni} = x_{ni}/c$, $i = 1, 2, 3, \ldots$. The corresponding eigenfunctions are $J_n(\lambda_{ni} r) = 0$.

(d) $u(r,\theta,t) = \sum_{i=1}^{n} (A_{0i} \cos a\alpha_{0i} t + B_{0i} \sin a\alpha_{0i} t) J_0(\alpha_{0i} r)$

$$+ \sum_{n=1}^{\infty} \sum_{i=1}^{\infty} \Big[(A_{ni} \cos a\alpha_{ni} t + B_{ni} \sin a\alpha_{ni} t) \cos n\theta$$

$$+ (C_{ni} \cos a\alpha_{ni} t + D_{ni} \sin a\alpha_{ni} t) \sin n\theta \Big] J_n(\alpha_{ni} r)$$

14.3 | Spherical Coordinates

3. The coefficients are given by

$$A_n = \frac{2n+1}{2c^n} \int_0^\pi \cos\theta\, P_n(\cos\theta) \sin\theta\, d\theta$$

$$= \frac{2n+1}{2c^n} \int_0^\pi P_1(\cos\theta) P_n(\cos\theta) \sin\theta\, d\theta \qquad \boxed{x = \cos\theta, \ dx = -\sin\theta\, d\theta}$$

$$= \frac{2n+1}{2c^n} \int_{-1}^1 P_1(x) P_n(x)\, dx$$

Since $P_n(x)$ and $P_m(x)$ are orthogonal for $m \neq n$, $A_n = 0$ for $n \neq 1$ and

$$A_1 = \frac{2(1)+1}{2c^1} \int_{-1}^1 P_1(x) P_1(x)\, dx = \frac{3}{2c} \int_{-1}^1 x^2\, dx = \frac{1}{c}.$$

Thus

$$u(r,\theta) = \frac{r}{c} P_1(\cos\theta) = \frac{r}{c} \cos\theta.$$

6. Referring to Example 1 in the text we have

$$R(r) = c_1 r^n \qquad \text{and} \qquad \Theta(\theta) = P_n(\cos\theta).$$

Now $\Theta(\pi/2) = 0$ implies that n is odd, so

$$u(r,\theta) = \sum_{n=0}^\infty A_{2n+1} r^{2n+1} P_{2n+1}(\cos\theta).$$

From

$$u(c,\theta) = f(\theta) = \sum_{n=0}^\infty A_{2n+1} c^{2n+1} P_{2n+1}(\cos\theta)$$

we see that

$$A_{2n+1} c^{2n+1} = (4n+3) \int_0^{\pi/2} f(\theta) \sin\theta\, P_{2n+1}(\cos\theta)\, d\theta.$$

Thus

$$u(r,\theta) = \sum_{n=0}^\infty A_{2n+1} r^{2n+1} P_{2n+1}(\cos\theta)$$

where

$$A_{2n+1} = \frac{4n+3}{c^{2n+1}} \int_0^{\pi/2} f(\theta) \sin\theta\, P_{2n+1}(\cos\theta)\, d\theta.$$

9. Checking the hint, we find

$$\frac{1}{r}\frac{\partial^2}{\partial r^2}(ru) = \frac{1}{r}\frac{\partial}{\partial r}\left[r\frac{\partial u}{\partial r} + u\right] = \frac{1}{r}\left[r\frac{\partial^2 u}{\partial r^2} + \frac{\partial u}{\partial r} + \frac{\partial u}{\partial r}\right] = \frac{\partial^2 u}{\partial r^2} + \frac{2}{r}\frac{\partial u}{\partial r}.$$

The partial differential equation then becomes

$$\frac{\partial^2}{\partial r^2}(ru) = r\frac{\partial u}{\partial t}.$$

Now, letting $ru(r,t) = v(r,t) + \psi(r)$, since the boundary condition is nonhomogeneous, we obtain

$$\frac{\partial^2}{\partial r^2}[v(r,t) + \psi(r)] = r\frac{\partial}{\partial t}\left[\frac{1}{r}v(r,t) + \psi(r)\right]$$

or

$$\frac{\partial^2 v}{\partial r^2} + \psi''(r) = \frac{\partial v}{\partial t}.$$

This differential equation will be homogeneous if $\psi''(r) = 0$ or $\psi(r) = c_1 r + c_2$. Now

$$u(r,t) = \frac{1}{r}v(r,t) + \frac{1}{r}\psi(r) \qquad \text{and} \qquad \frac{1}{r}\psi(r) = c_1 + \frac{c_2}{r}.$$

Since we want $u(r,t)$ to be bounded as r approaches 0, we require $c_2 = 0$. Then $\psi(r) = c_1 r$. When $r = 1$

$$u(1,t) = v(1,t) + \psi(1) = v(1,t) + c_1 = 100,$$

and we will have the homogeneous boundary condition $v(1,t) = 0$ when $c_1 = 100$. Consequently, $\psi(r) = 100r$. The initial condition

$$u(r,0) = \frac{1}{r}v(r,0) + \frac{1}{r}\psi(r) = \frac{1}{r}v(r,0) + 100 = 0$$

implies $v(r,0) = -100r$. We are thus led to solve the new boundary-value problem

$$\frac{\partial^2 v}{\partial r^2} = \frac{\partial v}{\partial t}, \quad 0 < r < 1, \quad t > 0,$$

$$v(1,t) = 0, \quad \lim_{r \to 0}\frac{1}{r}v(r,t) < \infty,$$

$$v(r,0) = -100r.$$

Letting $v(r,t) = R(r)T(t)$ and using the separation constant $-\lambda$ we obtain

$$R'' + \lambda R = 0 \qquad \text{and} \qquad T' + \lambda T = 0.$$

Using $\lambda = \alpha^2 > 0$ we then have

$$R(r) = c_3\cos\alpha r + c_4\sin\alpha r \qquad \text{and} \qquad T(t) = c_5 e^{-\alpha^2 t}.$$

The boundary conditions are equivalent to $R(1) = 0$ and $\lim_{r \to 0} R(r)/r < \infty$. Since

$$\lim_{r \to 0} \frac{\cos \alpha r}{r}$$

does not exist we must have $c_3 = 0$. Then $R(r) = c_4 \sin \alpha r$, and $R(1) = 0$ implies $\alpha = n\pi$ for $n = 1, 2, 3, \ldots$. Thus

$$v_n(r, t) = A_n e^{-n^2 \pi^2 t} \sin n\pi r$$

for $n = 1, 2, 3, \ldots$. Using the condition $\lim_{r \to 0} R(r)/r < \infty$ it is easily shown that there are no eigenvalues for $\lambda = 0$, nor does setting the common constant to $-\lambda = \alpha^2$ when separating variables lead to any solutions. Now, by the superposition principle,

$$v(r, t) = \sum_{n=1}^{\infty} A_n e^{-n^2 \pi^2 t} \sin n\pi r.$$

The initial condition $v(r, 0) = -100r$ implies

$$-100r = \sum_{n=1}^{\infty} A_n \sin n\pi r.$$

This is a Fourier sine series and so

$$A_n = 2 \int_0^1 (-100r \sin n\pi r)\, dr = -200 \left[-\frac{r}{n\pi} \cos n\pi r \Big|_0^1 + \int_0^1 \frac{1}{n\pi} \cos n\pi r\, dr \right]$$

$$= -200 \left[-\frac{\cos n\pi}{n\pi} + \frac{1}{n^2 \pi^2} \sin n\pi r \Big|_0^1 \right] = -200 \left[-\frac{(-1)^n}{n\pi} \right] = \frac{(-1)^n 200}{n\pi}.$$

A solution of the problem is thus

$$u(r, t) = \frac{1}{r} v(r, t) + \frac{1}{r} \psi(r) = \frac{1}{r} \sum_{n=1}^{\infty} (-1)^n \frac{20}{n\pi} e^{-n^2 \pi^2 t} \sin n\pi r + \frac{1}{r}(100r)$$

$$= \frac{200}{\pi r} \sum_{n=1}^{\infty} \frac{(-1)^n}{n} e^{-n^2 \pi^2 t} \sin n\pi r + 100.$$

12. Proceeding as in Example 1 we obtain

$$\Theta(\theta) = P_n(\cos \theta) \quad \text{and} \quad R(r) = c_1 r^n + c_2 r^{-(n+1)}$$

so that

$$u(r, \theta) = \sum_{n=0}^{\infty} (A_n r^n + B_n r^{-(n+1)}) P_n(\cos \theta).$$

To satisfy $\lim_{r \to \infty} u(r, \theta) = -Er \cos \theta$ we must have $A_n = 0$ for $n = 2, 3, 4, \ldots$. Then

$$\lim_{r \to \infty} u(r, \theta) = -Er \cos \theta = A_0 \cdot 1 + A_1 r \cos \theta,$$

so $A_0 = 0$ and $A_1 = -E$. Thus

$$u(r, \theta) = -Er \cos \theta + \sum_{n=0}^{\infty} B_n r^{-(n+1)} P_n(\cos \theta).$$

Now

$$u(c, \theta) = 0 = -Ec \cos \theta + \sum_{n=0}^{\infty} B_n c^{-(n+1)} P_n(\cos \theta)$$

so

$$\sum_{n=0}^{\infty} B_n c^{-(n+1)} P_n(\cos \theta) = Ec \cos \theta$$

and

$$B_n c^{-(n+1)} = \frac{2n+1}{2} \int_0^{\pi} Ec \cos \theta \, P_n(\cos \theta) \sin \theta \, d\theta.$$

Now $\cos \theta = P_1(\cos \theta)$ so, for $n \neq 1$,

$$\int_0^{\pi} \cos \theta \, P_n(\cos \theta) \sin \theta \, d\theta = 0$$

by orthogonality. Thus $B_n = 0$ for $n \neq 1$ and

$$B_1 = \frac{3}{2} Ec^3 \int_0^{\pi} \cos^2 \theta \sin \theta \, d\theta = Ec^3.$$

Therefore,

$$u(r, \theta) = -Er \cos \theta + Ec^3 r^{-2} \cos \theta.$$

Chapter 14 in Review

3. The conditions $\Theta(0) = 0$ and $\Theta(\pi) = 0$ applied to $\Theta = c_1 \cos \alpha \theta + c_2 \sin \alpha \theta$ give $c_1 = 0$ and $\alpha = n$, $n = 1, 2, 3, \ldots$, respectively. Thus we have the Fourier sine-series coefficients

$$A_n = \frac{2}{\pi} \int_0^{\pi} u_0(\pi\theta - \theta^2) \sin n\theta \, d\theta = \frac{4u_0}{n^3 \pi}[1 - (-1)^n].$$

Thus

$$u(r, \theta) = \frac{4u_0}{\pi} \sum_{n=1}^{\infty} \frac{1 - (-1)^n}{n^3} r^n \sin n\theta.$$

6. Two of the boundary conditions are $u(r, 0) = 0$ and $u_\theta(r, \pi) = 0$ which imply $\Theta(0) = 0$ and $\Theta'(\pi) = 0$. For $\lambda = [(2n-1)/2]^2$, $n = 1, 2, 3, \ldots$, we have $\Theta(\theta) = c_2 \sin[(2n-1)/2]\theta$. Assuming a bounded solution at $r = 0$ we have $R(r) = c_3 r^n$, so

$$u(r, \theta) = \sum_{n=1}^{\infty} A_n r^n \sin\left(\frac{2n-1}{2}\theta\right) \quad \text{and} \quad f(\theta) = \sum_{n=1}^{\infty} A_n c^n \sin\left(\frac{2n-1}{2}\theta\right).$$

This is not a Fourier series but it is an orthogonal series expansion of f, so

$$A_n = c^{-n} \frac{\displaystyle\int_0^\pi f(\theta) \sin\left(\frac{2n-1}{2}\theta\right) d\theta}{\displaystyle\int_0^\pi \sin^2\left(\frac{2n-1}{2}\theta\right) d\theta}.$$

9. The boundary-value problem in polar coordinates is

$$\frac{\partial^2 u}{\partial r^2} + \frac{1}{r}\frac{\partial u}{\partial r} + \frac{1}{r^2}\frac{\partial^2 u}{\partial \theta^2} = 0, \quad 0 < \theta < 2\pi, \quad 1 < r < 2,$$

$$u(1,\theta) = \sin^2\theta, \quad \left.\frac{\partial u}{\partial r}\right|_{r=2} = 0, \quad 0 < \theta < 2\pi.$$

Letting $u(r,\theta) = R(r)\Theta(\theta)$ and separating variables we obtain

$$r^2 R'' + rR' - \lambda R = 0 \quad \text{and} \quad \Theta'' + \lambda\Theta = 0.$$

For $\lambda = 0$ we have $\Theta'' = 0$ and $r^2 R'' + rR' = 0$. This gives $\Theta = c_1 + c_2\theta$ and $R = c_3 + c_4 \ln r$. The periodicity assumption (as mentioned in Example 1 of Section 14.1 in the text) implies $c_2 = 0$, while the boundary condition $R'(2) = 0$ implies $c_4 = 0$. Thus, for $\lambda = 0$, $u = c_1 c_3 = A_0$. Now, for $\lambda = \alpha^2$, the differential equations become $\Theta'' + \alpha^2\Theta = 0$ and $r^2 R'' + rR' - \alpha^2 R = 0$. The corresponding solutions are $\Theta = c_5 \cos\alpha\theta + c_6 \sin\alpha\theta$ and $R = c_7 r^\alpha + c_8 r^{-\alpha}$. In this case the periodicity assumption implies $\alpha = n$, $n = 1, 2, 3, \ldots$, while the boundary condition $R'(2) = 0$ implies $R = c_7(r^n + 4^n r^{-n})$. The product of the solutions is $u_n = (r^n + 4^n r^{-n})(A_n \cos n\theta + B_n \sin n\theta)$ and the superposition principle implies

$$u(r,\theta) = A_0 + \sum_{n=1}^\infty (r^n + 4^n r^{-n})(A_n \cos n\theta + B_n \sin n\theta).$$

Using the boundary condition at $r = 1$ we have

$$\sin^2\theta = \frac{1}{2}(1 - \cos 2\theta) = A_0 + \sum_{n=1}^\infty (1 + 4^n)(A_n \cos n\theta + B_n \sin n\theta).$$

From this we conclude that $B_n = 0$ for all integers n, $A_0 = \frac{1}{2}$, $A_1 = 0$, $A_2 = -\frac{1}{34}$, and $A_m = 0$ for $m = 3, 4, 5, \ldots$. Therefore

$$u(r,\theta) = A_0 + A_2 \cos 2\theta = \frac{1}{2} - \frac{1}{34}(r^2 + 16r^{-2})\cos 2\theta = \frac{1}{2} - \left(\frac{1}{34}r^2 + \frac{8}{17}r^{-2}\right)\cos 2\theta.$$

12. Letting $\lambda = \alpha^2 > 0$ and proceeding in the usual manner we find

$$u(r,t) = \sum_{n=1}^\infty A_n \cos a\alpha_n t J_0(\alpha_n r)$$

where the eigenvalues $\lambda_n = \alpha_n^2$ are determined by $J_0(\alpha) = 0$. Then the initial condition gives

$$u_0 J_0(x_k r) = \sum_{n=1}^{\infty} A_n J_0(\alpha_n r)$$

and so

$$A_n = \frac{2}{J_1^2(\alpha_n)} \int_0^1 r \left(u_0 J_0(x_k r) \right) J_0(\alpha_n r) \, dr.$$

But $J_0(\alpha) = 0$ implies that the eigenvalues are the positive zeros of J_0, that is, $\alpha_n = x_n$ for $n = 1, 2, 3, \ldots$. Therefore

$$A_n = \frac{2u_0}{J_1^2(\alpha_n)} \int_0^1 r J_0(\alpha_k r) J_0(\alpha_n r) \, dr = 0, \quad n \neq k$$

by orthogonality. For $n = k$,

$$A_k = \frac{2u_0}{J_1^2(\alpha_k)} \int_0^1 r J_0^2(\alpha_k) \, dr = u_0$$

by (9) of Section 12.6. Thus the solution $u(r, t)$ reduces to one term when $n = k$, and

$$u(r, t) = u_0 \cos a\alpha_k t J_0(\alpha_k r) = u_0 \cos a x_k t J_0(x_k r).$$

15. Referring to Example 1 in Section 14.3 of the text we have

$$u(r, \theta) = \sum_{n=0}^{\infty} A_n r^n P_n(\cos \theta).$$

For $x = \cos \theta$

$$u(1, \theta) = \begin{cases} 100, & 0 < \theta < \pi/2 \\ -100, & \pi/2 < \theta < \pi \end{cases} = g(x).$$

From Problem 22 in Exercise 12.6 we have

$$u(r, \theta) = 100 \left[\frac{3}{2} r P_1(\cos \theta) - \frac{7}{8} r^3 P_3(\cos \theta) + \frac{11}{16} r^5 P_5(\cos \theta) + \cdots \right].$$

18. Letting $u(r, t) = R(r)T(t)$ and the separation constant be $-\lambda = -\alpha^2$ we obtain

$$r R'' + R' + \alpha^2 r R = 0$$
$$T' + \alpha^2 T = 0,$$

with solutions

$$R(r) = c_1 J_0(\alpha r) + c_2 Y_0(\alpha r)$$
$$T(t) = c_3 e^{-\alpha^2 t}.$$

Now the boundary conditions imply

$$R(a) = 0 = c_1 J_0(\alpha a) + c_2 Y_0(\alpha a)$$
$$R(b) = 0 = c_1 J_0(\alpha b) + c_2 Y_0(\alpha b)$$

so that

$$c_2 = -\frac{c_1 J_0(\alpha a)}{Y_0(\alpha a)}$$

and

$$c_1 J_0(\alpha b) - \frac{c_1 J_0(\alpha a)}{Y_0(\alpha a)} Y_0(\alpha b) = 0$$

or

$$Y_0(\alpha a) J_0(\alpha b) - J_0(\alpha a) Y_0(\alpha b) = 0.$$

This equation defines α_n for $n = 1, 2, 3, \ldots$. Now

$$R(r) = c_1 J_0(\alpha r) - c_1 \frac{J_0(\alpha a)}{Y_0(\alpha a)} Y_0(\alpha r) = \frac{c_1}{Y_0(\alpha a)} \left[Y_0(\alpha a) J_0(\alpha r) - J_0(\alpha a) Y_0(\alpha r) \right]$$

and

$$u_n(r, t) = A_n \left[Y_0(\alpha_n a) J_0(\alpha_n r) - J_0(\alpha_n a) Y_0(\alpha_n r) \right] e^{-\alpha_n^2 t} = A_n u_n(r) e^{-\alpha_n^2 t}.$$

Thus

$$u(r, t) = \sum_{n=1}^{\infty} A_n u_n(r) e^{-\alpha_n^2 t}.$$

From the initial condition

$$u(r, 0) = f(r) = \sum_{n=1}^{\infty} A_n u_n(r)$$

we obtain

$$A_n = \frac{\int_a^b r f(r) u_n(r)\, dr}{\int_a^b r u_n^2(r)\, dr}.$$

21. Using $-\lambda$ as the separation constant leads to the equations

$$rR'' + R' + \lambda r R = 0 \quad \text{and} \quad Z'' - \lambda Z = 0.$$

With $\lambda = \alpha^2$ we get $R(r) = c_1 J_0(\alpha r) + c_2 Y_0(\alpha r)$ and $Z(z) = c_3 e^{-\alpha z} + c_4 e^{\alpha z}$. Boundedness as $r \to 0$ nd $z \to \infty$ prompts us to take $c_2 = c_4 = 0$. Therefore, using $R(1) = 0$ the eigenvalues are $\lambda_n = \alpha_n^2$ where the α_n are defined by $J_0(\alpha) = 0$. Therefore we have

$$u(r, z) = \sum_{n=1}^{\infty} A_n e^{-\alpha_n z} J_0(\alpha_n r)$$

At $z = 0$ and using the results of Section 12.6 we get

$$u(r,0) = 100 = \sum_{n=1}^{\infty} A_n J_0(\alpha_n r)$$

$$A_n = \frac{200}{J_1^2(\alpha_n)} \int_0^1 r J_0(\alpha_n r)\, dr = \frac{200}{\alpha_n J_1(\alpha_n)}.$$

The final solution is therefore

$$u(r,z) = \sum_{n=1}^{\infty} \frac{200}{\alpha_n J_1(\alpha_n)}\, e^{-\alpha_n z} J_0(\alpha_n r).$$

Chapter 15

Integral Transforms

15.1 | **Error Function**

3. By the first translation theorem,

$$\mathscr{L}\left\{e^{t}\operatorname{erf}(\sqrt{t}\,)\right\} = \mathscr{L}\left\{\operatorname{erf}(\sqrt{t}\,)\right\}\Bigg|_{s\to s-1} = \frac{1}{s\sqrt{s+1}}\Bigg|_{s\to s-1} = \frac{1}{\sqrt{s}\,(s-1)}.$$

6. Using the hints we get

$$\frac{1}{1+\sqrt{s+1}} = \frac{1}{1+\sqrt{s+1}}\cdot\frac{1-\sqrt{s+1}}{1-\sqrt{s+1}} = -\frac{1}{s}+\frac{\sqrt{s+1}}{s} = -\frac{1}{s}+\frac{\sqrt{s+1}}{s}\cdot\frac{\sqrt{s+1}}{\sqrt{s+1}}$$

$$= -\frac{1}{s}+\frac{s+1}{s\sqrt{s+1}} = -\frac{1}{s}+\frac{1}{\sqrt{s+1}}+\frac{1}{s\sqrt{s+1}}.$$

Using the result of Problem 2, together with formula 56 from the table of Laplace transforms, we get

$$\mathscr{L}^{-1}\left\{\frac{1}{1+\sqrt{s+1}}\right\} = -\mathscr{L}^{-1}\left\{\frac{1}{s}\right\} + \mathscr{L}^{-1}\left\{\frac{1}{\sqrt{s}}\Bigg|_{s\to s+1}\right\} + \mathscr{L}^{-1}\left\{\frac{1}{s\sqrt{s+1}}\right\}$$

$$= -1 + e^{-t}\frac{1}{\sqrt{\pi t}} + \operatorname{erf}\left(\sqrt{t}\,\right) = \frac{e^{-t}}{\sqrt{\pi t}} - \left(1 - \operatorname{erf}\left(\sqrt{t}\,\right)\right)$$

$$= \frac{e^{-t}}{\sqrt{\pi t}} - \operatorname{erfc}\left(\sqrt{t}\,\right).$$

9. Taking the Laplace transform of both sides of the equation we obtain

$$\mathscr{L}\{y(t)\} = \mathscr{L}\{1\} - \mathscr{L}\left\{\int_0^t \frac{y(\tau)}{\sqrt{t-\tau}}\, d\tau\right\}$$

$$Y(s) = \frac{1}{s} - Y(s)\frac{\sqrt{\pi}}{\sqrt{s}}$$

$$\frac{\sqrt{s}+\sqrt{\pi}}{\sqrt{s}}\, Y(s) = \frac{1}{s}$$

$$Y(s) = \frac{1}{\sqrt{s}\,(\sqrt{s}+\sqrt{\pi}\,)}.$$

Thus

$$y(t) = \mathscr{L}^{-1}\left\{\frac{1}{\sqrt{s}\,(\sqrt{s}+\sqrt{\pi}\,)}\right\} = e^{\pi t}\operatorname{erfc}(\sqrt{\pi t}\,). \qquad \boxed{\text{By entry 5 in Table 15.1.1}}$$

12. Since $f(x) = e^{-x^2}$ is an even function,

$$\int_{-a}^a e^{-u^2}\, du = 2\int_0^a e^{-u^2}\, du.$$

Therefore,

$$\int_{-a}^a e^{-u^2}\, du = \sqrt{\pi}\,\operatorname{erf}(a).$$

15. $\operatorname{erf}(x) + \operatorname{erfc}(x) = 1$

$$\operatorname{erfc}(x) = 1 - \operatorname{erf}(x)$$

$$\operatorname{erfc}(-x) = 1 - \operatorname{erf}(-x)$$

Since $\operatorname{erf}(x)$ is an odd function $\operatorname{erf}(-x) = -\operatorname{erf}(x)$, so

$$\operatorname{erfc}(-x) = 1 - (-\operatorname{erf}(x)) = 1 + \operatorname{erf}(x).$$

15.2 | Laplace Transform

3. The solution of

$$a^2\frac{d^2U}{dx^2} - s^2 U = 0$$

is in this case

$$U(x,s) = c_1 e^{-(x/a)s} + c_2 e^{(x/a)s}.$$

Since $\lim_{x\to\infty} u(x,t) = 0$ we have $\lim_{x\to\infty} U(x,s) = 0$. Thus $c_2 = 0$ and

$$U(x,s) = c_1 e^{-(x/a)s}.$$

If $\mathscr{L}\{u(0,t)\} = \mathscr{L}\{f(t)\} = F(s)$ then $U(0,s) = F(s)$. From this we have $c_1 = F(s)$ and

$$U(x,s) = F(s)e^{-(x/a)s}.$$

Hence, by the second translation theorem,

$$u(x,t) = f\left(t - \frac{x}{a}\right)\mathscr{U}\left(t - \frac{x}{a}\right).$$

6. Transforming the partial differential equation gives

$$\frac{d^2U}{dx^2} - s^2U = -\frac{\omega}{s^2 + \omega^2}\sin \pi x.$$

Using undetermined coefficients we obtain

$$U(x,s) = c_1 \cosh sx + c_2 \sinh sx + \frac{\omega}{(s^2 + \pi^2)(s^2 + \omega^2)}\sin \pi x.$$

The transformed boundary conditions $U(0,s) = 0$ and $U(1,s) = 0$ give, in turn, $c_1 = 0$ and $c_2 = 0$. Therefore

$$U(x,s) = \frac{\omega}{(s^2 + \pi^2)(s^2 + \omega^2)}\sin \pi x$$

and

$$u(x,t) = \omega \sin \pi x \mathscr{L}^{-1}\left\{\frac{1}{(s^2 + \pi^2)(s^2 + \omega^2)}\right\}$$

$$= \frac{\omega}{\omega^2 - \pi^2}\sin \pi x \mathscr{L}^{-1}\left\{\frac{1}{\pi}\frac{\pi}{s^2 + \pi^2} - \frac{1}{\omega}\frac{\omega}{s^2 + \omega^2}\right\}$$

$$= \frac{\omega}{\pi(\omega^2 - \pi^2)}\sin \pi t \sin \pi x - \frac{1}{\omega^2 - \pi^2}\sin \omega t \sin \pi x.$$

9. Transforming the partial differential equation gives

$$\frac{d^2U}{dx^2} - s^2U = -sxe^{-x}.$$

Using undetermined coefficients we obtain

$$U(x,s) = c_1 e^{-sx} + c_2 e^{sx} - \frac{2s}{(s^2 - 1)^2}e^{-x} + \frac{s}{s^2 - 1}xe^{-x}.$$

The transformed boundary conditions $\lim_{x \to \infty} U(x,s) = 0$ and $U(0,s) = 0$ give, in turn, $c_2 = 0$ and $c_1 = 2s/(s^2 - 1)^2$. Therefore

$$U(x,s) = \frac{2s}{(s^2 - 1)^2}e^{-sx} - \frac{2s}{(s^2 - 1)^2}e^{-x} + \frac{s}{s^2 - 1}xe^{-x}.$$

From entries (13) and (26) in the Table of Laplace transforms we obtain

$$u(x,t) = \mathscr{L}^{-1}\left\{\frac{2s}{(s^2 - 1)^2}e^{-sx} - \frac{2s}{(s^2 - 1)^2}e^{-x} + \frac{s}{s^2 - 1}xe^{-x}\right\}$$

$$= 2(t - x)\sinh(t - x)\mathscr{U}(t - x) - te^{-x}\sinh t + xe^{-x}\cosh t.$$

12. We use

$$U(x, s) = c_1 e^{-\sqrt{s}\,x} + c_2 e^{\sqrt{s}\,x} + \frac{u_1 x}{s}.$$

The condition $\lim\limits_{x\to\infty} u(x, t)/x = u_1$ implies $\lim\limits_{x\to\infty} U(x, s)/x = u_1/s$, so we define $c_2 = 0$. Then

$$U(x, s) = c_1 e^{-\sqrt{s}\,x} + \frac{u_1 x}{s}.$$

From $U(0, s) = u_0/s$ we obtain $c_1 = u_0/s$. Hence

$$U(x, s) = u_0 \frac{e^{-\sqrt{s}\,x}}{s} + \frac{u_1 x}{s}$$

and

$$u(x, t) = u_0 \mathscr{L}^{-1}\left\{\frac{e^{-x\sqrt{s}}}{s}\right\} + u_1 x \mathscr{L}^{-1}\left\{\frac{1}{s}\right\} = u_0 \operatorname{erfc}\left(\frac{x}{2\sqrt{t}}\right) + u_1 x.$$

15. We use

$$U(x, s) = c_1 e^{-\sqrt{s}\,x} + c_2 e^{\sqrt{s}\,x}.$$

The condition $\lim\limits_{x\to\infty} u(x, t) = 0$ implies $\lim\limits_{x\to\infty} U(x, s) = 0$, so we define $c_2 = 0$. Hence

$$U(x, s) = c_1 e^{-\sqrt{s}\,x}.$$

The transform of $u(0, t) = f(t)$ is $U(0, s) = F(s)$. Therefore

$$U(x, s) = F(s) e^{-\sqrt{s}\,x}$$

and

$$u(x, t) = \mathscr{L}^{-1}\left\{F(s) e^{-x\sqrt{s}}\right\} = \frac{x}{2\sqrt{\pi}} \int_0^t \frac{f(t - \tau) e^{-x^2/4\tau}}{\tau^{3/2}}\, d\tau.$$

18. The solution of the transformed equation

$$\frac{d^2 U}{dx^2} - sU = -100$$

by undetermined coefficients is

$$U(x, s) = c_1 e^{\sqrt{s}\,x} + c_2 e^{-\sqrt{s}\,x} + \frac{100}{s}.$$

From the fact that $\lim\limits_{x\to\infty} U(x, s) = 100/s$ we see that $c_1 = 0$. Thus

$$U(x, s) = c_2 e^{-\sqrt{s}\,x} + \frac{100}{s}. \tag{1}$$

Now the transform of the boundary condition at $x = 0$ is

$$U(0, s) = 20\left[\frac{1}{s} - \frac{1}{s}e^{-s}\right].$$

It follows from (**1**) that

$$\frac{20}{s} - \frac{20}{s}e^{-s} = c_2 + \frac{100}{s} \qquad \text{or} \qquad c_2 = -\frac{80}{s} - \frac{20}{s}e^{-s}$$

and so

$$U(x, s) = \left(-\frac{80}{s} - \frac{20}{s}e^{-s}\right)e^{-\sqrt{s}\,x} + \frac{100}{s}$$

$$= \frac{100}{s} - \frac{80}{s}e^{-\sqrt{s}\,x} - \frac{20}{s}e^{-\sqrt{s}\,x}e^{-s}.$$

Thus

$$u(x, t) = 100\mathscr{L}^{-1}\left\{\frac{1}{s}\right\} - 80\mathscr{L}^{-1}\left\{\frac{e^{-\sqrt{s}\,x}}{s}\right\} - 20\mathscr{L}^{-1}\left\{\frac{e^{-\sqrt{s}\,x}}{s}e^{-s}\right\}$$

$$= 100 - 80\,\mathrm{erfc}\left(x/2\sqrt{t}\right) - 20\,\mathrm{erfc}\left(x/2\sqrt{t-1}\right)\mathscr{U}(t-1).$$

21. Transforming the partial differential equation gives

$$\frac{d^2U}{dx^2} = \left(s^2 + 2s + 1\right)U \qquad \text{or} \qquad \frac{d^2U}{dx^2} - (s+1)^2\,U = 0$$

The general solution of the last equation is

$$U(x, s) = c_1 e^{-(s+1)x} + c_2 e^{(s+1)x}.$$

To guarantee that $u(x, t)$ is bounded as $x \to \infty$ we define $c_2 = 0$ and so $U(x, s) = c_1 e^{-(s+1)x}$. Now the transform of the boundary condition $u(0, t) = \sin 3t$ is

$$U(0, s) = \mathscr{L}\left\{\sin 2t\right\} = \frac{3}{s^2 + 9} = c_1$$

so that

$$U(x, s) = \frac{3}{s^2 + 9}e^{-(s+1)x} = e^{-x}\left(e^{-xs}\frac{3}{s^2 + 9}\right)$$

$$u(x, t) = e^{-x}\mathscr{L}^{-1}\left(e^{-xs}\frac{3}{s^2 + 9}\right).$$

By the inverse form of the second translation theorem we then have,

$$u(x, t) = e^{-x}\sin 3\,(t - x)\mathscr{U}(t - x).$$

24. The transform of the partial differential equation is

$$k\frac{d^2U}{dx^2} - hU + h\frac{u_m}{s} = sU - u_0$$

or

$$k\frac{d^2U}{dx^2} - (h+s)U = -h\frac{u_m}{s} - u_0.$$

By undetermined coefficients we find

$$U(x,s) = c_1 e^{\sqrt{(h+s)/k}\,x} + c_2 e^{-\sqrt{(h+s)/k}\,x} + \frac{hu_m + u_0 s}{s(s+h)}.$$

The transformed boundary conditions are $U'(0,s) = 0$ and $U'(L,s) = 0$. These conditions imply $c_1 = 0$ and $c_2 = 0$. By partial fractions we then get

$$U(x,s) = \frac{hu_m + u_0 s}{s(s+h)} = \frac{u_m}{s} - \frac{u_m}{s+h} + \frac{u_0}{s+h}.$$

Therefore,

$$u(x,t) = u_m \mathscr{L}^{-1}\left\{\frac{1}{s}\right\} - u_m \mathscr{L}^{-1}\left\{\frac{1}{s+h}\right\} + u_0 \mathscr{L}^{-1}\left\{\frac{1}{s+h}\right\} = u_m - u_m e^{-ht} + u_0 e^{-ht}.$$

27. We use

$$U(x,s) = c_1 e^{-\sqrt{RCs+RG}\,x} + c_2 e^{\sqrt{RCs+RG}} + \frac{Cu_0}{Cs+G}.$$

The condition $\lim_{x\to\infty} \partial u/\partial x = 0$ implies $\lim_{x\to\infty} dU/dx = 0$, so we define $c_2 = 0$. Applying $U(0,s) = 0$ to

$$U(x,s) = c_1 e^{-\sqrt{RCsRG}\,x} + \frac{Cu_0}{Cs+G}$$

gives $c_1 = -Cu_0/(Cs+G)$. Therefore

$$U(x,s) = -Cu_0 \frac{e^{-\sqrt{RCs+RG}\,x}}{Cs+G} + \frac{Cu_0}{Cs+G}$$

and

$$u(x,t) = u_0 \mathscr{L}^{-1}\left\{\frac{1}{s+G/C}\right\} - u_0 \mathscr{L}^{-1}\left\{\frac{e^{-x\sqrt{RC}\sqrt{s+G/C}}}{s+G/C}\right\}$$

$$= u_0 e^{-Gt/C} - u_0 e^{-Gt/C}\,\mathrm{erfc}\left(\frac{x\sqrt{RC}}{2\sqrt{t}}\right)$$

$$= u_0 e^{-Gt/C}\left[1 - \mathrm{erfc}\left(\frac{x}{2}\sqrt{\frac{RC}{t}}\right)\right]$$

$$= u_0 e^{-Gt/C}\,\mathrm{erf}\left(\frac{x}{2}\sqrt{\frac{RC}{t}}\right).$$

30. The boundary-value problem in Problem 29 in the case when $v_0 = a$ the solution of the transformed equation is

$$U(x, s) = c_1 e^{-(s/a)x} + c_2 e^{(s/a)x} - \frac{F_0}{2as} x e^{-(s/a)x}.$$

The usual analysis then leads to $c_1 = 0$ and $c_2 = 0$. Therefore

$$U(x, s) = -\frac{F_0}{2as} x e^{-(s/a)x}$$

and

$$u(x, t) = -\frac{xF_0}{2a} \mathscr{L}^{-1}\left\{\frac{e^{-(x/a)s}}{s}\right\} = -\frac{xF_0}{2a} \mathscr{U}\left(t - \frac{x}{a}\right).$$

33. Using the Laplace transform with respect to t of the partial differential equation gives

$$\frac{d^2 U}{dr^2} + \frac{2}{r}\frac{dU}{dr} = sU - \overbrace{u(r, 0)}^{0},$$

where $\mathscr{L}\{u(r, t)\} = U(r, s)$. The ordinary differential equation is equivalent to

$$rU'' + 2U' - srU = 0.$$

If we let $v(r, s) = rU(r, s)$ then differentiation with respect to r gives $v'' = rU'' + 2U'$. The transformed ordinary differential equation becomes

$$v'' - sv = 0$$

$$v = c_1 e^{-\sqrt{s}r} + c_2 e^{\sqrt{s}r}$$

$$U(r, s) = c_1 \frac{e^{-\sqrt{s}r}}{r} + c_2 \frac{e^{\sqrt{s}r}}{r}.$$

The usual argument about boundedness as $r \to \infty$ implies $c_2 = 0$. Therefore $U(r, s) = c_1 e^{-\sqrt{s}r}/r$. Now the Laplace transform of $u(1, t) = 100$ is

$$U(1, s) = \frac{100}{s} = c_1 e^{-\sqrt{s}}$$

so

$$c_1 = 100 \frac{e^{\sqrt{s}}}{s}$$

and

$$U(r, s) = \frac{100}{r} \frac{e^{-\sqrt{s}(r-1)}}{s}.$$

Thus

$$u(r, t) = \frac{100}{r} \mathscr{L}^{-1}\left\{\frac{e^{-(r-1)\sqrt{s}}}{s}\right\}.$$

From the third entry in Table 15.1.1 we get

$$u(r, t) = \frac{100}{r} \operatorname{erfc}\left(\frac{r-1}{2\sqrt{t}}\right).$$

36. If $U(r, s) = \mathscr{L}\{u(r,t)\}$, then the Laplace transform of the partial differential equation is

$$\frac{d^2U}{dr^2} + \frac{2}{r}\frac{dU}{dr} = sU - u_0 \qquad \text{or} \qquad r\frac{d^2U}{dr^2} + 2\frac{dU}{dr} - srU = -u_0r.$$

As in Problems 31 and 32, the substitution $v(r,s) = rU(r,s)$ gives

$$v'' - sv = -u_0r$$

and so

$$U(r,s) = c_1\frac{e^{-\sqrt{s}\,r}}{r} + c_2\frac{e^{\sqrt{s}\,r}}{r} + \frac{u_0}{s}.$$

Since $\lim\limits_{r\to\infty} u(r,t) = u_0$ then $\lim\limits_{r\to\infty} U(r,s) = u_0/s$ and so we must take $c_2 = 0$. Therefore

$$U(r,s) = c_1\frac{e^{-\sqrt{s}\,r}}{r} + \frac{u_0}{s}.$$

The transform of the boundary condition at $r = 1$ is

$$\frac{dU}{dr}\bigg|_{r=1} = U(1,s)$$

$$c_1\left[\frac{r\left(-\sqrt{s}\,e^{-\sqrt{s}\,r}\right) - e^{-\sqrt{s}\,r}}{r^2}\right]_{r=1} = \left[c_1\frac{e^{-\sqrt{s}\,r}}{r} + \frac{u_0}{s}\right]_{r=1}$$

$$c_1\left(-\sqrt{s} - 2\right)e^{-\sqrt{s}} = \frac{u_0}{s}$$

$$c_1 = -u_0\frac{e^{\sqrt{s}}}{s\left(\sqrt{s} + 2\right)}$$

$$U(r,s) = \frac{u_0}{s} - \frac{u_0}{r} \cdot \frac{e^{-\sqrt{s}(r-1)}}{s\left(\sqrt{s} + 2\right)}.$$

Then by entry #67 in the table in Appendix C, with $a = r - 1$ and $b = 2$, we have

$$u(r,t) = \mathscr{L}^{-1}\left\{\frac{u_0}{s}\right\} - \frac{u_0}{2r}\mathscr{L}^{-1}\left\{\frac{2e^{-\sqrt{s}(r-1)}}{s\left(\sqrt{s} + 2\right)}\right\}$$

$$= u_0 - \frac{u_0}{2r}\left[\operatorname{erfc}\left(\frac{r-1}{2\sqrt{t}}\right) - e^{2r-2+4t}\operatorname{erfc}\left(2\sqrt{t} + \frac{r-1}{2\sqrt{t}}\right)\right]$$

$$= u_0 - \frac{u_0}{2r}\operatorname{erfc}\left(\frac{r-1}{2\sqrt{t}}\right) + \frac{u_0e^{2r-2+4t}}{2r}\operatorname{erfc}\left(2\sqrt{t} + \frac{r-1}{2\sqrt{t}}\right).$$

15.3 | Fourier Integral

3. From formulas (5) and (6) in the text,

$$A(\alpha) = \int_0^3 x \cos \alpha x \, dx = \left. \frac{x \sin \alpha x}{\alpha} \right|_0^3 - \frac{1}{\alpha} \int_0^3 \sin \alpha x \, dx$$

$$= \frac{3 \sin 3\alpha}{\alpha} + \left. \frac{\cos \alpha x}{\alpha^2} \right|_0^3 = \frac{3\alpha \sin 3\alpha + \cos 3\alpha - 1}{\alpha^2}$$

and

$$B(\alpha) = \int_0^3 x \sin \alpha x \, dx = -\left. \frac{x \cos \alpha x}{\alpha} \right|_0^3 + \frac{1}{\alpha} \int_0^3 \cos \alpha x \, dx$$

$$= -\frac{3 \cos 3\alpha}{\alpha} + \left. \frac{\sin \alpha x}{\alpha^2} \right|_0^3 = \frac{\sin 3\alpha - 3\alpha \cos 3\alpha}{\alpha^2} \, .$$

Hence

$$f(x) = \frac{1}{\pi} \int_0^\infty \frac{(3\alpha \sin 3\alpha + \cos 3\alpha - 1) \cos \alpha x + (\sin 3\alpha - 3\alpha \cos 3\alpha) \sin \alpha x}{\alpha^2} \, d\alpha$$

$$= \frac{1}{\pi} \int_0^\infty \frac{3\alpha(\sin 3\alpha \cos \alpha x - \cos 3\alpha \sin \alpha x) + \cos 3\alpha \cos \alpha x + \sin 3\alpha \sin \alpha x - \cos \alpha x}{\alpha^2} \, d\alpha$$

$$= \frac{1}{\pi} \int_0^\infty \frac{3\alpha \sin \alpha(3 - x) + \cos \alpha(3 - x) - \cos \alpha x}{\alpha^2} \, d\alpha.$$

6. From formulas (5) and (6) in the text,

$$A(\alpha) = \int_{-1}^1 e^x \cos \alpha x \, dx = \frac{e(\cos \alpha + \alpha \sin \alpha) - e^{-1}(\cos \alpha - \alpha \sin \alpha)}{1 + \alpha^2}$$

$$= \frac{2(\sinh 1) \cos \alpha - 2\alpha(\cosh 1) \sin \alpha}{1 + \alpha^2}$$

and

$$B(\alpha) = \int_{-1}^1 e^x \sin \alpha x \, dx = \frac{e(\sin \alpha - \alpha \cos \alpha) - e^{-1}(-\sin \alpha - \alpha \cos \alpha)}{1 + \alpha^2}$$

$$= \frac{2(\cosh 1) \sin \alpha - 2\alpha(\sinh 1) \cos \alpha}{1 + \alpha^2} \, .$$

Hence

$$f(x) = \frac{1}{\pi} \int_0^\infty [A(\alpha) \cos \alpha x + B(\alpha) \sin \alpha x] \, d\alpha.$$

9. The function is even. Thus from formula (9) in the text

$$A(\alpha) = \int_0^\pi x \cos \alpha x \, dx = \left. \frac{x \sin \alpha x}{\alpha} \right|_0^\pi - \frac{1}{\alpha} \int_0^\pi \sin \alpha x \, dx$$

$$= \frac{\pi\alpha \sin \pi\alpha}{\alpha} + \left. \frac{1}{\alpha^2} \cos \alpha x \right|_0^\pi = \frac{\pi\alpha \sin \pi\alpha + \cos \pi\alpha - 1}{\alpha^2}.$$

Hence from formula (8) in the text

$$f(x) = \frac{2}{\pi} \int_0^\infty \frac{(\pi\alpha \sin \pi\alpha + \cos \pi\alpha - 1) \cos \alpha x}{\alpha^2} \, d\alpha.$$

12. The function is odd. Thus from formula (11) in the text

$$B(\alpha) = \int_0^\infty x e^{-x} \sin \alpha x \, dx.$$

Now recall

$$\mathscr{L}\{t \sin kt\} = -\frac{d}{ds} \mathscr{L}\{\sin kt\} = 2ks/(s^2 + k^2)^2.$$

If we set $s = 1$ and $k = \alpha$ we obtain

$$B(\alpha) = \frac{2\alpha}{(1 + \alpha^2)^2}.$$

Hence from formula (10) in the text

$$f(x) = \frac{4}{\pi} \int_0^\infty \frac{\alpha \sin \alpha x}{(1 + \alpha^2)^2} \, d\alpha.$$

15. For the cosine integral,

$$A(\alpha) = \int_0^\infty x e^{-2x} \cos \alpha x \, dx.$$

But we know

$$\mathscr{L}\{t \cos kt\} = -\frac{d}{ds} \frac{s}{(s^2 + k^2)} = \frac{s^2 - k^2}{(s^2 + k^2)^2}.$$

If we set $s = 2$ and $k = \alpha$ we obtain

$$A(\alpha) = \frac{4 - \alpha^2}{(4 + \alpha^2)^2}.$$

Hence

$$f(x) = \frac{2}{\pi} \int_0^\infty \frac{(4 - \alpha^2) \cos \alpha x}{(4 + \alpha^2)^2} \, d\alpha.$$

For the sine integral,

$$B(\alpha) = \int_0^\infty x e^{-2x} \sin \alpha x \, dx.$$

From Problem 12, we know

$$\mathscr{L}\{t \sin kt\} = \frac{2ks}{(s^2 + k^2)^2}.$$

If we set $s = 2$ and $k = \alpha$ we obtain

$$B(\alpha) = \frac{4\alpha}{(4 + \alpha^2)^2}.$$

Hence

$$f(x) = \frac{8}{\pi} \int_0^\infty \frac{\alpha \sin \alpha x}{(4 + \alpha^2)^2} \, d\alpha.$$

18. From the formula for sine integral of $f(x)$ we have

$$f(x) = \frac{2}{\pi} \int_0^\infty \left(\int_0^\infty f(x) \sin \alpha x \, dx \right) \sin \alpha x \, dx$$

$$= \frac{2}{\pi} \left[\int_0^1 1 \cdot \sin \alpha x \, d\alpha + \int_1^\infty 0 \cdot \sin \alpha x \, d\alpha \right]$$

$$= \frac{2}{\pi} \frac{(-\cos \alpha x)}{x} \bigg|_0^1 = \frac{2}{\pi} \frac{1 - \cos x}{x}.$$

15.4   Fourier Transforms

For the boundary-value problems in this section it is sometimes useful to note that the identities

$$e^{i\alpha} = \cos \alpha + i \sin \alpha \qquad and \qquad e^{-i\alpha} = \cos \alpha - i \sin \alpha$$

imply

$$e^{i\alpha} + e^{-i\alpha} = 2 \cos \alpha \qquad and \qquad e^{i\alpha} - e^{-i\alpha} = 2i \sin \alpha.$$

3. Using the Fourier sine transform, the partial differential equation becomes

$$\frac{dU}{dt} + k\alpha^2 U = k\alpha u_0.$$

The general solution of this linear equation is

$$U(\alpha, t) = ce^{-k\alpha^2 t} + \frac{u_0}{\alpha}.$$

But $U(\alpha, 0) = 0$ implies $c = -u_0/\alpha$ and so

$$U(\alpha, t) = u_0 \frac{1 - e^{-k\alpha^2 t}}{\alpha}$$

and

$$u(x, t) = \frac{2u_0}{\pi} \int_0^\infty \frac{1 - e^{-k\alpha^2 t}}{\alpha} \sin \alpha x \, d\alpha.$$

6. Since the domain of x is $(0, \infty)$ and the condition at $x = 0$ involves $\partial u/\partial x$ we use the Fourier cosine transform:

$$-k\alpha^2 U(\alpha, t) - ku_x(0, t) = \frac{dU}{dt}$$

$$\frac{dU}{dt} + k\alpha^2 U = kA$$

$$U(\alpha, t) = ce^{-k\alpha^2 t} + \frac{A}{\alpha^2}.$$

Since

$$\mathscr{F}\{u(x, 0)\} = U(\alpha, 0) = 0$$

we find $c = -A/\alpha^2$, so that

$$U(\alpha, t) = A\frac{1 - e^{-k\alpha^2 t}}{\alpha^2}.$$

Applying the inverse Fourier cosine transform we obtain

$$u(x, t) = \mathscr{F}_C^{-1}\{U(\alpha, t)\} = \frac{2A}{\pi} \int_0^\infty \frac{1 - e^{-k\alpha^2 t}}{\alpha^2} \cos \alpha x \, d\alpha.$$

9. (a) Using the Fourier transform we obtain

$$U(\alpha, t) = c_1 \cos \alpha a t + c_2 \sin \alpha a t.$$

If we write

$$\mathscr{F}\{u(x, 0)\} = \mathscr{F}\{f(x)\} = F(\alpha)$$

and

$$\mathscr{F}\{u_t(x, 0)\} = \mathscr{F}\{g(x)\} = G(\alpha)$$

we first obtain $c_1 = F(\alpha)$ from $U(\alpha, 0) = F(\alpha)$ and then $c_2 = G(\alpha)/\alpha a$ from $dU/dt\Big|_{t=0} = G(\alpha)$. Thus

$$U(\alpha, t) = F(\alpha) \cos \alpha a t + \frac{G(\alpha)}{\alpha a} \sin \alpha a t$$

and

$$u(x, t) = \frac{1}{2\pi} \int_{-\infty}^{\infty} \left(F(\alpha) \cos \alpha a t + G(\alpha) \alpha a \sin \alpha a t\right) e^{-i\alpha x} \, d\alpha.$$

(b) If $g(x) = 0$ then $c_2 = 0$ and

$$u(x, t) = \frac{1}{2\pi} \int_{-\infty}^{\infty} F(\alpha) \cos \alpha a t \, e^{-i\alpha x} \, d\alpha$$

$$= \frac{1}{2\pi} \int_{-\infty}^{\infty} F(\alpha) \left(\frac{e^{\alpha a t i} + e^{-\alpha a t i}}{2} \right) e^{-i\alpha x} \, d\alpha$$

$$= \frac{1}{2} \left[\frac{1}{2\pi} \int_{-\infty}^{\infty} F(\alpha) e^{-i(x-at)\alpha} \, d\alpha + \frac{1}{2\pi} \int_{-\infty}^{\infty} F(\alpha) e^{-i(x+at)\alpha} \, d\alpha \right]$$

$$= \frac{1}{2} \left[f(x - at) + f(x + at) \right].$$

12. Since the boundary condition at $y = 0$ now involves $u(x, 0)$ rather than $u'(x, 0)$, we use the Fourier sine transform. The transform of the partial differential equation is then

$$\frac{d^2 U}{dx^2} - \alpha^2 U + \alpha u(x, 0) = 0 \qquad \text{or} \qquad \frac{d^2 U}{dx^2} - \alpha^2 U = -\alpha.$$

The solution of this differential equation is

$$U(x, \alpha) = c_1 \cosh \alpha x + c_2 \sinh \alpha x + \frac{1}{\alpha}.$$

The transforms of the boundary conditions at $x = 0$ and $x = \pi$ in turn imply that $c_1 = 1/\alpha$ and

$$c_2 = \frac{\cosh \alpha \pi}{\alpha \sinh \alpha \pi} - \frac{1}{\alpha \sinh \alpha \pi} + \frac{\alpha}{(1 + \alpha^2) \sinh \alpha \pi}.$$

Hence

$$U(x, \alpha) = \frac{1}{\alpha} - \frac{\cosh \alpha x}{\alpha} + \frac{\cosh \alpha \pi}{\alpha \sinh \alpha \pi} \sinh \alpha x - \frac{\sinh \alpha x}{\alpha \sinh \alpha \pi} + \frac{\alpha \sinh \alpha x}{(1 + \alpha^2) \sinh \alpha \pi}$$

$$= \frac{1}{\alpha} - \frac{\sinh \alpha (\pi - x)}{\alpha \sinh \alpha \pi} - \frac{\sinh \alpha x}{\alpha (1 + \alpha^2) \sinh \alpha \pi}.$$

Taking the inverse transform it follows that

$$u(x, y) = \frac{2}{\pi} \int_{0}^{\infty} \left(\frac{1}{\alpha} - \frac{\sinh \alpha (\pi - x)}{\alpha \sinh \alpha \pi} - \frac{\sinh \alpha x}{\alpha (1 + \alpha^2) \sinh \alpha \pi} \right) \sin \alpha y \, d\alpha.$$

15. We use the Fourier sine transform with respect to x to obtain

$$U(\alpha, y) = c_1 \cosh \alpha y + c_2 \sinh \alpha y.$$

The transforms of $u(x, 0) = f(x)$ and $u(x, 2) = 0$ give, in turn, $U(\alpha, 0) = F(\alpha)$ and $U(\alpha, 2) = 0$. The first condition gives $c_1 = F(\alpha)$ and the second condition then yields

$$c_2 = -\frac{F(\alpha) \cosh 2\alpha}{\sinh 2\alpha}.$$

Hence

$$U(\alpha, y) = F(\alpha) \cosh \alpha y - \frac{F(\alpha) \cosh 2\alpha \sinh \alpha y}{\sinh 2\alpha}$$

$$= F(\alpha) \frac{\sinh 2\alpha \cosh \alpha y - \cosh 2\alpha \sinh \alpha y}{\sinh 2\alpha}$$

$$= F(\alpha) \frac{\sinh \alpha (2 - y)}{\sinh 2\alpha}$$

and

$$u(x, y) = \frac{2}{\pi} \int_0^\infty F(\alpha) \frac{\sinh \alpha (2 - y)}{\sinh 2\alpha} \sin \alpha x \, d\alpha.$$

18. We solve the three boundary-value problems:

Using separation of variables we find the solution of the first problem is

$$u_1(x, y) = \sum_{n=1}^\infty A_n e^{-ny} \sin nx \quad \text{where} \quad A_n = \frac{2}{\pi} \int_0^\pi f(x) \sin nx \, dx.$$

Using the Fourier sine transform with respect to y gives the solution of the second problem:

$$u_2(x, y) = \frac{200}{\pi} \int_0^\infty \frac{(1 - \cos \alpha) \sinh \alpha (\pi - x)}{\alpha \sinh \alpha \pi} \sin \alpha y \, d\alpha.$$

Also, the Fourier sine transform with respect to y gives the solution of the third problem:

$$u_3(x, y) = \frac{2}{\pi} \int_0^\infty \frac{\alpha \sinh \alpha x}{(1 + \alpha^2) \sinh \alpha \pi} \sin \alpha y \, d\alpha.$$

The solution of the original problem is

$$u(x, y) = u_1(x, y) + u_2(x, y) + u_3(x, y).$$

21. Using the Fourier transform with respect to x gives

$$U(\alpha, y) = c_1 \cosh \alpha y + c_2 \sinh \alpha y.$$

The transform of the boundary condition $\partial u / \partial y \big|_{y=0} = 0$ is $dU/dy \big|_{y=0} = 0$. This condition gives $c_2 = 0$. Hence

$$U(\alpha, y) = c_1 \cosh \alpha y.$$

Now by the given information the transform of the boundary condition $u(x, 1) = e^{-x^2}$ is $U(\alpha, 1) = \sqrt{\pi}\, e^{-\alpha^2/4}$. This condition then gives $c_1 = \sqrt{\pi}\, e^{-\alpha^2/4} \cosh \alpha$. Therefore

$$U(\alpha, y) = \sqrt{\pi}\, \frac{e^{-\alpha^2/4} \cosh \alpha y}{\cosh \alpha}$$

and

$$U(x, y) = \frac{1}{2\sqrt{\pi}} \int_{-\infty}^{\infty} \frac{e^{-\alpha^2/4} \cosh \alpha y}{\cosh \alpha} e^{-i\alpha x}\, d\alpha = \frac{1}{2\sqrt{\pi}} \int_{-\infty}^{\infty} \frac{e^{-\alpha^2/4} \cosh \alpha y}{\cosh \alpha} \cos \alpha x\, d\alpha$$

$$= \frac{1}{\sqrt{\pi}} \int_{0}^{\infty} \frac{e^{-\alpha^2/4} \cosh \alpha y}{\cosh \alpha} \cos \alpha x\, d\alpha.$$

24. The problem is

$$\frac{\partial^2 u}{\partial x^2} + \frac{\partial^2 u}{\partial y^2} = 0, \quad -\infty < x < \infty, \quad y > 0$$

$$u(x, 0) = \begin{cases} u_0, & |x| < 1 \\ 0, & |x| > 1 \end{cases}.$$

Using the Fourier transform with respect to x and $\mathscr{F}\{u(x, y)\} = U(\alpha, y)$, the transform of the PDE is

$$\frac{d^2 U}{dy^2} - \alpha^2 U = 0 \qquad \text{and thus} \qquad U(\alpha, y) = c_1 e^{-\alpha y} + c_2 e^{\alpha y}.$$

For $\alpha > 0$ boundedness as $y \to \infty$ implies $c_2 = 0$, so $U(\alpha, y) = c_1 e^{-\alpha y}$. For $\alpha < 0$ boundedness as $y \to \infty$ implies $c_1 = 0$, so $U(\alpha, y) = c_2 e^{\alpha y}$. Now the Fourier transform of the boundary condition is

$$\mathscr{F}\{u(x, 0)\} = U(\alpha, 0) = \int_{-1}^{1} u_0 e^{i\alpha x}\, dx = \int_{-1}^{1} u_0 \left(\cos \alpha x + i \sin \alpha x\right) dx$$

$$= u_0 \int_{-1}^{1} \cos \alpha x\, dx + i u_0 \overbrace{\int_{-1}^{1} \sin \alpha x\, dx}^{0 \text{ since integrand is odd}} = u_0 \frac{\sin \alpha x}{\alpha}\Bigg|_{-1}^{1} = 2u_0 \frac{\sin \alpha}{\alpha}.$$

Therefore

$$c_1 = 2u_0 \frac{\sin \alpha}{\alpha} \qquad \text{and} \qquad c_2 = 2u_0 \frac{\sin \alpha}{\alpha}$$

and so

$$U(\alpha, y) = \begin{cases} 2u_0 \dfrac{\sin \alpha}{\alpha} e^{-\alpha y}, & \alpha > 0 \\[2mm] 2u_0 \dfrac{\sin \alpha}{\alpha} e^{\alpha y}, & \alpha < 0 \end{cases} \qquad \text{or equivalently} \quad U(\alpha, y) = 2u_0 \frac{\sin \alpha}{\alpha} e^{-|\alpha| y}.$$

Then

$$u(x, y) = \frac{2u_0}{2\pi} \int_{-\infty}^{\infty} e^{-|\alpha| y} \frac{\sin \alpha}{\alpha} e^{-i\alpha x}\, d\alpha = \frac{u_0}{\pi} \int_{-\infty}^{\infty} e^{-|\alpha| y} \frac{\sin \alpha}{\alpha} \left(\cos \alpha x - i \sin \alpha x\right) d\alpha.$$

Note that

$$i \int_{-\infty}^{\infty} e^{-|\alpha|y} \frac{\sin \alpha}{\alpha} \sin \alpha x \, d\alpha = 0$$

since the integrand is an odd function of α. Hence

$$u(x,y) = \frac{u_0}{\pi} \int_{-\infty}^{\infty} e^{-|\alpha|y} \frac{\sin \alpha}{\alpha} \cos \alpha x \, d\alpha.$$

Finally, we can integrate with respect to α on the interval $[0, \infty)$ because the integrand is an even function of α:

$$u(x,y) = \frac{2u_0}{\pi} \int_0^{\infty} e^{-\alpha y} \frac{\sin \alpha}{\alpha} \cos \alpha x \, d\alpha.$$

15.5   Finite Fourier Transforms

3. The finite cosine transform of $f(x) = e^{-2x}$, $[0, \pi]$ is

$$\mathscr{F}_{cn} \{f(x)\} = \int_0^{\pi} e^{-2x} \cdot \cos nx \, dx = \left[\frac{ne^{-2x} \sin nx}{n^2 + 4} - \frac{2e^{-2x} \cos nx}{n^2 + 4} \right]_0^{\pi}$$

$$= \frac{2 - 2(-1)^n e^{-2\pi}}{n^2 + 4} = F(n)$$

$$F(0) = \int_0^{\pi} e^{-2x} \, dx = -\frac{1}{2} e^{-2x} \Big|_0^{\pi} = \frac{1}{2} \left(1 - e^{-2\pi} \right).$$

The inverse transform is

$$\mathscr{F}_{cn}^{-1} \{F(n)\} = \frac{1}{2\pi} \left(1 - e^{-2\pi} \right) + \frac{4}{\pi} \sum_{n=1}^{\infty} \frac{1 - (-1)^n e^{-2\pi}}{n^2 + 4} \cos nx = f(x).$$

6. Because of the two boundary conditions, we use the finite sine transform; we transform with respect to y. With $p = \pi$, we write

$$\mathscr{F}_{sn} \{u(x,y)\} = \int_0^{\pi} u(x,y) \sin n\pi x \, dx = U(x,n).$$

Using

$$\mathscr{F}_{sn} \left\{ \frac{\partial^2 u}{\partial y^2} \right\} = -n^2 U(x,n) + n \left[\overbrace{u(x,0)}^{0} - (-1)^n \overbrace{u(x,\pi)}^{0} \right].$$

Therefore the transform of the PDE is

$$\mathscr{F}_{sn} \left\{ \frac{\partial^2 u}{\partial x^2} \right\} + \mathscr{F}_{sn} \left\{ \frac{\partial^2 u}{\partial y^2} \right\} = \mathscr{F}_{sn} \{0\} \qquad \text{or} \qquad \frac{d^2 U}{dx^2} - n^2 U = 0.$$

Because $U(x, n)$ is defined on a finite interval $[0, \pi]$ we write the general solution of the last ODE in terms of hyperbolic functions:

$$U(x, n) = c_1 \cosh nx + c_2 \sinh nx.$$

The transforms of the two remaining boundary conditions are

$$\mathscr{F}_{sn}\{u(0, y)\} = U(0, n) = \mathscr{F}_{sn}\{0\} = 0$$

$$\mathscr{F}_{sn}\{u(\pi, y)\} = U(\pi, n) = \mathscr{F}_{sn}\{u_0\} = u_0 \int_0^\pi \sin ny \, dy = u_0 \frac{1 - (-1)^n}{n}.$$

Applying $U(0, n) = 0$ to the solution of the ODE implies $c_1 = 0$ and so $U(x, n) = c_2 \sinh nx$. Then applying $U(\pi, n) = u_0 \dfrac{1 - (-1)^n}{n}$ to $U(x, n) = c_2 \sinh nx$ implies $c_2 = u_0 \dfrac{1 - (-1)^n}{n \sinh n\pi}$. Therefore,

$$U(x, n) = u_0 \frac{1 - (-1)^n}{n \sinh n\pi} \cdot \frac{\sinh nx}{\sinh n\pi}.$$

Then from (4) of this section, the inverse of $U(x, n)$ is

$$u(x, y) = \frac{2}{\pi} \sum_{n=1}^{\infty} U(x, n) \sin ny \qquad \text{or} \qquad u(x, y) = \frac{2u_0}{\pi} \sum_{n=1}^{\infty} \frac{1 - (-1)^n}{n} \cdot \frac{\sinh nx}{\sinh n\pi} \cdot \sin ny.$$

9. Because of the two boundary conditions, we use the finite sine transform with $p = \pi$ and write

$$\mathscr{F}_{sn}\{u(x, t)\} = \int_0^\pi u(x, t) \sin nx \, dx = U(n, t).$$

From (6) in the text,

$$-n^2 U(n, t) = \frac{d^2 U}{dt^2}$$

$$\frac{d^2 U}{dt^2} + n^2 U(n, t) = 0.$$

The solution of the ODE is

$$U(n, t) = c_1 \cos nt + c_2 \sin nt.$$

The finite sine transforms of the initial conditions are then

$$U(n, 0) = F(n) = c_1 \qquad \text{and} \qquad \frac{dU}{dt}\bigg|_{t=0} = G(n) = nc_2$$

where $F(n)$ and $G(n)$ denote, respectively, the finite Fourier sine transforms of $f(x)$ and $g(x)$. Thus,

$$U(n, t) = F(n) \cos nt + \frac{G(n)}{n} \sin nt.$$

Then from (4) of this section, the inverse of $U(n,t)$ is

$$u(x,t) = \frac{2}{\pi} \sum_{n=1}^{\infty} U(n,t) \sin nx$$

or

$$u(x,t) = \frac{2}{\pi} \sum_{n=1}^{\infty} \left(F(n) \cos nt + \frac{G(n)}{n} \sin nt \right) \sin nx,$$

where

$$F(n) = \int_0^{\pi} f(x) \sin nx \, dx \quad \text{and} \quad G(n) = \int_0^{\pi} g(x) \sin nx \, dx.$$

12. Because of the two boundary conditions, we use the finite sine transform; we transform with respect to y. With $p = \pi$, we write

$$\mathscr{F}_{sn}\{u(x,y)\} = \int_0^{\pi} u(x,y) \sin nx \, dx = U(n,y).$$

From (6) in the text,

$$\mathscr{F}_{sn}\left\{ \frac{\partial^2 u}{\partial x^2} \right\} = -n^2 U(n,y).$$

Hence the transformed PDE is

$$-n^2 U(n,y) + \frac{d^2 U}{dy^2} = 0$$

$$\frac{d^2 U}{dy^2} - n^2 U(n,y) = 0.$$

$$U(n,y) = c_1 \cosh ny + c_2 \sinh ny.$$

The transforms of the two boundary conditions are

$$\mathscr{F}_{sn}\{u(x,0)\} = \mathscr{F}_{sn}\{u_0\} = U(n,0) = u_0 \frac{1-(-1)^n}{n}$$

$$c_1 = u_0 \frac{1-(-1)^n}{n}$$

$$U(n,y) = u_0 \frac{1-(-1)^n}{n} \cosh ny + c_2 \sinh ny$$

and

$$\mathscr{F}_{sn}\{u(x,\pi)\} = \mathscr{F}_{sn}\{u_0\} = U(n,\pi) = u_0 \frac{1-(-1)^n}{n}$$

$$U(n,\pi) = u_0 \frac{1-(-1)^n}{n} \cosh n\pi + c_2 \sinh n\pi = u_0 \frac{1-(-1)^n}{n}$$

$$c_2 = u_0 \frac{1-(-1)^n}{n} \cdot \frac{1 - \cosh n\pi}{\sin n\pi}.$$

Thus

$$U(n, y) = u_0 \frac{1 - (-1)^n}{n} \left[\frac{\sinh n\pi \cosh ny - \cosh n\pi \sinh ny + \sinh ny}{\sinh n\pi} \right].$$

The inverse of $U(n, y)$ is

$$u(x, y) = \frac{2u_0}{\pi} \sum_{n=1}^{\infty} \frac{1 - (-1)^n}{n} \left[\frac{\sinh n\pi \cosh ny - \cosh n\pi \sinh ny + \sinh ny}{\sinh n\pi} \right] \sin nx$$

or

$$u(x, y) = \frac{2u_0}{\pi} \sum_{n=1}^{\infty} \frac{1 - (-1)^n}{n} \left[\frac{\sinh n(\pi - y) + \sinh ny}{\sinh n\pi} \right] \sin nx.$$

15. Because of the two boundary conditions, we use the finite sine transform; we transform with respect to y. With $p = \pi$, we write

$$\mathscr{F}_{sn}\{u(x, y)\} = \int_0^{\pi} u(x, y) \sin nx \, dx = U(n, y).$$

From (6) in the text,

$$\mathscr{F}_{sn}\left\{ \frac{\partial^2 u}{\partial x^2} \right\} = -n^2 U(n, y) + n \left[\overbrace{u(0, y)}^{0} - (-1)^n \overbrace{u(\pi, y)}^{1} \right] = -n^2 U(n, y) - n(-1)^n.$$

Hence the transformed PDE is

$$-n^2 U(n, y) - n(-1)^n + \frac{d^2 U}{dy^2} = 0$$

$$\frac{d^2 U}{dy^2} - n^2 U(n, y) = n(-1)^n.$$

$$U(n, y) = c_1 \cosh ny + c_2 \sinh ny - \frac{(-1)^n}{n}.$$

The transforms of the two boundary conditions are

$$U'(n, 0) = 0 = c_2 n \qquad\qquad U(n, 1) = 0 = c_1 \cosh n - \frac{(-1)^n}{n}$$

$$c_2 = 0 \qquad\qquad c_1 = \frac{(-1)^n}{n \cosh n}.$$

Thus

$$U(n, y) = \frac{(-1)^n \cosh ny}{n \cosh n} - \frac{(-1)^n}{n}.$$

The inverse of $U(n, y)$ is

$$u(x, y) = \frac{2}{\pi} \sum_{n=1}^{\infty} \left[\frac{(-1)^n \cosh ny}{n \cosh n} - \frac{(-1)^n}{n} \right] \sin nx.$$

18. We use the finite sine transform with $p = \pi$ and write

$$\mathscr{F}_{sn}\{u(x,t)\} = \int_0^\pi u(x,t) \sin nx \, dx = U(n,t).$$

The transform of the PDE is

$$\mathscr{F}_{sn}\left\{\frac{\partial^4 u}{\partial x^4}\right\} = n^4 U(n,t) - n^3 \left[u(0,t) - (-1)^n u(\pi,t)\right] + n\left[\frac{\partial^2 u}{\partial x^2}\Big|_{x=0} - (-1)^n \frac{\partial^2 u}{\partial x^2}\Big|_{x=\pi}\right]$$

$$= n^4 U(n,t).$$

Hence the transformed PDE is

$$n^4 U(n,t) + \alpha^2 \frac{d^2 U}{dt^2} = \frac{\beta\pi}{n}(-1)^{n+1} \qquad \text{or} \qquad \frac{d^2 U}{dt^2} + \frac{n^4}{\alpha^2}U(n,t) = \frac{\beta\pi}{n\alpha^2}(-1)^{n+1}.$$

The solution of this ODE is

$$U(n,t) = c_1 \cos\frac{n^2}{\alpha}t + c_2 \sin\frac{n^2}{\alpha}t + \frac{\beta\pi}{n^5}(-1)^{n+1}.$$

The transforms of the initial condition are

$$U(n,0) = 0 \qquad \text{and} \qquad \frac{dU}{dt}\Big|_{t=0} = 0.$$

These conditions yield

$$c_1 + \frac{\beta\pi}{n^5}(-1)^{n+1} = 0 \qquad \text{or} \qquad c_1 = -\frac{\beta\pi}{n^5}(-1)^{n+1}$$

and

$$c_2 = 0.$$

Thus

$$U(n,t) = \frac{\beta\pi}{n^5}(-1)^{n+1}\left[1 - \cos\frac{n^2}{\beta}t\right].$$

The inverse of $U(n,t)$ is

$$u(x,t) = 2\beta \sum_{n=1}^\infty \frac{(-1)^{n+1}}{n^5}\left[1 - \cos\frac{n^2}{\alpha}t\right]\sin nx.$$

15.6 | Fast Fourier Transform

3. By the sifting property,

$$\mathscr{F}\{\delta(x)\} = \int_{-\infty}^\infty \delta(x)e^{i\alpha x}\,dx = e^{i\alpha 0} = 1.$$

6. Using a CAS we find

$$\mathscr{F}\{g(x)\} = \frac{1}{2}[\text{sign}(A - \alpha) + \text{sign}(A + \alpha)]$$

where $\text{sign}(t) = 1$ if $t > 0$ and $\text{sign}\, t = -1$ if $t < 0$. Thus

$$\mathscr{F}\{g(x)\} = \begin{cases} 1, & -A < \alpha < A \\ 0, & \text{elsewhere.} \end{cases}$$

Chapter 15 in Review

---

3. The Laplace transform gives

$$U(x, s) = c_1 e^{-\sqrt{s+h}\,x} + c_2 e^{\sqrt{s+h}\,x} + \frac{u_0}{s+h}.$$

The condition $\lim_{x \to \infty} \partial u / \partial x = 0$ implies $\lim_{x \to \infty} dU/dx = 0$ and so we define $c_2 = 0$. Thus

$$U(x, s) = c_1 e^{-\sqrt{s+h}\,x} + \frac{u_0}{s+h}.$$

The condition $U(0, s) = 0$ then gives $c_1 = -u_0/(s+h)$ and so

$$U(x, s) = \frac{u_0}{s+h} - u_0 \frac{e^{-\sqrt{s+h}\,x}}{s+h}.$$

With the help of the first translation theorem we then obtain

$$u(x, t) = u_0 \mathscr{L}^{-1} \left\{ \frac{1}{s+h} \right\} - u_0 \mathscr{L}^{-1} \left\{ \frac{e^{-\sqrt{s+h}\,x}}{s+h} \right\} = u_0 e^{-ht} - u_0 e^{-ht} \operatorname{erfc} \left(\frac{x}{2\sqrt{t}} \right)$$

$$= u_0 e^{-ht} \left[1 - \operatorname{erfc} \left(\frac{x}{2\sqrt{t}} \right) \right] = u_0 e^{-ht} \operatorname{erf} \left(\frac{x}{2\sqrt{t}} \right).$$

6. The Laplace transform and undetermined coefficients give

$$U(x, s) = c_1 \cosh sx + c_2 \sinh sx + \frac{s-1}{s^2 + \pi^2} \sin \pi x.$$

The conditions $U(0, s) = 0$ and $U(1, s) = 0$ give, in turn, $c_1 = 0$ and $c_2 = 0$. Thus

$$U(x, s) = \frac{s-1}{s^2 + \pi^2} \sin \pi x$$

and

$$u(x, t) = \sin \pi x \, \mathscr{L}^{-1} \left\{ \frac{s}{s^2 + \pi^2} \right\} - \frac{1}{\pi} \sin \pi x \, \mathscr{L}^{-1} \left\{ \frac{\pi}{s^2 + \pi^2} \right\}$$

$$= (\sin \pi x) \cos \pi t - \frac{1}{\pi} (\sin \pi x) \sin \pi t.$$

9. We solve the two problems

$$\frac{\partial^2 u_1}{\partial x^2} + \frac{\partial^2 u_1}{\partial y^2} = 0, \quad x > 0, \quad y > 0,$$

$$u_1(0, y) = 0, \quad y > 0,$$

$$u_1(x, 0) = \begin{cases} 100, & 0 < x < 1 \\ 0, & x > 1 \end{cases}$$

and

$$\frac{\partial^2 u_2}{\partial x^2} + \frac{\partial^2 u_2}{\partial y^2} = 0, \quad x > 0, \quad y > 0,$$

$$u_2(0, y) = \begin{cases} 50, & 0 < y < 1 \\ 0, & y > 1 \end{cases}$$

$$u_2(x, 0) = 0.$$

Using the Fourier sine transform with respect to x we find

$$u_1(x, y) = \frac{200}{\pi} \int_0^\infty \left(\frac{1 - \cos \alpha}{\alpha} \right) e^{-\alpha y} \sin \alpha x \, d\alpha.$$

Using the Fourier sine transform with respect to y we find

$$u_2(x, y) = \frac{100}{\pi} \int_0^\infty \left(\frac{1 - \cos \alpha}{\alpha} \right) e^{-\alpha x} \sin \alpha y \, d\alpha.$$

The solution of the problem is then

$$u(x, y) = u_1(x, y) + u_2(x, y).$$

12. Using the Laplace transform gives

$$U(x, s) = c_1 \cosh \sqrt{s}\, x + c_2 \sinh \sqrt{s}\, x.$$

The condition $u(0, t) = u_0$ transforms into $U(0, s) = u_0/s$. This gives $c_1 = u_0/s$. The condition $u(1, t) = u_0$ transforms into $U(1, s) = u_0/s$. This implies that $c_2 = u_0(1 - \cosh \sqrt{s}\,)/s \sinh \sqrt{s}$. Hence

$$U(x, s) = \frac{u_0}{s} \cosh \sqrt{s}\, x + u_0 \left[\frac{1 - \cosh \sqrt{s}}{s \sinh \sqrt{s}} \right] \sinh \sqrt{s}\, x$$

$$= u_0 \left[\frac{\sinh \sqrt{s}\, \cosh \sqrt{s}\, x - \cosh \sqrt{s}\, \sinh \sqrt{s}\, x + \sinh \sqrt{s}\, x}{s \sinh \sqrt{s}} \right]$$

$$= u_0 \left[\frac{\sinh \sqrt{s}\, (1 - x) + \sinh \sqrt{s}\, x}{s \sinh \sqrt{s}} \right]$$

$$= u_0 \left[\frac{\sinh \sqrt{s}\, (1 - x)}{s \sinh \sqrt{s}} + \frac{\sinh \sqrt{s}\, x}{s \sinh \sqrt{s}} \right]$$

and from Problem 8 in Exercises 15.1

$$u(x, t) = u_0 \left[\mathscr{L}^{-1} \left\{ \frac{\sinh \sqrt{s}\, (1 - x)}{s \sinh \sqrt{s}} \right\} + \mathscr{L}^{-1} \left\{ \frac{\sinh \sqrt{s}\, x}{s \sinh \sqrt{s}} \right\} \right]$$

$$= u_0 \sum_{n=0}^\infty \left[\operatorname{erf} \left(\frac{2n + 2 - x}{2\sqrt{t}} \right) - \operatorname{erf} \left(\frac{2n + x}{2\sqrt{t}} \right) \right]$$

$$+ u_0 \sum_{n=0}^\infty \left[\operatorname{erf} \left(\frac{2n + 1 + x}{2\sqrt{t}} \right) - \operatorname{erf} \left(\frac{2n + 1 - x}{2\sqrt{t}} \right) \right].$$

15. Using the Fourier transform with respect to x we obtain

$$\frac{d^2U}{dy^2} - \alpha^2 U = 0.$$

Since $0 < y < 1$ is a finite interval we use the general solution

$$U(\alpha, y) = c_1 \cosh \alpha y + c_2 \sinh \alpha y.$$

The boundary condition at $y = 0$ transforms into $U'(\alpha, 0) = 0$, so $c_2 = 0$ and $U(\alpha, y) = c_1 \cosh \alpha y$. Now denote the Fourier transform of f as $F(\alpha)$. Then $U(\alpha, 1) = F(\alpha)$ so $F(\alpha) = c_1 \cosh \alpha$ and

$$U(\alpha, y) = F(\alpha) \frac{\cosh \alpha y}{\cosh \alpha}.$$

Taking the inverse Fourier transform we obtain

$$u(x, y) = \frac{1}{2\pi} \int_{-\infty}^{\infty} F(\alpha) \frac{\cosh \alpha y}{\cosh \alpha} e^{-i\alpha x} \, d\alpha.$$

But

$$F(\alpha) = \int_{-\infty}^{\infty} f(t) e^{i\alpha t} \, dt,$$

and so

$$u(x, y) = \frac{1}{2\pi} \int_{-\infty}^{\infty} \left(\int_{-\infty}^{\infty} f(t) e^{i\alpha t} \, dt \right) \frac{\cosh \alpha y}{\cosh \alpha} e^{-i\alpha x} \, d\alpha$$

$$= \frac{1}{2\pi} \int_{-\infty}^{\infty} \int_{-\infty}^{\infty} f(t) e^{i\alpha(t-x)} \frac{\cosh \alpha y}{\cosh \alpha} \, dt \, d\alpha$$

$$= \frac{1}{2\pi} \int_{-\infty}^{\infty} \int_{-\infty}^{\infty} f(t)(\cos \alpha(t-x) + i \sin \alpha(t-x)) \frac{\cosh \alpha y}{\cosh \alpha} \, dt \, d\alpha$$

$$= \frac{1}{2\pi} \int_{-\infty}^{\infty} \int_{-\infty}^{\infty} f(t) \cos \alpha(t-x) \frac{\cosh \alpha y}{\cosh \alpha} \, dt \, d\alpha$$

$$= \frac{1}{\pi} \int_{0}^{\infty} \int_{-\infty}^{\infty} f(t) \cos \alpha(t-x) \frac{\cosh \alpha y}{\cosh \alpha} \, dt \, d\alpha,$$

since the imaginary part of the integrand is an odd function of α followed by the fact that the remaining integrand is an even function of α.

18. The Laplace transform with respect to t of the partial differential equation gives

$$\frac{d^2U}{dx^2} - sU = -50 \quad \text{so} \quad U(x, s) = c_1 e^{-\sqrt{s}x} + c_2 e^{\sqrt{s}x} + \frac{50}{s}.$$

The boundary condition

$$\lim_{x \to \infty} u(x, t) = 50 \quad \text{implies} \quad \lim_{x \to \infty} U(x, s) = \frac{50}{s}$$

so we take $c_2 = 0$. Thus

$$U(x, s) = c_1 e^{-\sqrt{s}x} + \frac{50}{s}.$$

The transform of the boundary condition at $x = 0$ is

$$U(0, s) = \frac{100}{s}e^{-5s} - \frac{100}{s}e^{-10s}.$$

Since

$$\frac{100}{s}e^{-5s} - \frac{100}{s}e^{-10s} = c_1 + \frac{50}{s}$$

we have

$$c_1 = -\frac{50}{s} + \frac{100}{s}e^{-5s} - \frac{100}{s}e^{-10s}$$

and

$$U(x, s) = \left(-\frac{50}{s} + \frac{100}{s}e^{-5s} - \frac{100}{s}e^{-10s}\right)e^{-\sqrt{s}x} + \frac{50}{s}.$$

Using the inverse form of the second translation theorem along with entry 3 of Table 15.1.1 yields

$$u(x, t) = 50 - 50\operatorname{erfc}\left(\frac{x}{2\sqrt{t}}\right) + 100\operatorname{erfc}\left(\frac{x}{2\sqrt{t-5}}\right)\mathscr{U}(t-5) - 100\operatorname{erfc}\left(\frac{x}{2\sqrt{t-10}}\right)\mathscr{U}(t-10)$$

or

$$u(x, t) = \begin{cases} 50 - 50\operatorname{erfc}\left(\frac{x}{2\sqrt{t}}\right), & 0 < t < 5 \\ 50 - 50\operatorname{erfc}\left(\frac{x}{2\sqrt{t}}\right) + 100\operatorname{erfc}\left(\frac{x}{2\sqrt{t-5}}\right), & 5 < t < 10 \\ 50 - 50\operatorname{erfc}\left(\frac{x}{2\sqrt{t}}\right) + 100\operatorname{erfc}\left(\frac{x}{2\sqrt{t-5}}\right) - 100\operatorname{erfc}\left(\frac{x}{2\sqrt{t-10}}\right), & t > 10. \end{cases}$$

21. The Laplace transform with respect to t of the partial differential equation gives

$$\frac{d^2U}{dx^2} - sU = 0 \quad \text{so} \quad U(x, s) = c_1 e^{-\sqrt{s}x} + c_2 e^{\sqrt{s}x}.$$

The boundary condition $\lim_{x\to\infty} u(x, t) = 0$ implies $\lim_{x\to\infty} U(x, s) = 0$, so we take $c_2 = 0$. Then

$$U(x, s) = c_1 e^{-\sqrt{s}x}.$$

The transform of the boundary condition at $x = 0$ implies

$$\left.\frac{dU}{dx}\right|_{x=0} = -\frac{1}{s}.$$

Also,

$$U'(0, s) = -\sqrt{s}\,c_1 = -\frac{1}{s}$$

so

$$c_1 = \frac{1}{s\sqrt{s}}$$

and

$$U(x, s) = \frac{1}{s\sqrt{s}} e^{-\sqrt{s}\,x} x.$$

From entry 4 of Table 15.1.1 with the identification $a = x$ we get

$$u(x, t) = 2\sqrt{\frac{t}{\pi}} e^{-x^2/4t} - x\,\mathrm{erfc}\left(\frac{x}{2\sqrt{t}}\right).$$

Had we used the convolution theorem in the inverse of

$$U(x, \alpha) = \frac{1}{s} \cdot \frac{1}{\sqrt{s}} e^{-\sqrt{s}\,x}$$

then from the inverse form of the convolution theorem, (5) of Section 4.4,

$$u(x, t) = \frac{1}{\sqrt{\pi}} \int_0^t \frac{e^{-x^2/4(t-\tau)}}{\sqrt{t-\tau}}\, d\tau.$$

24. Using the finite cosine transform with $p = 1$, we write

$$\mathscr{F}_{cn}\{u(x, t)\} = \int_0^1 u(x, t)\cos n\pi x\, dx = U(n, t).$$

From property (5) in Section 15.5,

$$\mathscr{F}_{cn}\left\{\frac{\partial^2 u}{\partial x^2}\right\} = -n^2\pi^2 U(n, t) - \left.\frac{\partial u}{\partial x}\right|_{x=0} + (-1)^n \left.\frac{\partial u}{\partial x}\right|_{x=1}$$

$$= -n^2\pi^2 U(n, t).$$

The transform of the PDE is

$$-n^2\pi^2 U(n, t) = \frac{d^2 U}{dt^2} + U \qquad \text{or} \qquad \frac{d^2 U}{dt^2} + \left(n^2\pi^2 + 1\right) U = 0.$$

For $n = 0$, the solution of the ODE is

$$\frac{d^2 U}{dt^2} + U = 0$$

$$U(0, t) = c_1 \cos t + c_2 \sin t.$$

For $n > 0$, the solution of the ODE is

$$\frac{d^2 U}{dt^2} + \left(n^2\pi^2 + 1\right) U = 0$$

$$U(n, t) = c_1 \cos\sqrt{n^2\pi^2 + 1}\, t + c_2 \sin\sqrt{n^2\pi^2 + 1}\, t.$$

Applying the boundary conditions for $n = 0$, we have

$$\mathscr{F}_{cn}\{u(x, 0)\} = U(0, 0) = 0 = c_1$$

$$U(0, t) = c_2 \sin t$$

and

$$\mathscr{F}_{cn}\{u_t(x, 0)\} = U'(0, 0) = \mathscr{F}_{cn}\{x\} = \int_0^1 x\,dx = \frac{1}{2} = c_2$$

$$U(0, t) = \frac{1}{2}\sin t.$$

Then for $n > 0$,

$$\mathscr{F}_{cn}\{u(x, 0)\} = U(n, 0) = 0 = c_1$$

$$U(n, t) = c_2 \sin \sqrt{n^2\pi^2 + 1}\,t$$

and

$$\mathscr{F}_{cn}\{u_t(x, 0)\} = U'(n, 0) = \mathscr{F}_{cn}\{x\} = \int_0^1 x \cos n\pi x\,dx = \frac{(-1)^n - 1}{n^2\pi^2} = c_2\sqrt{n^2\pi^2 + 1}$$

$$c_2 = \frac{(-1)^n - 1}{n^2\pi^2\sqrt{n^2\pi^2 + 1}}.$$

Thus

$$U(n, t) = \frac{(-1)^n - 1}{n^2\pi^2\sqrt{n^2\pi^2 + 1}} \sin \sqrt{n^2\pi^2 + 1}\,t.$$

The inverse of $U(n, t)$ is

$$u(x, t) = \frac{2}{1}\left[\frac{1}{2}U(0, t) + \sum_{n=1}^{\infty} U(n, t) \cos n\pi x\right]$$

or

$$u(x, t) = \frac{1}{2}\sin t + \frac{2}{\pi^2}\sum_{n=1}^{\infty} \frac{(-1)^n - 1}{n^2\sqrt{n^2\pi^2 + 1}} \sin\sqrt{n^2\pi^2 + 1}\,t \cos n\pi x.$$

Chapter 16

Numerical Solutions of Partial Differential Equations

16.1 Laplace's Equation

3. The figure shows the values of $u(x, y)$ along the boundary. We need to determine u_{11}, u_{21}, u_{12}, and u_{22}. By symmetry $u_{11} = u_{21}$ and $u_{12} = u_{22}$. The system is

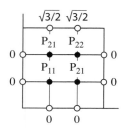

$$u_{21} + u_{12} + 0 + 0 - 4u_{11} = 0$$
$$0 + u_{22} + u_{11} + 0 - 4u_{21} = 0$$
$$u_{22} + \sqrt{3}/2 + 0 + u_{11} - 4u_{12} = 0$$
$$0 + \sqrt{3}/2 + u_{12} + u_{21} - 4u_{22} = 0$$

or

$$3u_{11} + u_{12} = 0$$
$$u_{11} - 3u_{12} = -\frac{\sqrt{3}}{2}.$$

Solving we obtain $u_{11} = u_{21} = \sqrt{3}/16$ and $u_{12} = u_{22} = 3\sqrt{3}/16$.

6. The coefficients of the unknowns are the same as shown above in Problem 5. The constant terms are 7.5, 5, 20, 10, 0, 15, 17.5, 5, 27.5. We use 32.5 as the initial guess for each variable. Then $u_{11} = 21.92$, $u_{21} = 28.30$, $u_{31} = 38.17$, $u_{12} = 29.38$, $u_{22} = 33.13$, $u_{32} = 44.38$, $u_{13} = 22.46$, $u_{23} = 30.45$, and $u_{33} = 46.21$.

16.2 Heat Equation

3. We identify $c = 1$, $a = 2$, $T = 1$, $n = 8$, and $m = 40$. Then $h = 2/8 = 0.25$, $k = 1/40 = 0.025$, and $\lambda = 2/5 = 0.4$.

TIME	X = 0.25	X = 0.50	X = 0.75	X = 1.00	X = 1.25	X = 1.50	X = 1.75
0.000	1.0000	1.0000	1.0000	1.0000	0.0000	0.0000	0.0000
0.025	0.7074	0.9520	0.9566	0.7444	0.2545	0.0371	0.0053
0.050	0.5606	0.8499	0.8685	0.6633	0.3303	0.1034	0.0223
0.075	0.4684	0.7473	0.7836	0.6191	0.3614	0.1529	0.0462
0.100	0.4015	0.6577	0.7084	0.5837	0.3753	0.1871	0.0684
0.125	0.3492	0.5821	0.6428	0.5510	0.3797	0.2101	0.0861
0.150	0.3069	0.5187	0.5857	0.5199	0.3778	0.2247	0.0990
0.175	0.2721	0.4652	0.5359	0.4901	0.3716	0.2329	0.1078
0.200	0.2430	0.4198	0.4921	0.4617	0.3622	0.2362	0.1132
0.225	0.2186	0.3809	0.4533	0.4348	0.3507	0.2358	0.1160
0.250	0.1977	0.3473	0.4189	0.4093	0.3378	0.2327	0.1166
0.275	0.1798	0.3181	0.3881	0.3853	0.3240	0.2275	0.1157
0.300	0.1643	0.2924	0.3604	0.3626	0.3097	0.2208	0.1136
0.325	0.1507	0.2697	0.3353	0.3412	0.2953	0.2131	0.1107
0.350	0.1387	0.2495	0.3125	0.3211	0.2808	0.2047	0.1071
0.375	0.1281	0.2313	0.2916	0.3021	0.2666	0.1960	0.1032
0.400	0.1187	0.2150	0.2725	0.2843	0.2528	0.1871	0.0989
0.425	0.1102	0.2002	0.2549	0.2675	0.2393	0.1781	0.0946
0.450	0.1025	0.1867	0.2387	0.2517	0.2263	0.1692	0.0902
0.475	0.0955	0.1743	0.2236	0.2368	0.2139	0.1606	0.0858
0.500	0.0891	0.1630	0.2097	0.2228	0.2020	0.1521	0.0814
0.525	0.0833	0.1525	0.1967	0.2096	0.1906	0.1439	0.0772
0.550	0.0779	0.1429	0.1846	0.1973	0.1798	0.1361	0.0731
0.575	0.0729	0.1339	0.1734	0.1856	0.1696	0.1285	0.0691
0.600	0.0683	0.1256	0.1628	0.1746	0.1598	0.1214	0.0653
0.625	0.0641	0.1179	0.1530	0.1643	0.1506	0.1145	0.0617
0.650	0.0601	0.1106	0.1438	0.1546	0.1419	0.1080	0.0582
0.675	0.0564	0.1039	0.1351	0.1455	0.1336	0.1018	0.0549
0.700	0.0530	0.0976	0.1270	0.1369	0.1259	0.0959	0.0518
0.725	0.0497	0.0917	0.1194	0.1288	0.1185	0.0904	0.0488
0.750	0.0467	0.0862	0.1123	0.1212	0.1116	0.0852	0.0460
0.775	0.0439	0.0810	0.1056	0.1140	0.1050	0.0802	0.0433
0.800	0.0413	0.0762	0.0993	0.1073	0.0989	0.0755	0.0408
0.825	0.0388	0.0716	0.0934	0.1009	0.0931	0.0711	0.0384
0.850	0.0365	0.0674	0.0879	0.0950	0.0876	0.0669	0.0362
0.875	0.0343	0.0633	0.0827	0.0894	0.0824	0.0630	0.0341
0.900	0.0323	0.0596	0.0778	0.0841	0.0776	0.0593	0.0321
0.925	0.0303	0.0560	0.0732	0.0791	0.0730	0.0558	0.0302
0.950	0.0285	0.0527	0.0688	0.0744	0.0687	0.0526	0.0284
0.975	0.0268	0.0496	0.0647	0.0700	0.0647	0.0495	0.0268
1.000	0.0253	0.0466	0.0609	0.0659	0.0608	0.0465	0.0252

(x, y)	exact	approx	abs error
(0.25, 0.1)	0.3794	0.4015	0.0221
(1, 0.5)	0.1854	0.2228	0.0374
(1.5, 0.8)	0.0623	0.0755	0.0132

6. **(a)** We identify $c = 15/88 \approx 0.1705$, $a = 20$, $T = 10$, $n = 10$, and $m = 10$. Then $h = 2$, $k = 1$, and $\lambda = 15/352 \approx 0.0426$.

TIME	X = 2	X = 4	X = 6	X = 8	X = 10	X = 12	X = 14	X = 16	X = 18
0	30.0000	30.0000	30.0000	30.0000	30.0000	30.0000	30.0000	30.0000	30.0000
1	28.7216	30.0000	30.0000	30.0000	30.0000	30.0000	30.0000	30.0000	28.7216
2	27.5521	29.9455	30.0000	30.0000	30.0000	30.0000	30.0000	29.9455	27.5521
3	26.4800	29.8459	29.9977	30.0000	30.0000	30.0000	29.9977	29.8459	26.4800
4	25.4951	29.7089	29.9913	29.9999	30.0000	29.9999	29.9913	29.7089	25.4951
5	24.5882	29.5414	29.9796	29.9995	30.0000	29.9995	29.9796	29.5414	24.5882
6	23.7515	29.3490	29.9618	29.9987	30.0000	29.9987	29.9618	29.3490	23.7515
7	22.9779	29.1365	29.9373	29.9972	29.9998	29.9972	29.9373	29.1365	22.9779
8	22.2611	28.9082	29.9057	29.9948	29.9996	29.9948	29.9057	28.9082	22.2611
9	21.5958	28.6675	29.8670	29.9912	29.9992	29.9912	29.8670	28.6675	21.5958
10	20.9768	28.4172	29.8212	29.9862	29.9985	29.9862	29.8212	28.4172	20.9768

(b) We identify $c = 15/88 \approx 0.1705$, $a = 50$, $T = 10$, $n = 10$, and $m = 10$. Then $h = 5$, $k = 1$, and $\lambda = 3/440 \approx 0.0068$.

TIME	X = 5	X = 10	X = 15	X = 20	X = 25	X = 30	X = 35	X = 40	X = 45
0	30.0000	30.0000	30.0000	30.0000	30.0000	30.0000	30.0000	30.0000	30.0000
1	29.7955	30.0000	30.0000	30.0000	30.0000	30.0000	30.0000	30.0000	29.7955
2	29.5937	29.9986	30.0000	30.0000	30.0000	30.0000	30.0000	29.9986	29.5937
3	29.3947	29.9959	30.0000	30.0000	30.0000	30.0000	30.0000	29.9959	29.3947
4	29.1984	29.9918	30.0000	30.0000	30.0000	30.0000	30.0000	29.9918	29.1984
5	29.0047	29.9864	29.9999	30.0000	30.0000	30.0000	29.9999	29.9864	29.0047
6	28.8136	29.9798	29.9998	30.0000	30.0000	30.0000	29.9998	29.9798	28.8136
7	28.6251	29.9720	29.9997	30.0000	30.0000	30.0000	29.9997	29.9720	28.6251
8	28.4391	29.9630	29.9995	30.0000	30.0000	30.0000	29.9995	29.9630	28.4391
9	28.2556	29.9529	29.9992	30.0000	30.0000	30.0000	29.9992	29.9529	28.2556
10	28.0745	29.9416	29.9989	30.0000	30.0000	30.0000	29.9989	29.9416	28.0745

(c) We identify $c = 50/27 \approx 1.8519$, $a = 20$, $T = 10$, $n = 10$, and $m = 10$. Then $h = 2$, $k = 1$, and $\lambda = 25/54 \approx 0.4630$.

TIME	X = 2	X = 4	X = 6	X = 8	X = 10	X = 12	X = 14	X = 16	X = 18
0	18.0000	32.0000	42.0000	48.0000	50.0000	48.0000	42.0000	32.0000	18.0000
1	16.1481	30.1481	40.1481	46.1481	48.1481	46.1481	40.1481	30.1481	16.1481
2	15.1536	28.2963	38.2963	44.2963	46.2963	44.2963	38.2963	28.2963	15.1536
3	14.2226	26.8414	36.4444	42.4444	44.4444	42.4444	36.4444	26.8414	14.2226
4	13.4801	25.4452	34.7764	40.5926	42.5926	40.5926	34.7764	25.4452	13.4801
5	12.7787	24.2258	33.1491	38.8258	40.7407	38.8258	33.1491	24.2258	12.7787
6	12.1622	23.0574	31.6460	37.0842	38.9677	37.0842	31.6460	23.0574	12.1622
7	11.5756	21.9895	30.1875	35.4385	37.2238	35.4385	30.1875	21.9895	11.5756
8	11.0378	20.9636	28.8232	33.8340	35.5707	33.8340	28.8232	20.9636	11.0378
9	10.5230	20.0070	27.5043	32.3182	33.9626	32.3182	27.5043	20.0070	10.5230
10	10.0420	19.0872	26.2620	30.8509	32.4400	30.8509	26.2620	19.0872	10.0420

(d) We identify $c = 260/159 \approx 1.6352$, $a = 100$, $T = 10$, $n = 10$, and $m = 10$. Then $h = 10$, $k = 1$, and $\lambda = 13/795 \approx 00164$.

TIME	X = 10	X = 20	X = 30	X = 40	X = 50	X = 60	X = 70	X = 80	X = 90
0	8.0000	16.0000	24.0000	32.0000	40.0000	32.0000	24.0000	16.0000	8.0000
1	8.0000	16.0000	23.6075	31.3459	39.2151	31.6075	23.7384	15.8692	8.0000
2	8.0000	15.9936	23.2279	30.7068	38.4452	31.2151	23.4789	15.7384	7.9979
3	7.9999	15.9812	22.8606	30.0824	37.6900	30.8229	23.2214	15.6076	7.9937
4	7.9996	15.9631	22.5050	29.4724	36.9492	30.4312	22.9660	15.4769	7.9874
5	7.9990	15.9399	22.1606	28.8765	36.2228	30.0401	22.7125	15.3463	7.9793
6	7.9981	15.9118	21.8270	28.2945	35.5103	29.6500	22.4610	15.2158	7.9693
7	7.9967	15.8791	21.5037	27.7261	34.8117	29.2610	22.2112	15.0854	7.9575
8	7.9948	15.8422	21.1902	27.1709	34.1266	28.8733	21.9633	14.9553	7.9439
9	7.9924	15.8013	20.8861	26.6288	33.4548	28.4870	21.7172	14.8253	7.9287
10	7.9894	15.7568	20.5911	26.0995	32.7961	28.1024	21.4727	14.6956	7.9118

9. (a) We identify $c = 15/88 \approx 0.1705$, $a = 20$, $T = 10$, $n = 10$, and $m = 10$. Then $h = 2$, $k = 1$, and $\lambda = 15/352 \approx 0.0426$.

TIME	X = 2.00	X = 4.00	X= 6.00	X = 8.00	X = 10.00	X = 12.00	X = 14.00	X = 16.00	X = 18.00
0.00	30.0000	30.0000	30.0000	30.0000	30.0000	30.0000	30.0000	30.0000	30.0000
1.00	28.7733	29.9749	29.9995	30.0000	30.0000	30.0000	29.9998	29.9916	29.5911
2.00	27.6450	29.9037	29.9970	29.9999	30.0000	30.0000	29.9990	29.9679	29.2150
3.00	26.6051	29.7938	29.9911	29.9997	30.0000	29.9999	29.9970	29.9313	28.8684
4.00	25.6452	29.6517	29.9805	29.9991	30.0000	29.9997	29.9935	29.8839	28.5484
5.00	24.7573	29.4829	29.9643	29.9981	29.9999	29.9994	29.9881	29.8276	28.2524
6.00	23.9347	29.2922	29.9421	29.9963	29.9997	29.9988	29.9807	29.7641	27.9782
7.00	23.1711	29.0836	29.9134	29.9936	29.9995	29.9979	29.9711	29.6945	27.7237
8.00	22.4612	28.8606	29.8782	29.9899	29.9991	29.9966	29.9594	29.6202	27.4870
9.00	27.7999	28.6263	29.8362	29.9848	29.9985	29.9949	29.9454	29.5421	27.2666
10.00	21.1829	28.3831	29.7878	29.9783	29.9976	29.9927	29.9293	29.4610	27.0610

(b) We identify $c = 15/88 \approx 0.1705$, $a = 50$, $T = 10$, $n = 10$, and $m = 10$. Then $h = 5$, $k = 1$, and $\lambda = 3/440 \approx 0.0068$.

TIME	X = 5.00	X = 10.00	X= 15.00	X = 20.00	X = 25.00	X = 30.00	X = 35.00	X = 40.00	X = 45.00
0.00	30.0000	30.0000	30.0000	30.0000	30.0000	30.0000	30.0000	30.0000	30.0000
1.00	29.7968	29.9993	30.0000	30.0000	30.0000	30.0000	30.0000	29.9998	29.9323
2.00	29.5964	29.9973	30.0000	30.0000	30.0000	30.0000	30.0000	29.9991	29.8655
3.00	29.3987	29.9939	30.0000	30.0000	30.0000	30.0000	30.0000	29.9980	29.7996
4.00	29.2036	29.9893	29.9999	30.0000	30.0000	30.0000	30.0000	29.9964	29.7345
5.00	29.0112	29.9834	29.9998	30.0000	30.0000	30.0000	29.9999	29.9945	29.6704
6.00	28.8212	29.9762	29.9997	30.0000	30.0000	30.0000	29.9999	29.9921	29.6071
7.00	28.6339	29.9679	29.9995	30.0000	30.0000	30.0000	29.9998	29.9893	29.5446
8.00	28.4490	29.9585	29.9992	30.0000	30.0000	30.0000	29.9997	29.9862	29.4830
9.00	28.2665	29.9479	29.9989	30.0000	30.0000	30.0000	29.9996	29.9827	29.4222
10.00	28.0864	29.9363	29.9986	30.0000	30.0000	30.0000	29.9995	29.9788	29.3621

(c) We identify $c = 50/27 \approx 1.8519$, $a = 20$, $T = 10$, $n = 10$, and $m = 10$. Then $h = 2$, $k = 1$, and $\lambda = 25/54 \approx 0.4630$.

TIME	X = 2.00	X = 4.00	X= 6.00	X = 8.00	X = 10.00	X = 12.00	X = 14.00	X = 16.00	X = 18.00
0.00	18.0000	32.0000	42.0000	48.0000	50.0000	48.0000	42.0000	32.0000	18.0000
1.00	16.4489	30.1970	40.1562	46.1502	48.1531	46.1773	40.3274	31.2520	22.9449
2.00	15.3312	28.5350	38.3477	44.3130	46.3327	44.4671	39.0872	31.5755	24.6930
3.00	14.4219	27.0429	36.6090	42.5113	44.5759	42.9362	38.1976	31.7478	25.4131
4.00	13.6381	25.6913	34.9606	40.7728	42.9127	41.5716	37.4340	31.7086	25.6986
5.00	12.9409	24.4545	33.4091	39.1182	41.3519	40.3240	36.7033	31.5136	25.7663
6.00	12.3088	23.3146	31.9546	37.5566	39.8880	39.1565	35.9745	31.2134	25.7128
7.00	11.7294	22.2589	30.5939	36.0884	38.5109	38.0470	35.2407	30.8434	25.5871
8.00	11.1946	21.2785	29.3217	34.7092	37.2109	36.9834	34.5032	30.4279	25.4167
9.00	10.6987	20.3660	28.1318	33.4130	35.9801	35.9591	33.7660	29.9836	25.2181
10.00	10.2377	19.5150	27.0178	32.1929	34.8117	34.9710	33.0338	29.5224	25.0019

(d) We identify $c = 260/159 \approx 1.6352$, $a = 100$, $T = 10$, $n = 10$, and $m = 10$. Then $h = 10$, $k = 1$, and $\lambda = 13/795 \approx 00164$.

TIME	X = 10.00	X = 20.00	X= 30.00	X = 40.00	X = 50.00	X = 60.00	X = 70.00	X = 80.00	X = 90.00
0.00	8.0000	16.0000	24.0000	32.0000	40.0000	32.0000	24.0000	16.0000	8.0000
1.00	8.0000	16.0000	24.0000	31.9979	39.7425	31.9979	24.0000	16.0026	8.3218
2.00	8.0000	16.0000	23.9999	31.9918	39.4932	31.9918	24.0000	16.0102	8.6333
3.00	8.0000	16.0000	23.9997	31.9820	39.2517	31.9820	24.0001	16.0225	8.9350
4.00	8.0000	16.0000	23.9993	31.9687	39.0176	31.9687	24.0002	16.0392	9.2272
5.00	8.0000	16.0000	23.9987	31.9520	38.7905	31.9521	24.0003	16.0599	9.5103
6.00	8.0000	15.9999	23.9978	31.9323	38.5701	31.9324	24.0005	16.0845	9.7846
7.00	8.0000	15.9999	23.9966	31.9097	38.3561	31.9098	24.0008	16.1126	10.0506
8.00	8.0000	15.9998	23.9951	31.8844	38.1483	31.8846	24.0012	16.1441	10.3084
9.00	8.0000	15.9997	23.9931	31.8566	37.9463	31.8569	24.0017	16.1786	10.5585
10.00	8.0000	15.9996	23.9908	31.8265	37.7499	31.8270	24.0023	16.2160	10.8012

12. We identify $c = 1$, $a = 1$, $T = 1$, $n = 5$, and $m = 20$. Then $h = 0.2$, $k = 0.04$, and $\lambda = 1$. The values below were obtained using *Excel*, which carries more than 12 significant digits. In order to see evidence of instability use $0 \le t \le 2$.

TIME	X = 0.2	X = 0.4	X = 0.6	X = 0.8
0.00	0.5878	0.9511	0.9511	0.5878
0.04	0.3633	0.5878	0.5878	0.3633
0.08	0.2245	0.3633	0.3633	0.2245
0.12	0.1388	0.2245	0.2245	0.1388
0.16	0.0858	0.1388	0.1388	0.0858
0.20	0.0530	0.0858	0.0858	0.0530
0.24	0.0328	0.0530	0.0530	0.0328
0.28	0.0202	0.0328	0.0328	0.0202
0.32	0.0125	0.0202	0.0202	0.0125
0.36	0.0077	0.0125	0.0125	0.0077
0.40	0.0048	0.0077	0.0077	0.0048
0.44	0.0030	0.0048	0.0048	0.0030
0.48	0.0018	0.0030	0.0030	0.0018
0.52	0.0011	0.0018	0.0018	0.0011
0.56	0.0007	0.0011	0.0011	0.0007
0.60	0.0004	0.0007	0.0007	0.0004
0.64	0.0003	0.0004	0.0004	0.0003
0.68	0.0002	0.0003	0.0003	0.0002
0.72	0.0001	0.0002	0.0002	0.0001
0.76	0.0001	0.0001	0.0001	0.0001
0.80	0.0000	0.0001	0.0001	0.0000
0.84	0.0000	0.0000	0.0000	0.0000
0.88	0.0000	0.0000	0.0000	0.0000
0.92	0.0000	0.0000	0.0000	0.0000
0.96	0.0000	0.0000	0.0000	0.0000
1.00	0.0000	0.0000	0.0000	0.0000

TIME	X = 0.2	X = 0.4	X = 0.6	X = 0.8
1.04	0.0000	0.0000	0.0000	0.0000
1.08	0.0000	0.0000	0.0000	0.0000
1.12	0.0000	0.0000	0.0000	0.0000
1.16	0.0000	0.0000	0.0000	0.0000
1.20	−0.0001	0.0001	−0.0001	0.0001
1.24	0.0001	−0.0002	0.0002	−0.0001
1.28	−0.0004	0.0006	−0.0006	0.0004
1.32	0.0010	−0.0015	0.0015	−0.0010
1.36	−0.0025	0.0040	−0.0040	0.0025
1.40	0.0065	−0.0106	0.0106	−0.0065
1.44	−0.0171	0.0277	−0.0277	0.0171
1.48	0.0448	−0.0724	0.0724	−0.0448
1.52	−0.1172	0.1897	−0.1897	0.1172
1.56	0.3069	−0.4965	0.4965	−0.3069
1.60	−0.8034	1.2999	−1.2999	0.8034
1.64	2.1033	−3.4032	3.4032	−2.1033
1.68	−5.5064	8.9096	−8.9096	5.5064
1.72	14.416	−23.326	23.326	−14.416
1.76	−37.742	61.067	−61.067	37.742
1.80	98.809	−159.88	159.88	−98.809
1.84	−258.68	418.56	−418.56	258.685
1.88	677.24	−1095.8	1095.8	−677.245
1.92	−1773.1	2868.9	−2868.9	1773.1
1.96	4641.9	−7510.8	7510.8	−4641.9
2.00	−12153	19663	−19663	12153

16.3 | Wave Equation

3. (a) Identifying $h = 1/5$ and $k = 0.5/10 = 0.05$ we see that $\lambda = 0.25$.

TIME	X = 0.2	X = 0.4	X = 0.6	X = 0.8
0.00	0.5878	0.9511	0.9511	0.5878
0.05	0.5808	0.9397	0.9397	0.5808
0.10	0.5599	0.9059	0.9059	0.5599
0.15	0.5256	0.8505	0.8505	0.5256
0.20	0.4788	0.7748	0.7748	0.4788
0.25	0.4206	0.6806	0.6806	0.4206
0.30	0.3524	0.5701	0.5701	0.3524
0.35	0.2757	0.4460	0.4460	0.2757
0.40	0.1924	0.3113	0.3113	0.1924
0.45	0.1046	0.1692	0.1692	0.1046
0.50	0.0142	0.0230	0.0230	0.0142

(b) Identifying $h = 1/5$ and $k = 0.5/20 = 0.025$ we see that $\lambda = 0.125$.

TIME	X = 0.2	X = 0.4	X = 0.6	X = 0.8
0.00	0.5878	0.9511	0.9511	0.5878
0.03	0.5860	0.9482	0.9482	0.5860
0.05	0.5808	0.9397	0.9397	0.5808
0.08	0.5721	0.9256	0.9256	0.5721
0.10	0.5599	0.9060	0.9060	0.5599
0.13	0.5445	0.8809	0.8809	0.5445
0.15	0.5257	0.8507	0.8507	0.5257
0.18	0.5039	0.8153	0.8153	0.5039
0.20	0.4790	0.7750	0.7750	0.4790
0.23	0.4513	0.7302	0.7302	0.4513
0.25	0.4209	0.6810	0.6810	0.4209
0.28	0.3879	0.6277	0.6277	0.3879
0.30	0.3527	0.5706	0.5706	0.3527
0.33	0.3153	0.5102	0.5102	0.3153
0.35	0.2761	0.4467	0.4467	0.2761
0.38	0.2352	0.3806	0.3806	0.2352
0.40	0.1929	0.3122	0.3122	0.1929
0.43	0.1495	0.2419	0.2419	0.1495
0.45	0.1052	0.1701	0.1701	0.1052
0.48	0.0602	0.0974	0.0974	0.0602
0.50	0.0149	0.0241	0.0241	0.0149

6. We identify $c = 24944.4$, $k = 0.00010022$ seconds $= 0.10022$ milliseconds, and $\lambda = 0.25$. Time in the table is expressed in milliseconds.

TIME	X = 10	X = 20	X = 30	X = 40	X = 50
0.00000	0.2000	0.2667	0.2000	0.1333	0.0667
0.10022	0.1958	0.2625	0.2000	0.1333	0.0667
0.20045	0.1836	0.2503	0.1997	0.1333	0.0667
0.30067	0.1640	0.2307	0.1985	0.1333	0.0667
0.40089	0.1384	0.2050	0.1952	0.1332	0.0667
0.50111	0.1083	0.1744	0.1886	0.1328	0.0667
0.60134	0.0755	0.1407	0.1777	0.1318	0.0666
0.70156	0.0421	0.1052	0.1615	0.1295	0.0665
0.80178	0.0100	0.0692	0.1399	0.1253	0.0661
0.90201	−0.0190	0.0340	0.1129	0.1184	0.0654
1.00223	−0.0435	0.0004	0.0813	0.1077	0.0638
1.10245	−0.0626	−0.0309	0.0464	0.0927	0.0610
1.20268	−0.0758	−0.0593	0.0095	0.0728	0.0564
1.30290	−0.0832	−0.0845	−0.0278	0.0479	0.0493
1.40312	−0.0855	−0.1060	−0.0639	0.0184	0.0390
1.50334	−0.0837	−0.1237	−0.0974	−0.0150	0.0250
1.60357	−0.0792	−0.1371	−0.1275	−0.0511	0.0069
1.70379	−0.0734	−0.1464	−0.1533	−0.0882	−0.0152
1.80401	−0.0675	−0.1515	−0.1747	−0.1249	−0.0410
1.90424	−0.0627	−0.1528	−0.1915	−0.1595	−0.0694
2.00446	−0.0596	−0.1509	−0.2039	−0.1904	−0.0991
2.10468	−0.0585	−0.1467	−0.2122	−0.2165	−0.1283
2.20491	−0.0592	−0.1410	−0.2166	−0.2368	−0.1551
2.30513	−0.0614	−0.1349	−0.2175	−0.2507	−0.1772
2.40535	−0.0643	−0.1294	−0.2154	−0.2579	−0.1929
2.50557	−0.0672	−0.1251	−0.2105	−0.2585	−0.2005
2.60580	−0.0696	−0.1227	−0.2033	−0.2524	−0.1993
2.70602	−0.0709	−0.1219	−0.1942	−0.2399	−0.1889
2.80624	−0.0710	−0.1225	−0.1833	−0.2214	−0.1699
2.90647	−0.0699	−0.1236	−0.1711	−0.1972	−0.1435
3.00669	−0.0678	−0.1244	−0.1575	−0.1681	−0.1115
3.10691	−0.0649	−0.1237	−0.1425	−0.1348	−0.0761
3.20713	−0.0617	−0.1205	−0.1258	−0.0983	−0.0395
3.30736	−0.0583	−0.1139	−0.1071	−0.0598	−0.0042
3.40758	−0.0547	−0.1035	−0.0859	−0.0209	0.0279
3.50780	−0.0508	−0.0889	−0.0617	0.0171	0.0552
3.60803	−0.0460	−0.0702	−0.0343	0.0525	0.0767
3.70825	−0.0399	−0.0478	−0.0037	0.0840	0.0919
3.80847	−0.0318	−0.0221	0.0297	0.1106	0.1008
3.90870	−0.0211	0.0062	0.0648	0.1314	0.1041
4.00892	−0.0074	0.0365	0.1005	0.1464	0.1025
4.10914	0.0095	0.0680	0.1350	0.1558	0.0973
4.20936	0.0295	0.1000	0.1666	0.1602	0.0897
4.30959	0.0521	0.1318	0.1937	0.1606	0.0808
4.40981	0.0764	0.1625	0.2148	0.1581	0.0719
4.51003	0.1013	0.1911	0.2291	0.1538	0.0639
4.61026	0.1254	0.2164	0.2364	0.1485	0.0575
4.71048	0.1475	0.2373	0.2369	0.1431	0.0532
4.81070	0.1659	0.2526	0.2315	0.1379	0.0512
4.91093	0.1794	0.2611	0.2217	0.1331	0.0514
5.01115	0.1867	0.2620	0.2087	0.1288	0.0535

Chapter 16 in Review

3. (a)

TIME	X = 0.0	X = 0.2	X = 0.4	X = 0.6	X = 0.8	X = 1.0
0.00	0.0000	0.2000	0.4000	0.6000	0.8000	0.0000
0.01	0.0000	0.2000	0.4000	0.6000	0.5500	0.0000
0.02	0.0000	0.2000	0.4000	0.5375	0.4250	0.0000
0.03	0.0000	0.2000	0.3844	0.4750	0.3469	0.0000
0.04	0.0000	0.1961	0.3609	0.4203	0.2922	0.0000
0.05	0.0000	0.1883	0.3346	0.3734	0.2512	0.0000

(b)

TIME	X = 0.0	X = 0.2	X = 0.4	X = 0.6	X = 0.8	X = 1.0
0.00	0.0000	0.2000	0.4000	0.6000	0.8000	0.0000
0.01	0.0000	0.2000	0.4000	0.6000	0.8000	0.0000
0.02	0.0000	0.2000	0.4000	0.6000	0.5500	0.0000
0.03	0.0000	0.2000	0.4000	0.5375	0.4250	0.0000
0.04	0.0000	0.2000	0.3844	0.4750	0.3469	0.0000
0.05	0.0000	0.1961	0.3609	0.4203	0.2922	0.0000

(c) The table in part **(b)** is the same as the table in part **(a)** shifted downward one row.

Chapter 17

Functions of a Complex Variable

17.1 | **Complex Numbers**

3. $i^{42} = \left(i^2\right)^{21} = (-1)^{21} = -1$

6. $-4i$

9. $7 - 13i$

12. $-7 + 8i$

15. $-5 + 12i$

18. $\dfrac{i}{1+i} \cdot \dfrac{1-i}{1-i} = \dfrac{i+1}{2} = \dfrac{1}{2} + \dfrac{1}{2}\,i$

21. $\dfrac{9+7i}{1+i} \cdot \dfrac{1-i}{1-i} = \dfrac{16-2i}{2} = 8 - i$

24. $\dfrac{4+3i}{3+4i} \cdot \dfrac{3-4i}{3-4i} = \dfrac{24-7i}{25} = \dfrac{24}{25} - \dfrac{7}{25}\,i$

27. $20 + 23i + \dfrac{1}{2-i} \cdot \dfrac{2+i}{2+i} = 20 + 23i + \dfrac{2}{5} + \dfrac{1}{5}\,i = \dfrac{102}{5} + \dfrac{116}{5}\,i$

30. $\dfrac{1}{6+8i} \cdot \dfrac{6-8i}{6-8i} = \dfrac{6-8i}{84} = \dfrac{1}{14} - \dfrac{2}{21}\,i$

33. $\dfrac{x}{x^2 + y^2}$

36. 0

39. $2x + 2yi = -9 + 2i$ implies $2x = -9$ and $2y = 2$. Hence $z = -\frac{9}{2} + i$.

42. $x^2 - y^2 - 4x + (-2xy - 4y)i = 0 + 0i$ implies $x^2 - y^2 - 4x = 0$ and $y(-2x - 4) = 0$. If $y = 0$ then $x(x - 4) = 0$ and so $z = 0$ and $z = 4$. If $-2x - 4 = 0$ or $x = -2$ then $12 - y^2 = 0$ or $y = \pm 2\sqrt{3}$. This gives $z = -2 + 2\sqrt{3}\,i$ and $z = -2 - 2\sqrt{3}\,i$.

45. $|10 + 8i| = \sqrt{164}$ and $|11 - 6i| = \sqrt{157}$. Hence $11 - 6i$ is closer to the origin.

48. By the triangle inequality, $|z + 6 + 8i| \leq |z| + |6 + 8i|$. On the circle, $|z| = 2$ and so $|z + 6 + 8i| \leq 2 + \sqrt{100} = 12$.

17.2 | Powers and Roots

3. $3 \left(\cos \dfrac{3\pi}{2} + i \sin \dfrac{3\pi}{2} \right)$

6. $5\sqrt{2} \left(\cos \dfrac{7\pi}{4} + i \sin \dfrac{7\pi}{4} \right)$

9. $\dfrac{3\sqrt{2}}{2} \left(\cos \dfrac{5\pi}{4} + i \sin \dfrac{5\pi}{4} \right)$

12. $z = -8 + 8i$

15. $z_1 z_2 = 8 \left[\cos \left(\dfrac{\pi}{8} + \dfrac{3\pi}{8} \right) + i \sin \left(\dfrac{\pi}{8} + \dfrac{3\pi}{8} \right) \right] = 8i$

$\dfrac{z_1}{z_2} = \dfrac{1}{2} \left[\cos \left(\dfrac{\pi}{8} - \dfrac{3\pi}{8} \right) + i \sin \left(\dfrac{\pi}{8} - \dfrac{3\pi}{8} \right) \right] = \dfrac{\sqrt{2}}{4} - \dfrac{\sqrt{2}}{4} i$

18. $\left[4\sqrt{2} \left(\cos \dfrac{\pi}{4} + i \sin \dfrac{\pi}{4} \right) \right] \left[\sqrt{2} \left(\cos \dfrac{3\pi}{4} + i \sin \dfrac{3\pi}{4} \right) \right]$

$\qquad = 8 \left[\cos \left(\dfrac{\pi}{4} + \dfrac{3\pi}{4} \right) + i \sin \left(\dfrac{\pi}{4} + \dfrac{3\pi}{4} \right) \right] = -8$

21. $2^9 \left[\cos \dfrac{9\pi}{3} + i \sin \dfrac{9\pi}{3} \right] = -512$

24. $(2\sqrt{2})^4 \left[\cos \dfrac{8\pi}{3} + i \sin \dfrac{8\pi}{3} \right] = -32 + 32\sqrt{3}\, i$

27. $8^{1/3} = 2 \left[\cos \dfrac{2k\pi}{3} + i \sin \dfrac{2k\pi}{3} \right], \quad k = 0, 1, 2$

$w_0 = 2[\cos 0 + i \sin 0] = 2; \quad w_1 = 2 \left[\cos \dfrac{2\pi}{3} + i \sin \dfrac{2\pi}{3} \right] = -1 + \sqrt{3}\, i$

$w_2 = 2 \left[\cos \dfrac{4\pi}{3} + i \sin \dfrac{4\pi}{3} \right] = -1 - \sqrt{3}\, i$

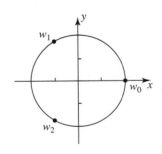

30. $(-1+i)^{1/3} = 2^{1/6} \left[\cos \left(\dfrac{\pi}{4} + \dfrac{2k\pi}{3} \right) + i \sin \left(\dfrac{\pi}{4} + \dfrac{2k\pi}{3} \right) \right], \quad k = 0, 1, 2$

$w_0 = 2^{1/6} \left[\cos \dfrac{\pi}{4} + i \sin \dfrac{\pi}{4} \right] = \dfrac{1}{\sqrt[3]{2}} + \dfrac{1}{\sqrt[3]{2}} i = 0.7937 + 0.7937i$

$w_1 = 2^{1/6} \left[\cos \dfrac{11\pi}{12} + i \sin \dfrac{11\pi}{12} \right] = -1.0842 + 0.2905i$

$w_2 = 2^{1/6} \left[\cos \dfrac{19\pi}{12} + i \sin \dfrac{19\pi}{12} \right] = 0.2905 - 1.0842i$

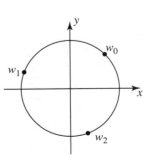

33. The solutions are the four fourth roots of -1;

$$w_k = \cos \dfrac{\pi + 2k\pi}{4} + i \sin \dfrac{\pi + 2k\pi}{4}, \quad k = 0, 1, 2, 3.$$

We have

$$w_1 = \cos\frac{\pi}{4} + i\sin\frac{\pi}{4} = \frac{\sqrt{2}}{2} + \frac{\sqrt{2}}{2}i \qquad\qquad w_3 = \cos\frac{5\pi}{4} + i\sin\frac{5\pi}{4} = -\frac{\sqrt{2}}{2} - \frac{\sqrt{2}}{2}i$$

$$w_2 = \cos\frac{3\pi}{4} + i\sin\frac{3\pi}{4} = -\frac{\sqrt{2}}{2} + \frac{\sqrt{2}}{2}i \qquad w_4 = \cos\frac{7\pi}{4} + i\sin\frac{7\pi}{4} = \frac{\sqrt{2}}{2} - \frac{\sqrt{2}}{2}i.$$

36. $$\frac{\left[8\left(\cos\dfrac{3\pi}{8} + i\sin\dfrac{3\pi}{8}\right)\right]^3}{\left[2\left(\cos\dfrac{\pi}{16} + i\sin\dfrac{\pi}{16}\right)\right]^{10}} = \frac{2^9}{2^{10}}\left[\cos\left(\frac{9\pi}{8} - \frac{10\pi}{16}\right) + i\left(\frac{9\pi}{8} - \frac{10\pi}{16}\right)\right]$$

$$= \frac{1}{2}\left(\cos\frac{\pi}{2} + i\sin\frac{\pi}{2}\right) = \frac{1}{2}i$$

39. (a) $\mathrm{Arg}(z_1) = \pi$, $\mathrm{Arg}(z_2) = \dfrac{\pi}{2}$, $\mathrm{Arg}(z_1 z_2) = -\dfrac{\pi}{2}$, $\mathrm{Arg}(z_1) + \mathrm{Arg}(z_2) = \dfrac{3\pi}{2} \neq \mathrm{Arg}(z_1 z_2)$

(b) $\mathrm{Arg}(z_1/z_2) = -\dfrac{\pi}{2}$, $\mathrm{Arg}(z_1) - \mathrm{Arg}(z_2) = \pi - \dfrac{\pi}{2} = \dfrac{\pi}{2} \neq \mathrm{Arg}(z_1/z_2)$

17.3 Sets in the Complex Plane

3.
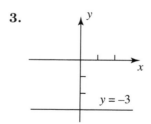

6.
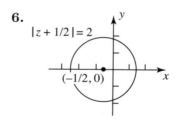

9.
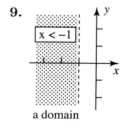

a domain

12.
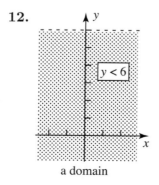

a domain

15.
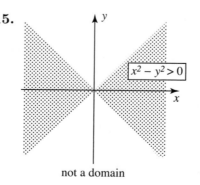

not a domain

18.
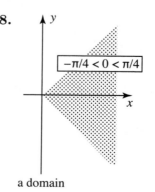

a domain

21.

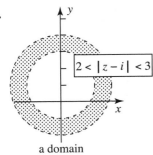

$$2 < |z - i| < 3$$

a domain

24. $|\text{Re}(z)| = |x|$ is the same as $\sqrt{x^2}$ and $|z| = \sqrt{x^2 + y^2}$. Since $y^2 \geq 0$ the inequality $\sqrt{x^2} \leq \sqrt{x^2 + y^2}$ is true for all complex numbers.

17.4 Functions of a Complex Variable

3. $x = 0$ gives $u = -y^2$, $v = 0$. Since $-y^2 \leq 0$ for all real values of y, the image is the origin and the negative u-axis.

6. $y = -x$ gives $u = 0$, $v = -2x^2$. Since $-x^2 \leq 0$ for all real values of x, the image is the origin and the negative v-axis.

9. $f(z) = (x^2 - y^2 - 3x) + i(2xy - 3y + 4)$ **12.** $f(z) = (x^4 - 6x^2y^2 + y^4) + i(4x^3y - 4xy^3)$

15. (a) $f(0 + 2i) = -4 + i$ **(b)** $f(2 - i) = 3 - 9i$ **(c)** $f(5 + 3i) = 1 + 86i$

18. (a) $f\left(0 + \dfrac{\pi}{4}i\right) = \dfrac{\sqrt{2}}{2} + \dfrac{\sqrt{2}}{2}i$ **(b)** $f(-1 - \pi i) = -e^{-1}$

(c) $f\left(3 + \dfrac{\pi}{3}i\right) = \dfrac{1}{2}e^3 + \dfrac{\sqrt{3}}{2}e^3 i$

21. $\displaystyle\lim_{z \to i} \frac{z^4 - 1}{z - i} = \lim_{z \to i} \frac{(z^2 - 1)(z - i)(z + i)}{z - i} = -4i$

24. Along the line $x = 1$, $\displaystyle\lim_{z \to 1} \frac{x + y - 1}{z - 1} = \lim_{y \to 0} \frac{y}{iy} = \frac{1}{i} = -i$, whereas along the x-axis,

$\displaystyle\lim_{z \to 1} \frac{x + y - 1}{z - 1} = \lim_{x \to 1} \frac{x - 1}{x - 1} = 1.$

27. $f'(z) = 12z^2 - (6 + 2i)z - 5$

30. $f'(z) = (z^5 + 3iz^3)(4z^2 + 3iz^2 + 4z - 6i) + (z^4 + iz^3 + 2z^2 - 6iz)(5z^4 + 9iz^2)$

33. $f'(z) = \dfrac{(2z + i)3 - (3z - 4 + 8i)2}{(2z + 1)^2} = \dfrac{8 - 13i}{(2z + i)^2}$

36. $0,\ 2 - 5i$

39. We have
$$\lim_{\Delta z \to 0} \frac{\overline{z + \Delta z} - \overline{z}}{\Delta z} = \lim_{\Delta z \to 0} \frac{\overline{\Delta z}}{\Delta z}.$$

If we let $\Delta z \to 0$ along a horizontal line then $\Delta z = \Delta x$, $\overline{\Delta z} = \Delta x$, and
$$\lim_{\Delta z \to 0} \frac{\overline{\Delta z}}{\Delta z} = \lim_{\Delta x \to 0} \frac{\Delta x}{\Delta x} = 1.$$

If we let $\Delta z \to 0$ along a vertical line then $\Delta z = i\Delta y$, $\overline{\Delta z} = -i\Delta y$, and
$$\lim_{\Delta z \to 0} \frac{\overline{\Delta z}}{\Delta z} = \lim_{\Delta y \to 0} \frac{-i\Delta y}{i\Delta y} = -1.$$

Since these two limits are not equal, $f(z) = \overline{z}$ cannot be differentiable at any z.

42. The system $\dfrac{dx}{dt} = -y,\ \dfrac{dy}{dt} = x$ can be solved as in Section 3.11. We obtain
$$x(t) = c_1 \cos t + c_2 \sin t,\ y(t) = c_1 \sin t - c_2 \cos t.$$

45. If $y = \frac{1}{2}x^2$ the equations $u = x^2 - y^2$, $v = 2xy$ give $u = x^2 - \frac{1}{4}x^4$, $v = x^3$. With the aid of a computer, the graph of these parametric equations is shown.

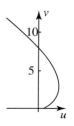

17.5 | Cauchy–Riemann Equations

3. $u = x,\ v = 0;\quad \dfrac{\partial u}{\partial x} = 1,\ \dfrac{\partial v}{\partial y} = 0.$ Since $1 \neq 0$, f is not analytic at any point.

6. $u = x^2 - y^2,\ v = -2xy;\quad \dfrac{\partial u}{\partial x} = 2x,\ \dfrac{\partial v}{\partial y} = -2x;\quad \dfrac{\partial u}{\partial y} = -2y,\ -\dfrac{\partial v}{\partial x} = 2y$

The Cauchy–Riemann equations hold only at $(0,0)$. Since there is no neighborhood about $z = 0$ within which f is differentiable we conclude f is nowhere analytic.

9. $u = e^x \cos y,\ v = e^x \sin y;\quad \dfrac{\partial u}{\partial x} = e^x \cos y = \dfrac{\partial v}{\partial y};\quad \dfrac{\partial u}{\partial y} = -e^x \sin y = -\dfrac{\partial v}{\partial x}.$

f is analytic for all z.

12. $u = 4x^2 + 5x - 4y^2 + 9,\ v = 8xy + 5y - 1;\quad \dfrac{\partial u}{\partial x} = 8x + 5 = \dfrac{\partial v}{\partial y},\ \dfrac{\partial u}{\partial y} = -8y = -\dfrac{\partial v}{\partial x}.$

f is analytic for all z.

15. $\dfrac{\partial u}{\partial x} = 3 = b = \dfrac{\partial v}{\partial y};\quad \dfrac{\partial u}{\partial y} = -1 = -a = -\dfrac{\partial v}{\partial x}.$

f is analytic for all z when $b = 3$, $a = 1$.

18. $u = 3x^2y^2$, $v = -6x^2y^2$; $\dfrac{\partial u}{\partial x} = 6xy^2$, $\dfrac{\partial v}{\partial y} = -12x^2y$; $\dfrac{\partial u}{\partial y} = 6x^2y$, $-\dfrac{\partial v}{\partial x} = 12xy^2$

u and v are continuous and have continuous first partial derivatives. The Cauchy–Riemann equations are satisfied whenever $6xy(y + 2x) = 0$ and $6xy(x - 2y) = 0$. The point satisfying $y + 2x = 0$ and $x - 2y = 0$ is $z = 0$. The points that satisfy $6xy = 0$ are the points along the y-axis ($x = 0$) or along the x-axis ($y = 0$). The function f is differentiable but not analytic on either axis; there is no neighborhood about any point $z = x$ or $z = iy$ within which f is differentiable.

21. Since f is entire,

$$f'(z) = \frac{\partial u}{\partial x} + i\frac{\partial v}{\partial x} = e^x \cos y + ie^x \sin y = f(z).$$

24. $\dfrac{\partial^2 u}{\partial x^2} = 0$, $\dfrac{\partial^2 u}{\partial y^2} = 0$ gives $\dfrac{\partial^2 u}{\partial x^2} + \dfrac{\partial^2 u}{\partial y^2} = 0$. Thus u is harmonic. Now $\dfrac{\partial u}{\partial x} = 2 - 2y = \dfrac{\partial v}{\partial y}$

implies $v = 2y - y^2 + h(x)$, $\dfrac{\partial u}{\partial y} = -2x = -\dfrac{\partial v}{\partial x} = -h'(x)$ implies $h'(x) = 2x$ or $h(x) = x^2 + C$.

Therefore $f(z) = 2x - 2xy + i(2y - y^2 + x^2 + C)$.

27. $\dfrac{\partial^2 u}{\partial x^2} = \dfrac{2y^2 - 2x^2}{(x^2 + y^2)^2}$, $\dfrac{\partial^2 u}{\partial y^2} = \dfrac{2x^2 - 2y^2}{(x^2 + y^2)^2}$ gives $\dfrac{\partial^2 u}{\partial x^2} + \dfrac{\partial^2 u}{\partial y^2} = 0$. Thus u is harmonic. Now

$\dfrac{\partial u}{\partial x} = \dfrac{2x}{x^2 + y^2} = \dfrac{\partial v}{\partial y}$ implies $v = 2\tan^{-1}\dfrac{y}{x} + h(x)$, $\dfrac{\partial u}{\partial y} = \dfrac{2y}{x^2 + y^2} = -\dfrac{\partial v}{\partial x} = \dfrac{2y}{x^2 + y^2} - h'(x)$

implies $h'(x) = 0$ or $h(x) = C$. Therefore $f(z) = \log_e(x^2 + y^2) + i\left(\tan^{-1}\dfrac{y}{x} + C\right)$, $z \neq 0$.

30. $f(x) = \dfrac{x}{x^2 + y^2} - i\dfrac{y}{x^2 + y^2}$. The level curves $u(x, y) = c_1$ and $v(x, y) = c_2$ are the family of circles $x = c_1(x^2 + y^2)$ and $-y = c_2(x^2 + y^2)$, with the exception that $(0, 0)$ is not on the circumference of any circle.

17.6 | Exponential and Logarithmic Functions

3. $e^{-1+\frac{\pi}{4}i} = e^{-1}\cos\dfrac{\pi}{4} + ie^{-1}\sin\dfrac{\pi}{4} = e^{-1}\left(\dfrac{\sqrt{2}}{2} + \dfrac{\sqrt{2}}{2}i\right)$

6. $e^{-\pi+\frac{3\pi}{2}i} = e^{-\pi}\cos\dfrac{3\pi}{2} + ie^{-\pi}\sin\dfrac{3\pi}{2} = -e^{-\pi}i$

9. $e^{5i} = \cos 5 + i\sin 5 = 0.2837 - 0.9589i$

12. $e^{5+\frac{5\pi}{2}i} = e^5\cos\dfrac{5\pi}{2} + ie^5\sin\dfrac{5\pi}{2} = e^5i$

15. $e^{z^2} = e^{x^2-y^2+2xyi} = e^{x^2-y^2}\cos 2xy + ie^{x^2-y^2}\sin 2xy$

18. $\dfrac{e^{z_1}}{e^{z_2}} = \dfrac{e^{x_1}\cos y_1 + ie^{x_1}\sin y_1}{e^{x_2}\cos y_2 + ie^{x_2}\sin y_2} = \dfrac{(e^{x_1}\cos y_1 + ie^{x_1}\sin y_1)(e^{x_2}\cos y_2 - ie^{x_2}\sin y_2)}{e^{2x_2}}$

$= e^{x_1 - x_2}[(\cos y_1 \cos y_2 + \sin y_1 \sin y_2) + i(\sin y_1 \cos y_2 - \cos y_1 \sin y_2)]$

$= e^{x_1 - x_2}[\cos(y_1 - y_2) + i\sin(y_1 - y_2)] = e^{x_1 - x_2 + i(y_1 - y_2)} = e^{(x_1 + iy_1) - (x_2 + iy_2)} = e^{z_1 - z_2}$

21. $u = e^x \cos y, \quad v = -e^x \sin y; \quad \dfrac{\partial u}{\partial x} = e^x \cos y, \quad \dfrac{\partial v}{\partial y} = -e^x \cos y; \quad \dfrac{\partial u}{\partial y} = -e^x \sin y,$

$-\dfrac{\partial v}{\partial x} = e^x \sin y$

Since the Cauchy–Riemann equations are not satisfied at any point, f is nowhere analytic.

24. $\ln(-ei) = \log_e e + i\left(-\dfrac{\pi}{2} + 2n\pi\right) = 1 + \left(-\dfrac{\pi}{2} + 2n\pi\right)i$

27. $\ln(\sqrt{2} + \sqrt{6}\,i) = \log_e 2\sqrt{2} + i\left(\dfrac{\pi}{3} + 2n\pi\right) = 1.0397 + \left(\dfrac{\pi}{3} + 2n\pi\right)i$

30. $\mathrm{Ln}(-e^3) = \log_e e^3 + \pi i = 3 + \pi i$

33. $\mathrm{Ln}(1 + \sqrt{3}\,i)^5 = \mathrm{Ln}(16 - 16\sqrt{3}\,i) = \log_e 32 - \dfrac{\pi}{3}i = 3.4657 - \dfrac{\pi}{3}i$

36. $\dfrac{1}{z} = \ln(-1) = \log_e 1 + i(\pi + 2n\pi) = (2n+1)\pi i$ and so $z = -\dfrac{i}{(2n+1)\pi}$.

39. $(-i)^{4i} = e^{4i\ln(-i)} = e^{4i[\log_e 1 + i(-\frac{\pi}{2} + 2n\pi)]} = e^{(2-8n)\pi}$

42. $(1-i)^{2i} = e^{2i\ln(1-i)} = e^{2i[\log_e \sqrt{2} + i(-\frac{\pi}{4} + 2n\pi)]} = e^{\frac{\pi}{2} - 4n\pi}[\cos(\log_e 2) + i\sin(\log_e 2)]$

$= e^{-4n\pi}[3.7004 + 3.0737i]$

45. If $z_1 = i$ and $z_2 = -1 + i$ then

$$\mathrm{Ln}(z_1 z_2) = \mathrm{Ln}(-1 - i) = \log_e \sqrt{2} - \dfrac{3\pi}{4}i,$$

whereas

$$\mathrm{Ln}\, z_1 + \mathrm{Ln}\, z_2 = \dfrac{\pi}{2}i + \left(\log_e \sqrt{2} + \dfrac{3\pi}{4}i\right) = \log_e \sqrt{2} + \dfrac{5\pi}{4}i.$$

48. (a) $(i^i)^2 = (e^{i\ln i})^2 = [e^{-(\frac{\pi}{2} + 2n\pi)}]^2 = e^{-(\pi + 4n\pi)}$ and $i^{2i} = e^{2i\ln i} = e^{-(\pi + 4n\pi)}$

(b) $(i^2)^i = (-1)^i = e^{i\ln(-1)} = e^{-(\pi + 2n\pi)}$, whereas $i^{2i} = e^{-(\pi + 4n\pi)}$

17.7 | Trigonometric and Hyperbolic Functions

3. $\sin\left(\dfrac{\pi}{4} + i\right) = \sin\dfrac{\pi}{4}\cosh(1) + i\cos\dfrac{\pi}{4}\sinh(1) = 1.0911 + 0.8310i$

6. $\cot\left(\dfrac{\pi}{2} + 3i\right) = \dfrac{\cos\left(\frac{\pi}{2} + 3i\right)}{\sin\left(\frac{\pi}{2} + 3i\right)} = \dfrac{-i\sinh(3)}{\cosh(3)} = -0.9951i$

9. $\cosh(\pi i) = \cos(i(\pi i)) = \cos(-\pi) = \cos\pi = -1$

12. $\cosh(2+3i) = \cosh(2)\cos(3) + i\sinh(2)\sin(3) = -3.7245 + 0.5118i$

15. $\dfrac{e^{iz} - e^{-iz}}{2i} = 2$ gives $e^{2(iz)} - 4ie^{iz} - 1 = 0$. By the quadratic formula, $e^{iz} = 2i \pm \sqrt{3}\,i$ and so

$$iz = \ln\left[(2 \pm \sqrt{3}\,)i\right]$$

$$z = -i\left[\log_e(2 \pm \sqrt{3}\,) + \left(\frac{\pi}{2} + 2n\pi\right)i\right] = \frac{\pi}{2} + 2n\pi - i\log_e(2 \pm \sqrt{3}\,), \quad n = 0, \pm1., \pm2, \dots.$$

18. $\dfrac{e^z - e^{-z}}{2} = -1$ gives $e^{2z} + 2e^z - 1 = 0$. By the quadratic formula, $e^z = -1 \pm \sqrt{2}$, and so

$$z = \ln(-1 \pm \sqrt{2}\,)$$

$$z = \log_e(\sqrt{2} - 1) + 2n\pi i \quad \text{or} \quad z = \log_e(\sqrt{2} + 1) + (\pi + 2n\pi)i,$$

$n = 0, \pm1, \pm2, \dots.$

21. $\cos z = \cosh 2$ implies $\cos x \cosh y - i\sin x \sinh y = \cosh 2 + 0i$ and so we must have $\cos x \cosh y = \cosh 2$ and $\sin x \sinh y = 0$. The last equation has solutions $x = n\pi$, $n = 0, \pm1,$ $\pm2, \dots$, or $y = 0$. For $y = 0$ the first equation becomes $\cos x = \cosh 2$. Since $\cosh 2 > 1$ this equation has no solutions. For $x = n\pi$ the first equation becomes $(-1)^n \cosh y = \cosh 2$. Since $\cosh y > 0$ we see n must be even, say, $n = 2k$, $k = 0, \pm1, \pm2, \dots$. Now $\cosh y = \cosh 2$ implies $y = \pm2$. Solutions of the original equation are then

$$z = 2k\pi \pm 2i, \quad k = 0, \pm1, \pm2, \dots.$$

24. $\sinh z = \dfrac{e^{x+iy} - e^{-x-iy}}{2} = \dfrac{1}{2}(e^x e^{iy} - e^{-x}e^{-iy}) = \dfrac{1}{2}[e^x(\cos y + i\sin y) - e^{-x}(\cos y - i\sin y)]$

$$= \left(\frac{e^x - e^{-x}}{2}\right)\cos y + i\left(\frac{e^x + e^{-x}}{2}\right)\sin y = \sinh x \cos y + i\cosh x \sin y$$

27. $|\cosh z|^2 = \cosh^2 x \cos^2 y + \sinh^2 x \sin^2 y = (1 + \sinh^2 x)\cos^2 y + \sinh^2 x \sin^2 y$

$$= \cos^2 y + \sinh^2 x(\cos^2 y + \sin^2 y) = \cos^2 y + \sinh^2 x$$

30. $\tan z = \dfrac{\sin z}{\cos z} = \dfrac{\sin z \cos z}{|\cos z|^2} = \dfrac{[\sin x \cosh y + i\cos x \sinh y][\cos x \cosh y + i\sin x \sinh y]}{\cos^2 x + \sinh^2 y}$

$$= \frac{(\sin x \cos x \cosh^2 y - \sin x \cos x \sinh^2 y)}{\cos^2 x + \sinh^2 y} + i\,\frac{\cos^2 x \sinh y \cosh y + \sin^2 x \sinh y \cosh y}{\cos^2 x + \sinh^2 y}$$

$$= \frac{\sin x \cos x(\cosh^2 y - \sinh^2 y)}{\cos^2 x + \sinh^2 y} + i\sin y \cosh y(\cos^2 x + \sin^2 x)\cos^2 x + \sinh^2 y$$

$$= \frac{\sin x \cos x}{\cos^2 x + \sinh^2 y} + i\,\frac{\sinh y \cosh y}{\cos^2 x + \sinh^2 y} = \frac{\sin 2x}{2(\cos^2 x + \sinh^2 y)} + i\,\frac{\sinh 2y}{2(\cos^2 x + \sinh^2 y)}$$

But

$$2\cos^2 x + 2\sinh^2 y = (2\cos^2 x - 1) + (2\sinh^2 y + 1) = \cos 2x + \cosh 2y.$$

Therefore $\tan z = u + iz$ where

$$u = \frac{\sin 2x}{\cos 2x + \cosh 2y}, \qquad v = \frac{\sinh 2y}{\cos 2x + \cosh 2y}.$$

17.8   Inverse Trigonometric and Hyperbolic Functions

3. $\sin^{-1} 0 = -i \ln (\pm 1) = \begin{cases} 2n\pi + i \log_e 1 \\ (2n+1)\pi + i \log_e 1 \end{cases} = \begin{cases} 2n\pi \\ (2n+1)\pi \end{cases} = n\pi, \quad n = 0, \pm 1, \pm 2, \ldots$

6. $\cos^{-1} 2i = -i \ln [(2 \pm \sqrt{5}\,)i] = \begin{cases} 2n\pi - \dfrac{\pi}{2} + i \log_e (2 + \sqrt{5}\,) \\[2mm] 2n\pi + \dfrac{\pi}{2} - i \log_e (2 + \sqrt{5}\,) \end{cases}, \quad n = 0, \pm 1, \pm 2, \ldots$

9. $\tan^{-1} 1 = \dfrac{i}{2} \ln \dfrac{i+1}{i-1} = \dfrac{i}{2} \ln (-i) = -n\pi + \dfrac{\pi}{4} + \dfrac{i}{2} \log_e 1 = \dfrac{\pi}{4} - n\pi, \quad n = 0, \pm 1, \pm 2, \ldots$

Note that this can also be written as $\tan^{-1} 1 = \frac{\pi}{4} + n\pi, \quad n = 0, \pm 1, \pm 2, \ldots .$

12. $\cosh^{-1} i = \ln [(1 + \pm\sqrt{2}\,)i] = \begin{cases} \log_e (1 + \sqrt{2}\,) + \left(\dfrac{\pi}{2} + 2n\pi \right) i \\[2mm] \log_e (\sqrt{2} - 1) + \left(-\dfrac{\pi}{2} + 2n\pi \right) i \end{cases}, \quad n = 0, \pm 1, \pm 2, \ldots$

Chapter 17 in Review

3. $-7/25$

6. The closed annular region between the circles $|z + 2| = 1$ and $|z + 2| = 3$. These circles have center at $z = -2$.

9. $z = \ln (2i) = \log_e 2 + i \left(\dfrac{\pi}{2} + 2n\pi \right), \quad n = 0, \pm 1, \pm 2, \ldots$

12. $f(-1 + i) = -33 + 26i$ **15.** $\operatorname{Ln} (-ie^3) = \log_e e^3 + \left(-\dfrac{\pi}{2} \right) i = 3 - \dfrac{\pi}{2} i$

18. $-\dfrac{1}{13} - \dfrac{17}{13} i$

21. The region satisfying $xy \le 1$ is shown in the figure.

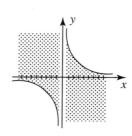

24. The region satisfying $y < x$ is shown in the figure.

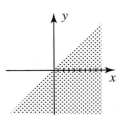

27. The four fourth roots of $1 - i$ are given by

$$w_R = 2^{1/8} \left[\cos \left(-\frac{\pi}{16} + \frac{k\pi}{2} \right) + i \sin \left(-\frac{\pi}{16} + \frac{k\pi}{2} \right) \right], \quad n = 0, \ 1, \ 2, \ 3$$

$$w_0 = 2^{1/8} \left[\cos \left(-\frac{\pi}{16} \right) + i \sin \left(-\frac{\pi}{16} \right) \right] = 1.0696 - 0.2127i$$

$$w_1 = 2^{1/8} \left[\cos \frac{7\pi}{16} + i \sin \frac{7\pi}{16} \right] = 0.2127 + 1.0696i$$

$$w_2 = 2^{1/8} \left[\cos \frac{15\pi}{16} + i \sin \frac{15\pi}{16} \right] = -1.0696 + 0.2127i$$

$$w_3 = 2^{1/8} \left[\cos \frac{23\pi}{16} + i \sin \frac{23\pi}{16} \right] = -0.2127 - 1.0696i$$

30. $\text{Im}(z - 3\bar{z}) = 4y$, $z\text{Re}(z^2) = (x^3 - xy^2) + i(x^2y - y^3)$. Thus,

$$f(z) = (4y + x^3 - xy^2 - 5x) + i(x^2y - y^3 - 5y).$$

33. $z = z^{-1}$ gives $z^2 = 1$ or $(z - 1)(z + 1) = 0$. Thus $z = \pm 1$.

36. $z^2 = \bar{z}^2$ gives $xy = -xy$ or $xy = 0$. This implies $x = 0$ or $y = 0$. All real numbers $(y = 0)$ and all pure imaginary numbers $(x = 0)$ satisfy the equation.

39. $\text{Ln}\,(1 + i)(1 - i) = \text{Ln}\,(2) = \log_e 2$; $\text{Ln}\,(1 + i) = \log_e \sqrt{2} + \frac{\pi}{4}i$; $\text{Ln}\,(1 - i) = \log_e \sqrt{2} - \frac{\pi}{4}i$.

Therefore,
$$\text{Ln}\,(1 + i) + \text{Ln}\,(1 - i) = 2\log_e \sqrt{2} = \log_e 2 = \text{Ln}\,(1 + i)(1 - i).$$

Chapter 18

Integration in the Complex Plane

18.1 | Contour Integrals

3. $\displaystyle\int_C z^2\, dz = (3+2i)^3 \int_{-2}^{2} t^2\, dt = \frac{16}{3}(3+2i)^3 = -48 + \frac{736}{3}\, i$

6. $\displaystyle\int_C |z|^2\, dz = \int_{1}^{2}\left(2t^5 + \frac{2}{t}\right) dt - i\int_{1}^{2}\left(t^2 + \frac{1}{t^4}\right) dt = 21 + \ln 4 - \frac{21}{8}\, i$

9. Using $y = -x + 1$, $0 \le x \le 1$, $z = x + (-x+1)i$, $dz = (1-i)\, dx$,

$$\int_C (x^2 + iy^3)\, dz = (1-i)\int_{1}^{0} [x^2 + (1-x)^3 i]\, dx = -\frac{7}{12} + \frac{1}{12}\, i.$$

12. $\displaystyle\int_C \sin z\, dz = \int_{C_1} \sin z\, dz + \int_{C_2} \sin z\, dz$ where C_1 and C_2 are the line segments $y = 0$,

$0 \le x \le 1$, and $x = 1$, $0 \le y \le 1$, respectively. Now

$$\int_{C_1} \sin z\, dz = \int_{0}^{1} \sin x\, dx = 1 - \cos 1$$

$$\int_{C_2} \sin z\, dz = i\int_{0}^{1} \sin(1 + iy)\, dy = \cos 1 - \cos(1+i).$$

Thus

$$\int_C \sin z\, dz = (1 - \cos 1) + (\cos 1 - \cos(1+i)) = 1 - \cos(1+i)$$

$$= (1 - \cos 1 \cosh 1) + i\sin 1 \sinh 1 = 0.1663 + 0.9889i.$$

15. We have $\displaystyle\oint_C ze^z\, dz = \int_{C_1} ze^z\, dz + \int_{C_2} ze^z\, dz + \int_{C_3} ze^z\, dz + \int_{C_4} ze^z\, dz$

On C_1, $y = 0$, $0 \le x \le 1$, $z = x$, $dz = dx$,

$$\int_{C_1} ze^z\, dz = \int_{0}^{1} xe^x\, dx = xe^x - e^x \Big|_{0}^{1} = 1.$$

On C_2, $x = 1$, $0 \leq y \leq 1$, $z = 1 + iy$, $dz = i\,dy$,

$$\int_{C_2} ze^z\,dz = i\int_0^1 (1 + iy)e^{1+iy}\,dy = ie^{i+1}.$$

On C_3, $y = 1$, $0 \leq x \leq 1$, $z = x + i$, $dz = dx$,

$$\int_{C_3} ze^z\,dz = \int_1^0 (x + i)e^{x+i}\,dx = (i - 1)e^i - ie^{1+i}.$$

On C_4, $x = 0$, $0 \leq y \leq 1$, $z = iy$, $dz = i\,dy$,

$$\int_{C_4} ze^z\,dz = -\int_1^0 ye^{iy}\,dy = (1 - i)e^i - 1.$$

Thus

$$\oint_C ze^z\,dz = 1 + ie^{i+1} + (i - 1)e^i - ie^{1+i} + (1 - i)e^i - 1 = 0.$$

18. We have
$$\oint_C (2z - 1)\,dz = \int_{C_1} (2z - 1)\,dz + \int_{C_2} (2z - 1)\,dz + \int_{C_3} (2z - 1)\,dz$$

On C_1, $y = 0$, $0 \leq x \leq 1$, $z = x$, $dz = dx$,

$$\int_{C_1} (2z - 1)\,dz = \int_0^1 (2x - 1)\,dx = 0.$$

On C_2, $x = 1$, $0 \leq y \leq 1$, $z = 1 + iy$, $dz = i\,dy$,

$$\int_{C_2} (2z - 1)\,dz = -2\int_0^1 y\,dy + i\int_0^1 dy = -1 + i.$$

On C_3, $y = x$, $z = x + ix$, $dz = (1 + i)\,dx$,

$$\int_{C_3} (2z - 1)\,dz = (1 + i)\int_1^0 (2x - 1 + 2ix)\,dx = 1 - i.$$

Thus

$$\oint_C (2z - 1)\,dz = 0 - 1 + i + 1 - i = 0.$$

21. On C, $y = -x + 1$, $0 \leq x \leq 1$, $z = x + (-x + 1)i$, $dz = (1 - i)\,dx$,

$$\int_C (z^2 - z + 2)\,dz = (1 - i)\int_0^1 [x^2 - (1 - x)^2 - x + 2 + (3x - 2x^2 - 1)i]\,dx = \frac{4}{3} - \frac{5}{3}i.$$

24. On C, $x = \sin t$, $y = \cos t$, $0 \leq t \leq \pi/2$ or $z = ie^{-it}$, $dz = e^{-it}\,dt$,

$$\int_C (z^2 - z + 2)\,dz = \int_0^{\pi/2} (-e^{-2it} - ie^{-it} + 2)e^{-it}\,dt = \int_0^{\pi/2} (-e^{-3it} - ie^{-2it} + 2e^{-it})\,dt$$

$$= -\frac{1}{3}ie^{-3\pi i/2} + \frac{1}{2}e^{-\pi i} + 2ie^{-\pi i/2} + \frac{1}{3}i - \frac{1}{2} - 2i = \frac{4}{3} - \frac{5}{3}i.$$

27. The length of the line segment from $z = 0$ to $z = 1+i$ is $\sqrt{2}$. In addition, on this line segment

$$|z^2 + 4| \leq |z|^2 + 4 \leq |1 + i|^2 + 4 = 6.$$

Thus $\left| \displaystyle\int_C (z^2 + 4)\, dz \right| \leq 6\sqrt{2}$.

30. With $z_k^* = z_k$,

$$\int_C z\, dz = \lim_{\|P\| \to 0} \sum_{k=1}^{n} z_k(z_k - z_{k-1})$$

$$= \lim_{\|P\| \to 0} [(z_1^2 - z_1 z_0) + (z_2^2 - z_2 z_1) + \cdots + (z_n^2 - z_n z_{n-1})]. \tag{1}$$

With $z_k^* = z_{k-1}$,

$$\int_C z\, dz = \lim_{\|P\| \to 0} \sum_{k=1}^{n} z_{k-1}(z_k - z_{k-1})$$

$$= \lim_{\|P\| \to 0} [(z_0 z_1 - z_0^2) + (z_1 z_2 - z_1^2) + \cdots + (z_{n-1} z_n - z_{n-1}^2)]. \tag{2}$$

Adding **(1)** and **(2)** gives

$$2\int_C z\, dz = \lim_{\|P\| \to 0} (z_n^2 - z_0^2) \quad \text{or} \quad \int_C z\, dz = \frac{1}{2}(z_n^2 - z_0^2).$$

33. For $f(z) = 2z$, $\overline{f(z)} = 2\bar{z}$, so on $z = e^{it}$, $\bar{z} = e^{-it}$, $dz = ie^{it}\, dt$, and

$$\oint_C \overline{f(z)}\, dz = \int_0^{2\pi} (e^{-it})(ie^{it}\, dt) = 2i \int_0^{2\pi} dt = 4\pi i.$$

Thus circulation $= \text{Re}\left(\oint_C \overline{f(z)}\, dz\right) = 0$, and net flux $= \text{Im}\left(\oint_C \overline{f(z)}\, dz\right) = 4\pi$.

18.2   Cauchy–Goursat Theorem

3. $f(z) = \dfrac{z}{2z + 3}$ is discontinuous at $z = -3/2$ but is analytic within and on the circle $|z| = 1$.

6. $f(z) = \dfrac{e^z}{2z^2 + 11z + 15}$ is discontinuous at $z = -5/2$ and at $z = -3$ but is analytic within and on the circle $|z| = 1$.

9. By the principle of deformation of contours we can choose the more convenient circular contour C_1 defined by $|z| = 1$. Thus

$$\oint_C \frac{1}{z}\, dz = \oint_{C_1} \frac{1}{z}\, dz = 2\pi i$$

by (4) of Section 18.2.

12. By Theorem 18.2.1 and (4) of Section 18.2,

$$\oint_C \left(z + \frac{1}{z^2}\right) dz = \oint_C \frac{1}{z} dz + \oint_C \frac{1}{z^2} dz = 0 + 0 = 0.$$

15. By partial fractions, $\displaystyle\oint_C \frac{2z+1}{z(z+1)} dz = \oint_C \frac{1}{z} dz + \oint_C \frac{1}{z+1} dz.$

(a) By Theorem 18.2.1 and (4) of Section 18.2,

$$\oint_C \frac{1}{z} dz + \oint_C \frac{1}{z+1} dz = 2\pi i + 0 = 2\pi i.$$

(b) By writing $\displaystyle\oint_C = \oint_{C_1} + \oint_{C_2}$ where C_1 and C_2 are the circles $|z| = 1/2$ and $|z + 1| = 1/2$,

respectively, we have by Theorem 18.2.1 and (4) of Section 18.2,

$$\oint_C \frac{1}{z} dz + \oint_C \frac{1}{z+1} dz = \oint_{C_1} \frac{1}{z} dz + \oint_{C_1} \frac{1}{z+1} dz + \oint_{C_2} \frac{1}{z} dz + \oint_{C_2} \frac{1}{z+1} dz$$

$$= 2\pi i + 0 + 0 + 2\pi i = 4\pi i.$$

(c) Since $f(z) = \dfrac{2z+1}{z(z+1)}$ is analytic within and on C it follows from Theorem 18.2.1 that

$$\oint_C \frac{2z+1}{z^2+z} dz = 0.$$

18. (a) By writing $\displaystyle\oint_C = \oint_{C_1} + \oint_{C_2}$ where C_1 and C_2 are the circles $|z + 2| = 1$ and $|z - 2i| = 1$,

respectively, we have by Theorem 18.2.1 and (4) of Section 18.2,

$$\oint_C \left(\frac{3}{z+2} - \frac{1}{z-2i}\right) dz = \oint_{C_1} \frac{3}{z+2} dz - \oint_{C_1} \frac{1}{z-2i} dz + \oint_{C_2} \frac{3}{z+2} dz - \oint_{C_2} \frac{1}{z-2i} dz$$

$$= 3(2\pi i) - 0 + 0 - 2\pi i = 4\pi i.$$

(b) By Theorem 18.2.1 and (4) of Section 18.2,

$$\int_C \frac{3}{z+2} dz - \int_C \frac{1}{z-2i} dz = 0 - 2\pi i = -2\pi i$$

21. We have

$$\oint_C \frac{8z-3}{z^2-z} dz = \oint_{C_1} \frac{8z-3}{z^2-z} dz - \oint_{C_2} \frac{8z-3}{z^2-z} dz$$

where C_1 and C_2 are the closed portions of the curve C enclosing $z = 0$ and $z = 1$, respectively. By partial fractions, Theorem 18.2.1, and (4) of Section 18.2,

$$\oint_{C_1} \frac{8z-3}{z^2-z} dz = 5\oint_{C_1} \frac{1}{z-1} dz + 3\oint_{C_1} \frac{1}{z} dz = 5(0) + 3(2\pi i) = 6\pi i$$

$$\oint_{C_1} \frac{8z-3}{z^2-z} dz = 5\oint_{C_2} \frac{1}{z-1} dz + 3\oint_{C_2} \frac{1}{z} dz = 5(2\pi i) + 3(0) = 10\pi i.$$

Thus

$$\oint_C \frac{8z-3}{z^2-z}\,dz = 6\pi i - 10\pi i = -4\pi i.$$

24. Write

$$\oint_C (z^2 + z + \text{Re}(z))\,dz = \oint_C (z^2 + z)\,dz + \oint_C \text{Re}(z)\,dz.$$

By Theorem 18.2.1, $\oint_C (z^2 + z)\,dz = 0$. However, since $\text{Re}(z) = x$ is not analytic,

$$\oint_C x\,dz = \oint_{C_1} x\,dz + \oint_{C_2} x\,dz + \oint_{C_3} x\,dz$$

where C_1 is $y = 0$, $0 \le x \le 1$, C_2 is $x = 1$, $0 \le y \le 2$, and C_3 is $y = 2x$, $0 \le x \le 1$. Thus,

$$\oint_C x\,dz = \int_0^1 x\,dx + i\int_0^2 dy + (1+2i)\int_1^0 x\,dx = \frac{1}{2} + 2i - \frac{1}{2}(1+2i) = i.$$

18.3   Independence of Path

3. The given integral is independent of the path. Thus

$$\int_C 2z\,dz = \int_{-2+7i}^{2-i} 2z\,dz = z^2 \Big|_{-2+7i}^{2-i} = 48 + 24i.$$

6. $\displaystyle\int_{-2i}^1 (3z^2 - 4z + 5i)\,dz = z^3 - 2z^2 + 5iz \Big|_{-2i}^1 = -19 - 3i$

9. $\displaystyle\int_{-i/2}^{1-i} (2z+1)^2\,dz = \frac{1}{6}(2z+1)^3 \Big|_{-i/2}^{1-i} = -\frac{7}{6} - \frac{22}{3}i$

12. $\displaystyle\int_{1-i}^{1+2i} ze^{z^2}\,dz = \frac{1}{2}e^{z^2}\Big|_{1-i}^{1+2i} = \frac{1}{2}[e^{-3+4i} - e^{-2i}] = \frac{1}{2}(e^{-3}\cos 4 - \cos 2) + \frac{1}{2}(e^{-3}\sin 4 + \sin 2)i =$
$0.1918 + 0.4358i$

15. $\displaystyle\int_{\pi i}^{2\pi i} \cosh z\,dz = \sinh z \Big|_{\pi i}^{2\pi i} = \sinh 2\pi i - \sinh \pi i = i\sin 2\pi - i\sin\pi = 0$

18. $\displaystyle\int_{1+i}^{4+4i} \frac{1}{z}\,dz = \text{Ln}\,z \Big|_{1+i}^{4+4i} = \text{Ln}\,(4+4i) - \text{Ln}\,(1+i)$

$$= \log_e 4\sqrt{2} + \frac{\pi}{4}i - \left(\log_e \sqrt{2} + \frac{\pi}{4}i\right) = \log_e 4 = 1.3863$$

21. Integration by parts gives

$$\int e^z \cos z\,dz = \frac{1}{2}e^z(\cos z + \sin z) + C$$

and so

$$\int_\pi^i e^z \cos z\, dz = \frac{1}{2}\, e^z (\cos z + \sin z)\Big|_\pi^i = \frac{1}{2}[e^i(\cos i + \sin i) - e^\pi(\cos \pi + \sin \pi)]$$

$$= \frac{1}{2}[(\cos 1 \cosh 1 - \sin 1 \sinh 1 + e^\pi) + i(\cos 1 \sinh 1 + \sin 1 \cosh 1)$$

$$= 11.4928 + 0.9667i.$$

24. Integration by parts gives

$$\int z^2 e^z\, dz = z^2 e^z - 2z e^z + 2e^z + C$$

and so

$$\int_0^{\pi i} z^2 e^z\, dz = e^z(z^2 - 2z + 2)\Big|_0^{\pi i} = e^{\pi i}(-\pi^2 - 2\pi i + 2) - 2 = \pi^2 - 4 + 2\pi i.$$

18.4   Cauchy's Integral Formulas

3. By Theorem 18.4.1 with $f(z) = e^z$,

$$\oint_C \frac{e^z}{z - \pi i}\, dz = 2\pi i e^{\pi i} = -2\pi i.$$

6. By Theorem 18.4.1 with $f(z) = \frac{1}{3}\cos z$,

$$\oint_C \frac{\frac{1}{3}\cos z}{z - \frac{\pi}{3}}\, dz = 2\pi i \left(\frac{1}{3}\cos\frac{\pi}{3}\right) = \frac{\pi}{3}i.$$

9. By Theorem 18.4.1 with $f(z) = \frac{z^2 + 4}{z - i}$,

$$\oint_C \frac{\frac{z^2 + 4}{z - i}}{z - 4i}\, dz = 2\pi i \left(-\frac{12}{3i}\right) = -8\pi.$$

12. By Theorem 18.4.2 with $f(z) = z$, $f'(z) = 1$, $f''(z) = 0$, and $f'''(z) = 0$,

$$\oint_C \frac{z}{(z - (-i))^4}\, dz = \frac{2\pi i}{3!}(0) = 0.$$

15. (a) By Theorem 18.4.1 with $f(z) = \frac{2z + 5}{z - 2}$,

$$\oint_C \frac{\frac{2z + 5}{z - 2}}{z}\, dz = 2\pi i \left(-\frac{5}{2}\right) = -5\pi i.$$

(b) Since the circle $|z - (-1)| = 2$ encloses only $z = 0$, the value of the integral is the same as in part **(a)**.

(c) From Theorem 18.4.1 with $f(z) = \dfrac{2z+5}{z}$,

$$\oint_C \frac{\frac{2z+5}{z}}{z-2}\, dz = 2\pi i\left(\frac{9}{2}\right) = 9\pi i.$$

(d) Since the circle $|z - (-2i)| = 1$ encloses neither $z = 0$ nor $z = 2$ it follows from the Cauchy–Goursat Theorem, Theorem 18.2.1, that

$$\oint_C \frac{2z+5}{z(z-2)}\, dz = 0.$$

18. (a) By Theorem 18.4.2 with $f(z) = \dfrac{1}{z-4}$, $f'(z) = -\dfrac{1}{(z-4)^2}$, and $f''(z) = \dfrac{2}{(z-4)^3}$,

$$\oint_C \frac{\frac{1}{z-4}}{z^3}\, dz = \frac{2\pi i}{2!}\left(\frac{2}{-64}\right) = -\frac{\pi}{32}i.$$

(b) By the Cauchy–Goursat Theorem, Theorem 18.2.1,

$$\oint_C \frac{1}{z^3(z-4)}\, dz = 0.$$

21. We have

$$\oint_C \frac{1}{z^3(z-1)^2}\, dz = \oint_{C_1} \frac{\frac{1}{(z-1)^2}}{z^3}\, dz + \oint_{C_2} \frac{\frac{1}{z^3}}{(z-1)^2}\, dz$$

where C_1 and C_2 are the circles $|z| = 1/3$ and $|z-1| = 1/3$, respectively. By Theorem 18.4.2,

$$\oint_{C_1} \frac{\frac{1}{(z-1)^2}}{z^3}\, dz = \frac{2\pi i}{2!}(6) = 6\pi i, \qquad \oint_{C_2} \frac{\frac{1}{z^3}}{(z-1)^2}\, dz = \frac{2\pi i}{1!}(-3) = -6\pi i.$$

Thus

$$\oint_C \frac{1}{z^3(z-1)^2}\, dz = 6\pi i - 6\pi i = 0.$$

24. We have

$$\oint_C \frac{e^{iz}}{(z^2+1)^2}\, dz = \oint_{C_1} \frac{\frac{e^{iz}}{(z+i)^2}}{(z-i)^2}\, dz - \oint_{C_2} \frac{\frac{e^{iz}}{(z-i)^2}}{(z-(-i))^2}\, dz$$

where C_1 and C_2 are the closed portions of the curve C enclosing $z = i$ and $z = -i$, respectively. By Theorem 18.4.2,

$$\oint_{C_1} \frac{\frac{e^{iz}}{(z+i)^2}}{(z-i)^2}\, dz = \frac{2\pi i}{1!}\left(\frac{-4e^{-1}}{-8i}\right) = \pi e^{-1}, \qquad \oint_{C_2} \frac{\frac{e^{iz}}{(z-i)^2}}{(z-(-i))^2}\, dz = \frac{2\pi i}{1!}\left(\frac{0}{8i}\right) = 0.$$

Thus

$$\oint_C \frac{e^{iz}}{(z^2+1)^2}\, dz = \pi e^{-1}.$$

Chapter 18 in Review

3. True   **6.** $\pi(-16 + 8i)$

9. True (Use partial fractions and write the given integral as two integrals.)

12. 12π

15. $\displaystyle\int_C |z^2|\, dz = \int_0^2 (t^4 + t^2)\, dt + 2i \int_0^2 (t^5 + t^3)\, dt = \frac{136}{15} + \frac{88}{3} i$

18. $\displaystyle\int_{3i}^{1-i} (4z - 6)\, dz = 2z^2 - 6z \Big|_{3i}^{1-i} = 12 + 20i$

21. On $|z| = 1$, let $z = e^{it}$, $dz = ie^{it}\, dt$, so that

$$\oint_C (z^{-2} + z^{-1} + z + z^2)\, dz = i \int_0^{2\pi} (e^{-2it} + e^{-it} + e^{it} + e^{2it}) e^{it}\, dt$$

$$= -e^{-it} + it + \frac{1}{2} e^{2it} + \frac{1}{3} e^{3it} \Big|_0^{2\pi} = 2\pi i.$$

24. By Theorem 18.4.2 with $f(z) = \dfrac{\cos z}{z - 1}$ and $f'(z) = \dfrac{\sin z - \cos z - z \sin z}{(z-1)^2}$,

$$\oint_C \frac{\frac{\cos z}{z-1}}{z^2}\, dz = \frac{2\pi i}{1!}\left(\frac{-1}{1}\right) = -2\pi i.$$

27. Using the principle of deformation of contours we choose C to be the more convenient circular contour $|z + i| = \frac{1}{4}$. On this circle $z = -i + \frac{1}{4} e^{it}$ and $dz = \frac{1}{4} ie^{it}\, dt$. Thus

$$\oint_C \frac{z}{z+i}\, dz = i \int_0^{2\pi} \left(\frac{1}{4} e^{it} - i\right) dt = 2\pi.$$

30. We have

$$\left| \int_C \operatorname{Ln}(z + 1)\, dz \right| \le |\text{max of } \operatorname{Ln}(z + 1) \text{ on } C| \cdot 2,$$

where 2 is the length of the line segment. Now

$$|\operatorname{Ln}(z + 1)| \le |\log_e (z + 1)| + |\operatorname{Arg}(z + 1)|.$$

But $\max \operatorname{Arg}(z + 1) = \pi/4$ when $z = i$ and $\max|z + 1| = \sqrt{10}$ when $z = 2 + i$. Thus,

$$\left| \int_C \operatorname{Ln}(z + 1)\, dz \right| \le \left(\frac{1}{2} \log_e 10 + \frac{\pi}{4}\right) 2 = \log_e 10 + \frac{\pi}{2}.$$

Chapter 19

Series and Residues

| **Sequences and Series**

3. 0, 2, 0, 2, 0

6. Converges. To see this write the general term as $\left(\dfrac{2}{5}\right)^n \dfrac{1 + n2^{-n}i}{1 + 3n5^{-n}i}$.

9. Diverges. To see this write the general term as $\sqrt{n}\left(1 + \dfrac{1}{\sqrt{n}}i^n\right)$.

12. Write $z_n = \left(\dfrac{1}{4} + \dfrac{1}{4}i\right)^n$ in polar form as $z_n = \left(\dfrac{\sqrt{2}}{4}\right)^n \cos n\theta + i\left(\dfrac{\sqrt{2}}{4}\right)^n \sin n\theta$. Now

$$\text{Re}(z_n) = \left(\frac{\sqrt{2}}{4}\right)^n \cos n\theta \to 0 \text{ as } n \to \infty \quad \text{and} \quad \text{Im}(z_n) = \left(\frac{\sqrt{2}}{4}\right)^n \sin n\theta \to 0 \text{ as } n \to \infty$$

since $\sqrt{2}/4 < 1$.

15. We identify $a = 1$ and $z = 1 - i$. Since $|z| = \sqrt{2} > 1$ the series is divergent.

18. We identify $a = 1/2$ and $z = i$. Since $|z| - 1$ the series is divergent.

21. From

$$\lim_{n \to \infty} \left| \frac{\dfrac{1}{(1-2i)^{n+2}}}{\dfrac{1}{(1-2i)^{n+1}}} \right| = \frac{1}{|1-2i|} = \frac{1}{\sqrt{5}}$$

we see that the radius of convergence is $R = \sqrt{5}$. The circle of convergence is $|z - 2i| = \sqrt{5}$.

24. From

$$\lim_{n \to \infty} \left| \frac{\dfrac{1}{(n+1)^2(3+4i)^{n+1}}}{\dfrac{1}{n^2(3+4i)^n}} \right| = \lim_{n \to \infty} \left(\frac{n}{n+1}\right)^2 \frac{1}{|3+4i|} = \frac{1}{5}$$

we see that the radius of convergence is $R = 5$. The circle of convergence is $|z + 3i| = 5$.

27. From

$$\lim_{n\to\infty} \sqrt[n]{\left|\frac{1}{5^{2n}}\right|} = \lim_{n\to\infty} \frac{1}{25} = \frac{1}{25}$$

we see that the radius of convergence is $R = 25$. The circle of convergence is $|z - 4 - 3i| = 25$.

30. (a) The circle of convergence is $|z| = 1$. Since the series of absolute values

$$\sum_{k=1}^{\infty}\left|\frac{z^k}{k^2}\right| = \sum_{k=1}^{\infty}\frac{|z|^k}{k^2} = \sum_{k=1}^{\infty}\frac{1}{k^2}$$

converges, the given series is absolutely convergent for every z on $|z| = 1$. Since absolute convergence implies convergence, the given series converges for all z on $|z| = 1$.

(b) The circle of convergence is $|z| = 1$. On the circle, $n|z|^n \to \infty$ as $n \to \infty$. This implies $nz^n \nrightarrow 0$ as $n \to \infty$. Thus by Theorem 19.1.3 the series is divergent for every z on the circle $|z| = 1$.

19.2 Taylor Series

3. Differentiating $\dfrac{1}{1+2z} = 1 - 2z + 2^2z^2 - 2^3z^3 + \cdots$ gives $\dfrac{-2}{(1+2z)^2} = -2 + 2\cdot 2^2 z - 3\cdot 2^3 z^2 + \cdots$. Thus

$$\frac{1}{(1+2z)} = 1 - 2\cdot(2z) + 3\cdot(2z)^2 - \cdots = \sum_{k=1}^{\infty}(-1)^{k-1}k(2z)^{k-1} \text{ where } R = \frac{1}{2}.$$

6. Replacing z in $e^z = \displaystyle\sum_{k=0}^{\infty}\frac{z^k}{k!}$ by $-z^2$ and multiplying the result by z gives

$$ze^{-z^2} = \sum_{k=0}^{\infty}\frac{(-1)^k}{k!}z^{2k+1} \text{ where } R = \infty.$$

9. Replacing z in $\cos z = \displaystyle\sum_{k=0}^{\infty}(-1)^k\frac{z^{2k}}{(2k)!}$ by $z/2$ gives $\cos\dfrac{z}{2} = \displaystyle\sum_{k=0}^{\infty}\frac{(-1)^k}{(2k)!}\left(\frac{z}{2}\right)^{2k}$ where $R = \infty$.

12. Using the identity $\cos z = \dfrac{1}{2}(1 + \cos 2z)$ and the series $\cos z = \displaystyle\sum_{k=0}^{\infty}(-1)^k\frac{z^{2k}}{(2k)!}$ gives

$$\cos^2 z = \frac{1}{2} + \frac{1}{2}\sum_{k=0}^{\infty}(-1)^k\frac{(2z)^{2k}}{(2k)!} = 1 + \sum_{k=1}^{\infty}(-1)^k\frac{2^{2k-1}}{(2k)!}z^{2k} \text{ where } R = \infty.$$

15. Using (5) of Section 19.1,

$$\frac{1}{3-z} = \frac{1}{3-2i-(z-2i)} = \frac{1}{3-2i} \cdot \frac{1}{1-\dfrac{z-2i}{3-2i}}$$

$$= \frac{1}{3-2i}\left[1 + \frac{z-2i}{3-2i} + \frac{(z-2i)^2}{(3-2i)^2} + \frac{(z-2i)^3}{(3-2i)^3} + \cdots\right]$$

$$= \frac{1}{3-2i} + \frac{z-2i}{(3-2i)^2} + \frac{(z-2i)^2}{(3-2i)^3} + \frac{(z-2i)^3}{(3-2i)^4} + \cdots$$

$$= \sum_{k=0}^{\infty} \frac{(z-2i)^k}{(3-2i)^{k+1}} \quad \text{where } R = \sqrt{13}.$$

18. Using (5) of Section 19.1,

$$\frac{1+z}{1-z} = -1 + \frac{2}{1-z} = -1 + \frac{2}{1-i-(z-i)} = -1 + \frac{2}{1-i} \cdot \frac{1}{1-\dfrac{z-i}{1-i}}$$

$$= -1 + \frac{2}{1-i}\left[1 + \frac{z-i}{1-i} + \frac{(z-i)^2}{(1-i)^2} + \frac{(z-i)^3}{(1-i)^3} + \cdots\right]$$

$$= -1 + \frac{2}{1-i} + \frac{2(z-i)}{(1-i)^2} + \frac{2(z-i)^2}{(1-i)^3} + \frac{2(z-i)^3}{(1-i)^4} + \cdots$$

$$= -1 + \sum_{k=0}^{\infty} \frac{2(z-i)^k}{(1-i)^{k+1}} \quad \text{where } R = \sqrt{2}.$$

21. Using $e^z = e^{3i} \cdot e^{z-3i}$ and (12) of Section 19.2, $e^z = e^{3i} \sum_{k=0}^{\infty} \frac{(z-3i)^k}{k!}$ where $R = \infty$.

24. Using (7) of Section 19.2, $e^{1/(1+z)} = e - ez + \frac{3e}{2}z^2 - \cdots$.

27. The distance from $2 + 5i$ to i is $|2 + 5i - i| = |2 + 4i| = 2\sqrt{5}$.

30. The series are

$$f(z) = \sum_{k=0}^{\infty} (-1)^k \frac{(z-3)^k}{3^{k+1}} \quad \text{where } R = 3$$

and

$$f(z) = \sum_{k=0}^{\infty} (-1)^k \frac{(z-1-i)^k}{(1+i)^{k+1}} \quad \text{where } R = \sqrt{2}.$$

33. From $e^z \approx 1 + z + \frac{z^2}{2}$ we obtain

$$e^{(1+i)/10} \approx 1 + \frac{1+i}{10} + \frac{(1+i)^2}{100} = 1.1 + 0.12i.$$

36. $e^{iz} = \displaystyle\sum_{k=0}^{\infty} \frac{(iz)^k}{k!} = 1 + i\frac{z}{1!} - \frac{z^2}{2!} - i\frac{z^3}{3!} + \frac{z^4}{4!} + i\frac{z^5}{5!} - \frac{z^6}{6!} - i\frac{z^7}{7!} + \cdots$

$$= \left(1 - \frac{z^2}{2!} + \frac{z^4}{4!} - \frac{z^6}{6!} + \cdots\right) + i\left(\frac{z}{1!} - \frac{z^3}{3!} + \frac{z^5}{5!} - \frac{z^7}{7!} + \cdots\right) = \cos z + i\sin z$$

19.3 | Laurent Series

3. $f(z) = 1 - \dfrac{1}{1!z^2} + \dfrac{1}{2!z^4} - \dfrac{1}{3!z^6} + \cdots$

6. $f(z) = z\left(1 - \dfrac{1}{2!z^2} + \dfrac{1}{4!z^4} - \dfrac{1}{6!z^6} + \cdots\right) = z - \dfrac{1}{2!z} + \dfrac{1}{4!z^3} - \dfrac{1}{6!z^5} + \cdots$

9. $f(z) = \dfrac{1}{z-3} \cdot \dfrac{1}{3+z-3}$

$$= \frac{1}{3(z-3)} \cdot \frac{1}{1 + \dfrac{z-3}{3}} = \frac{1}{3(z-3)}\left[1 - \frac{z-3}{3} + \frac{(z-3)^2}{3^2} - \frac{(z-3)^3}{3^3} + \cdots\right]$$

$$= \frac{1}{3(z-3)} - \frac{1}{3^2} + \frac{z-3}{3^3} - \frac{(z-3)^2}{3^4} + \cdots$$

12. $f(z) = \dfrac{1}{3}\left[\dfrac{1}{z-3} - \dfrac{1}{z}\right] = \dfrac{1}{3}\left[\dfrac{1}{-4+z+1} - \dfrac{1}{z+1-1}\right]$

$$= \frac{1}{3}\left[-\frac{1}{4}\cdot\frac{1}{1 - \dfrac{z+1}{4}} - \frac{1}{z+1}\cdot\frac{1}{1 - \dfrac{1}{z+1}}\right]$$

$$= \frac{1}{3}\left[-\frac{1}{4}\left(1 + \frac{z+1}{4} + \frac{(z+1)^2}{4^2} + \frac{(z+1)^3}{4^3} + \cdots\right)\right.$$

$$\left. - \frac{1}{z+1}\left(1 + \frac{1}{z+1} + \frac{1}{(z+1)^2} + \frac{1}{(z+1)^3} + \cdots\right)\right]$$

$$= \cdots - \frac{1}{(z+1)^2} - \frac{1}{z+1} - \frac{1}{12} - \frac{z+1}{3\cdot 4^2} - \frac{(z+1)^2}{3\cdot 4^3} - \cdots$$

15. $f(z) = \dfrac{1}{z-1} \cdot \dfrac{-1}{1-(z-1)} = \dfrac{-1}{z-1}[1 + (z-1) + (z-1)^2 + (z-1)^3 + \cdots]$

$$= -\frac{1}{z-1} - 1 - (z-1) - (z-1)^2 - \cdots$$

18. $f(z) = \dfrac{1}{3(z+1)} + \dfrac{2}{3} \cdot \dfrac{1}{(z+1)-3} = \dfrac{1}{3(z+1)} + \dfrac{2}{3(z+1)} \cdot \dfrac{1}{1 - \dfrac{3}{z+1}}$

$\qquad = \dfrac{1}{3(z+1)} + \dfrac{2}{3(z+1)} \left(1 + \dfrac{3}{z+1} + \dfrac{3^2}{(z+1)^2} + \dfrac{3^3}{(z+1)^3} + \cdots\right)$

$\qquad = \dfrac{1}{z+1} + \dfrac{2}{(z+1)^2} + \dfrac{2 \cdot 3}{(z+1)^3} + \dfrac{2 \cdot 3^2}{(z+1)^4} + \cdots$

21. $f(z) = \dfrac{1}{z}(1-z)^{-2} = \dfrac{1}{z}\left(1 + (-2)(-z) + \dfrac{(-2)(-3)}{z!}(-z)^2 + \dfrac{(-2)(-3)(-4)}{3!}(-z)^3 + \cdots\right)$

$\qquad = \dfrac{1}{z} + 2 + 3z + 4z^2 + \cdots$

24. $f(z) = \dfrac{1}{(z-3)^3} \cdot \dfrac{-1}{1-(z-1)} = \dfrac{-1}{(z-1)^3}\left[1 + (z-1) + (z-1)^2 + (z-1)^3 + \cdots\right]$

$\qquad = -\dfrac{1}{(z-1)^3} - \dfrac{1}{(z-1)^2} - \dfrac{1}{z-1} - 1 - (z-1) - \cdots$

27. $f(z) = z + \dfrac{2}{z-2} = 1 + (z-1) + \dfrac{2}{-1+z-1} = 1 + (z-1) + \dfrac{2}{z-1} \cdot \dfrac{1}{1 - \dfrac{1}{z-1}}$

$\qquad = 1 + (z-1) + \dfrac{2}{z-1}\left(1 + \dfrac{1}{z-1} + \dfrac{1}{(z-1)^2} + \dfrac{1}{(z-1)^3} + \cdots\right)$

$\qquad = \cdots + \dfrac{2}{(z-1)^2} + \dfrac{2}{z-1} + 1 + (z-1)$

19.4 | Zeros and Poles

3. Since $f(-2+i) = f'(-2+i) = 0$ and $f''(z) = 2$ for all z, $z = -2+i$ is a zero of order two.

6. Write $f(z) = (z^2+9)/z = (z-3i)(z+3i)/z$ to see that $3i$ and $-3i$ are zeros of f. Now $f'(z) = 1 - 9/z^2$ and $f'(3i) = f'(-3i) = 2 \neq 0$. This indicates that each zero is of order one.

9. From $\qquad f(z) = z(1 - \cos z^2) = z\left(-\dfrac{z^4}{2!} + \dfrac{z^8}{4!} - \cdots\right) = z^5\left(-\dfrac{1}{2!} + \dfrac{z^4}{4!} - \cdots\right)$

we see that $z = 0$ is a zero of order five.

12. From the series $e^z = -\displaystyle\sum_{k=0}^{\infty} \dfrac{(z-\pi i)^k}{k!}$ centered at πi and

$\qquad f(z) = 1 - \pi i + z + e^z = 1 - \pi i + z + \left(-1 - \dfrac{z-\pi i}{1!} - \dfrac{(z-\pi i)^2}{2!} - \dfrac{(z-\pi i)^3}{3!} - \cdots\right)$

$\qquad = -\dfrac{(z-\pi i)^2}{2!} - \dfrac{(z-\pi i)^3}{3!} - \cdots = (z-\pi i)^2\left(-\dfrac{1}{2!} - \dfrac{z-\pi i}{3!} - \cdots\right)$

we see that $z = \pi i$ is a zero of order two.

15. From $f(z) = \dfrac{1+4i}{(z+2)(z+i)^4}$ and Theorem 19.4.1 we see that -2 is a simple pole and $-i$ is a pole of order four.

18. From $z^2 \sin \pi z = z^3 \left(\pi - \dfrac{\pi^3 z^2}{3!} + \cdots \right)$ we see $z = 0$ is a zero of order three. From $f(z) = \dfrac{\cos \pi z}{z^2 \sin \pi z}$ and Theorem 19.4.1 we see 0 is a pole of order three. The numbers n, $n = \pm 1, \pm 2, \ldots$ are simple poles.

21. From $1 - e^z = 1 - \left(1 + \dfrac{z}{1!} + \dfrac{z^2}{2!} + \cdots \right) = z \left(-1 - \dfrac{z}{2!} - \cdots \right)$ we see that $z = 0$ is a zero of order one. By periodicity of e^z it follows that $z = 2n\pi i$, $n = 0, \pm 1, \pm 2, \ldots$ are zeros of order one. From $f(z) = \dfrac{1}{1 - e^z}$ and Theorem 19.4.1 we see that the numbers $2n\pi i$, $n = 0, \pm 1, \pm 2, \ldots$ are simple poles.

24. The function is rewritten as

$$\frac{\cos z - \cos 2z}{z^6} = \frac{1}{z^6} \left[\left(1 - \frac{z^2}{2!} + \frac{z^4}{4!} - \frac{z^6}{6!} + \frac{z^8}{8!} \cdots \right) - \left(1 - \frac{(2z)^2}{2!} + \frac{(2z)^4}{4!} - \frac{(2z)^6}{6!} + \frac{(2z)^8}{8!} \cdots \right) \right]$$

$$= \frac{1}{z^6} \left[1 - \frac{z^2}{2!} + \frac{z^4}{4!} - \frac{z^6}{6!} + \frac{z^8}{8!} \cdots - 1 + \frac{(2z)^2}{2!} - \frac{(2z)^4}{4!} + \frac{(2z)^6}{6!} - \frac{(2z)^8}{8!} \cdots \right]$$

$$= \frac{1}{z^6} \left[\frac{(2^2 - 1)z^2}{2!} - \frac{(2^4 - 1)z^4}{4!} + \frac{(2^6 - 1)z^6}{6!} - \frac{(2^8 - 1)z^8}{8!} \cdots \right]$$

$$= \frac{(2^2 - 1)}{2! z^4} - \frac{(2^4 - 1)}{4! z^2} + \frac{(2^6 - 1)}{6!} - \frac{(2^8 - 1)}{8!} z^2 \cdots$$

Therefore $z = 0$ is a pole of order 4.

19.5 | Residue Theorem

3. $f(z) = -\dfrac{3}{z} - \dfrac{1}{z-2} = -\dfrac{3}{z} + \dfrac{1}{2} \cdot \dfrac{1}{1 - \dfrac{z}{2}} = -\dfrac{3}{z} + \dfrac{1}{2} \left(1 + \dfrac{z}{2} + \dfrac{z^2}{2^2} + \dfrac{z^3}{2^3} + \cdots \right)$

$$= -\frac{3}{z} + \frac{1}{2} + \frac{z}{2^2} + \frac{z^2}{2^3} + \cdots$$

$\text{Res}(f(z), 0) = -3$

6. $f(z) = \dfrac{e^{-2}}{(z-2)^2} e^{-(z-2)} = \dfrac{e^{-2}}{(z-2)^2} \left(1 - \dfrac{z-2}{1!} + \dfrac{(z-2)^2}{2!} - \dfrac{(z-2)^3}{3!} + \cdots \right)$

$$= \frac{e^{-2}}{(z-2)^2} - \frac{e^{-2}}{z-2} + \frac{e^{-2}}{2} - \frac{e^{-2}(z-2)}{3!} + \cdots$$

$\text{Res}(f(z), 2) = -e^{-2}$

9. $\text{Res}(f(z), 1) = \lim_{z \to 1} (z-1) \dfrac{1}{z^2(z+2)(z-1)} = \lim_{z \to 1} \dfrac{1}{z^2(z+2)} = \dfrac{1}{3}$

$\text{Res}(f(z), -2) = \lim_{z \to -2} (z+2) \dfrac{1}{z^2(z+2)(z-1)} = \lim_{z \to -2} \dfrac{1}{z^2(z-1)} = -\dfrac{1}{12}$

$\text{Res}(F(z), 0) = \dfrac{1}{1!} \lim_{z \to 0} \dfrac{d}{dz} \left[z^2 \cdot \dfrac{1}{z^2(z+2)(z-1)} \right] = \lim_{z \to 0} \dfrac{-2z-1}{(z+2)^2(z-1)^2} = -\dfrac{1}{4}$

12. $\text{Res}(f(z), -3) = \lim_{z \to -3} (z+3) \cdot \dfrac{2z-1}{(z-1)^4(z+3)} = \lim_{z \to -3} \dfrac{2z-1}{(z-1)^4} = -\dfrac{7}{256}$

$\text{Res}(f(z), 1) = \dfrac{1}{3!} \lim_{z \to 1} \dfrac{d^3}{dz^3} \left[(z-1)^4 \cdot \dfrac{2z-1}{(z-1)^4(z+3)} \right] = \dfrac{1}{6} \lim_{z \to 1} \dfrac{-42}{(z+3)^4} = -\dfrac{7}{256}$

15. Using $\dfrac{d}{dz} \cos z = -\sin z$ and the result in (4) in the text,

$$\text{Res}\left(f(z), (2n+1)\dfrac{\pi}{2} \right) = \dfrac{1}{-\sin z} \bigg|_{z=(2n+1)\frac{\pi}{2}} = \dfrac{1}{-\sin(2n+1)\frac{\pi}{2}} = (-1)^{n+1}.$$

18. (a) $\displaystyle\oint_C \dfrac{z+1}{z^2(z-2i)} \, dz = 2\pi i \, \text{Res}(f(z), 0) = \pi \left(-1 + \dfrac{1}{2} i \right)$

(b) $\displaystyle\oint_C \dfrac{z+1}{z^2(z-2i)} \, dz = 2\pi i \, \text{Res}(f(z), 2i) = \pi \left(1 - \dfrac{1}{2} i \right)$

(c) $\displaystyle\oint_C \dfrac{z+1}{z^2(z-2i)} \, dz = 2\pi i [\text{Res}(f(z), 0) + \text{Res}(f(z), 2i)] = 2\pi i \left[\dfrac{1}{4} + \dfrac{1}{2} i + \left(-\dfrac{1}{4} - \dfrac{1}{2} i \right) \right] = 0$

21. $\displaystyle\oint_C \dfrac{1}{z^2 + 4z + 13} \, dz = 2\pi i \, \text{Res}(f(z), -2+3i) = \dfrac{\pi}{3}$

24. $\displaystyle\oint_C \dfrac{z}{(z+1)(z^2+1)} \, dz = 2\pi i [\text{Res}(f(z), i) + \text{Res}(f(z), -i)] = 2\pi i \left[\dfrac{1}{4} - \dfrac{1}{4} i + \dfrac{1}{4} + \dfrac{1}{4} i \right] = \pi i$

27. $\displaystyle\oint_C \dfrac{\tan z}{z} \, dz = 2\pi i \, \text{Res}\left(f(z), \dfrac{\pi}{2} \right) = -4i$. Note: $z = 0$ is not a pole. See Example 1, Section 19.4.

30. $\displaystyle\oint_C \dfrac{2z-1}{z^2(z^3+1)} \, dz = 2\pi i \left[\text{Res}(f(z), 0) + \text{Res}(f(z), -1) + \text{Res}\left(f(z), \dfrac{1}{2} + \dfrac{\sqrt{3}}{2} i \right) \right]$

$$= 2\pi i \left[2 + (-1) + \left(-\dfrac{1}{2} - \dfrac{1}{6}\sqrt{3}\, i \right) \right] = \pi \left(\dfrac{\sqrt{3}}{3} + i \right)$$

19.6   Evaluation of Real Integrals

3. $\displaystyle\int_0^{2\pi} \dfrac{\cos\theta}{3 + \sin\theta} \, d\theta = \oint_C \dfrac{z^2+1}{z(z^2 + 6iz - 1)} \, dz = 2\pi i [\text{Res}(f(z), 0) + \text{Res}(f(z), -3 + 2\sqrt{2}\, i)] = 0$

6. $\displaystyle\int_0^\pi \frac{d\theta}{1+\sin^2\theta} = \frac{1}{2}\int_0^{2\pi} \frac{d\theta}{1+\sin^2\theta} = -\frac{2}{i}\oint_C \frac{z}{z^4-6z^2+1}\,dz$

$$= \left(-\frac{2}{i}\right) 2\pi i[\operatorname{Res}(f(z),\sqrt{3-2\sqrt{2}}\,) + \operatorname{Res}(f(z),-\sqrt{3-2\sqrt{2}}\,)] = \frac{\pi}{\sqrt{2}}$$

9. We use $\cos 2\theta = (z^2+z^{-2})/2$.

$$\int_0^{2\pi} \frac{\cos 2\theta}{5-4\cos\theta}\,d\theta = \frac{i}{2}\oint_C \frac{z^4+1}{z^2(2z^2-5z+2)}\,dz = \left(\frac{i}{2}\right) 2\pi i\left[\operatorname{Res}(f(z),0) + \operatorname{Res}\left(f(z),\frac{1}{2}\right)\right]$$

$$= \frac{\pi}{6}$$

12. $\displaystyle\int_{-\infty}^\infty \frac{1}{x^2-2x+25}\,dx = 2\pi i\,\operatorname{Res}(f(z),1+2\sqrt{6}\,i) = \frac{\pi}{2\sqrt{6}}$

15. $\displaystyle\int_{-\infty}^\infty \frac{1}{(x^2+1)^3}\,dx = 2\pi i\,\operatorname{Res}(f(z),i) = \frac{3\pi}{8}$

18. $\displaystyle\int_{-\infty}^\infty \frac{dx}{(x^2+1)^2(x^2+9)} = 2\pi i[\operatorname{Res}(f(z),i) + \operatorname{Res}(f(z),3i)] = \frac{5\pi}{96}$

21. $\displaystyle\int_{-\infty}^\infty \frac{e^{ix}}{x^2+1}\,dx = 2\pi i\,\operatorname{Res}(f(z),i) = \pi e^{-1}.$

Therefore, $\displaystyle\int_{-\infty}^\infty \frac{\cos x}{x^2+1}\,dx = \operatorname{Re}\left(\int_{-\infty}^\infty \frac{e^{ix}}{x^2+1}\,dx\right) = \pi e^{-1}.$

24. $\displaystyle\int_{-\infty}^\infty \frac{e^{ix}}{(x^2+4)^2}\,dx = 2\pi i\,\operatorname{Res}(f(z),2i) = \frac{3e^{-2}}{16}\,\pi;$

$$\int_{-\infty}^\infty \frac{\cos x}{(x^2+4)^2}\,dx = \operatorname{Re}\left(\int_{-\infty}^\infty \frac{e^{ix}}{(x^2+4)^2}\,dx\right) = \frac{3e^{-2}}{16}\,\pi.$$

Therefore, $\displaystyle\int_0^\infty \frac{\cos x}{(x^2+4)^2}\,dx = \frac{1}{2}\left(\frac{3e^{-2}}{16}\,\pi\right) = \frac{3e^{-2}}{32}\pi.$

27. $\displaystyle\int_{-\infty}^\infty \frac{e^{2ix}}{x^4+1}\,dx = 2\pi i\left[\operatorname{Res}\left(f(z),\frac{1}{\sqrt{2}}+\frac{1}{\sqrt{2}}\,i\right) + \operatorname{Res}\left(f(z),-\frac{1}{\sqrt{2}}+\frac{1}{\sqrt{2}}\,i\right)\right]$

$$= 2\pi i\left[\left(-\frac{\sqrt{2}}{8}-\frac{\sqrt{2}}{8}\,i\right) e^{(-\sqrt{2}+\sqrt{2}\,i)} + \left(\frac{\sqrt{2}}{8}-\frac{\sqrt{2}}{8}\,i\right) e^{(-\sqrt{2}-\sqrt{2}\,i)}\right]$$

$$= \pi e^{-\sqrt{2}}\left[\frac{\sqrt{2}}{2}\cos\sqrt{2} + \frac{\sqrt{2}}{2}\sin\sqrt{2}\right]$$

$$\int_{-\infty}^\infty \frac{\cos 2x}{x^4+1}\,dx = \operatorname{Re}\left(\int_{-\infty}^\infty \frac{e^{2ix}}{x^4+1}\,dx\right) = \pi e^{-\sqrt{2}}\left[\frac{\sqrt{2}}{2}\cos\sqrt{2} + \frac{\sqrt{2}}{2}\sin\sqrt{2}\right]$$

Therefore, $\displaystyle\int_0^\infty \frac{\cos 2x}{x^4+1}\,dx = \pi e^{-\sqrt{2}}\frac{\sqrt{2}}{4}(\cos\sqrt{2} + \sin\sqrt{2}\,).$

30. $\displaystyle\int_{-\infty}^{\infty} \frac{xe^{ix}}{(x^2+1)(x^2+4)}\,dx = 2\pi i[\mathrm{Res}(f(z),i) + \mathrm{Res}(f(z),2i)] = 2\pi i\left[\frac{1}{6}e^{-1} - \frac{1}{6}e^{-2}\right]$

$$= \frac{\pi}{3}(e^{-1} - e^{-2})i$$

$\displaystyle\int_{-\infty}^{\infty} \frac{x\sin x}{(x^2+1)(x^2+4)}\,dx = \mathrm{Im}\left(\int_{-\infty}^{\infty} \frac{xe^{ix}}{(x^2+1)(x^2+4)}\,dx\right) = \frac{\pi}{3}(e^{-1} - e^{-2})$

Therefore, $\displaystyle\int_{0}^{\infty} \frac{x\sin x}{(x^2+1)(x^2+4)}\,dx = \frac{1}{2}\left[\frac{\pi}{3}(e^{-1} - e^{-2})\right] = \frac{\pi}{6}(e^{-1} - e^{-2})$.

33. $\displaystyle\int_{0}^{\pi} \frac{d\theta}{(a+\cos\theta)^2} = \frac{1}{2}\int_{0}^{2\pi} \frac{d\theta}{(a+\cos\theta)^2}$

$$= \frac{2}{i}\oint_C \frac{z}{(z^2+2az+1)^2}\,dz \quad (C \text{ is } |z|=1) \;= \frac{2}{i}\oint_C \frac{z}{(z-r_1)^2(z-r_2)^2}\,dz$$

where $r_1 = -a + \sqrt{a^2-1}$, $r_2 = -a - \sqrt{a^2-1}$. Now

$$\oint_C \frac{z}{(z-r_1)^2(z-r_2)^2}\,dz = 2\pi i\,\mathrm{Res}(f(z),r_1) = 2\pi i\,\frac{a}{4(\sqrt{a^2-1})^3} = \frac{a\pi}{2(\sqrt{a^2-1})^3}\,i.$$

Thus,

$$\int_{0}^{\pi} \frac{d\theta}{(a+\cos\theta)^2} = \frac{2}{i}\cdot\frac{a\pi}{2(\sqrt{a^2-1})^3}\,i = \frac{a\pi}{(\sqrt{a^2-1})^3}.$$

When $a=2$ we obtain

$$\int_{0}^{\pi} \frac{d\theta}{(2+\cos\theta)^2} = \frac{2\pi}{(\sqrt{3})^3} \qquad \text{and so} \qquad \int_{0}^{2\pi} \frac{d\theta}{(2+\cos\theta)^2} = \frac{4\pi}{3\sqrt{3}}.$$

36. Using the Fourier sine transform with respect to y the partial differential equation becomes $\dfrac{d^2U}{dx^2} - \alpha^2 U = 0$ and so

$$U(x,\alpha) = c_1 \cosh\alpha x + c_2 \sinh\alpha x.$$

The boundary condition $u(0,y)$ becomes $U(0,\alpha) = 0$ and so $c_1 = 0$. Thus $U(x,\alpha) = c_2 \sinh\alpha x$. Now to evaluate

$$U(\pi,\alpha) = \int_{0}^{\infty} \frac{2y}{y^4+4}\sin\alpha y\,dy = \int_{-\infty}^{\infty} \frac{y}{y^4+4}\sin\alpha y\,dy$$

we use the contour integral $\displaystyle\int_C \frac{ze^{i\alpha z}}{z^4+4}\,dz$ and

$\displaystyle\int_{-\infty}^{\infty} \frac{xe^{i\alpha x}}{x^4+4}\,dx = 2\pi i[\mathrm{Res}(f(z),1+i) + \mathrm{Res}(f(z),-1+i)] = 2\pi i\left[-\frac{1}{8}ie^{(-1+i)\alpha} + \frac{1}{8}ie^{(-1-i)\alpha}\right]$

$$= \frac{\pi}{2}(e^{-\alpha}\sin\alpha)i$$

$$\int_{-\infty}^{\infty} \frac{x\sin\alpha x}{x^4+4}\,dx = \mathrm{Im}\left(\int_{-\infty}^{\infty} \frac{xe^{i\alpha x}}{x^4+4}\,dx\right) = \frac{\pi}{2}e^{-\alpha}\sin\alpha.$$

Finally, $U(\pi, \alpha) = \dfrac{\pi}{2} e^{-\alpha} \sin \alpha = c_2 \sinh \alpha \pi$ gives $c_2 = \dfrac{\pi}{2} \dfrac{e^{-\alpha} \sin \alpha}{\sinh \alpha \pi}$. Hence

$U(x, \alpha) = \dfrac{\pi}{2} \dfrac{e^{-\alpha} \sin \alpha}{\sinh \alpha \pi} \sinh \alpha x$ and

$$u(x, y) = \int_0^\infty \frac{e^{-\alpha} \sin \alpha}{\sinh \alpha \pi} \sinh \alpha x \sin \alpha y \, d\alpha.$$

Chapter 19 in Review

3. False   **6.** True   **9.** $1/\pi$ **12.** False

15. $f(z) = \dfrac{1}{z^4} \left[1 - \left(1 + \dfrac{iz}{1!} + \dfrac{i^2 z^2}{2!} + \dfrac{i^3 z^3}{3!} + \dfrac{i^4 z^4}{4!} + \cdots \right) \right] = -\dfrac{i}{z^3} + \dfrac{1}{2! z^2} + \dfrac{i}{3! z} - \dfrac{1}{4!} - \dfrac{iz}{5!} + \cdots$

18. $\dfrac{1 - \cos z^2}{z^5} = \dfrac{1}{z^5} \left[1 - \left(1 - \dfrac{z^4}{2!} + \dfrac{z^8}{4!} - \dfrac{z^{12}}{6!} + \dfrac{z^{16}}{8!} - \cdots \right) \right] = \dfrac{1}{2! z} - \dfrac{z^3}{4!} + \dfrac{z^7}{6!} - \dfrac{z^{11}}{8!} + \cdots$

21. $\displaystyle\oint_C \dfrac{2z + 5}{z(z + 2)(z - 1)^4} \, dz = 2\pi i [\mathrm{Res}(f(z), 0) + \mathrm{Res}(f(z), -2)] = \dfrac{404}{81} \pi i$

24. $\displaystyle\oint_C \dfrac{z + 1}{\sinh z} \, dz = 2\pi i [\mathrm{Res}(f(z), 0) + \mathrm{Res}(f(z), \pi i)] = 2\pi i [1 + (-\pi i - 1)] = 2\pi^2$

27. $\displaystyle\oint_C \dfrac{1}{z(e^z - 1)} \, dz = 2\pi i \, \mathrm{Res}(f(z), 0) = -\pi i.$ Note: $z = 0$ is a pole of order two, and so

$$\mathrm{Res}(f(z), 0) = \lim_{z \to 0} \frac{d}{dz} z^2 \cdot \frac{1}{z^2 \left(1 + \dfrac{z}{2!} + \dfrac{z^2}{3!} + \cdots \right)} = \lim_{z \to 0} - \frac{\left(\dfrac{1}{2!} + \dfrac{2z}{3!} + \cdots \right)}{\left(1 + \dfrac{z}{2!} + \dfrac{z^2}{3!} + \cdots \right)^2} = -\frac{1}{2}.$$

30. $\displaystyle\oint_C \csc \pi z \, dz = 2\pi i [\mathrm{Res}(f(z), 0) + \mathrm{Res}(f(z), 1) + \mathrm{Res}(f(z), 2)] = 2\pi i \left[\dfrac{1}{\pi} + \left(-\dfrac{1}{\pi} \right) + \dfrac{1}{\pi} \right] = 2i$

33. $\displaystyle\int_0^{2\pi} \dfrac{\cos^2 \theta}{2 + \sin \theta} \, d\theta = \dfrac{1}{2} \oint_C \dfrac{z^4 + 2z^2 + 1}{z^2 (z^2 + 4iz - 1)} \, dz$ (C is $|z| = 1$)

$$= \pi i [\mathrm{Res}(f(z), 0) + \mathrm{Res}(f(z), (-2 + \sqrt{3})i)]$$

$$= \pi i [-4i + 2\sqrt{3} i] = (4 - 2\sqrt{3})\pi$$

[Note: The answer in the text is correct but not simplified.]

36. We have

$$C e^{-a^2 z^2} e^{ibz} \, dz = \int_{-r}^r + \int_{C_1} + \int_{C_2} + \int_{C_3} = 0$$

by the Cauchy–Goursat Theorem. Therefore,

$$\int_{-r}^r = -\int_{C_1} - \int_{C_2} - \int_{C_3}.$$

Let C_1 and C_3 denote the vertical sides of the rectangle. By the ML-inequality, $\displaystyle\int_{C_1} \to 0$ and $\displaystyle\int_{C_3} \to 0$ as $r \to \infty$. On C_2, $z = x + \dfrac{b}{2a^2}\, i$, $-r \le x \le r$, $dz = dx$,

$$\int_{-\infty}^{\infty} e^{-ax^2} e^{ibx}\, dx = -\int_{\infty}^{-\infty} e^{-a^2\left(x+\frac{b}{2a^2}i\right)^2} e^{ib\left(x+\frac{b}{2a^2}i\right)}\, dx = \int_{-\infty}^{\infty} e^{-a^2x^2} e^{-b^2/4a^2}\, dx$$

$$\int_{-\infty}^{\infty} e^{-ax^2}\left(\cos bx + i \sin bx\right) dx = e^{-b^2/4a^2} \int_{-\infty}^{\infty} e^{-a^2x^2}\, dx.$$

Using the given value of $\displaystyle\int_{-\infty}^{\infty} e^{-a^2x^2}\, dx$ and equating real and imaginary parts gives

$$\int_{-\infty}^{\infty} e^{-ax^2} \cos bx\, dx = \frac{\sqrt{\pi}}{a}\, e^{-b^2/4a^2} \qquad \text{and so} \qquad \int_{0}^{\infty} e^{-ax^2} \cos bx\, dx = \frac{\sqrt{\pi}}{2a}\, e^{-b^2/4a^2}.$$

Chapter 20

Conformal Mappings

20.1 | **Complex Functions as Mappings**

3. For $w = z^2$, $u = x^2 - y^2$ and $v = 2xy$. If $xy = 1$, $v = 2$ and so the hyperbola $xy = 1$ is mapped onto the line $v = 2$.

6. If $\theta = \pi/4$, then $v = \theta = \pi/4$. In addition $u = \log_e r$ will vary from $-\infty$ to ∞. The image is therefore the horizontal line $v = \pi/4$.

9. For $w = e^z$, $u = e^x \cos y$ and $v = e^x \sin y$. Therefore if $e^x \cos y = 1$, $u = 1$. The curve $e^x \cos y = 1$ is mapped into the line $u = 1$. Since $v = \dfrac{\sin y}{\cos y} = \tan y$, v varies from $-\infty$ to ∞ and the image is the line $u = 1$.

12. For $w = \dfrac{1}{z}$, $u = \dfrac{x}{x^2 + y^2}$ and $v = \dfrac{-y}{x^2 + y^2}$. The line $y = 0$ is mapped to the line $v = 0$, and, from Problem 2, the line $y = 1$ is mapped onto the circle $\left| w + \frac{1}{2} i \right| = \frac{1}{2}$. Since $f(\frac{1}{2} i) = -2i$, the region $0 \le y \le 1$ is mapped onto the points in the half-plane $v \le 0$ which are on or outside the circle $\left| w + \frac{1}{2} i \right| = \frac{1}{2}$. (The image does not include the point $w = 0$.)

15. The mapping $w = z + 4i$ is a translation which maps the circle $|z| = 1$ to a circle of radius $r = 1$ and with center $w = 4i$. This circle may be described by $|w - 4i| = 1$.

18. Since $w = (1 + i)z = \sqrt{2}\, e^{i\pi/4} z$, the mapping is the composite of a rotation through $45°$ and a magnification by $\alpha = \sqrt{2}$. The image of the first quadrant is therefore the angular wedge $\pi/4 \le \operatorname{Arg} w \le 3\pi/4$.

21. We first let $z_1 = z - i$ to map the region $1 \le y \le 4$ to the region $0 \le y_1 \le 3$. We then let $w = e^{-i\pi/2} z_1$ to rotate this strip through $-90°$. Therefore $w = -i(z - i) = -iz - 1$ maps $1 \le y \le 4$ to the strip $0 \le u \le 3$.

24. The mapping $w = iz$ will rotate the strip $-1 \le x \le 1$ through $90°$ so that the strip $-1 \le v \le 1$ results.

27. By Example 1, Section 20.1, $z_1 = e^z$ maps the strip $0 \le y \le \pi$ onto the upper half-plane $y_1 \ge 0$, or $0 \le \operatorname{Arg} z_1 \le \pi$. The power function $w = z_1^{3/2}$ changes the opening of this wedge by a factor of 3/2 so the wedge $0 \le \operatorname{Arg} w \le 3\pi/2$ results. The composite of these two mappings is $w = (e^z)^{3/2} = e^{3z/2}$.

30. The mapping $z_1 = -(z - \pi i)$ lowers R by π units in the vertical direction and then rotates the resulting region through $180°$. The image region R_1 is upper half-plane $y_1 \ge 0$. By Example 1, Section 20.1, $w = \operatorname{Ln} z_1$ maps R_1 onto the strip $0 \le v \le \pi$. The composite of these two mappings is $w = \operatorname{Ln}(\pi i - z)$.

20.2 | Conformal Mappings

3. $f'(z) = 1 + e^z$ and $1 + e^z = 0$ for $z = \pm i \pm 2n\pi i$. Therefore f is conformal except for $z = \pi i \pm 2n\pi i$.

6. The function $f(z) = \pi i - \frac{1}{2}[\operatorname{Ln}(z+1) + \operatorname{Ln}(z-1)]$ is analytic except on the branch cut $x - 1 \le 0$ or $x \le 1$, and

$$f'(z) = -\frac{1}{2}\left(\frac{1}{z+1} + \frac{1}{z-1}\right) = -\frac{z}{z^2 - 1}$$

is non-zero for $z \ne 0, \pm 1$. Therefore f is conformal except for $z = x$, $x \le 1$.

9. $f(z) = (\sin z)^{1/4}$ is the composite of $z_1 = \sin z$ and $w = z_1^{1/4}$. The region $-\pi/2 \le x \le \pi/2$, $y \ge 0$ is mapped to the upper half-plane $y_2 \ge 0$ by $z_1 = \sin z$ (see Example 2) and the power function $w = z_1^{1/4}$ maps this upper half-plane to the angular wedge $0 \le \operatorname{Arg} w \le \pi/4$. The real interval $[-\pi/2, \pi/2]$ is first mapped to $[-1, 1]$ and then to the union of the line segments from $e^{i\frac{\pi}{4}}$ to 0 and 0 to 1. See the figures below.

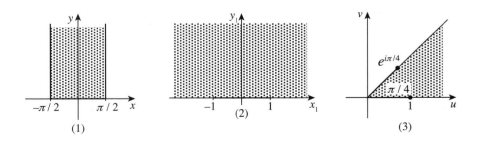

12. Using C-3 $w = e^z$ maps R onto the target region R'. The image of AB is shown in the figure.

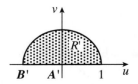

15. Using H-6, $z_1 = \dfrac{e^{\pi/z} + e^{-\pi/z}}{e^{\pi/z} - e^{-\pi/z}}$ maps R onto the upper half-plane

$y_1 \geq 0$, and $w = z_1^{1/2}$ maps this half-plane onto the target region R'.
Therefore

$$w = \left(\frac{e^{\pi/z} + e^{-\pi/z}}{e^{\pi/z} - e^{-\pi/z}} \right)^{1/2}$$

and the image of AB is shown in the figure.

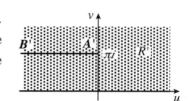

18. Using E-9, $z_1 = \cosh z$ maps R onto the upper half-plane $y_1 \geq 0$.
Using M-7, $w = z_1 + \operatorname{Ln} z_1 + 1$ maps this half-plane onto the
target region R'. Therefore $w = \cosh z + \operatorname{Ln}(\cosh z) + 1$ and the
image of AB is shown in the figure.

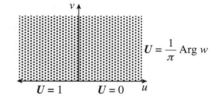

*In Exercises 19–22, we find a conformal mapping $w = f(z)$ that
maps the given region R onto the upper half-plane $v \geq 0$ and
transfers the boundary conditions so that the resulting Dirichlet
problem is as shown in the figure.*

$U = \dfrac{1}{\pi} \operatorname{Arg} w$

$U = 1 \qquad U = 0$

21. $f(z) = i\,\dfrac{1-z}{1+z}$, using H-1, and so $u = U(f(z)) = \dfrac{1}{\pi} \operatorname{Arg}\left(\dfrac{1-z}{1+z}\right)$. The solution may also be

written as $u(x,y) = \dfrac{1}{\pi} \tan^{-1}\left(\dfrac{1 - x^2 - y^2}{2y}\right)$.

*In Exercises 23–26, we find a conformal mapping $w = f(z)$ that
maps the given region R onto the upper half-plane $v \geq 0$ and
transfers the boundary conditions so that the resulting Dirichlet
problem is as shown in the figure.*

$U = 0$ -1 $U = c_0$ 1 $U = 0$

$U = \dfrac{c_0}{\pi}\,[\operatorname{Arg}(w-1) - \operatorname{Arg}(w+1)]$

24. The mapping $z_1 = z^2$ maps R onto the region R_1 defined by $y_1 \geq 0$, $|z_1| \geq 1$ and shown in

H-3, and $w = \dfrac{1}{2}\left(z_1 + \dfrac{1}{z_1}\right)$ maps R_1 onto the upper half-plane $v \geq 0$. Letting $c_0 = 5$,

$$u = \frac{5}{\pi} \left[\operatorname{Arg}\left(\frac{1}{2}\left[z^2 + \frac{1}{z^2} \right] - 1 \right) - \operatorname{Arg}\left(\frac{1}{2}\left[z^2 + \frac{1}{z^2} \right] + 1 \right) \right].$$

27. (a) If $u = \dfrac{\partial^2 \phi}{\partial x^2} + \dfrac{\partial^2 \phi}{\partial y^2}$,

$$\frac{\partial^2 u}{\partial x^2} + \frac{\partial^2 u}{\partial y^2} = \frac{\partial^4 \phi}{\partial x^4} + 2\frac{\partial^4 \phi}{\partial x^2 \partial y^2} + \frac{\partial^4 \phi}{\partial y^4} = 0$$

since ϕ is assumed to be biharmonic.

(b) If $g = u + iv$, then $\phi = \text{Re}(\bar{z}g(z)) = xu + yv$.

$$\frac{\partial^2 \phi}{\partial x^2} = 3\frac{\partial u}{\partial x} + x\frac{\partial^2 u}{\partial x^2} + y\frac{\partial^2 v}{\partial x^2}$$

$$\frac{\partial^2 \phi}{\partial y^2} = 2\frac{\partial v}{\partial y} + x\frac{\partial^2 u}{\partial y^2} + y\frac{\partial^2 v}{\partial y^2}.$$

Since u and v are harmonic and $\dfrac{\partial u}{\partial x} = \dfrac{\partial v}{\partial y}$,

$$\frac{\partial^2 \phi}{\partial x^2} + \frac{\partial^2 \phi}{\partial y^2} = 2\frac{\partial u}{\partial x} + 2\frac{\partial v}{\partial y} = 4\frac{\partial u}{\partial x}.$$

Now $u_1 = \dfrac{\partial u}{\partial x}$ is also harmonic and so $\dfrac{\partial^2 u_1}{\partial x^2} + \dfrac{\partial^2 u_1}{\partial y^2} = 0$. But

$$\frac{\partial^2 u_1}{\partial x^2} + \frac{\partial^2 u_1}{\partial y^2} = \frac{1}{4}\left[\frac{\partial^4 \phi}{\partial x^4} + 2\frac{\partial^4 \phi}{\partial x^2 \partial y^2} + \frac{\partial^4 \phi}{\partial y^4}\right]$$

and so ϕ is biharmonic.

20.3 | Linear Fractional Transformations

3. (a) For $T(z) = \dfrac{z+1}{z-1}$, $T(0) = -1$, $T(1) = \infty$, and $T(\infty) = 1$.

(b) The circle $|z| = 1$ passes through the pole at $z = 1$ and so the image is a line. Since $T(-1) = 0$ and $T(i) = -i$, the image is the line $u = 0$. If $|z - 1| = 1$,

$$|w - 1| = \left|\frac{z+1}{z-1} - 1\right| = \frac{2}{|z-1|} = 2$$

and so the image is the circle $|w - 1| = 2$ in the w-plane.

(c) Since $T(0) = -1$, the image of the disk $|z| \leq 1$ is the half-plane $u \leq 0$.

6. $S^{-1}(T(z)) = \dfrac{az+b}{cz+d}$ where

$$\begin{bmatrix} a & b \\ c & d \end{bmatrix} = \text{adj}\left(\begin{bmatrix} 2 & 1 \\ 1 & 1 \end{bmatrix}\right)\begin{bmatrix} i & 0 \\ 1 & -2i \end{bmatrix} = \begin{bmatrix} -1+i & 2i \\ 2-i & -4i \end{bmatrix}.$$

Therefore,

$$S^{-1}(T(z)) = \frac{(-1+i)z + 2i}{(2-i)z - 4i} \quad \text{and} \quad S^{-1}(w) = \frac{w-1}{-w+2}.$$

9. $T(z) = \dfrac{(z - z_1)(z_2 - z_3)}{(z - z_3)(z_2 - z_1)}$ maps z_1, z_2, z_3 to 0, 1, ∞. Therefore,

$$T(z) = \frac{(z+1)(-2)}{(z-2)(1)} = -2\frac{z+1}{z-2} \quad \text{maps} \quad -1, 0, 2 \text{ to } 0, 1, \infty.$$

12. As in Exercise 11, $z = \dfrac{(w - 1 - i)(-1 + i)}{(w - 1 + i)(-1 - i)}$ and, solving for w, $w = \dfrac{2z - 2}{(1 + i)z - 1 + i}$ maps $0, 1, \infty$ to $1 + i, 0, 1 - i$.

15. Using the cross-ratio formula (7),

$$S(w) = \frac{(w + 1)(-3)}{(w - 3)(1)} = \frac{(z - 1)(2i)}{(z + i)(i - 1)} = T(z).$$

We can solve for w to obtain

$$w = 3\,\frac{(1 + i)z + (1 - i)}{(-3 + 5i)z - 3 - 5i}.$$

Alternatively we can apply the matrix method to compute $w = S^{-1}(T(z))$.

18. The mapping $T(z) = \dfrac{1}{2}\dfrac{z + 1}{z}$ maps $-1, 1, 0$ to $0, 1, \infty$ and maps each of the two circles in R to lines since both circles pass through the pole at $z = 0$. Since $T(\frac{1}{2} + \frac{1}{2}i) = 1 - i$ and $T(1) = 1$, the circle $|z - \frac{1}{2}| = \frac{1}{2}$ is mapped onto the line $u = 1$. Likewise, the circle $|z + \frac{1}{2}| = \frac{1}{2}$ is mapped onto the line $u = 0$. The transferred boundary conditions are shown in the figure and $U(u, v) = u$ is the solution. The solution to the Dirichlet problem in Figure 20.3.6 is

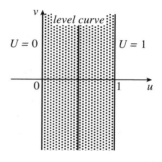

$$u = U(T(z)) = \mathrm{Re}\left(\frac{1}{2}\frac{z + 1}{z}\right) = \frac{1}{2} + \frac{1}{2}\frac{x}{x^2 + y^2}.$$

The level curves $u = c$ are the circles with centers on the x-axis which pass through the origin. The level curve $u = \frac{1}{2}$, however, is the vertical line $x = 0$.

21. $T_2(T_1(z)) = \dfrac{a_2 T_1(z) + b_2}{c_2 T_1(z) + d_2} = \dfrac{a_2 \dfrac{a_1 z + b_1}{c_1 z + d_1} + b_2}{c_2 \dfrac{a_1 z + b_1}{c_1 z + d_1} + d_2} = \dfrac{a_1 a_2 z + a_2 b_1 + b_2 c_1 z + b_2 d_1}{a_1 c_2 z + b_1 c_2 + c_1 d_2 z + d_1 d_2}$

$$= \frac{(a_1 a_2 + c_1 b_2)z + (b_1 a_2 + d_1 b_2)}{(a_1 c_2 + c_1 d_2)z + (b_1 c_2 + d_1 d_2)}$$

20.4 Schwarz–Christoffel Transformations

3. $\arg f'(t) = -\dfrac{1}{2}\,\mathrm{Arg}\,(t + 1) + \dfrac{1}{2}\,\mathrm{Arg}\,(t - 1) = \begin{cases} 0, & t < -1 \\ \pi/2, & -1 < t < 1 \\ 0, & t > 1 \end{cases}$

and $\alpha_1 = \pi/2$ and $\alpha_2 = 3\pi/2$. Since $f(-1) = 0$, the image of the upper half-plane is the region shown in the figure.

6. Since $\alpha_1 = \pi/3$ and $\alpha_2 = \pi/2$, $\alpha_1/\pi - 1 = -2/3$ and $\alpha_2/\pi - 1 = -1/2$ and so $f'(z) = A(z+1)^{-2/3}z^{-1/2}$ for some constant A.

9. Since $\alpha_1 = \alpha_2 = \pi/2$, $f'(z) = A(z+1)^{-1/2}(z-1)^{-1/2} = A/(z^2-1)^{1/2}$. Therefore, $f(z) = A\cosh^{-1} z + B$. But $f(-1) = \pi i$ and $f(1) = 0$. Since $\cosh^{-1} 1 = 0$, $B = 0$. Since $\cosh^{-1}(-1) = \pi i$, $\pi i = A(\pi i)$ and so $A = 1$. Hence $f(z) = \cosh^{-1} z$.

12. From (3), $f'(z) = Az^{-3/4}(z-1)^{(\alpha_2/\pi)-1}$. But $\alpha_2 \to \pi$ as $\theta \to 0$. This suggests that we examine $f'(z) = Az^{-3/4}$. Therefore, $f(z) = A_1 z^{1/4} + B_1$. But $f(0) = 0$ and $f(1) = 1$ so that $B_1 = 0$ and $A_1 = 1$. Hence $f(z) = z^{1/4}$ and we recognize that this power function maps the upper half-plane onto the wedge $0 \le \operatorname{Arg} w \le \pi/4$.

20.5 | Poisson Integral Formulas

3. The harmonic function

$$u_1 = \frac{5}{\pi}[\pi - \operatorname{Arg}(z-1)] = \begin{cases} 5, & x > 1 \\ 0, & x < 1 \end{cases}, \quad \text{and} \quad u_2 = -\frac{1}{\pi}\operatorname{Arg}\left(\frac{z+1}{z+2}\right) + \frac{1}{\pi}\operatorname{Arg}\left(\frac{z}{z+1}\right)$$

from (3) satisfies all boundary conditions except that $u_2 = 0$ for $x > 1$. Therefore $u = u_1 + u_2$ is the solution to the given Dirichlet problem.

6. From Theorem 20.5.1,

$$u(x,y) = \frac{y}{\pi}\int_{-\infty}^{\infty}\frac{\cos t}{(x-t)^2+y^2}\,dt = \frac{y}{\pi}\int_{-\infty}^{\infty}\frac{\cos(x-s)}{s^2+y^2}\,ds$$

letting $s = x - t$. But $\cos(x-s) = \cos x \cos x + \sin x \sin s$. It follows that

$$u(x,y) = \frac{y\cos x}{\pi}\int_{-\infty}^{\infty}\frac{\cos s}{s^2+y^2}\,ds + \frac{y\sin x}{\pi}\int_{-\infty}^{\infty}\frac{\sin s}{s^2+y^2}\,ds = \frac{y\cos x}{\pi}\left(\frac{\pi e^{-y}}{y}\right)$$

$$= e^{-y}\cos x, \quad y > 0.$$

9. Using H-1, $f(z) = i\dfrac{1-z}{1+z}$ maps R onto the upper half-plane R'. The corresponding Dirichlet problem in R' is shown in the figure. From (3),

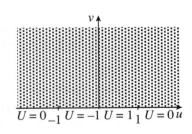

$$U = -\frac{1}{\pi}\operatorname{Arg}\left(\frac{w}{w+1}\right) + \frac{1}{\pi}\operatorname{Arg}\left(\frac{w-1}{w}\right)$$

$$= \frac{1}{\pi}\left[\operatorname{Arg}\left(\frac{w-1}{w}\right) - \operatorname{Arg}\left(\frac{w}{w+1}\right)\right].$$

The harmonic function $u = U(f(z))$ may be simplified to

$$u = \frac{1}{\pi}\left[\operatorname{Arg}\left(\frac{(1-i)z-(1+i)}{1-z}\right) - \operatorname{Arg}\left(\frac{1-z}{-(1+i)z+1-i}\right)\right]$$

and is the solution to the original Dirichlet problem in R.

12. From Theorem 20.5.2, $u(x,y) = \dfrac{1}{2\pi}\displaystyle\int_{-\pi}^{\pi} e^{-|t|}\dfrac{1-|z|^2}{|e^{it}-z|^2}\,dt.$ Therefore,

$$u(0,0) = \frac{1}{2\pi}\int_{-\pi}^{\pi} e^{-|t|}\,dt = \frac{1}{\pi}\int_{0}^{\pi} e^{-t}\,dt = \frac{1}{\pi}(1-e^{-\pi}).$$

With the aid of Simpson's Rule, $u(0.5,0) = 0.5128$ and $u(-0.5,0) = 0.1623$.

15. For $u(e^{i\theta}) = \sin\theta + \cos\theta$, the Fourier series solution (6) reduces to

$$u(r,\theta) = r\sin\theta + r\cos\theta \quad\text{or}\quad u(x,y) = y+x.$$

The corresponding system of level curves is shown in the figure.

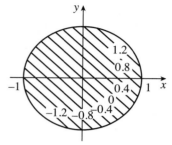

20.6 Applications

3. $g(z) = \dfrac{x}{x^2+y^2} - \dfrac{y}{x^2+y^2}i = \dfrac{1}{z}$ is analytic for $z \neq 0$ and so

div $\mathbf{F} = 0$ and curl $\mathbf{F} = \mathbf{0}$ by Theorem 20.6.1. A complex potential is $G(z) = \operatorname{Ln} z$ and

$$\phi(x,y) = \operatorname{Re}(G(z)) = \frac{1}{2}\log_e(x^2+y^2).$$

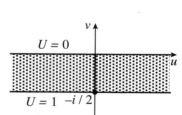

The equipotential lines $\phi(x,y) = c$ are circles $x^2+y^2 = e^{2c}$ and are shown in the figure.

6. The function $f(z) = \dfrac{1}{z}$ maps the original region R to the strip $-\frac{1}{2} \leq v \leq 0$ (see Example 2, Section 20.1). The boundary conditions transfer as shown in the figure. $U = -2v$ is the solution in the horizontal strip and so

$$\phi(x,y) = -2\operatorname{Im}\left(\frac{1}{z}\right) = \frac{2y}{x^2+y^2}$$

is the potential in the original region R. The equipotential lines $\dfrac{2y}{x^2+y^2} = c$ may be written

as $x^2 + \left(y+\dfrac{1}{c}\right)^2 = \left(\dfrac{1}{c}\right)^2$ for $c \neq 0$ and are circles. If $c = 0$, we obtain the line $y = 0$. Note

that $\phi(x,y) = \operatorname{Re}\left(\dfrac{2i}{z}\right)$ and so $G(z) = \dfrac{2i}{z}$ is a complex potential. The corresponding vector field is

$$\mathbf{F} = \overline{G'(z)} = \frac{2i}{\bar{z}^2} = \left(\frac{-4xy}{(x^2+y^2)^2}, \frac{2(x^2-y^2)}{(x^2+y^2)^2}\right).$$

9. (a) $\psi(x,y) = \text{Im}(z^4) = 4xy(x^2 - y^2)$ and so $\psi(x,y) = 0$ when $y = x$ and $y = 0$.

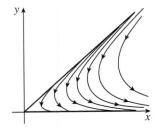

(b) $\mathbf{V} = \overline{G'(z)} = \overline{4z^3} = 4(x^3 - 3xy^2, y^3 - 3x^2 y)$

(c) In polar coordinates $r^4 \sin 4\theta = c$ or $r = (c \csc 4\theta)^{1/4}$, for $0 < \theta < \pi/4$, are the streamlines. See the figure.

12. (a) The image of R under $w = i \sin^{-1} z$ is the horizontal strip (see E-6) $-\pi/2 \le v \le \pi/2$ and

$$\psi(x,y) = \text{Im}(i \sin^{-1} z) = \begin{cases} \pi/2, & x \ge 1 \\ -\pi/2, & x \le -1 \end{cases}.$$

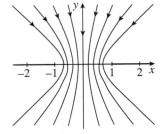

Each piece of boundary is therefore a streamline.

(b) $\mathbf{V} = \overline{G'(z)} = \overline{\dfrac{i}{(1-z^2)^{1/2}}} = \dfrac{-i}{(1-\bar{z}^2)^{1/2}}$

(c) The streamlines are the images of the lines $v = b$, $-\pi/2 < b < \pi/2$ under $z = -i \sin w$ and are therefore hyperbolas. See Example 2, Section 20.2, and the figure. Note that at $z = 0$, $v = -i$ and the flow is downward.

15. (a) For $f(z) = \pi i - \frac{1}{2}[\text{Ln}\,(z+1) + \text{Ln}\,(z-1)]$

$$f(t) = \pi i - \frac{1}{2}[\log_e |t+1| + \log_e |t-1| + i \,\text{Arg}\,(t+1) + i \,\text{Arg}\,(t-1)]$$

and so $\text{Im}(f(t)) = \begin{cases} 0, & t < -1 \\ \pi/2, & -1 < t < 1 \\ \pi, & t > 1 \end{cases}$. Hence $\text{Im}(G(z)) = \psi(x,y) = 0$ on the boundary of R.

(b) $x = -\dfrac{1}{2}[\log_e |t+1+ic| + \log_e |t-1+ic|], \quad y = \pi - \dfrac{1}{2}[\text{Arg}\,(t+1+ic) + \text{Arg}\,(t-1+ic)]$ for $c > 0$

(c)

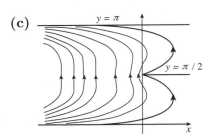

18. (a) For $f(z) = 2(z+1)^{1/2} + \text{Ln}\left(\dfrac{(z+1)^{1/2} - 1}{(z+1)^{1/2} + 1}\right)$,

$$f(t) = 2(t+1)^{1/2} + \text{Ln}\,\frac{(t+1)^{1/2} - 1}{(t+1)^{1/2} + 1}.$$

If we write $(t+1)^{1/2} = |t+1|^{1/2}e^{(i/2)\operatorname{Arg}(t+1)}$, we may conclude that

$$\operatorname{Im}(f(t)) = \begin{cases} 0, & t > 0 \\ \pi, & -1 < t < 0 \end{cases} \quad \text{and} \quad \operatorname{Re}(f(t)) = 0 \text{ for } t < -1.$$

Therefore $\operatorname{Im}(G(z)) = \psi(x,y) = 0$ on the boundary of R.

(b) $x = \operatorname{Re}\left[2(t+ic+1)^{1/2} + \operatorname{Ln}\dfrac{(t+ic+1)^{1/2} - 1}{(t+ic+1)^{1/2} + 1}\right]$

$y = \operatorname{Im}\left[2(t+ic+1)^{1/2} + \operatorname{Ln}\dfrac{(t+ic+1)^{1/2} - 1}{(t+ic+1)^{1/2} + 1}\right] \quad \text{for } c > 0$

(c)

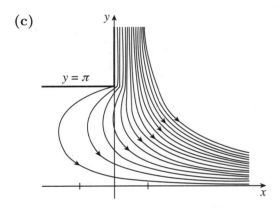

21. $f(z) = z^2$ maps the first quadrant onto the upper half-plane and $f(\xi_0) = f(1) = 1$. Therefore $G(z) = \operatorname{Ln}(z^2 - 1)$ is the complex potential, and so

$$\psi(x,y) = \operatorname{Arg}(z^2 - 1) = \tan^{-1}\left(\frac{2xy}{x^2 - y^2 - 1}\right)$$

is the streamline function where $\tan^{-1}$ is chosen to be between 0 and π. If $\psi(x,y) = c$, then $x^2 + Bxy - y^2 - 1 = 0$ where $B = -2\cot c$. Each hyperbola in the family passes through $(1,0)$ and the boundary of the first quadrant satisfies $\psi(x,y) = 0$.

24. (a) $\mathbf{V} = \dfrac{a+ib}{\bar{z}} = \left(\dfrac{ax-by}{x^2+y^2}, \dfrac{bx+ay}{x^2+y^2}\right)$ and since $(x'(t), y'(t)) = \mathbf{V}$, the path of the particle satisfies

$$\frac{dx}{dt} = \frac{ax-by}{x^2+y^2}, \qquad \frac{dy}{dt} = \frac{bx+ay}{x^2+y^2}.$$

(b) Switching to polar coordinates,

$$\frac{dr}{dt} = \frac{1}{r}\left(x\frac{dx}{dt} + y\frac{dy}{dt}\right) = \frac{1}{r}\left(\frac{ax^2 - bxy}{r^2} + \frac{bxy + ay^2}{r^2}\right) = \frac{a}{r}$$

$$\frac{d\theta}{dt} = \frac{1}{r^2}\left(-y\frac{dx}{dt} + x\frac{dy}{dt}\right) = \frac{1}{r^2}\left(\frac{-axy + by^2}{r^2} + \frac{bx^2 + axy}{r^2}\right) = \frac{b}{r^2}.$$

Therefore $\dfrac{dr}{d\theta} = \dfrac{a}{b}r$ and so $r = ce^{a\theta/b}$.

(c) $\dfrac{dr}{dt} = \dfrac{a}{r} < 0$ if and only if $a < 0$, and in this case r is decreasing and the curve spirals inward. $\dfrac{d\theta}{dt} = \dfrac{b}{r^2} < 0$ if and only if $b < 0$, and in this case θ is decreasing and the curve is traversed clockwise.

Chapter 20 in Review

3. The wedge $0 \le \operatorname{Arg} w \le 2\pi/3$. See Figure 20.1.6 in the text.

6. A line, since $|z - 1| = 1$ passes through the pole at $z = 2$.

9. False. $\overline{g(z)} = P - iQ$ is analytic. See Theorem 20.6.1.

12. First use $z_1 = z^2$ to map the first quadrant onto the upper half-plane $y_1 \ge 0$, and segment AB to the negative real axis. We then use $w = \frac{1}{\pi}(z_1 + \operatorname{Ln} z_1 + 1)$ to map this half-plane onto the target region R'. The composite transformation is

$$w = \frac{1}{\pi}[z^2 + \operatorname{Ln}(z^2) + 1]$$

and the image of AB is the ray extending to the left from $w = i$ along the line $v = 1$.

15. The inversion $w = 1/z$ maps R onto the horizontal strip $-1 \le v \le -1/2$ and the transferred boundary conditions are shown in the figure. The solution in the strip is $U = 2v + 2$ and so

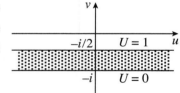

$$u = U\left(\frac{1}{z}\right) = 2\operatorname{Im}\left(\frac{1}{z}\right) + 2 = \frac{-2y}{x^2 + y^2} + 2$$

is the solution to the original Dirichlet problem in R.

18. (a) From Theorem 20.5.1,

$$u(x, y) = \frac{y}{\pi}\int_{-\infty}^{\infty} \frac{\sin t}{(x-t)^2 + y^2}\, dt = \frac{y}{\pi}\int_{-\infty}^{\infty} \frac{\sin(x-s)}{s^2 + y^2}\, ds \quad \text{(letting } s = x - t\text{)}.$$

But $\sin(x - s) = \sin x \cos s - \cos x \sin s$. We now proceed as in the solution to Problem 6, Section 20.5 to show that $u(x, y) = e^{-y}\sin x$.

(b) For $u(e^{i\theta}) = \sin\theta$, the Fourier series solution (6) in Section 20.5 reduces to $u(r, \theta) = r\sin\theta$.